MHCC WITHDRAWN

Environmental Impacts of Aquaculture

Sheffield Biological Sciences

A series which provides an accessible source of information at research and professional level in chosen sectors of the biological sciences.

Series Editors:

Professor Jeremy A. Roberts, Plant Science Division, School of Biological Sciences, University of Nottingham.
Professor Peter N.R. Usherwood, Molecular Toxicology Division, School of Biological Sciences, University of Nottingham.

Titles in the series:

Biology of Farmed Fish
Edited by K.D. Black and A.D. Pickering.

Stress Physiology in Animals
Edited by P.H.M. Balm.

Leaf Development and Canopy Growth
Edited by B. Marshall and J.A. Roberts.

Seed Technology and its Biological Basis
Edited by M. Black and J.D. Bewley.

Herbicides and their Mechanisms of Action
Edited by A.H. Cobb and R.C. Kirkwood.

Environmental Impacts of Aquaculture
Edited by K.D. Black.

Environmental Impacts of Aquaculture

Edited by

KENNETH D. BLACK
Dunstaffnage Marine Laboratory
Oban, UK

CRC Press

First published 2001
Copyright © 2001 Sheffield Academic Press

Published by
Sheffield Academic Press Ltd
Mansion House, 19 Kingfield Road
Sheffield S11 9AS, UK

ISBN 1-84127-041-5

Published in the U.S.A. and Canada (only) by
CRC Press LLC
2000 Corporate Blvd., N.W.
Boca Raton, FL 33431, U.S.A.
Orders from the U.S.A. and Canada (only) to CRC Press LLC

U.S.A. and Canada only:
ISBN 0-8493-0501-2

All rights reserved. No part of this publication may be reproduced, stored in a retrieval system or transmitted in any form or by any means, electronic, mechanical, photocopying or otherwise, without the prior permission of the copyright owner.

This book contains information obtained from authentic and highly regarded sources. Reprinted material is quoted with permission, and sources are indicated. Reasonable efforts have been made to publish reliable data and information, but the author and the publisher cannot assume responsibility for the validity of all materials or for the consequences of their use.

Trademark Notice: Product or corporate names may be trademarks or registered trademarks, and are used only for identification and explanation, without intent to infringe.

Printed on acid-free paper in Great Britain by
Antony Rowe Ltd., Chippenham, Wiltshire

British Library Cataloguing-in-Publication Data:
A catalogue record for this book is available from the British Library

Library of Congress Cataloging-in-Publication Data:
Environmental impacts of aquaculture / edited by Kenneth D. Black.
     p. cm.-- (Sheffield biological sciences)
  Includes bibliographical references (p.).
  ISBN 0-8493-0501-2 (alk. paper)
  1. Aquaculture--Environmental aspects.   I. Black, Kenneth D.   II. Series.

SH135 .E58 2000
639.8--dc21

00-029762

Preface

When I started researching the impacts of aquaculture in 1991, I was advised by a senior colleague that I should get out of this area of science as soon as possible, as there was little left to learn and I could expect no continued funding. This advice reminds me of the oft-quoted opinion of a computer company executive questioned on the future of the personal computer in the 1970s: 'They will never catch on'. A great deal of research on the environmental effects of aquaculture has been published over the last decade and there is a great deal of consequence left to learn. This volume is a review of important recent developments in the context of what was previously known.

The worldwide expansion of interest in the impacts of aquaculture has been fuelled by two main factors:

- The massive increase in world aquaculture production through the 1990s, which has seen the development of new culture species, increased production of established species, increased geographical spread of culture and changes to the practices and technologies applied to culture. Each of these has required research into environmental aspects.

- Increasing pressure on regulators from consumers, the public, non-governmental organisations, pressure groups and scientific expert groups. This has driven regulators to seek a more robust understanding of the environmental consequences of aquaculture, in order to enlighten and refine regulations and codes of practice.

It has become increasingly recognised amongst the research community that research on the impacts of aquaculture can provide real insights into processes of more general relevance and applicability. For example, the marine culture of carnivorous species in coastal waters provides an opportunity for studying ecological and biogeochemical processes across extremely strong carbon and nutrient gradients. Such farms produce strong environmental signals where the source terms, i.e. nutrient or pollutant input, can be clearly defined, and they therefore represent excellent sites for hypothesis testing by experiment—an area which has traditionally been neglected by oceanographers in favour of the collection of large sets of data. Such farms are often located within areas of restricted water exchange which further constrain the boundary conditions for modelling and experiment—the familiar natural mesocosm concept.

In this volume, an attempt has been made to cover the impacts of the main culture types. Given the diversity of aquaculture techniques and practices, there is no claim of encyclopaedic coverage, but close attention has been paid to processes and principles of wide applicability. On the other hand, classic examples of the problems faced by aquaculture are exemplified by cage culture of salmonids and by the production of tropical shrimps, and aspects of the extensive research carried out on these systems can be found in several of the chapters. These aspects illuminate different perspectives rather than overlap. The book is ordered such that chapters dealing with specific types of impact occur near the beginning of the book, with a gradual move towards generality.

In the first chapter, the impacts of marine cage culture are reviewed. Although the key processes were identified in the 1970s and 1980s, much detail has been added in recent years. Chapter 2 reviews progress on the impacts of freshwater cage culture, but concerns are raised that research progress has been lacking in this area - particularly in regard to impacts on sediments. Chapter 3 reviews aspects of shellfish culture, both in temperate and tropical systems. This is complemented by chapter 4, which looks at tropical systems in greater depth, paying particular concern to the major issues surrounding the impacts of shrimp culture. Developments in the technologies for containing effluents in tank culture are reviewed in chapter 4. As environmental regulation becomes more rigorous and/or better enforced, it is likely that recirculation systems will play a large part in future developments, as they afford almost total control of effluents. The concept of confinement is also explored in chapter 6, which examines the issues surrounding genetic pollution, including potential impacts associated with the development of transgenic culture species. The ability to generalise research findings and test process understanding is crucial and chapter 7 reviews progress in modelling relevant to the impacts of aquaculture. In chapter 8, the wider place of aquaculture in integrated resource management is discussed in terms of progress towards coastal zone management. The book is completed in chapter 9 with a discussion of issues relating to the sustainability of aquaculture.

Sustainability is a recurring theme throughout the volume. Much remains to be learned on this complex subject, which is likely to become an ever more important topic of debate and research as we look beyond the local scale at the wider or systemic effects of aquaculture. Linked with this is the concept of environmental carrying capacity on the regional scale and the place of aquaculture within a threatened world fishery. This book will go some way to informing this debate and will aid the identification of gaps in our understanding.

I am grateful to the chapter authors for the professional way in which they have dealt with the task they undertook and by writing what I believe to be clear, well structured and informed texts. As in the companion volume in this series, *Biology of Farmed Fish,* I am happy to have the opportunity of thanking Dr Graeme MacKintosh, Publisher for the Sciences and Technology at Sheffield Academic Press, for his enthusiasm and wise words during the development of the project.

Kenny Black
Oban

Contributors

Dr K.D. Black	Dunstaffnage Marine, Laboratory, Oban, Argyll PA34 4AD, UK
Ms D.J. Brister	Department of Fisheries and Wildlife, University of Minnesota, 200 Hodson Hall, 1980 Folwell Avenue, St Paul, MN 55108, USA
Ms P.-S. Choo	Fisheries Research Institute, 11960 Batu Maung, Penang, Malaysia
Mr C.J. Cromey	Dunstaffnage Marine Laboratory, Oban, Argyll PA34 4AD, UK
Dr D.W. Donnan	Scottish Natural Heritage, 2/5 Anderson Place, Edinburgh EH6 5NP, UK
Dr I.R. Elberizon	Department of Biological Sciences, Heriot-Watt University, Edinburgh EH14 4AS, UK
Dr M.J. Kaiser	School of Ocean Sciences, University of Wales–Bangor, Menai Bridge, Gwynedd LL59 5EY, UK
Professor A.R. Kapuscinski	Department of Fisheries and Wildlife, University of Minnesota, 200 Hodson Hall, 1980 Folwell Avenue, St Paul, MN 55108, USA
Dr L.A. Kelly	Department of Civil and Offshore Engineering, Heriot-Watt University, Edinburgh EH14 4AS, UK
Dr T.H. Pearson	Scottish Environmental Advisory Service, Oban, Argyll PA34 4AD, UK
Dr W. Silvert	Instituto Nacional de Investigação das Pescas, Departmento de Ambiente Aquático, Avenida de Brasília, s/n, 1449-006 Lisboa, Portugal
Dr U. Waller	Institut für Meereskunde Kiel, Düsternbrooker Weg 20, D-24105 Kiel, Germany

Contents

1 The environmental impacts of marine fish cage culture   1
T. H. PEARSON and K. D. BLACK

1.1 Introduction	1
1.2 Background	2
1.3 Enrichment of the natural environment	3
1.3.1 Nutrient enrichment	3
1.3.2 The nitrogen budget of cage farming	5
1.3.3 The phosphorus budget of cage farming	7
1.3.4 The silicon budget of cage farming	7
1.3.5 Organic enrichment	8
1.3.6 Carbon	16
1.3.7 Seasonality in benthic response to sedimentary enrichment	18
1.3.8 Benthic recovery	18
1.4 Ecological impacts on wild populations	20
1.5 Impacts in differing environments	23
1.5.1 Mediterranean	23
1.5.2 Tropical culture	24
1.6 Minimisation of impacts	25
1.7 Research needs	26
1.8 Conclusions	26
References	27

2 Freshwater finfish cage culture   32
L. A. KELLY and I. R. ELBERIZON

2.1 Introduction	32
2.2 The impacts of cage aquaculture in freshwaters	34
2.2.1 Introduction	34
2.2.2 Impacts of cage aquaculture at a site	34
2.2.3 The wider impacts of cage aquaculture in freshwater	38
2.3 The environmental management of cage aquaculture in freshwater	41
2.3.1 Introduction	41
2.3.2 Controlling and treating waste outputs from cages	42
2.3.3 Management of freshwater lakes for and by aquaculture	44
2.4 Conclusions	45
References	47

3 Ecological effects of shellfish cultivation   51
M. J. KAISER

3.1 Introduction	51
3.2 Cultivation of molluscs	52

	3.3	Seed collection		54
		3.3.1 Subtidal dredging		54
		3.3.2 Intertidal collection		55
		3.3.3 Use of collectors and cultch		55
		3.3.4 Hatchery seed		56
		3.3.5 Introductions of alien species		57
	3.4	On-growing		58
		3.4.1 Intertidal cultivation		58
		3.4.2 Suspended cultivation		62
		3.4.3 Stock enhancement		63
	3.5	Harvesting		63
	3.6	Crustacean farming		65
		3.6.1 Disease outbreaks		66
		3.6.2 Environmental issues		68
		3.6.3 Mangrove destruction		68
		3.6.4 Stock enhancement		70
	3.7	Summary		70
	References			71
4	**Environmental effects of warm water culture in ponds/lagoons**			**76**
	P.-S. CHOO			
	4.1	Introduction		76
	4.2	Culture systems and their environmental effects		77
		4.2.1 Freshwater culture systems		78
		4.2.2 Shrimp farming		81
	4.3	Culture of exotic species		88
		4.3.1 Interaction with local flora and fauna		88
		4.3.2 Disease transmission		89
		4.3.3 Habitat alteration		89
	4.4	Feral animals		89
	4.5	Mitigation measures		90
	References			92
5	**Tank culture and recirculating systems**			**99**
	U. WALLER			
	5.1	Introduction: the benefits of closed and semi-closed systems		99
	5.2	General design of rearing units		102
	5.3	Environmental control		106
		5.3.1 Light regime		106
		5.3.2 Temperature regime		107
		5.3.3 Swimming activity		107
		5.3.4 Rearing density		109
	5.4	Water supply		110
		5.4.1 Through-flow systems		112
		5.4.2 Recirculation systems		114
	References			120

6 Genetic impacts of aquaculture — 128
A. R. KAPUSCINSKI and D. J. BRISTER

- 6.1 Introduction — 128
- 6.2 Genetically-engineered organisms — 129
- 6.3 Assessing biosafety — 129
- 6.4 Mitigating risks — 133
- 6.5 Interbreeding with natural populations — 136
  - 6.5.1 Panmictic populations *vs* genetically distinct populations — 137
  - 6.5.2 Increased vulnerability to environmental change due to loss of genetic differences between populations — 138
  - 6.5.3 Decreased production and fitness of wild populations due to outbreeding depression — 139
  - 6.5.4 Escapees from domesticated aquacultural stocks increase the hazard of outbreeding depression — 141
  - 6.5.5 Estimating the probability of interbreeding — 143
  - 6.5.6 Interactions with naturalized populations — 144
- 6.6 Conclusion — 145
- Acknowledgments — 146
- Glossary — 147
- References — 148

7 Modelling impacts — 154
W. SILVERT and C. J. CROMEY

- 7.1 Introduction — 154
- 7.2 Why model? — 154
- 7.3 What are models? — 155
- 7.4 What is a system? — 156
- 7.5 Scale issues — 157
- 7.6 Modular approaches — 159
- 7.7 Modelling individual fish — 160
- 7.8 Physical transport modelling — 162
- 7.9 Primary *vs* secondary effects — 165
- 7.10 Water quality models — 166
- 7.11 Models of benthic loading — 168
  - 7.11.1 Modelling primary deposition — 168
  - 7.11.2 After the fall — 171
- 7.12 Disease transmission — 173
- 7.13 Pharmaceuticals — 174
- 7.14 Tools to improve modelling — 175
- 7.15 Providing management advice — 175
  - 7.15.1 Decision support systems — 176
- 7.16 The three Ms — 177
- 7.17 Summary — 178
- 7.18 Additional resources — 178
- Acknowledgements — 179
- References — 179

8 Aquaculture in the age of integrated coastal management (ICM) 182
D. DONNAN

8.1	Introduction	182
8.2	The need for an integrated approach to the management of aquaculture	183
	8.2.1 The loss of mangrove habitat to tropical shrimp and fish farming	183
	8.2.2 Interactions between farmed salmon and wild fish stocks in Europe	184
8.3	Integrated coastal management (ICM)	185
	8.3.1 United Nations Conference on Environment and Development 1992	185
	8.3.2 FAO Code of Conduct for Sustainable Fisheries	186
8.4	Progress towards the management of aquaculture in an ICM environment	188
8.5	The implications of marine areas for aquaculture planning	191
8.6	Practical considerations for managing aquaculture within an ICM environment	194
	8.6.1 Estimating environmental capacity	194
	8.6.2 Mapping coastal resources	195
8.7	Conclusion	195
References		196

9 Sustainability of aquaculture 199
K. D. BLACK

9.1	Introduction	199
9.2	Assessing sustainability	200
9.3	Impact types and sustainability	202
	9.3.1 Wild seed collection	202
	9.3.2 Feedstocks	202
	9.3.3 Soluble wastes and water resources	203
	9.3.4 Solid wastes	206
	9.3.5 Waste chemicals	206
	9.3.6 Containment of cultured species, disease and parasites	207
	9.3.7 Fisheries	208
9.4	Benefits of aquaculture	208
	9.4.1 Environmental benefits	208
	9.4.2 Socioeconomic benefits	209
9.5	Growth and global carrying capacity	210
9.6	Conclusion and future predictions	210
Acknowledgement		211
References		211

Index 213

1 The environmental impacts of marine fish cage culture

T.H. Pearson and K.D. Black

1.1 Introduction

Culture of fish in cages has the potential to cause both onshore and offshore impacts on the surrounding environment, with a severity scaled to the size and intensity of the farming operation. Such impacts include: distortion of the local ecosystem; short- and long-term and near-field and far-field effects of eutrophication; contamination by xenobiotics; cross-transmission of parasites and pathogens; and aesthetic deterioration in coastal areas. Such environmental effects must be minimised, both to maintain the most favourable conditions for farm stocks in the cages and to preserve environmental amenity for other users. Thus, it is necessary to assess the environmental carrying capacity for fish cages at all farm sites and to institute regulations to monitor, control and harmonise farming practices with environmental stability.

As awareness of the environmental consequences of caged fish culture has grown with the growth of the practice, so have the scope and sophistication of regulatory controls on this industry. It is self-evident that the success of such controls is dependent on an accurate assessment of the environmental risks associated with any type and intensity of cage farming. It is equally evident that, although there are certain environmental trends common to all such impacts, there are site-specific factors associated with local topographic, hydrographic and climatic conditions that must be considered in applying any regulatory controls. In this chapter, the environmental impacts of carbon and nutrient enrichment from intensive marine cage culture are reviewed. In addition, some of the ecological effects of cage culture, including those from the transfer of parasites and pathogens, are examined. Genetic impacts are reviewed by Kapuscinski and Brister in Chapter 6 of this volume. Coastal zone management aspects are examined by Donnan in Chapter 8.

The most obvious changes caused by intensive marine culture of carnivorous fish in cages are those to the local environment. These impacts, predominantly on the benthos, may be long-lived at sites in relatively quiescent waters, persisting for many years after the culture activity has ceased, or, at more dynamic sites, may be relatively short-lived. Much of the research activity into impacts has concentrated, understandably, on the immediate local environment and over relatively short time-scales. However, it is likely that with the continued expansion of marine cage farming across many parts of the world, the focus will change to long-term, far-field, systemic interactions. This is particularly true of nutrient discharges, where, in essence, nutrients are extracted from the open seas in the form of wild fish that are subsequently converted into meal for consumption by cultured fish, followed by excretion into the more valuable and more vulnerable coastal zone.

A goal of environmental research in this area has long been the definition of carrying capacity, such that the scale of the aquacultural activity can be matched to the assimilative capacity of the local environment. While this is relatively straightforward for benthic impacts in temperate environments, great difficulties arise in the pelagic environment. Regulators, therefore, generally require a relatively cautious approach to be taken, based on the precautionary principle, which may be challenged by farmers.

Guidelines, assessment and regulation form a feedback loop with monitoring. This process is much better developed in some countries than others, and is poorly advanced in many warm water counties where aquaculture is rapidly expanding. There is a risk that impacts in lower latitude areas will be assumed to be comparable to those described from the more intensively studied higher latitude areas and that the necessary research will not be carried out.

The goal of the regulatory sequence must be a sustainable marine cage culture activity, with positive socioeconomic benefits and negligible impacts on the environment. This topic is discussed further by Black in Chapter 9 of this volume.

1.2 Background

In a bibliographic review, Munday *et al.* (1992) surveyed the existing literature on environmental impacts of aquaculture and found, among other things, that the majority of such research originated in developed western countries and concentrated on the major cultured species. Gowen and Rosenthal (1993), in attempting to draw conclusions applicable to aquaculture in developing countries, reviewed and discussed some of these results. Wu (1995) reviewed the literature on marine cage culture, essentially reaching similar conclusions to previous reviewers. These suggest that the degree of impact from effluent wastes is dependent on husbandry parameters, including species, culture method and feed type, and on the nature of the receiving environment in terms of physics, chemistry and biology. The major impacts are on the seabed, generally within a localised area around the farm and, while chemical treatments or food additives may have a significant local impact, they are unlikely to pose a significant threat to the regional environment at current levels of use. This last view is widely held within the research community. However, it begs the question of the ultimate carrying capacity of aquatic systems for fish before significant systemic effects become apparent.

The interactions between aquaculture and biodiversity were reviewed by Beveridge *et al.* (1994), who argued that aquaculture impacts are usually negative and only rarely positive. These arise from the consumption of a set of resources and their transformation into a desirable product, with the simultaneous production of a waste stream that includes food, faeces, respiratory products, chemicals, microbiota, parasites and escaped fish. They concluded that while fish culture may indeed be a necessary activity in some cases due to over-exploitation of natural stocks, intensive systems, which require a large input of fishmeal, must exacerbate pressures on wild fisheries in general. These authors regard the output of energy rich wastes from intensive aquaculture, discharges that frequently cause pollution problems, as poor use of a valuable dwindling resource.

1.3 Enrichment of the natural environment

1.3.1 Nutrient enrichment

Most marine culture of carnivorous species takes place in cages suspended in more or less open water. Cage structures are relatively cheap compared with equivalent land-based structures and, by being immersed in the receiving environment, cage culture avoids the need for expensive pumping of water to supply the oxygen requirements of the fish and to remove waste products. This does, however, mean that dissolved components are released directly into the marine environment in a highly biologically active form. The dissolved products include ammonia, phosphorus and dissolved organic carbon (DOC). The DOC component contains fractions rich in nitrogen (DON) and phosphorus (DOP). These waste products have a variety of sources: they may be directly excreted; they may be dissolved from the feed or from faecal particles; or they may be released from particles that have been deposited on the seabed around the cages. Lipids released from the diet may form a film on the water surface, which is often observed around cages after feeding.

The effect that dissolved wastes may have on the environment will depend on the speed at which these nutrients are diluted before being assimilated by the pelagic ecosystem. Cage structures are often located in areas of partially restricted exchange, as such locations generally provide shelter from extreme weather, thus protecting staff and equipment. In restricted exchange environments, it is generally useful to estimate flushing time (i.e. the time taken to exchange all or some part of the local water volume with new coastal water) in order to assess the risks of significantly increasing the nutrient concentrations in the immediate environment (hypernutrification). Strutton *et al.* (1996) have provided a simple expression for estimation of the flushing time of semi-enclosed bays or estuaries, with only limited data:

$$T_0 = \frac{h \Delta S}{S_1(E-R) - h\frac{dS_2}{dt}}$$

where, h is the mean depth of the bay, $(E - R)$ is the net evaporation rate, S_1 is the salinity of the open ocean and S_2 is the volume-averaged salinity of the bay.

Flushing time can also be calculated using the planar area and volumes of the body of water under investigation. Planar area is the area of the surface of the water body and total volume is determined by summing the volume of smaller subvolumes (= planar area × depth). Flushing time (T) in days can be determined from the following formula (Edwards and Sharples, 1986):

$$T = \frac{1.05 \, lwvolume}{0.7 \, tiderange \, (hwarea + lwarea)}$$

where, *lwvolume* and *lwarea* are low water volume and planar area at lowest astronomical tide (LAT), *hwarea* is planar area of the water body at MHWS (mean high water spring tides) and *tiderange* is the tidal range between MLWS (mean low water spring tides) and MHWS.

The equilibrium concentration of nitrogen originating from the farm can be estimated from the formula (Gowen and Ezzi, 1992):

Equilibrium concentration = nitrogen released/volume × flushing time

It is assumed that, where the flushing time is less than the typical generation time of phytoplankton, any increase in nutrient concentrations caused by the farm will not lead to measurable increases in the local phytoplankton biomass (eutrophication). However, flushing times are calculated for mean conditions and it is possible that certain events, or combinations of events (e.g. unusually persistent onshore winds combined with neap tides reducing exchange of nutrient rich surface waters), might lead to longer periods of water retention.

Smaller pelagic organisms, such as nanoflagellates (microzooplankton) and bacteria, have much shorter generation times. Heterotrophic bacterioplankton around fish farms will have access to highly-available organic nutrients, however, there has been little research on bacterial response and on any consequences that increased bacterial biomass might have on other compartments of the microzooplankton.

It is generally assumed that, in contrast to freshwaters, nitrogen is likely to be the nutrient limiting phytoplankton growth in marine waters. There have been suggestions, however, that this is not always the case and that alterations in the ratio of nitrogen to phosphorus might cause an effect on the phytoplankton (Arzul *et al.*, 1996). Phosphate limitation resulting from natural alterations in the N:P ratio has been implicated in, for instance, the formation of the toxic *Chrysochromulina polylepis* bloom in the Skaggerak (Smayda, 1990).

Wildish *et al.* (1993) found that mariculture of salmonids did not significantly affect the concentrations of nitrate, phosphate and silicate in the Bay of Fundy,

Nova Scotia, but there were local increases in the concentration of ammonia. In a more recent study in the Mediterranean, Pitta *et al.* (1998) reported elevated concentrations of phosphate and ammonium near three fish farms, but no significant effect on chlorophyll.

It has only rarely been possible to demonstrate any linkage between the nutrients produced from farming and a biological response, although many such linkages have been claimed. Beyond looking at purely local enrichment, it is normally not feasible to attribute wider-scale effects to nutrients from farms. However, in the brackish waters of the Finnish archipelago, in the northern Baltic Sea, increasing levels of eutrophication attributable to nutrients from rainbow trout *Oncorhyncus mykiss* farms have resulted in the rapid growth, in summer, of algal mats on the bottom sediments. These have caused bottom-water anoxia and strong reductions in local fish and benthic populations (Bonsdorff and co-workers, 1996, 1997). Moreover, in 1993, a dense and catastrophic bloom of *Alexandrium tamarense* in an Italian lagoon was linked to local intensive aquaculture activities (Sorokin *et al.*, 1996). The occurrence of nuisance or harmful algal blooms appears to be increasing on a worldwide scale (Anderson, 1997). A diatomaceous bloom on the west coast of Scotland was responsible for a major closure of the wild scallop fishery during 1999–2000, when scallops from a very large area tested positively for the toxin (domoic acid), which causes amnesiac shellfish poisoning (ASP). It was widely reported in the popular media that this toxic event was a consequence of ecological disturbance caused by fish farm nutrient effluents. No evidence has been presented to support this view and, given that the bloom occurred across a massive area of the Scottish West coast, including areas far distant from coastal aquaculture, it seems that there is little likelihood of this being the case.

1.3.2 The nitrogen budget of cage farming

Carbon inputs drive sedimentary biogeochemical processes but waste materials from fish farms also contain large amounts of other environmentally relevant elements that influence biological processes in the water column, namely nitrogen, phosphorus and silicon. In one of the few studies that have examined the processing of nitrogen through intensive cage farms in marine waters, Hall and co-workers (1992) found that 67–80% of the nitrogen added to the cage system is lost to the environment. The majority (50–60% of total nitrogen) is lost in dissolved form, either directly from the fish or by benthic flux from solid waste beneath the cages. The level of nitrogen (and phosphorus) in feeds has decreased as feed manufacture becomes more closely aligned with the dietary requirements of fish. In particular, modern diets tend to contain more lipid and less protein. This has resulted in a general reduction in feed conversion ratios. These currently approach 1:1 in western Europe, although more efficient feeding methods also play a part in this. The net effect is a reduction in nitrogen released to the environment. Enell (1995) reported a reduction from 132 to 55 kg N t^{-1} fish between 1974 and 1994, and estimated that nitrogen from aquaculture contributed only 0.5% of the atmospheric input to the Baltic region.

The excretion of nitrogenous compounds by marine fish and their associated toxicity to fish have been comprehensively reviewed by Handy and Poxton (1993). Marine fish can assimilate a large proportion of ingested nitrogen (up to 95%), but any reduction in assimilation efficiency may lead to significantly elevated levels of faecal nitrogen. Using a wide range of published information, Handy and Poxton (1993) produced worst- and best-case scenarios for loss of nitrogen to the environment, and concluded that 52–95% of nitrogen added in food is ultimately lost through a combination of food wastage and incomplete absorption and retention. Their worst-case scenario was, however, just that and it is questionable whether any culture operation could perform anywhere near as badly without going out of business.

Fish utilise dietary protein efficiently but a significant proportion is metabolised for energy purposes, thereby releasing large amounts of nitrogenous waste, mostly as ammonia but partly as urea. Excretion of nitrogenous wastes from sea bass, sea bream, turbot, rainbow trout and brown trout have been compared (Dosdat *et al.*, 1996). Ammonia and urea losses were similar for all species except turbot, which excreted relatively more ammonia and less urea. In addition, nitrogen losses from fish are dependent on temperature and on dietary protein content (Buttle *et al.*, 1996).

The nitrogen budget for the areolated grouper *Epinephelus areolatus* cultured under both laboratory and cage conditions has recently been reported (Leung *et al.*, 1999). These workers attempted to balance the following equation:

$$C = G + M + E + F$$

where, each term relates to a mass of nitrogen, and: C is consumption, G is nitrogen retained for growth, M represents losses from mortality, E is excretory loss and F loss through faeces. For a fish farm, the budget is represented by:

$$C = I - W$$

where, I is the total input to the farm and W is the food wasted.

Using a variety of techniques, Leung *et al.* (1999) were able to quantify each of these terms either by direct measurement or by difference. Figure 1.1 shows the budget derived for cage farming of this species using trash fish as feed. Excretion of ammonia was the greatest contributor to nitrogen loss, followed by feed wastage, with faecal nitrogen loss being relatively unimportant. Nitrogen loss from this subtropical species fed trash fish is around three times greater than for temperate species fed formulated diets. Trash diets are inherently wasteful as a consequence of their high nitrogen content and their tendency to break up and shed small unconsumed particles during feeding.

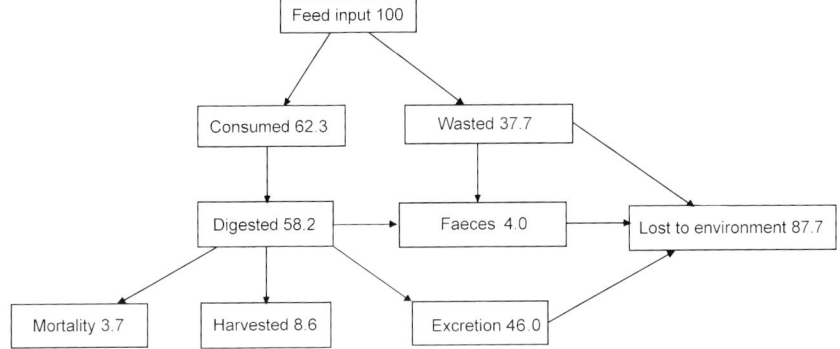

Figure 1.1 Nitrogen budget for areolated grouper, *Epiniphelus areolatus*, in cage culture (data from Leung *et al.*, 1999).

Benthic fluxes of ammonia have also received attention (Hargrave *et al.*, 1993). Uptake of oxygen and release of ammonia from sediments directly beneath cages can be 4–27 times higher than at control stations, reaching maximum values of 99 and 38 mmol m^{-2} day^{-1}, respectively.

1.3.3 The phosphorus budget of cage farming

Phosphorus in fish farm effluents has been the subject of considerable interest, especially with regard to freshwater and brackish systems, where it may be the limiting nutrient for plant growth, (see Kelly, Chapter 2 of this volume, for a review of freshwater culture nutrient impacts). In the marine environment, Holby and Hall (1991) showed environmental losses of phosphorus to be 19.6–22.4 kg t^{-1} fish produced, 34–41% of which was released in dissolved form, with the remainder lost by sedimentation. They estimated that 4–8% of the sedimented phosphorus was returned to the water column per year.

Karrakassis and co-workers (1999) attributed increases in benthic diatom biomass beneath farms in Greece to the periodic release of phosphate stored in sediments as the sediments oscillated between reduced and oxidised states. Under oxidising conditions, phosphate becomes bound as iron (III) phosphate, which precipitates near the sediment surface. Under reducing conditions, deeper in the sediment, iron (III) is reduced to iron (II), the phosphate of which is soluble.

1.3.4 The silicon budget of cage farming

Only one comprehensive study of silicon budgets has been published (Holby and Hall, 1994), which is perhaps surprising given the key role of this element in the metabolism of diatoms. Almost all of the silicon added in food to the marine trout farm investigated was lost to the environment but this only accounted for around 20% of the total silicon budget. The majority of silicon was thought to have arisen from diatoms associated with the farm itself, that is, the farm provided a good niche for organisms utilising silicon from sources external to the farm. Fluxes from the sediment to the overlying water were around 2.5 times higher than from local 'control' sediments. There is continuing concern that toxic dinoflagellate species may be promoted under conditions of nutrient imbalance (Berry, 1996). Therefore, it has been proposed (Holby and Hall, 1994) that, in order to maintain ratios of silicon relative to phosphorus and nitrogen, cage farms should be sited in areas which are well supplied with fresh sea water and that the deliberate addition of biogenic silica to feed formulations should be considered.

1.3.5 Organic enrichment

Sediments are diverse environments supporting a range of flora and fauna existing in a complex matrix, whose defining parameters include particle size, carbon (food) availability, oxygen concentration and redox potential. Biogeochemical processes in marine sediments are dominated by ambient hydrography (deposition, erosion and oxygen supply) and by the net input of carbon, which determines sedimentary oxygen demand and thus redox chemistry.

1.3.5.1 Sampling techniques and strategies
Because of their predominantly sedentary nature, benthic infauna (animals living within the sediments) are the populations most frequently selected for study in assessment of the effects of any environmental disturbance. Using standardised grabs or corers, quantitative samples of the sediments are taken from disturbed areas and at varying distances from those areas. All animals found in the samples are enumerated and identified and the composition and density of the populations are compared from area to area. Concomitant sediment samples and hydrographic records are taken for the measurement of a range of natural and disturbance-related physical and chemical parameters. These would commonly include: flow rate through the cages recorded over a full tidal cycle on spring and neap tides; grain size; pH (acidity); Eh (redox potential); total organic material (TOM) and/or total organic carbon (TOC) in the sediments; together with analyses for any chemicals used in cage treatments. Comparisons of the biological and environmental data are made and common trends with distance from the disturbance are sought. Such sampling surveys are repeated at intervals to assess trends over time. Sampling techniques of this type have now been standardised for a range of survey purposes (see, OsParCom, 1997; NS 9423, 1988; HelCom, 1999; SFT, 1999; Rumohr, 1999). Such standardised guideline manuals have now been issued for use in fish farm surveys in Norway (NS 9410, 2000) and Scotland (SEPA, 1998), and are being compiled for use in other European areas.

The strategies most frequently adopted for fish farm surveys are to establish sampling stations along transects orientated on the axis of the prevailing current flow through the cages and originating beside the cages. Samples are taken at intervals along the transects, for example at 0, 25, 50 and 100 m from the cages and at a reference station sited in an area beyond the influence of any cage effects and having similar sedimentary characteristics to those of the transect stations. The intensity of sampling in both time and space is, however, generally site specific and defined in consultation with the regulatory authorities. In Scotland, the sampling regime is varied in relation to the estimated carrying capacity of the site and the proposed stocking densities. Small farms in hydrodynamically energetic areas may only be required to undertake a small-scale, qualitative, visual survey (photo or video) of sedimentary conditions, or to take a single set of five replicate samples from under the cages during the time of maximum cage biomass. Large farms in quiescent areas, on the other hand, may be required to carry out a fully quantitative survey, at the time of peak biomass, of a wide range of physical, chemical and biological factors along up to four transects radiating from the cages. In such cases, an annual visual survey is also required. Full technical details for the implementation of the required sampling are given in SEPA (1998).

In Norway, a management system termed 'Modelling Ongrowing fish farms Monitoring' (MOM) has been adopted, which combines simulation modelling of potential environmental impacts with a monitoring programme of increasing elaboration, dependent on the model's predicted scale of impacts (Ervik, *et al.*, 1997; NS 9410, 2000). The monitoring has to ensure a farm's compliance with a set of required environmental quality standards (EQSs). Three classes of monitoring are defined. Class A, for farms predicted to suffer no appreciable environmental impact outside the immediate cage area, requires only a simplified assessment of sedimentation rates beneath the cages and does not need to conform to EQSs. Class B applies to most farms where some degree of impact is anticipated. It specifies a qualitative assessment of infauna, quantitative measurements of acidity and redox potential and a ranked scoring of a range of visually recorded sedimentary characteristics in samples taken locally around the cages. These measurements have to meet EQSs. Class C applies to those farms predicted to have a high impact level. These are required to undertake a long-term survey of benthic faunal changes along transects through the anticipated impact zones. Such surveys also have to meet set EQSs. The recommended techniques for carrying out each class of survey are set out in NS 9410 (2000).

1.3.5.2 Sedimentary microbiological processes
Various terminal electron acceptors are used by different bacterial communities in marine sediments. The oxygen concentration at any point in the sediment is dependent on the rate of its uptake, either to fuel aerobic metabolism, or to reoxidise reduced products released from deeper in the sediment. When the oxygen

demand caused by input of organic matter exceeds the oxygen diffusion rate from overlying waters, sediments become anoxic and anaerobic processes dominate. As sediments become more reducing with increasing distance from the water column interface, a range of microbiological processes become successively dominant, in the order:

- Aerobic respiration, ammonium oxidation (to nitrite) and nitrite oxidation (to nitrate)—these aerobic nitrifying processes are inhibited by sulphide and are, therefore, of limited importance in sediments beneath marine fish farms
- Denitrification (producing dinitrogen from nitrate)
- Nitrate reduction (producing ammonium from nitrate) and manganese reduction
- Iron reduction
- Sulphate reduction (producing hydrogen sulphide)
- Under the most reducing conditions, methanogenesis (producing methane)

To some extent, these processes may overlap spatially. In marine systems, sulphate reduction is the most important terminal anaerobic process for the degradation of organic material (Holmer and Kristensen, 1992) but is much less important in fresh water, due to the normally low sulphate concentration.

The redox potential (Eh) profile measured down the sediment column to a depth of 10–15 cm gives a useful guide to the relative degree of carbon enrichment in the sediments (Pearson and Stanley, 1979). Positive Eh values are indicative of aerobic conditions, whereas negative values are associated with anaerobic microbial processes. Under normal rates of detrital carbon input to sediments, the redox discontinuity level (RDL), that is, the point at which anaerobic processes become predominant, lies some centimetres below the surface. As carbon inputs increase, the RDL approaches ever closer to the surface as the biological oxygen demand (BOD) within the sediments increases. Eventually, under very high detrital inputs, the RDL coincides with the sediment/water interface, where, under low flow conditions, it might even rise into the water column. Since organic degradation rates are lower under anaerobic than in oxygenated conditions (2–3 times lower first order rate constants; Westrich and Berner, 1984), these extreme states are highly undesirable. Not only do they result in the elimination of the benthic infauna (see Section 1.3.5.3), which, through bioturbation, play a crucial role in oxygenating the upper sediment layers, but in slowing down carbon recycling, carbon accumulation is enhanced, thus perpetuating sedimentary anoxia. The regular monitoring of Eh levels in the sediments adjacent to fish cages can, therefore, give a good indication of the possible build up of such conditions.

Table 1.1 shows redox profiles from sediment cores taken from a highly-enriched area adjacent to a fish cage and from areas 25 and 1000 m from

THE ENVIRONMENTAL IMPACTS OF MARINE FISH CAGE CULTURE

Table 1.1 Redox potential (Eh in mV) profiles of sediment core samples taken from a sampling station beside a fish cage in a west of Scotland sea loch and from stations 25 and 1000 m from the cage

Position		Adjacent to fish cage		25 m from fish cage		1000 m from fish cage	
Core no.		1	2	1	2	1	2
Water	+10 mm	+85	+185	+383	+415	+427	+432
Surface	0 mm	−149	−96	+316	+198	+428	+416
	−5 mm	−174	−130	+234	+62	+451	+173
	−10 mm	−193	−149	+103	+23	+449	+158
	−20 mm	−198	−164	+99	+10	+235	+127
	−30 mm	−205	−173	+70	+5	+162	+132
	−40 mm	−211	−183	+27	−7	+152	+100
	−50 mm	−220	−192	−7	−25	+148	+56
	−60 mm	−225	−189	−40	−82	+117	+64
	−75 mm	−238	−211	−68	−103	+100	+34
	−100 mm	−240		−99	−134	+91	

All stations were at the same depth (38 m) and had the same type of sediment (>90% silt/clay).

the cage. In the latter case, which represents normal conditions in the loch, the RDL lies well below 10 cm in the core and all the values recorded are positive. At 25 m from the cage, the RDL lies between 40 and 50 mm. This indicates moderately enriched conditions at that point, whereas at the station beside the cages the RDL is at the sediment surface and all values are negative. This suggests highly enriched conditions. It is worth noting that the values recorded in the water above the sediment surface are all around 400 mV at the two more distant stations and represent the values expected for oxygen saturation, but at the cage station the bottom water values are much lower, indicating oxygen depletion. The sampling area in this case is hydrodynamically quiescent, thus regular sampling is necessary to allow farm managers to limit stock biomass if sedimentary conditions should deteriorate further. A useful summary of the impact of salmonid cage farming on benthic microbiology was given by Davies *et al.* (1996). A significant disruption to nitrogen cycling in the sediments immediately beneath fish cages, with inhibition of nitrification and denitrification, has recently been reported by McCaig *et al.* (1999). They demonstrated the dominance of a novel subgroup of nitrosammonas bacteria at the cage site, which is absent in the normal sediments. This complements the information of Hall and co-workers (1992) that pore water ammonium levels beneath fish cages was greatly enhanced compared to that outside the area (see Section 1.3.2). However, these authors found that the net flux of nitrogen from the sediments beneath the cages was only 11% of the total nitrogen input.

A study in the Bay of Fundy, Nova Scotia, by Hargrave and co-workers (1997) at sites beneath cage farms and at a comparable number of reference sites, showed that total sulphur and Eh in the surface sediment and benthic CO_2

release and O_2 uptake were sensitive indicators of benthic enrichment. However, surface sediment water content, grain size, pore water salinity and sulphate were less sensitive indicators. The biomass of deposit-feeding organisms was significantly higher at the cage sites but not the total macrofaunal biomass.

1.3.5.3 Benthic macrofaunal response to enrichment

The gross effects of wastes from intensive cage culture on marine benthic habitats and processes in northern European and other cool-temperate regions are now fairly well established (Brown et al., 1987; Weston, 1990; Holmer and Kristensen, 1992; Hargrave et al., 1993; Wildish et al., 1993; Findlay et al., 1995). Essentially, these follow the pattern of impacts from other organic pollutant sources (see Pearson and Rosenberg, 1978) but on a more reduced spatial scale. Recorded effects include: reducing sediments; hypoxia in the water overlying the sediment; increased sulphate reduction; and marked changes in benthic faunal and meiofaunal (Duplisea and Hargrave, 1996) assemblages in terms of species number, diversity, abundance and biomass.

Benthic macrofaunal populations in sediments receiving normal detrital inputs derived from planktonic production in the overlying water column are species rich, have a relatively low total abundance/species richness ratio, and include a wide range of higher taxa, body sizes and functional types, that is, they are highly diverse communities. As detrital inputs increase, diversity also initially increases because the enhanced food supply provides opportunities for the expansion of existing populations and the immigration of additional species. However, the concomitant changes in the physical and chemical conditions in the deeper levels of the sediments progressively eliminate the larger and deeper burrowing and longer-lived forms and favour smaller, more rapidly growing 'opportunistic' species. Eventually, if input levels continue to increase, the surface sediments become anoxic and only a small number of specialist taxa can survive, principally small annelid and nematode worms. However, under such conditions, populations of these species may reach very high levels. Where the RDL occurs close to the sediment surface, this may become covered in dense mats of sulphide oxidising bacteria, *Beggiatoa* sp. Ultimately, increasing levels of sedimentary BOD bring about deoxygenation of the lower levels of the overlying water column leading to the elimination of all macrobenthic fauna. This successional process is illustrated in Figure 1.2.

Although the succession is continuous along the enrichment gradient, a number of phases characterised by differing dominant taxa can be distinguished. Some of the taxa characterising such different successional states in western Scottish sea lochs are listed in Table 1.2. It must be stressed, however, that, with the exception of those at the higher end of the enrichment gradient, these are not necessarily bioindicators for such conditions in other geographical areas (see Section 1.3.5.4). The successional phases may also be described by reference to the changing relationships, along the enrichment gradient, of the

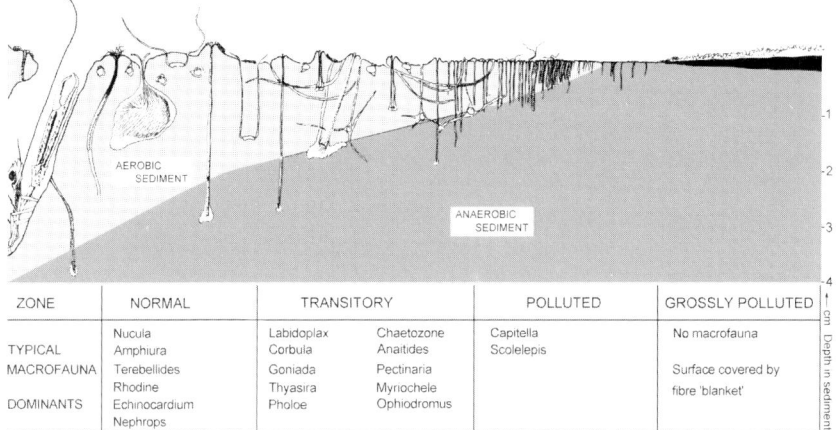

Figure 1.2 Diagrammatic representation of faunal succession along a gradient of organic enrichment. As carbon inputs to the sediments increase (from left to right on the diagram), the diverse normal community that includes, for example, molluscs, echinoderms, crustacea and polychaetes, is gradually simplified. Ultimately, at high levels of input, only high numbers of small, opportunistic worms occur before surface anoxia eliminates all macrofauna.

three most important population parameters, species richness (S), abundance (A) and biomass (B), of benthic communities. This is depicted in Figure 1.3. Thus, in grossly enriched areas, where the immediate bottom water is intermittently or fully anoxic and Eh levels are uniformly negative, no infaunal metazoans survive (zone A). In zone B, only a few vermiform taxa can survive, but often in very high numbers, and the RDL lies close to the sediment surface. Beyond this, zone C is an area where species richness and biomass increase rapidly to a maximum and positive Eh values are found at some centimetres depth in the sediment. In this zone, faunal densities are much lower but still well above background levels. Zone D is where levels of detrital input return to normal and the levels of all three parameters fall to their background levels. Here, the Eh values are highly positive throughout the sediment column. Four such zones of effect were reported by Brown *et al.* (1987): an azoic zone immediately under the cages; a highly-enriched zone 0–8 m from the cage edge, with a high biomass dominated by a large number of a few opportunistic species; a slightly-enriched transitional zone between 8 and ~25 m from the cages; and a normal zone beyond that. However, each of these zones may vary in spatial extent (or may be absent) depending on local hydrography, and benthic effects may be measurable at considerably greater distances than 25 m from cages.

The most significant environmental constraint on the development of sedimentary anoxia under conditions of progressive carbon enrichment is the flow rate of bottom water over any particular site. High flow rates bring a

Table 1.2 Descriptive model of benthic faunal succession during sedimentary recovery from loading with fish farm wastes

Degree of enrichment	Highly enriched	Moderately enriched	Lightly enriched	Normal community
Characteristic species	*Capitella capitata*	*Apistobranchus tullbergi*	*Scoloplos* sp.	*Glycera alba*
	Malacoceros fuliginosus	*Spio decorata*	*Thyasira ferruginea*	*Mugga wahbergi*
	Nematoda	*Mediomastus fuliginosus*	*Diplocirrus glaucus*	*Ophelina* sp.
	Ophryotrocha sp.	*Protodorvillea kefersteini*	*Sphaerosyllis tetralix*	*Perugia caeca*
	Pseudopolydora paucibranchiata	*Microspio* sp.	*Leptognathia brevirostris*	*Synelmis klatti*
		Prionospio fallax	*Abra alba*	*Owenia fusiformis*
		Scalibregma inflatum	*Glycera alba*	*Magelona* sp.
		Chaetozone setosa		*Amphiura filiformis*
		Caulleriella sp.		*Polycirrus plumosus*
		Cossura sp.		*Eumida* sp.
		Melinna palmata		*Ophelina* sp.
		Pholoe inornata		*Lanice conchilega*
		Cirratulidae		
		Abra alba		
Expected number of species present	5–10	25–40	40–45	30–35
Approximate time between stages (months)	0	9	18	21–24

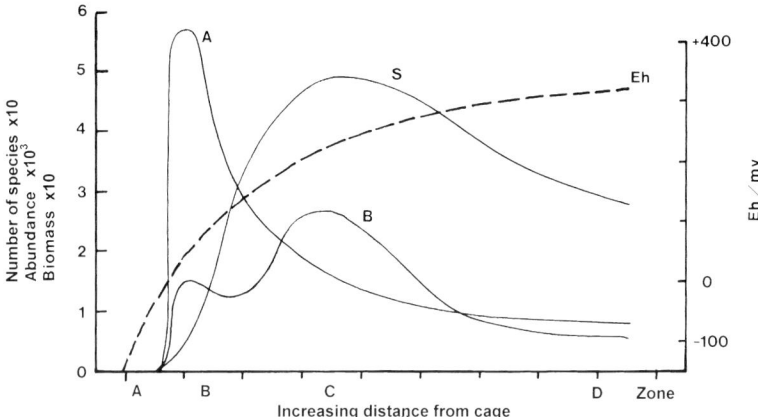

Figure 1.3 Idealised graph of changing macrobenthic population parameters and sedimentary redox levels with increasing distance from highly-enriched sediments beneath a sea cage. Note the changing scale of the population parameter values. See text for a description of the sequential changes from zone to zone. Key: A, the total abundance of all macrofauna present; B, the total biomass of all macrofauna; S, the total number of macrofaunal species present; Eh, the redox potential of the sediments at 40 mm depth in the sediment column.

continuous supply of oxygen to the sediment water interface, which permits the survival of infauna even when sedimentary boundary layers are anoxic. Conversely, low rates of bottom water renewal can lead to deoxygenation in the overlying water column as sedimentary BOD increases. Findlay and Watling (1997) examined the relationship between remineralisation rate, bioturbation by macrofauna and oxygen supply from the overlying water. They showed that where sediments suffer oxygen deficit for even relatively short periods of a few hours, for example, caused by slack water, large sections of the benthic macrofauna are eliminated. Thus, the degree of impact at a site can be estimated from knowledge of carbon input and from an examination of the prevailing hydrography.

1.3.5.4 Bioindicators of sedimentary enrichment

There is a considerable literature detailing species and higher taxa associated with various degrees of carbon enrichment in boreal marine sediments (e.g. Pearson and Rosenberg, 1978; Pearson *et al.*, 1983; Brown *et al.*, 1987; Weston 1990). The majority of such studies and compilations have shown that there are specific organisms or groups of organisms associated with particular levels of enrichment in any one area. However, only at the highest levels of enrichment-related sedimentary disturbance is there any ubiquity in the organisms recorded. Thus, throughout the world, highly polluted marine sediments are dominated by a very few opportunistic macrofaunal species, such as *Capitella* sp., often in high abundance. *Capitella* sp. (their taxonomy is not yet well established) are small deposit-feeding polychaete worms capable of both benthic and pelagic larval dispersion. They thrive in organic-rich sulphidic sediments and have low oxygen requirements (Chareonpanich *et al.*, 1994a). As such, they are well suited to the rapid colonisation of sediments where the original fauna has been impoverished, as occurs beneath marine fish cages. Other polychaete taxa that commonly occur under such conditions are the surface deposit-feeding members of the Spionid and Ophryotrochid groups, such as *Malacoceros* and *Ophryotrocha* in northern European waters. Large carnivorous nematode worms also frequently predominate in highly-enriched areas, but non-vermiform phyla, such as the Mollusca and Echinodermata, are completely excluded.

In moderately enriched sediments in boreal latitudes, polychaete groups continue to predominate but are largely comprised of other members of the Capitellid, Spionid and Cirratulid families. However, small bivalves of the Tellinid and Erinacean families can also become numerous under such conditions. As detrital inputs decline and conditions approach the normal ambient state, these intermediate populations are gradually replaced by the much more diverse but less dense communities typical of undisturbed conditions. It must be stressed that there are no universal bioindicators for moderate to low levels of sedimentary enrichment. Local conditions and populations govern the composition of benthic communities at these levels of disturbance.

1.3.6 Carbon

1.3.6.1 Carbon inputs

In open fish culture, fish faeces are lost to the environment. Intensive methods, where fish are fed from an external source, also lead to wastage of uneaten food (processed fishmeal, trash fish). Processed fish food is composed of a highly labile (digestible) mixture of protein, fat, carbohydrate and other minor components, such as vitamins and pigments. As it is undigested, uneaten food has a much greater capacity (weight for weight) than faecal material to impact the environment, both in terms of energy content and degradation rate. Current estimates of waste pelletised fish food from salmonid cage culture vary between 5 and 15%, but in fact this has rarely been objectively measured. There is evidence, however, that feed losses are being progressively reduced. A sediment-trap study at an Atlantic salmon farm, in Scotland in 1999, estimated food loss at 0–5% (C.J. Cromey, T.D. Nickell and K.D. Black, unpublished data). This has been forced by declining profits, which compel farmers to use more efficient feeding strategies, and, perhaps, also by the greater palatability of modern diets. In two sediment trap studies in 1993 and 1994 at a sea bass/bream cage site in Greece, an order of magnitude reduction in food wasted between annual sampling events was found, which was thought to be due, in part, to the introduction of new feeding strategies in response to the wastage observed during the initial study (MacDougall and Black, 1999).

Intensive culture in the developed world has come to rely on the use of pelletised fishmeal from high quality and, therefore, highly digestible sources. This has led to lower feed conversion ratios and less waste. Enell (1995) reported that the mean feed coefficient in the Nordic area in 1976 was 2.08, but had fallen to 1.25 in 1994 and was down to 1.0–1.1 in 1995 at Danish and Norwegian farms. Intensive culture in many other areas still relies on waste, or 'trash', fish of variable quality (Wu *et al.*, 1994). As mentioned above, such diets tend to cause greater benthic impacts due to lower digestibility and to wastage during the feeding activity itself.

Several published studies have examined the total amount of particulates released to sediments from fish cages (Hall *et al.*, 1990; Ye *et al.*, 1991; Findlay *et al.*, 1995). Estimates of wastes have varied between 29 and 71% of input carbon, depending on year (Hall *et al.*, 1990), and 78% of input carbon (Ye *et al.*, 1991). The variation in the sedimentation rate data presented in these studies (4.1–77.8 g C m^{-2} day^{-1}) may be attributed to various causes relating to input parameters. However, the difficulty in determining the precise position of sediment traps in relation to (partially mobile) cages may also play a part. A study using precisely-positioned sediment traps under cages at Scottish farms gave rates in the range 11–33 g solids m^{-2} day^{-1} (T.D. Nickell, personal communication). To put these input rates in context, typical sedimentation rates near long-sea sewage outfalls around the British coast are <1 g C m^{-2} day^{-1}, mainly because of their exposed, dispersive locations (Cromey *et al.*, 1998).

The isotopic ratio of organic carbon in sediments has been used as a tracer indicating the extent of dispersion of waste organic material (Ye *et al.*, 1991), as has the presence in sediments of lipids derived from fishmeal (Henderson *et al.*, 1998). In both of these studies, the influence of the farm was measurable at considerable distance from the cages.

1.3.6.2 Carbon recycling and benthic productivity in enriched sediments

Various studies have shown that microbiological metabolism in sediments is stimulated by organic wastes from fish farms. In a shallow marine cage farm in Kolding fjord, Denmark, microbiological metabolism in the sediments was measured at around 10 times (525–619 mmol CO_2 m^{-2} day^{-1}) that at a reference site, most of which could be accounted for by sulphate reduction (Holmer and Kristensen, 1992).

Studies on the production of macrobenthic populations in the vicinity of cage farms in the west of Scotland (Pearson *et al.*, 1995) have suggested that the production of infaunal benthos close to the cages is 4–6 times the background levels. This supports a greatly increased population of epibenthic predators, principally demersal fish and malacoceran crustacea, whose productivity is some 50% higher than that of similar predator populations on normal sediments. Thus, the cage farms make a significant contribution to the production of surrounding native populations.

1.3.7 Seasonality in benthic response to sedimentary enrichment

Several studies have shown that the micro- and macrobiological changes associated with sedimentary enrichment are influenced by the seasonal cycle. Holmer and Kristensen (1992, 1996) found that the sulphate reduction rate in surface sediments at a marine fish farm in shallow Danish waters varied from low levels in the early spring and late summer to peaks in early summer and winter, when pools of organic matter decreased. Methane production occurred concurrently with sulphate reduction in the late summer. Peaks in pore water solutes appeared in late summer but diminished as water temperatures fell in the autumn. Karakassis and co-workers (1998) also investigated seasonal changes in geochemical processes in the sediments around farm cages in Cephalonia Bay in the eastern Mediterranean, and reported increased accumulation of organic matter in summer followed by increased carbon mineralisation in the cooler winter months.

In contrast to these case histories, Johnsen (1996) reported no seasonal effects on the concentration of fatty acids, which act as markers for microbiological sedimentary activity, in the sediments adjacent to a fish cage in western Norway.

Hargrave and co-workers (1993, 1997) found that maximum ammonium release from the sediments coincided with highest water temperatures in July, around and beneath salmon cages in the Bay of Fundy, Nova Scotia. Peak oxygen uptake and sulphide accumulation in the sediments occurred in September. The latter was in excess of 100 mM sulphur in the pore water in that month, which could have been toxic to infaunal benthos and coincided with a reduction in the population of *Capitella* spp. In a similar study in coastal waters in Maine, Findlay *et al.* (1995) reported low levels of sedimentary organic matter beneath cages in spring and early summer, followed by increases which reached a peak in November, accompanied by the development of surface bacterial mats and a switch in the benthic communities to those associated with high levels of enrichment. Subsequent storm-related resuspension reversed this development. In the warmer waters of an embayment in southern Japan, with a 22-year history of cage aquaculture, Tsutsumi (1995) found that extremely reducing conditions developed in the bottom water throughout the bay in the summer months, resulting in the elimination of all bottom fauna. Recolonisation took place during the autumn and spring but only by an opportunist polychaete fauna. Molluscs, which had formerly dominated the benthos, were entirely absent and the total biomass of benthos had markedly decreased.

1.3.8 Benthic recovery

The rate at which sedimentary ecosystems recover following the removal of cages or the cessation of farming is of considerable interest, particularly as the fallowing of sites and rotation of cages has now become recommended practice in many areas where hypertrophic sedimentary conditions are a problem. Up to the present time, there have been few detailed studies of the time taken for sedimentary conditions to return to an undisturbed state following the fallowing of a cage site. Karakassis and co-workers (1999) reported on one such study undertaken over 2 years following the removal of cages at an intensive farm in Cephalonia Bay, Greece. They found that most geochemical variables in the sediment 10 m from the edge of the cage site had values similar to those found at the reference station 1.2 km from the site 11 months after cage removal. However, values at a station in the centre of the site were still showing large fluctuations, indicative of continuing enrichment, after 23 months. At this time, the benthic populations at the centre station were still dominated by small opportunistic polychaetes typical of high levels of enrichment, although the presence of some small bivalves and infaunal brittle stars suggested that the recovery process was continuing. At the 10 m station, the fauna showed a strong recovery at the end of the first year, but a regression to hypertrophic characteristics occurred during the second year. The communities were then again dominated by small polychaetes, although the bivalves, crustacea and echinoderms, which had recolonised that area during the first year, retained their presence. This secondary regression was attributed to increased carbon input from benthic algal production stimulated by nutrient regeneration from the enriched sediments during the recovery process in the summer months. Lu and Wu (1998) carried out an experimental study on the macrobenthic recolonistation of trays of defaunated, organically-enriched sediment collected from beneath a fish farm in Hong Kong. The trays were exposed to the subtidal area of a clean site with similar hydrographic characteristics. Recolonisation was rapid, with 144 individuals from 26 species per tray recorded within the first month. Molluscs made up half the species present but polychaetes comprised 80% of the total abundance. Maximum abundance in the trays was recorded after 3 months and species richness peaked after 4 months, followed in both cases by a decline to putative background levels. It was suggested that, because of such rapid recolonisation, cage farming in the area would be unlikely to have any long-term effects.

Little has been published concerning the recovery times of benthic populations around fish farms in northern boreal areas. However, at a number of fish cage farms in the west of Scotland, a two-year study has been undertaken of benthic faunal succession and sedimentary conditions during fallowing periods following harvesting of mature cage stocks. The results are summarised in Table 1.2, showing that the communities adjacent to the cages returned to normal 21–24 months after destocking. The switch from highly enriched (type B) to moderately enriched (type C) communities took 9 months, followed by a further 9 months to achieve a lightly-enriched status. Species numbers peaked at 18 months.

These various studies suggest that recovery rates are considerably higher in the warmer waters of the eastern Mediterranean and the South China Sea than in the boreal North Atlantic. However, it should be noted that local hydrographic conditions have a considerable influence on recovery rates in all areas. In quiescent areas, normalisation of the benthic community might be expected to take considerably longer that in more hydrodynamically energetic areas.

1.4 Ecological impacts on wild populations

The genetic impacts of escaped farmed species are reviewed by Kapuscinski and Brister (Chapter 6 of this volume). Ecological effects from the discharge of nutrient and carbon wastes are discussed above, but cage farms may also have ecological effects stemming from the release of parasites and pathogens. It is, however, difficult to find diseased animals in the marine environment—such animals quickly succumb to predation. It is, therefore, not easy to be confident about the frequency or significance of transfer of pathogens to wild stocks. An infamous example involved *Gyrodactylus salaris*, a monogenean parasite of salmon, which was transferred from resistant Baltic salmon populations to Norwegian populations lacking resistance, as a result of movements of farmed fish stocks in the mid-1970s. This resulted in the extinction of many wild populations (Johnsen and Jensen, 1996).

Sea lice infestations are endemic in most salmonid culture areas and, in recent years, declines in wild salmonid populations have led to the widespread belief that there is a link between farming and this decline. In Scotland, the main focus has been on wild sea trout, *Salmo trutta*, which have seen catastrophic population declines, particularly in the North-west where salmon culture is concentrated (Northcott and Walker, 1996). On their first visit to sea in the spring of the year following hatching, these animals may be confronted with very high standing stocks of infective sea lice larval stages and quickly become infested with up to several hundred lice (personal observation). Although these fish may choose to return to fresh water to avoid the parasite, it is likely that many are severely compromised. A burden of only 10 adult lice is sufficient to cause mortality, especially in fish that may not have fully developed their osmoregulatory system and are, therefore, already under stress (Jens Christian Holst, personal communication).

The position is less clear with wild Atlantic salmon, *Salmo salar*, which are also in decline in many areas where fish farms are common. Smolts of this species migrate directly to the ocean without remaining in the coastal or estuarine zone, as do sea trout. It was previously thought that these fish would not be exposed to the same degree of infestation, owing to the limited period of contact. However, it is now suggested that, particularly in long fjordic systems with several fish farms, salmon may receive sufficient infestation to compromise their survival. This hypothesis is not easy to test as it is rather difficult to catch salmon smolts in coastal waters, particularly in such a way as to protect the fish from skin/scale damage, which may remove any early lice stages present. However, researchers in Norway have recently made significant progress in this area, using a fishing net with an aquarium in the cod-end designed to minimise damage to the fish. The results from a cooperative research project between the Institute of Marine Research, Bergen, Norway and the University of Bergen indicate that more than 86% of the wild postsmolts of Atlantic salmon migrating out of the Sognefjord, and 48.5–81.5% of the postsmolts from the Nordfjord were killed as a direct consequence of sea lice infections during the spring of 1999. The surviving fish were probably weakened due to the infection. Only two fjords were investigated, but it seems probable that postsmolts from other fjords also experience the same problem (Jens Christian Holst, personal communication).

Although the relationship between sea lice infection and the decline of wild populations is striking, there is no conservative proof of a causal link, as the origin of lice on wild fish has not been clearly established. There is, however, genetic evidence which might support the case that at least some lice on wild fish are of farmed fish origin (Todd *et al.*, 1997).

It has now become a priority to reduce lice numbers on salmon farms, and regulations are in force, or anticipated, in the UK, Norway, Canada, Ireland and Chile. The main features of strategies to reduce lice numbers include:

- Regular monitoring of lice numbers
- Coordinated chemical treatments between farms sharing the same body of water
- Single generation sites
- Fallowing of management areas to break lice cycles
- Treatment of lice in the spring, when their numbers are low

In the past, fish farmers had access to only a few treatment agents, such as dichlorvos (an organophosphorus now banned for use on fish) and hydrogen peroxide. As a consequence of this limited chemical armoury, there have been many reports of reduced efficacy caused by resistance. There are, however, several new lice treatment agents on, or approaching, the market that should have a dramatic effect on reducing lice numbers.

Even with this greater access to efficacious sea lice treatment agents, it is doubtful if total lice numbers can be brought down to low enough levels to fully protect wild salmonids. This is a consequence of the continuously increasing numbers of fish entering culture; the numbers of farmed fish far exceeds the size of wild populations. Any decrease in lice numbers by a lowering of acceptable levels on farmed fish is likely to be counterbalanced by future increases in production. Given that there will always be economic and environmental constraints on the frequency of therapeutic application, it would appear that if lice from salmon farming are a major contributor to declines in wild populations, we will have to await a much more radical solution, for example, a totally effective vaccine.

With the exception of sea lice, there appears to be no significant transfer of parasites between farmed Atlantic salmon and wild populations—in fact, the reverse transfer is more apparent (McGeorge and Sommerville, 1996). Little research has been reported on the parasitic interactions of other cultured species and wild populations.

The potential for bacterial and viral diseases to be transmitted from farmed fish to wild is real. Furunculosis (caused by the bacteria, *Aeromonas salmonicida*) was believed to have been reintroduced to Norway via cultured-fish imports from Scotland in 1985, causing severe damage both to farmed and wild populations (Johnsen and Jensen, 1996).

During and since the major outbreak of infectious salmonid anaemia (ISA) in a number of Scottish fish farms in 1998–99, there have been several claims of a threat to wild populations. That ISA is present in wild populations is now confirmed (Scottish Executive Press Release, 4/11/99) but it is not clear whether this is a consequence of the outbreak in farmed stocks, nor is it clear what impact the disease will have on wild populations. The policy of complete eradication of this disease by compulsory slaughter must be beneficial to wild stocks, if there is a risk of disease transfer.

Escaped fish may also have ecological impacts. If the cultured species is exotic (but reproductively competent), once escaped large effects are possible in native populations. The classic example of an introduction gone wrong (in this case intentional and for commercial fishery purposes) is that of the Nile perch, a large piscivorous fish, which has decimated the native fish of Lake Victoria, Kenya (Gophen *et al.*, 1995; Goudswaard and Witte, 1997). The culture of endemic species (and local strains) must, therefore, be preferred on ecological grounds. Endemic species may also perform better than exotics due to their local adaptations, especially with regards to disease resistance.

Fish farming may also have a direct ecological effect on wild fish populations by providing food, either directly or through an increase in algal and zooplankton biomass, and by providing refuge for fish around cage structures. This has been studied more thoroughly in fresh water (reviewed by Phillips *et al.*, 1985), where populations of species including roach and native brown trout have been observed to increase as a consequence of rainbow trout cage culture. Little is known about fish populations around marine cages, although Carss (1990) has shown that wild fish may congregate in large numbers around Atlantic salmon cages in Scottish sea lochs. In addition to the species mentioned by Carss (1990), very large numbers of mackerel may congregate around such cages during the summer months, with gut-content analysis revealing food pellets (K.D. Black, personal observations). Large concentrations of a variety of wild and escaped fish have also been observed around cages in Greece (Papoutsoglou *et al.*, 1996; Mac Dougall and Black, 1999) and in France (Pergent *et al.*, 1999), where they appear to reduce impacts on the benthos. Although these effects may be important locally, nothing is known of their wider significance to wild fish populations and this remains an important area for study.

Published reports of interactions between birds and marine fish culture are relatively few, although several species are known to take fish from ponds and cages. In such cases, deterrence is relatively easy and shooting must be viewed as a management failure. Damage to bird populations from shooting around Scottish marine farms is unlikely to be significant (Furness, 1996). Mammals, such as seals, otters and mink, may also cause problems and proximity to known haunts of these animals is often considered when determining the siting of new culture operations. There are few data concerning the effects of shooting or entrapment on local mammal populations, as activities of this sort, even when licensed, are likely to be underreported. Farmed fish losses may, however, be locally significant, especially from seals.

1.5 Impacts in differing environments

Although most research has concentrated on marine culture in temperate zones, much of the major expansion in aquaculture is occurring in subtropical and tropical zones. It is relatively easy, therefore, to pick out those few research areas that have been studied and to identify those where further work is urgently required.

1.5.1 Mediterranean

Very little information has been published regarding impacts in the Mediterranean, where there is a rapidly expanding marine fish culture industry, especially in Greece. Due to its restricted nature, the Mediterranean has an insignificant tidal range and currents are, therefore, predominantly wind and density driven. Given the weak-to-moderate nature of the resulting flows in many areas it might be thought that benthic impacts would be extreme.

There are, however, important differences between eastern Mediterranean coastal sites and those in the north Atlantic. For example, water temperatures are higher, many areas are highly oligotrophic and they may be phosphorus rather than nitrogen limited (Krom *et al.*, 1991). Mac Dougall and Black (1999) examined the benthic impacts at a long-established European sea-bass, *Dicentrarchus labrax*, and gilt-head sea bream, *Sparus aurata*, fish farm in Greece. They presented evidence from an acoustic ground discrimination survey, supplemented by photographs from a remotely-deployed underwater camera. This supported the conclusion that the effects of farming on the benthic environment around fish cages are very much less than those typically experienced in comparatively eutrophic areas, for example in Scottish sea lochs.

From the limited information available in the Aegean, the impacts of aquaculture often appear to be qualitatively different from those described elsewhere (Papoutsoglou et al., 1996; Mac Dougall and Black, 1999). In this oligotrophic environment, sites are characterised by very large densities of wild and escaped fish outside the cages. These can be observed not only eating waste food pellets but also grazing on algae growing on the nets. It is hypothesised that these animals are responsible for consuming much of the particulate waste.

Karakassis et al. (1998, 1999) examined the biogeochemistry and biology of sediments beneath cages in Cephalonia Bay, western Greece. Thick, black, organic-rich sediments were sampled, which had low redox potentials and, in many respects, showed similar impacts to those typical of salmon farms in comparatively eutrophic, cool-temperate coastal areas. In a subsequent paper, Karakassis and co-workers (in press) have contrasted the impacts in Cephalonia Bay with two other sites in Greece and report lower impacts (positive redox potentials) at these sites, which are characterised by coarse sandy sediments as opposed to the silty muds found in Cephalonia Bay. Strong relationships between macrofaunal indices and distance from the cages are reported for all three sites.

Meadows of the seagrass, *Posidonia oceanica*, cover vast areas in shallow regions of the Mediterranean. They are regarded as the cornerstone of the littoral ecosystem, providing a wide variety of niches, accounting for the high diversity of these areas. A recent study (Pergent et al., 1999) examined the impact of cage farming on this protected species. It concluded that while the nutrient input from the farms studied resulted in an increase in leaf length and increases in the biomass of epiphytes and ichthyofauna, there was evidence of decreased meadow density and total disappearance beneath cages. Because of its ability to record environmental alterations, including nutrients, light and trace metals (by examination of dead leaf sheaths), it is proposed that *Posidonia* is a useful bioindicator for monitoring fish farm impacts in these environments.

1.5.2 Tropical culture

Much less work has been published on impacts in warmer climates (see Choo, Chapter 4 of this volume, for a discussion of impacts of crustacean culture in tropical areas). Wu et al. (1994) studied four small, shallow farms around Hong Kong and reported broadly similar results to Brown et al. (1987), but found that benthic impacts extended up to 1–1.5 km from the site. This covers a considerably larger area than that described by Brown et al. (1987), which may be due to the use of trash fish as feed in the Hong Kong study. Particles of waste fish may have a much lower sinking velocity than pellets and can, therefore, be dispersed across a greater area. Unfortunately, Wu et al. (1994) provided no data on the positioning of their sampling stations relative to the cages, and their method of determining the affected area is not clear. In their study of nitrogen budgets, Leung *et al.* (1999) predicted that from an annual production of only 3000 tonnes, 962 tonnes of nitrogen are released into the Hong Kong marine environment.

1.6 Minimisation of impacts

The use of filter-feeding shellfish species to reduce impacts of intensive culture while producing a secondary product has also been examined. Oysters, *Crassostrea gigas*, grow up to three times faster, have greater condition factors and faster post-spawn recovery when grown near marine Chinook salmon cage farms compared to control sites, as a consequence of the greater availability of particulate organic matter (Jones and Iwama, 1991). However, these results were not repeated with mussels, *Mytilus edulis*, where no increase in performance with proximity to the farm was found (Taylor *et al.*, 1992).

In several areas of the enclosed coastal seas of Japan, where intensive cage culture has been practised for many years, sediments around fish farms may become anoxic and azoic during the summer months. There is the potential for the rapid degradation of the accumulated organic material by seeding the seabed with large numbers of cultured *Capitella* sp. (Chareonpanich *et al.*, 1993). It was calculated that abundances of 59,000 individuals m^{-2} could treat organic material arriving at the sediment at the rate of $0.58 \, kg \, m^{-2} \, yr^{-1}$. This represents a considerably lower sedimentation rate than is common at many farms in Western Europe (up to $15 \, kg \, solids \, m^{-2} \, yr^{-1}$), where the abundance of *Capitella* sp. is typically up to 90,000 individuals m^{-2} (T. Nickell, personal communication). Thus, taking the calculations of Chareonpanich and co-workers (1994b), *Capitella* only directly process a fraction of the total organic input to these sites. However, the working of the sediments by these animals will render organic material more available to microbiota.

Various methods of reducing the discharge of feed from farms have been examined, including: using an acoustic feed detector on marine cage sites to reduce the loss of medicated pellets (Ervik *et al.*, 1994); using various sensors to detect when fish reduce feeding activity, linked to feed input controllers; and use of cages that are completely enclosed by a tarpaulin, supplied with water by a pump and with recovery of wastes for processing. None of these systems are in very widespread use at the moment, although it is feasible that a regulator might require their use in certain circumstances.

The practice of site rotation of marine cage sites, allowing the seabed to return to more normal conditions for several months or years before further farming, is becoming increasingly common (Pereira, 1997), and may become a condition of effluent discharge regulations in Scotland. This is seen by many as having positive benefits, both for the environment and future farm productivity. While short-term (a few weeks or months) fallowing to break disease cycles is generally regarded as having production advantages (Wheatley *et al.*, 1995), the benefits to the environment are less clear. Implicit in site rotation is that more sites are required than are used at any one time, thereby increasing the area of seabed over which contamination occurs. A more environmentally-sound alternative may be simply to limit the size of each farm, in terms of stocking density per unit area, to that where the seasonally-averaged inputs of organic material to the seabed are matched by the rate of benthic remineralisation, thus reaching some environmentally-stable steady state.

1.7 Research needs

At present, there is an urgent need to assess the environmental consequences of the use of the new generation of anti-sea-lice prophylactic treatments. These have the potential to affect many wild pelagic and benthic invertebrate species. The seriousness and extent of this threat is not yet known, although studies have recently been initiated in Scotland to assess their ecological impact on local populations around cage farms. The interactions between wild and cultured fish populations in relation to the transmission of parasites and pathogens also requires further detailed study in order to minimise the likelihood of cross-infections. In particular, it is essential that the contribution that farmed-origin sea lice are making to the current pressure on wild salmonids must be definitively determined, so that serious attempts can be made to protect such stocks prior to their local extinction.

Much further research is needed into the potential for the development of eutrophic problems in the vicinity of cage farms in the warmer waters of lower latitudes. It is possible that hypernutrification will have greater potential to influence pelagic ecosystem function in oligotrophic areas, such as parts of the Mediterranean and the Tropics.

Further methods of minimising environmental impact need to be devised. In this area, research should be encouraged into the establishment and cost-benefits of plural cultures, which include fish, shell fish and benthic detritivores on the same site.

1.8 Conclusions

In the past, the rapid growth of the salmonid farming industry, which outstripped the understanding of its environmental consequences, resulted in avoidable disturbances both to cage production rates and to the local environment. Recent advances in our knowledge of the environmental consequences of cage farming and in regulatory practices based on this knowledge have greatly improved the situation in many areas. However, much research and developmental work remains to be carried out to improve environmental best practice, both in farm management and in statutory control, particularly in the rapidly developing cage farm industries in the warmer low-latitude regions.

References

Anderson, D.M. (1997) Turning back the harmful red tide. *Nature*, **388** 513-514.
Arzul, G., Clement, A. and Pinier, A. (1996) Effects on phytoplankton growth of dissolved substances produced by fish farming. *Aquatic Living Resources*, **9** 95-102.
Berry, A.W. (1996) Aquaculture and sea loch nutrient ratios: a hypothesis, in *Aquaculture and Sea Lochs* (ed. K.D. Black), Scottish Association for Marine Science, Oban, pp. 7-15.
Beveridge, M.C.M., Ross, L.G. and Kelly, L.A. (1994) Aquaculture and biodiversity. *Ambio*, **23** 497-502.
Bonsdorff, E., Blomqvist, E.M., Mattila, J. and Norkko, A. (1996) Long-term changes and coastal eutrophication; examples from the Åland Islands and the Archipelago Sea, northern Baltic Sea. *Oceanologica Acta*, **20** 319-329.
Bonsdorff, E., Blomqvist, E.M., Mattila, J. and Norkko, A. (1997) Coastal eutrophication: causes, consequences and perspectives in the archipelago areas of the northern Baltic Sea. *Estuarine, Coastal and Shelf Science*, **44** (Suppl. A) 63-72.
Brown, J.R., Gowen, R.J. and McLusky, D.S. (1987) The effect of salmon farming on the benthos of a Scottish sea loch. *Journal of Experimental Marine Biology and Ecology*, **109** 39-51.
Buttle, L.G., Uglow, R.F. and Cowx, I.G. (1996) Temperature and dietary factors affecting the nitrogen efflux rates. *Aquaculture Research*, **27** 391-397.
Carss, D.N. (1990) Concentrations of wild and escaped fishes immediately adjacent to fish farm cages. *Aquaculture*, **90** 29-40.
Chareonpanich, C., Montani, S., Tsutsumi, H. and Matsuoka, S. (1993) Modification of chemical characteristics of organically-enriched sediment by *Capitella* sp. I. *Marine Pollution Bulletin*, **26** 375-379.
Chareonpanich, C., Montani, S., Tsutsumi, H. and Nakamura, H. (1994a) Estimation of oxygen-consumption of a deposit-feeding polychaete, *Capitella* sp. I. *Fisheries Science*, **60** 249-251.
Chareonpanich, C., Tsutsumi, H. and Montani, S. (1994b) Efficiency of the decomposition of organic matter, loaded on the sediment, as a result of the biological activity of *Capitella* sp. I. *Marine Pollution Bulletin*, **28** 314-318.
Cromey, C.J., Black, K.D., Edwards, A. and Jack, I.A. (1998) Modelling the deposition and biological effects of organic carbon from marine sewage discharges. *Estuarine, Coastal and Shelf Science*, **47** 295-308.
Cromey, C.J., Nickell, T.D. and Black, K.D. (Submitted) Modelling the deposition and biological effects of waste solids from marine cage farms. DEPOMOD.
Davies, I.M., Smith, P., Nickell, T.D. and Provost, P.G. (1996) Interactions of salmon farming and benthic microbiology in sea lochs, in *Aquaculture and Sea Lochs* (ed. K.D. Black), Scottish Association for Marine Science, Oban, pp. 27-32.
Dosdat, A., Servais, F., Metailler, R., Huelvan, C. and Desbruyeres, E. (1996) Comparison of nitrogenous losses in five teleost fish species. *Aquaculture*, **141** 107-127.
Duplisea, D.E. and Hargrave, B.T. (1996) Response of meiobenthic size-structure, biomass and respiration to sediment organic enrichment. *Hydrobiologia*, **339** 161-170.
Edwards, A. and Sharples, F. (1986) *Scottish Sea Lochs: A Catalogue*. Scottish Marine Biological Association, Oban.
Enell, M. (1995) Environmental impact of nutrients from Nordic fish farms. *Water Science and Technology*, **31** 61-71.

Ervik, A., Samuelsen, O.B., Juell, J.E. and Sveier, H. (1994) Reduced environmental impact of antibacterial agents applied in fish farms using the Liftup feed collector system or a hydroacoustic feed detector. *Diseases of Aquatic Organisms*, **19** 101-104.

Ervik, A., Hansen, P.K., Aure, J., Stigebrant, A., Johannessen, P. and Jahnsen, T. (1997) Regulating the local environmental impact of intensive marine fish farming. 1. The concept of the MOM system (modelling ongrowing fish farms monitoring). *Aquaculture*, **158** 85-94.

Findlay, R.H. and Watling, L. (1997) Prediction of benthic impact for salmon net-pens based on the balance of benthic oxygen supply and demand. *Marine Ecology Progress Series*, **155** 147-157.

Findlay, R.H., Watling, L. and Mayer, L.M. (1995) Environmental impact of salmon net-pen culture on marine benthic communities in Maine: a case study. *Estuaries*, **18** 145-179.

Furness, B.W. (1996) Interactions between seabirds and aquaculture in sea lochs, in *Aquaculture and Sea Lochs* (ed. K.D. Black), The Scottish Association for Marine Science, Oban, pp. 50-55.

Gophen, M., Ochumba, P.B.O. and Kaufman, L.S. (1995) Some aspects of perturbation in the structure and biodiversity of the ecosystem of Lake Victoria (East-Africa). *Aquatic Living Resources*, **8** 27-41.

Goudswaard, K.P.C. and Witte, F. (1997) The catfish fauna of Lake Victoria after the Nile perch upsurge. *Environmental Biology of Fishes*, **49** 21-43.

Gowen, R.J. and Ezzi, I.A. (1992) Assessment and prediction of the potential for hypernutrification and eutrophication associated with salmonid culture in Scottish coastal waters. Dunstaffnage Marine Laboratory, Oban, ISBN 0-9518959-0-7.

Gowen, R.J. and Rosenthal, H. (1993) The environmental consequences of intensive coastal aquaculture in developed countries: what lessons can be learnt?, in *ICLARM Conference Proceedings*, Vol. 31 (eds. R.S.V. Pullin, H. Rosenthal and J.L. Maclean), Phillipines, Manila, pp. 102-115.

Hall, P.O.J., Anderson, L.G., Holby, O., Kollberg, S. and Samuelsson, M.O. (1990) Chemical fluxes and mass balances in a marine fish cage farm. 1. Carbon. *Marine Ecology Progress Series*, **61** 61-73.

Hall, P.O.J., Holby, O., Kollberg, S. and Samuelsson, M.O. (1992) Chemical fluxes and mass balances in a marine fish cage farm. 4. Nitrogen. *Marine Ecology Progress Series*, **89** 81-91.

Handy, R.D. and Poxton, M.G. (1993) Nitrogen pollution in mariculture: toxicity and excretion of nitrogenous compounds by marine fish. *Reviews in Fish Biology and Fisheries*, **3** 205-241.

Hargrave, B.T., Duplisea, D.E., Pfeiffer, E. and Wildish, D.J. (1993) Seasonal changes in benthic fluxes of dissolved oxygen and ammonium associated with marine culture of Atlantic salmon. *Marine Ecology Progress Series*, **96** 249-257.

Hargrave, B.T., Phillips, G.A., Doucette, L.I., White, M.J., Milligan, T.G., Wildish, D.J. and Cranston, R.E. (1997) Assessing benthic impacts of organic enrichment from marine aquaculture. *Water, Air and Soil Pollution*, **99** 641-650.

HelCom (1999) Annex C-8 Soft Bottom Macrozoobenthos (http://www.helcom.fi/manual2/anxc8.html).

Henderson, R.J., Forrest, D.A.M., Black, K.D. and Park, M.T. (1998) Environmental distribution of lipids in sealoch sediments underlying marine fish cages. *Aquaculture*, **158** 69-83.

Holby, O. and Hall, P.O.J. (1991) Chemical fluxes and mass balances in a marine fish cage farm. 2. Phosphorus. *Marine Ecology Progress Series*, **70** 263-272.

Holby, O. and Hall, P.O.J. (1994) Chemical fluxes and mass balances in a marine fish cage farm. 3. Silicon. *Aquaculture*, **120** 305-318.

Holmer, M. and Kristensen, E. (1992) Impact of marine fish cage farming on metabolism and sulphate reduction of underlying sediments. *Marine Ecology Progress Series*, **80** 191-201.

Holmer, M. and Kristensen, E. (1996) Seasonality of sulphate reduction and pore water solutes in a marine fish farm sediment: the importance of temperature and sedimentary organic matter. *Biogeochemistry*, **32** 15-39.

ICES (in press) *Techniques in Marine Monitoring. Soft Bottom Macrofauna: Collection, Treatment and Quality Assurance of Samples* (ed. H. Rumohr). Preprint version, available for download from ICES home pages.

Johnsen, B.O. and Jensen, A.J. (1996) *Gyrodactylus* and furunculosis in Atlantic salmon rivers in Norway, in *The Role of Aquaculture in World Fisheries.* (ed. T.G. Heggeberget), Oxford & IBH Publishing Co., New Dehli, pp. 52-64.

Johnsen, R.I. (1996) Environmental and nutritional aspects in cultivation of Atlantic salmon (*Salmo salar*) evaluated by fatty acids. Dr Scient. thesis, Department of Chemistry, University of Bergen, ISBN 82-7406-015-66, 120 pp.

Jones, T.O. and Iwama, G.K. (1991) Polyculture of the Pacific oyster, *Crassostrea gigas* (Thunberg), with chinook salmon, *Oncorhynchus tshawytscha*. *Aquaculture*, **92** 313-322.

Karakassis, I., Tsapakis, M. and Hatziyanni, E. (1998) Seasonal variability in sediment profiles beneath fish farm cages in the Mediterranean. *Marine Ecology Progress Series*, **162** 243-252.

Karakassis, I., Hatziyanni, E., Tsapakis, M. and Plaiti, W. (1999) Benthic recovery following cessation of fish farming: a series of successes and catastrophes. *Marine Ecology Progress Series*, **184** 205-218.

Karakassis, I., Tsapakis, M. Hatziyanni, E., Papadopoulou, K.-N. and Plaiti, W. (In press) Impact of fish farming on the seabed in three Mediterranean coastal areas. *ICES Journal of Marine Science*.

Krom, M.D., Kress, N., Brenner, S. and Gordon, L.I. (1991) Phosphorus limitation of primary productivity in the eastern Mediterranean sea. *Limnology and Oceanography*, **36** 424-432.

Leung, K.M.Y., Chu, J.C.W. and Wu, R.S.S. (1999) Nitrogen budgets for the areolated grouper, *Epinephelus areolatus*, cultured under laboratory conditions and in open-sea cages. *Marine Ecology Progress Series*, **186** 271-281.

Lu, L. and Wu, R.S.S. (1998) Recolonisation and succession of marine macrobenthos in organically-enriched sediment deposited from fish farms. *Environmental Pollution*, **101** 241-251.

Mac Dougall, N. and Black, K.D. (1999) Determining sediment properties around a marine cage farm using acoustic ground discrimination: RoxAnn^{TM}. *Aquaculture Research*, **30** 1-8.

McCaig, A.E., Phillips, C.J., Stephen, J.R., Kowalchuk, J.R., Harvey, S.M., Herbert, R.A., Embley, T.M. and Prosser, J.I. (1999) Nitrogen cycling and community structure of proteobacterial beta-subgroup ammonia oxidising bacteria within polluted marine fish farm sediments. *Applied and Environmental Microbiology*, **65** 213-220.

McGeorge, J. and Somerville, C. (1996) The potential for interaction between parasites of wild salmonids, non-salmonids and farmed Atlantic salmon in Scottish sea lochs, in *Aquaculture and Sea Lochs* (ed. K.D. Black), Scottish Association for Marine Science, Oban, pp. 59-71.

Munday, B.W., Eleftheriou, A., Kentouri, M. and Divanach, P. (1992) *The Interaction of Aquaculture and the Environment—A Bibliographical Review.* Commisioners of the European Community, DGXIV/D/3, Brussels.

Northcott, S.J. and Walker, A.F. (1996) Farming salmon, saving sea trout: a cool look at a hot issue, in *Aquaculture and Sea Lochs* (ed. K.D. Black), The Scottish Association for Marine Science, Oban, pp. 72-81.

NS 9423 (1998) Water quality—guidelines for quantitative investigations of sublittoral soft-bottom benthic fauna in the marine environment. Norwegian General Standardizing Body (NAS), Oslo, 16 pp. (in Norwegian, English translation available from NAS).

NS 9410 (2000) Environmental monitoring of marine fish farms. Norwegian General Standardizing Body (NAS), Oslo, 16 pp. (in Norwegian, English translation available from NAS).

OsParCom (1997) Eutrophication and Monitoring Guidelines: Benthos. JAMP (Joint Assessment and Monitoring Programme), Oslo and Paris Commissions.

Papoutsoglou, S., Costello, M.J., Stamou, E. and Tziha, G. (1996) Environmental conditions at sea-cages, and ectoparasites on farmed European sea-bass, *Dicentrarchus labrax* (L.), and gilt-head sea-bream, *Sparus aurata* L., at two farms in Greece. *Aquaculture Research*, **27** 25-34.

Pearson, T.H. and Rosenberg, R. (1978) Macrobenthic succession in relation to organic enrichment and pollution of the marine environment. *Oceanography and Marine Biology Annual Review*, **16** 229-311.

Pearson, T.H. and Stanley, S.O. (1979) Comparative measurement of redox potential of marine sediments as a rapid means of assessing the effect of organic pollution. *Marine Biology*, **53** 371-379.

Pearson, T.H., Gray, J.S. and Johannessen, P.J. (1983) Objective selection of sensitive species indicative of pollution-induced change in benthic communities. 2. Data analyses. *Marine Ecology Progress Series*, **12** 237-255.

Pearson, T.H., Blackstock, J. and Duncan, J.A.R. (1995) Productivity of organically enriched ecosystems. *MAFF R and D Project, Code CSA1628/2567. Final Report*. Part 1: Text, 48 pp. Part 2: Annexes 1-4.

Pereira, P.M.F. (1997) Macrobenthic succession and changes in sediment biogeochemistry following marine fish farming. Ph.D. thesis, University of Stirling.

Pergent, G., Mendez, S., Pergent-Martini, C. and Pasqualini, V. (1999) Preliminary data on the impact of fish farming facilities on *Posidonia oceanica* meadows in the Mediterranean. *Oceanologica Acta*, **22** 95-107.

Phillips, M.J., Beveridge, M.C.M. and Ross, L.G. (1985) The environmental impact of salmonid cage culture on inland fisheries: present staus and future trends. *Journal of Fish Biology*, **27** 123-137.

Pitta, P., Karakassis, I., Tsapakis, M. and Zivanovic, S. (1998) Natural vs mariculture-induced variability in nutrients and plankton in the eastern Mediterranean. *Hydrobiologia*, **391** 181-194.

Rumohr, H. (1999) Soft bottom macrofauna: collection, treatment and quality assurance of samples. *Techniques in Marine Environmental Sciences*, ICES, Copenhagen, **27** 19.

SEPA (1998) Regulation and Monitoring of Marine Cage Fish Farming in Scotland. A Procedures Manual. Version 1. Scottish Environmental Protection Agency, Stirling. 11 Sections + 8 Annexes.

SFT (1999) Environmental monitoring of petroleum activities on the Norwegian Shelf; Guidelines. Norwegian State Pollution Control Authority (SFT) report 99:01, ISBN 82-7655-164-5; 123 pp.

Smayda, T.J. (1990) Novel and nuisance phytoplankton blooms in the sea: evidence for a global epidemic, in *Toxic Marine Phytoplankton* (eds. E. Graneli, B. Sundstom, L. Edler and D.M. Anderson), Elsevier, pp. 29-40.

Sorokin, Y.I., Sorokin, P.Y. and Ravagnan, G. (1996) On an extremely dense bloom of the dinoflagellate, *Alexandrium tamarense*, in lagoons of the Po River Delta: impact on the environment. *Journal of Sea Research*, **35** 251-255.

Strutton, P.G., Bye, J.A.T. and Mitchell, J.G. (1996) Determining coastal inlet flushing times: a practical expression for use in aquaculture and pollution management. *Aquaculture Research*, **27** 497-504.

Taylor, B.E., Jamieson, G. and Carefoot, T.H. (1992) Mussel culture in British Columbia: the influence of salmon farms on growth of *Mytilus edulis*. *Aquaculture*, **108** 51-66.

Todd, C.D., Walker, A.M., Wolff, K., Northcott, S.J., Walker, A.F., Ritchie, M.G., Hoskins, R., Abbott, R.J. and Hazon, N. (1997) Genetic differentiation of populations of the copepod sea louse, *Lepeophtheirus salmonis* (Kroyer), ectoparasitic on wild and farmed salmonids around the coasts of Scotland: evidence from RAPD markers. *Journal of Experimental Marine Biology and Ecology*, **210** 251-274.

Tsutsumi, H. (1995) Impact of fish net pen culture on the benthic environment of a cove in South Japan. *Estuaries*, **18** 108-115.

Weston, D.P. (1990) Quantitative examination of macrobenthic community change along an organic enrichment gradient. *Marine Ecology Progress Series*, **61** 233-244.

Westrich, J.T. and Berner, R.A. (1984) The role of sedimentary organic matter in bacterial sulphate reduction: the G model tested. *Limnology and Oceanography*, **29** 236-249.

Wheatley, S.B., McLoughlin, M.F., Menzies, F.D. and Goodall, E.A. (1995) Site management factors influencing mortality rates in Atlantic salmon (*Salmo salar* L.) during marine production. *Aquaculture*, **136** 3-4.

Wildish, D.J., Keizer, P.D., Wilson, A.J. and Martin, J.L. (1993) Seasonal changes of dissolved oxygen and plant nutrients in seawater near salmonid net pens in the macrotidal Bay of Fundy. *Canadian Journal of Fisheries and Aquatic Sciences*, **50** 303-311.

Wu, R.S.S. (1995) The environmental impact of marine fish culture—towards a sustainable future. *Marine Pollution Bulletin*, **31** 159-166.

Wu, R.S.S., Lam, K.S., Mackay, D.W., Lau, T.C. and Yam, V. (1994) Impact of marine fish farming on water quality and bottom sediment: a case study in the subtropical environment. *Marine Environmental Research*, **38** 115-145.

Ye, L.-X., Ritz, D.A., Fenton, G.E. and Lewis, M.E. (1991) Tracing the influence on sediments of organic waste from a salmonid farm using stable isotope analysis. *Journal of Experimental Marine Biology and Ecology*, **145** 161-174.

2 Freshwater finfish cage culture
L.A. Kelly and I.R. Elberizon

2.1 Introduction

Although the last decade has seen little change in our understanding of the physical impacts of cage aquaculture in freshwater, there has been a radical alteration in the way in which the management and socioeconomic consequences of such activities are addressed. Research, carried out mostly during the 1980s, produced a range of reports on the major physical environmental impacts of intensive freshwater cage aquaculture (e.g. Enell and Löf, 1983; Beveridge, 1984; Merican and Phillips, 1985). The results of these and other studies have been summarised in other texts (e.g. NCC, 1990; Pillay, 1992; Baird et al., 1996). The view prevails that, in common with other areas of aquaculture, cage farms operating at lower intensities are limited in their environmental impact. Biological impacts of intensive aquaculture have largely addressed an agenda that relates to both the security of indigenous fish stocks (Costa-Pierce et al., 1993) and the exploitation of commercial fish species as a food source for aquaculture production (Beveridge et al., 1994).

Freshwater cage culture takes place on a truly global scale, with production occurring on all the inhabited continents. One consequence of this breadth is that a wide variety of production methods exists. Systems used for culture include: net pens or 'hapas', where the bed of the lake or river forms the base of a net corral; small nursery cages (Vines et al., 1996); traditional floating cages; and fully enclosed cage systems. Technologies and materials employed to develop larger volume cage systems do not differ greatly from their marine counterparts. The range of species cultured within cages is similarly wide and includes salmonids, catfish and carp as the main part of the global production inventory. For further details, the reader is directed to Beveridge (1996), who provides a thorough review of methods and systems used in freshwater production.

The major new trends in freshwater cage production are to be found in Asia, where the bulk of all such production presently occurs. Throughout Asia, dam-building programmes by governments are creating large new areas of freshwater, but elsewhere in the world, demands upon existing freshwater bodies are increasing, against a fixed 'stock' of sites. Until relatively recently, intensive freshwater cage aquaculture was largely unknown in South East Asia, with the bulk of freshwater production instead associated with semi-intensive pond culture (De Silva et al., 1991; Bao-Tong, 1994). The current trend, however, is towards intensification and proliferation of cage production, with methodologies applied that were previously associated mostly with temperate fish culture (Li and Mathias, 1994).

Elsewhere in the world, many areas of freshwater are already in use by other stakeholders and the potential for increasing cage farming activities is severely limited, either through a scarcity of suitable new sites or legislative restriction. For example, in countries like Norway and Denmark, freshwater cage aquaculture is prohibited, whilst in Chile, a combination of scarcity of accessible areas, adverse public opinion and regulatory control is creating a shortage of sites (Lindbergh, 1999).

Our understanding of the term 'environmental impact' has, however, widened considerably during the 1990s and now addresses broader issues than the direct effects that human activities may have upon their immediate physical environment. In considering environmental impacts, concepts of 'sustainability' are now embraced. These include not only the immediate impacts of an activity but also the effects on resources used by that activity. As resources are required to maintain intensive production systems and are often obtained at distances remote from the point of manufacture, part of the total impact of the activity itself is displaced to the site of resource extraction. In addition, the economic effect of human activity may also impact upon population and people associated with a given activity, either by their physical proximity or employment. More recently, therefore, discussion of the environmental impacts of aquaculture has increasingly sought to express those concerns in more holistic terms.

This broad interpretation of both the nature and extent of freshwater cage production should not, however, deflect from the debate on the role of aquaculture within the freshwater environment. In a global context, the continuing pressures on standing freshwater resources are well documented (Figure 2.1; UNEP, 1994). At an international level, the emergence of legislation, such as the EU Water Framework Directive, further limits the potential for an 'isolationist' view of aquaculture. Water resource planning on an integrated catchment management basis precludes the treatment of fish farming as anything other than one of a number of potential users attempting to access these resources. Instead, many of the most pressing questions currently faced are those relating to how fish farming fits into the wider mosaic of water, and other, resource usage.

The view is, therefore, that allocation of water resources is a multi-user rather than individual-user problem. This has led to a broadening and deepening of the perspective of many researchers investigating the environmental issues that affect freshwater cage aquaculture at this time. An increased awareness of aquaculture as only one of many, often conflicting interests and users of the freshwater environment, is now to the fore (Muir, 1996; Remane, 1997).

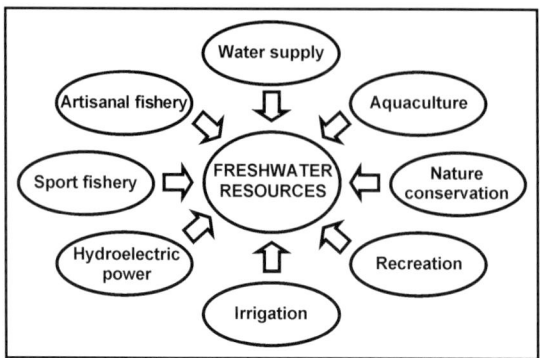

Figure 2.1 Pressures on freshwater resources worldwide.

2.2 The impacts of cage aquaculture in freshwaters

2.2.1 Introduction

The production of an aquaculture species requires the importation of materials and energy to a site and, in common with other forms of productive activity, together with the commodity generated, wastes and by-products also occur. Unlike its land-based counterpart, cage culture relies upon natural forces to deliver water and oxygen, both to sustain production and remove waste composed of fish metabolic outputs and production system losses. It is these wastes, which may also include live biological material, that form the basis of debate on the environmental impacts of cage culture in freshwater and are discussed below. The unique mode of waste output from cages (i.e. direct discharge to a recipient body with no intermediate stage), has considerable bearing upon the quantities and quality of wastes discharged. The limited potential that exists for treatment of waste materials produced is a key issue in the environmental concerns raised against cage production. In this context, the need for appropriate mitigation or containment measures remains strong. The restricted range of options presently available also has an important bearing upon the nature of management of such activities, and is discussed further in Section 3.

2.2.2 Impacts of cage aquaculture at a site

The generation of wastes from cages can be viewed as a simple input–output process. The main input streams are seed stock, feed and water, and the major outputs are composed of solid wastes (mostly faeces and feed), dissolved wastes and harvested biomass (Figure 2.2). Veterinary treatments used in cage farming include both chemical and antibiotic compounds, ranging from simple materials,

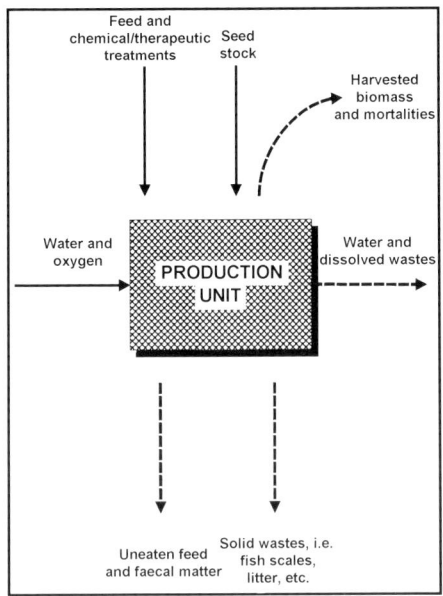

Figure 2.2 Schematic representation of major inputs (solid arrows) and outputs (dashed arrows) from cage aquaculture.

like common salt, through to complex compounds, such as malachite green. Although small in mass relative to the other waste streams, losses of therapeutic treatments from cage systems, as with discharges from their land-based counterparts, are of much concern with regard to their potential environmental impact (Smith et al., 1994a). It is rare, however, to find examples within the scientific literature that deal specifically with the residues of such treatments lost from freshwater cage farms.

The major impacts in the physical environment relating to cage aquaculture remain those of the potential threat of eutrophication, that is, the enrichment of an ecosystem with nutrient elements (Welch, 1992). In terms of cage aquaculture, this outcome is most likely to occur as a result of the products of feed given to fish and the process of feeding itself (Cho et al., 1991). The input of nitrogen (N), phosphorus (P) and carbon (C) in feed material given to the fish, and the subsequent losses which occur, have the potential to increase nutrient content within the water body as a whole. These losses may lead in turn to:

- transient (Lazur, 1995; Moylette, 1998) or longer term effects upon water column chemistry;
- deoxygenation and enrichment of sediments with organic wastes (Yokom et al., 1997);
- loss of benthic diversity (Cornel and Whoriskey, 1993);

- and, the formation of a potential long-term source of water column nutrient enrichment, through leaching of wastes, long after production ends (Kelly, 1993; Yokom *et al.*, 1997).

In temperate waters, it is normally estimated that approximately 60% of phosphorus and 30% of nitrogen discharged from freshwater cages is in solid form. Carbon is by far the largest part of the solid waste budget and is discharged mainly in solid form, but this can be solubilised from falling particles or sediment.

The process of eutrophication for any water body is, however, complex. Naturally occurring processes and conditions within a water body, as much as the imposed influence of an anthropogenic nutrient source, can influence the susceptibility, or otherwise, of a given lake to become eutrophic. Eutrophication, therefore, may be variable in its severity of occurrence under the influence of a nutrient source, such as nutrients released from fish cages.

The key factors that have been noted as influencing the likelihood of eutrophication occurring at a site as a result of the development of cage aquaculture are as follows:

- the form of nutrient limitation that exists in the freshwater body prior to production commencing;
- the quality of feed given and the management of feeding;
- the fish biomass generated and methods used in its production;
- the size of the lake and the rate at which water passes through the basin;
- the position of the cage site relative to natural obstacles (e.g. embayments) and the inflows and outflows of the lake basin;
- the depth of the lake basin;
- and, the response of algal species resident in the lake to the nutrient additions made by fish production.

Examples of eutrophication resulting solely from the impacts of waste outputs of aquaculture in freshwater have been documented, and increased biological productivity resulting from waste products and discharges have also been noted in laboratory studies (Kelly, 1993; Massik and Costello, 1995). Yokom *et al.* (1997) documented the progressive eutrophication and subsequent recovery of a pit mine lake in Minnesota, concurrently with the operation and abandonment of a fish farm project, from 1988–1992, with phosphorus levels in the water column increasing almost tenfold during the production period. The severity of impact in this instance was enhanced by a number of factors, including the high feed conversion ratios (1.5:1), which in turn result in higher levels of waste production. In terms of the physical characteristics of the lake, its eutrophication was further aided by slow water exchange and steep basin morphometry, both of which are factors that encourage retention of solid materials lost from the cages within the lake itself.

More frequently, however, aquaculture is identified either as a localised nutrient enrichment impact within a lake (Baffico and Pedrozo, 1996), or as one of a number of nutrient sources contributing to the overall load (Marsden *et al.*, 1995). Depending upon the nature and magnitude of the other nutrient sources, a fish farm may be considered to have more or less potential to influence the trophic status of the lake system. It follows, therefore, that the introduction of fish production has the potential to promote eutrophication of a water body, but the presence of cages alone does not guarantee its occurrence (Black *et al.*, 1997).

In freshwater sciences, it is normally assumed that phosphorus will be the nutrient element limiting algal growth (Welch, 1992). This assumption holds true for the large majority of freshwater lakes worldwide. It is clear, however, from a growing number of studies (e.g. Faafeng and Hessen, 1993; Pedrozo *et al.*, 1993) that algal growth in freshwater may, under certain circumstances, be limited by nitrogen rather than phosphorus concentration. Nitrogen limitation in freshwater lakes reportedly occurs with either very low ambient nutrient concentrations or in highly eutrophic systems (Downing and McCauley, 1992). These findings have important implications, not only for the way in which aquaculture is regulated but also with regard to the nature of sites that are currently most sought after for aquaculture development. In temperate regions, favoured sites for salmonid aquaculture typically have low nitrogen and phosphorus concentrations (Maitland *et al.*, 1994), whilst for some forms of fish culture (e.g. silver carp) in tropical areas, eutrophic sites are preferred. The former sites rely upon clean, oxygen-rich waters to sustain fish production, with all food sources artificially provided, whilst the latter sites passively supplement artificial food supplies through opportunistic consumption of products of the lake ecosystem (i.e. algae, zooplankton, etc.) in which the culture system is placed. Current emphasis of control and management of cage aquaculture in many parts of the world is directed towards the limitation of phosphorus discharges. It may, in certain cases, be appropriate to widen this scope to include the potential impacts of nitrogen discharges in order to protect the aquatic environment from potential degradation.

The literature is dominated by examples of temperate intensive aquaculture impacts, concentrating upon the effects of the intensive practices which are most prevalent in production in these regions (e.g. Cornel and Whoriskey, 1993). In some respects, this is to be expected given the relatively recent increase in the use of intensive cage systems in other climatic regions. However, in order to improve our ability to plan and manage non-temperate systems, our knowledge requires refinement and further applied research to determine how tropical lake systems respond to these impacts, together with a better understanding of the functioning of freshwater systems in warmer climates. In terms of aquaculture-specific research, the nature and volumes of inputs (Tacon *et al.*, 1995) and waste production from a wider variety of tropical fish species being cultured in differing production systems will also need to continue to improve (Kaushik, 1995).

In one of the few detailed studies so far undertaken of intensive cage aquaculture impacts in tropical freshwater, Troell and Berg (1997) observed substantial variations in the mode and quantity of waste losses from an intensive tilapia rearing operation based on Lake Kariba, Zimbabwe. Rapid degradation of wastes released from the cages led to them becoming impoverished with respect to carbon, nitrogen and phosphorus content. In studies from temperate regions, dissolution and degradation occur at much reduced rates. It was concluded, however, that the enhanced degradation observed could be due to a number of factors, including not only higher ambient water temperatures but also the influence of the atypical hydrodynamic regime in this artificially created and managed water body (Troell and Berg, 1997).

So far, little research has been published regarding gaseous phase inputs and outputs of nitrogen related to cage sites. Typically, studies assume atmospheric nitrogen fluxes to be zero, although in pond aquaculture operations, particularly in tropical zones, where water temperatures and pond surface area:pond water volume ratio is high, gaseous phase exchanges may have a significant influence upon nitrogen cycling (Boyd, 1995). For cages in lakes, the reduced surface area:water volume ratio will limit, but not prevent, the occurrence of such exchanges. Future studies in tropical regimes should seek to quantify the nature of this relationship.

As stated previously, a unique aspect of cage aquaculture as a nutrient source is that the generation of waste actually occurs directly within the water body. Other external nutrient sources, such as agricultural run-off, may be altered or in fact reduced in quantity en route to the lake, through sorption onto clay mineral surfaces, plant uptake, etc. For cage farms, however, there is no potential for this natural 'buffering' to occur, and therefore the impacts of such operations are considered to be directly linked to the environment in which they are maintained. The lack of potential for treatment and interception of outputs is commonly cited as one of the most environmentally undesirable aspects of cage production, and is discussed further in Section 3 (Cripps and Kelly, 1996; SNIFFER, 1998).

2.2.3 The wider impacts of cage aquaculture in freshwater

The changing nature of our appreciation of the impacts of cage aquaculture can be best evidenced by the trend towards examining the effects of cage production that occur well away from the site of the farm itself. The questions most actively being discussed in this context concern the sustainability of production systems. Middleton (1999) offers a threefold definition of sustainable development as: a) continuing to support human life; b) maintaining environmental quality and the long-term stock of biological resources; and c) the retention of existing resources for future generations. Therefore, sustainable development of production is defined by ecological, economic and social considerations and, in that respect, aquaculture is no different from any other form of human activity.

2.2.3.1 Ecological sustainability

The debate regarding the ecological sustainability of aquaculture has been one of the most keenly discussed and argued in recent times (see Folke *et al.*, 1994 and 1997; Black *et al.*, 1997; see also Black, Chapter 9 of this volume). Mostly, however, this debate has continued to focus upon coastal aquaculture, rather than its freshwater counterpart. The examination of aquaculture from the viewpoint of sustainability has centred largely upon applying the concept of the 'ecological footprint' to fish production. The 'footprint' of any form of human activity is the sum of the 'goods and services' it takes from the natural environment. These goods or services are used to maintain production and, depending upon the level of analysis, to assimilate wastes. The term 'footprint' is used as these goods or services are often expressed as an area, and then used as an accounting tool for ecological resources (Wackernagel and Rees, 1996) (Figure 2.3). These 'goods' or 'services', however, may take on many different forms, such as the amount of water, fuel energy or fishmeal required for fish production.

By undertaking this form of analysis, several new views of an activity become apparent. Firstly, it is possible to examine, on an equal basis, how differing activities may have an impact upon the same environment. Secondly, through understanding the linkages between the goods and services used, and the true

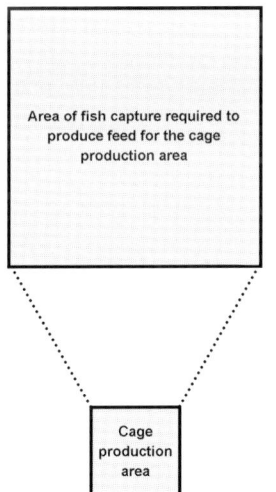

Figure 2.3 The 'ecological footprint' concept: fishmeal required for aquaculture production. The 'ecological footprint' of the production area is, therefore, the area of the larger upper box, that is, the area required to sustain production.

nature of the environmental impacts their use incurs, the key weaknesses in the production system may be revealed.

Kautsky *et al.* (1997) demonstrated the assessment of resource use for marine shrimp and freshwater tilapia farming in Colombia and Zimbabwe, respectively. In doing so, it was shown that intensive cage aquaculture in Lake Kariba, Zimbabwe, would require the support of primary production within the lake from an area equivalent to 10,000 times the size of the existing cages being used for cultivation (Kautsky *et al.*, 1997). In contrast, a semi-intensive system of pond cultivation, based close to Lake Kariba and using waste products from terrestrial production and traditional fisheries to stimulate production was estimated to be almost self-sustaining. This analysis highlights a continuing problem of cage aquaculture. Unlike pond farming, cages are far more open to the environment, and, as a result, input materials are easily lost. In ponds where recycling of these wastes occurs, the fertilising and growth potential of the inputs is maximised. Thus, resource use is often greater in cages than on land for the same amount of fish production (Kautsky *et al.*, 1997).

Critics of this form of approach have argued that it is simplistic in the evaluation of impacts as a whole (Black *et al.*, 1997). These arguments have echoes of those used when waste loads from aquaculture were evaluated in terms of 'population equivalents'. Simply put, the gross comparison of productivity in natural and artificial systems, such as a cage pen and a lake, is not a valid comparison. The complexity of natural systems, it is argued, is not adequately characterised by a simple 'accounting' of productivity. Arguments regarding the assessment of aquaculture waste in terms of population equivalents have proceeded along similar lines: the relative mass of nutrient output from a cage farm and a human community can be compared, but the form of the waste is substantially different. The particular physical and chemical nature of aquaculture waste (e.g. dilute, small solid particles, nutrient element ratios), it is argued, presents an erroneous image of the waste problem when compared directly to human waste outputs (Ackefors and Olburs, 1996). Therefore, the utility of the ecological footprint approach remains to be confirmed; however, it should be noted that, so far, these arguments are developed only in relation to marine rather than freshwater aquaculture projects.

2.2.3.2 Social and economic sustainability
Fish farming has been proposed, or actually used, as a tool for economic development in as many areas of the world as it has different forms. Primarily, it has been used to offer greater food security to undernourished populations. In addition, aquaculture has been applied to widen the economic base and, therefore, stabilise and strengthen populations of remote regions or marginalised social groups in developed countries (Buttner, 1997). Thus, aquaculture has the capacity to impact upon the human environment, and its impacts may be at least as important as those related to the strictly physical and biological impacts considered elsewhere.

The increase in area of artificial surface water reservoirs in South East Asia since 1980 has been described as 'Asia's final aquatic frontier' (Costa-Pierce, 1998). The opportunity which these newly created areas of surface waters present to fish farmers has contributed to a wider trend of increasing net-pen aquaculture production in the SE Asia region over the past 20 years. This trend has also coincided with an increase in the use of less traditional, more intensive methods of production (Li, 1994; Kestemont, 1995; Anderson and De Silva, 1997). In order to repay the greater investment required of such systems, producers mostly favour high-value species that are suitable for export (Tacon *et al.*, 1995). In areas where food security is a key issue, the potential for economic and social sustainability of such developments may, therefore, be questioned (Beveridge *et al.*, 1997).

In the Saguling and Cirata hydroelectric dams of West Java, Indonesia, development of aquaculture was intended to form part of a resettlement plan for families displaced as a consequence of dam development (Costa-Pierce, 1998). So far, success has been modest in terms of social improvement, although the dams offer great opportunity for fish production. At Saguling Dam, cage culture units expanded in number from 756 in 1988 to 4425 in 1995, with culture being based on *Cyprinus carpio* and *Oreochromis niloticus* production in net-pen units. High levels of fish mortality were reported to occur in response to seasonally low oxygen levels during overturn and worsening water quality conditions within the reservoir near to crowds of cage units (Djuangsih *et al.*, 1999). However, these adverse culture conditions appear to have resulted from a breakdown in planning control, partly due to the intended beneficiaries of the project being displaced by external commercial enterprises, and the local population becoming employees rather than owners. Costa-Pierce (1998) concluded that for cage aquaculture development at these dams to be sustainable in either a social or environmental sense, greater institutional control over both distribution of licenses and enforcement of production limits will be required. In addition, less intensive production methods would also take advantage of existing water column productivity, rather than contribute to waste accumulation. Therefore, whilst aquaculture may open up new and existing dams as a source of opportunity, the regulatory means by which to ensure the sustainable development of cage aquaculture remain elusive.

2.3 The environmental management of cage aquaculture in freshwater

2.3.1 Introduction

Cage farming in freshwater is traditionally viewed as having a simple, low-cost infrastructure by which fish may be produced, when compared with land-based systems. However, in environmental terms, there is a trade-off between the simplicity of the cage system and the ability to manage and control its processes and products. The major consequences of this choice in terms of environmental issues are the difficulties in controlling waste outputs and resultant limitations on aquaculture in the planning processes. The most effective means by which control over the impacts of cage aquaculture has been implemented is through the use of both external and internal management procedures, including careful selection of appropriate feed types and feeding methods (Gavine *et al.*, 1995; Anderson and De Silva, 1997; Cho and Bureau, 1997). It is of great importance, therefore, that initial waste generation is limited as far as possible. Careful siting of cages and a requirement to limit biomass production and the quality and quantity of feed given now commonly feature as licensing conditions, when approving new sites.

2.3.2 Controlling and treating waste outputs from cages

Cage aquaculture presents a unique problem in terms of wastewater treatment, as there is no fixed 'stream' of waste when compared with land-based systems. Instead, routes of material losses from a cage are modified by fluctuating natural factors, such as the water currents generated by wind in upper water layers, lake inflows and outflows, and by the circulating mass of fish themselves. Therefore, under normal conditions, solid wastes are lost through the sides and base of the cage (Figure 2.4). Whilst, for the purposes of planning, a cage may be considered as a 'point' source of pollution, the details of the waste outputs at a site are considerably more complex. Sediment trap studies recently undertaken by the authors at Scottish loch sites indicate that less than 3% of the total solid waste production passes through the base of the cage (Figure 2.4).

However, fall velocities of both faeces (0.015–0.03 m s^{-1}) and feed pellets (0.02–0.12 m s^{-1}) are low because, as a result of their low density and irregular shapes, they do not conform to traditional modelled estimations of sedimentation

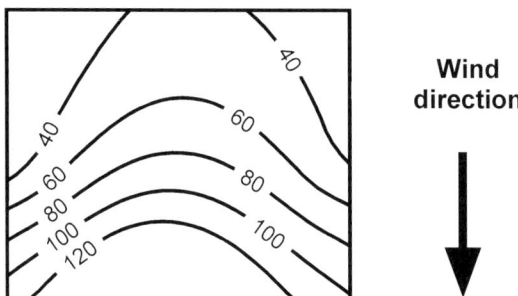

Figure 2.4 Wind-induced currents within large lakes can influence the pattern of solid waste deposition from cages at a variety of water column depths. Solid waste deposition (in mg m^{-2} day^{-1}) within a 5 × 6 m rainbow trout cage at 7.5 m depth was aligned with wind direction and prevailing current (authors' observations). The axes represent the perimeter of the net cage at this depth.

(Elberizon and Kelly, 1998). However, freshwater cages, particularly in large lakes and/or sites exposed to wind action, can be as prone to the effects of currents as their marine counterparts. The combination of currents, both from internal and external sources, affecting solid waste movement is sufficient to create initial deflection of wastes away from the cage itself, although their final deposition still often occurs close to the site (authors' observations).

Behmer et al. (1993) experimentally investigated the potential for capture of wastes within cage systems, using a semi-rigid design based upon modified plastic tanks. Whilst trapping of solids was reasonably successful during calm conditions (16.3% of total feed mass added), the effectiveness of the system declined rapidly under high currents (<5% of total feed mass in currents $\sim$0.5 m s^{-1}). This system differed considerably from either enclosed or netcage-based designs, and benefited from both side baffling and a width:depth ratio of approximately 1:1 (Behmer et al., 1993). Both features reduce the potential for ejection of solid particles through the side of the cage during their downward transit through the water column of the cage.

Dissolved wastes from cages are also lost through a combination of diffusive and mechanical processes. As with solid wastes, the rate at which the loss of these wastes occurs is also highly variable, both in space and time.

At sites already in existence, continued production may only be possible through adequate treatment and removal of wastes. To successfully effect treatment, presently available technologies that have the potential to control and minimise such waste outputs are fully enclosed cage systems of a type such as that shown in Figure 2.5. In such systems, 'enclosed cage' is a misnomer as the

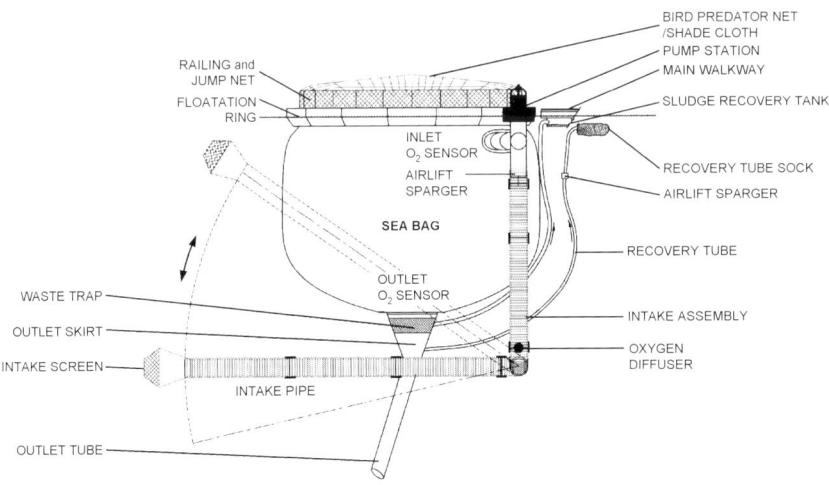

Figure 2.5 Schematic diagram of a fully-enclosed cage system (source: Future SEA Technologies, British Columbia, Canada).

net cage is replaced with a flexible plastic bag. Water is circulated artificially, entering the bag at the top and flowing out through a sump outlet in the base. This outlet presents the opportunity for treatment to occur. At this time, fully enclosed cage systems are mainly deployed in marine waters, although several have also been installed in freshwater sites (C. Bullen, Future Seas Technolgies, personal communication).

If such systems are to be adopted in freshwater, however, the following points need to be addressed:

- Wastewater treatment technology used in aquaculture (e.g. filtration) generally still performs poorly (Cripps and Kelly, 1996), and no opportunity exists for removal of soluble wastes on cage systems.
- Unlike marine sites, which due to the volumes required and costs involved (e.g. power to pump water) would be uneconomic on land, freshwater producers may be better able to obtain suitable alternative land-based sites, where feeding is less wasteful and wastewater treatment can be more easily applied.
- As freshwater cage aquaculture is traditionally low in infrastructure costs, the costs associated with enclosed systems may exceed the financial capacity of some producers. In addressing the issue of sustainable production from the perspective of water quality, may such systems be economically unsustainable?

2.3.3 Management of freshwater lakes for and by aquaculture

Where considerations of the effect of phosphorus additions to the water column are concerned, the use of the Dillon and Rigler (1974) model has gained widespread acceptance as a tool for predicting estimates of nutrient impacts from cage sites in freshwater lakes. However, this model remains, at best, a first estimate of reality when applied in assessments of aquaculture operations (Kelly, 1995). The simplicity of the model is in its application, but it in turn requires a level of data that is not often available to those undertaking the assessment.

The original model was developed by its authors through the collection of phosphorus input data from all sources in a range of lakes in Ontario (Dillon and Rigler, 1974). Given the often 'pioneering' nature of aquaculture developments in either isolated or newly formed lake resources, the quantity and quality of data available to input to such a model may be very limited. Therefore, any interpretation of phosphorus increases due to a cage farm estimated from the Dillon and Rigler model should be very carefully evaluated before proceeding to approval of a production site. Philosophically, the treatment of a cage farm as the sole source of nutrient input when applying this model does little to assist planning. Whole lake assessments of phosphorus budgets provide a far better interpretative framework for assessing developments, although estimation errors may remain substantial (Marsden *et al.*, 1995).

Aquaculture, although often associated with negative impacts upon lakes, can also be used as a form of management and restoration. Traditionally, this has been through the introduction, with or without enclosure, of fish species capable of grazing and thus reducing nuisance algal groups (e.g. bighead carp in Singapore; Kestemont, 1995). Starling and co-workers (1998, 1999) identified the use of silver carp in net-pen enclosures as being capable of reducing levels of algae in a man-made lake in Brazil. In essence, this parallels extensive cage culture practices in many tropical areas, where captive fish feed on organisms available *in situ*. However, by using fish for algal control, the outcome of the process can provide direct benefits both to the grower and to other users of the water body.

Controlled addition of phosphorus as fertiliser in lakes affected by low pH, in order to influence their base exchange, has been shown to provide an effective method of ameliorating the impact of what is termed 'acid rain' (George and Davison, 1998). This is achieved by a controlled increase in productivity in a freshwater lake system to the extent that base exchange increases as a consequence of biomass growth. Any suggestions that the wastes from cage farms may be capable of generating a direct benefit through indirect fertilisation of such waters are, however, largely unfounded. Firstly, the fish themselves must be tolerant of water acidity (pH <5.5), thus restricting the choice perhaps to non-indigenous species, which in turn brings its own problems (Beveridge *et al.*, 1994). Secondly, control over the input of phosphorus requires careful regulation to successfully promote productivity (George and Davison, 1998). For a fish farm, there are many difficulties associated with monitoring both the total input and timing of nutrient additions (Cho *et al.*, 1991; Kelly *et al.*, 1996). Variations in both quality of input (i.e. feed composition) and other factors, such as fish growth rate, all contribute to a highly variable pattern of nutrient output, which is unsuitable for this task. Finally, these experimental fertilisations have used phosphorus only. The wastes from fish farms, however, are primarily composed of carbon, nitrogen and phosphorus, which may influence both water chemistry and the development of plant biomass in directions that are not appropriate.

2.4 Conclusions

Little new information has become available to researchers in the past decade, largely as a result of the identification of the major impacts of cage aquaculture in freshwater during the late 1970s and 1980s. Within Europe and North America, it appears that many, although not all, of the questions regarding impacts associated with freshwater cage production have been addressed. At present, most concerns relating to the impacts of cage farming form part of wider assessments of water resource management, rather than addressing aquaculture alone.

That said, for those concerned with the environmental impact of freshwater cage aquaculture, the current trend for intensification of cage production in SE Asia has opened a series of new challenges. However, in both temperate and tropical regions a number of areas remain poorly understood and require attention in order to enhance our understanding of the nature of environmental impact from cage aquaculture in freshwaters:

- The fate of chemical treatments in freshwater lakes—The scientific literature remains poorly supplied with evidence regarding the fate of chemical treatments in freshwater cage farming, when compared with investigations in the marine environment (Smith *et al.*, 1994b; Weston, 1996). Most assessments of treatments in freshwater currently available relate to land-based systems and rarely consider the potential benthic impact within a lake.
- The behaviour and fate of waste outputs from cages in non-temperate regions—The study by Troell and Berg (1997) remains one of the few in tropical regions to have examined, in detail, the fluxes of waste from intensive cages. In order to meet the challenge of planning for increased cage production in tropical/subtropical zones, further data of this nature will need to be acquired.
- Methodologies for managing cage aquaculture—Use of models in the freshwater environment remains rooted to the application of Dillon and Rigler (1974), although this framework may well be inappropriate. The future would appear to be best secured by ensuring that aquaculture is properly addressed in catchment management planning activities. Additionally, the requirement for waste treatment from many cage systems will continue to be revisited. Matching both economic and environmental targets of production remains a key area of debate if cage farms in freshwater are to continue to operate in the long-term.
- Linking the environmental impacts of freshwater production—Current assessments of production impacts seldom combine the environmental impacts from the other phases of production, such as marine grow-out and processing. If we are concerned to reduce the environmental impacts from any form of aquaculture, including that of freshwater cage culture, then the use of life cycle assessment should be strongly considered. The piecemeal approach to environmental issues that the aquaculture literature currently presents does not assist in identifying those facets of production that are most harmful to the environment. Instead, an ever-increasing list of unlinked potential impacts is generated. If we are to plan for the future and realise a potentially sustainable production

(Bardach, 1997), then it is essential that we are able to identify and prioritise action for all impacts of activities contained within fish production.

References

Ackefors, H. and Olburs, C. (1996) Swedish aquaculture policy—a nightmare for the industry? *Aquaculture Europe*, **21** 6-13.
Anderson, T.A. and De Silva, S.S. (1997) Strategies for low-pollution feeds and feeding. *Aquaculture Asia*, **2** 18-25.
Baffico, G.D. and Pedrozo, F.L. (1996) Growth factors controlling periphyton production in a temperate reservoir in Patagonia used for fish farming. *Lakes and Reservoirs: Research and Management*, **9** 243-249.
Baird, D.J., Beveridge, M.C.M., Kelly, L.A. and Muir, J.F. (eds.) (1996) *Aquaculture and Water Resource Management*. Blackwell Science Ltd.
Bao-Tong, H. (1994) Cage culture development and its role in aquaculture in China. *Aquaculture and Fisheries Management*, **25** 305-310.
Bardach, J.E. (ed.) (1997) *Sustainable Aquaculture*. John Wiley and Sons, New York.
Behmer, D.J., Greil, D.W., Greil, D.C. and Fessell, B.P. (1993) Evaluation of cone-bottom cages for removal of solid wastes and phosphorus from pen-cultured rainbow trout. *The Progressive Fish Culturist*, **55** 255-260.
Beveridge, M.C.M. (1984) Cage and Pen Fish Farming: Carrying Capacity Models and Environmental Impact. FAO Fisheries Technical Paper 255, FAO, Rome.
Beveridge, M.C.M. (1996) *Cage Aquaculture*. 2nd Edition. Fishing News Books, Oxford.
Beveridge M.C.M., Ross L.G.R. and Kelly L.A. (1994) Aquaculture and biodiversity. *Ambio*, **23** 497-502.
Beveridge, M.C.M., Phillips, M.J. and Macintosh, D.J. (1997) Aquaculture and the environment: the supply of and demand for environmental goods and services by Asian aquaculture and the implications for sustainability. *Aquaculture Research*, **28** 797-807.
Black, E., Gowen, R., Rosenthal, H., Roth, E., Stechy, D. and Taylor, F.J.R. (1997) The costs of eutrophication from salmon farming: implications for policy—a comment. *Journal of Environmental Management*, **50** 105-109.
Boyd, C.E. (1995) *Bottom Soils, Sediment and Pond Aquaculture*. Chapman and Hall, New York.
Buttner J.K. (1997) First Nation people and Great Lakes aquaculture. *Aquaculture Magazine*, **23** 27-38.
Cho, C.Y., Hynes, J.D., Wood, K.R. and Yoshida, H.K. (1991) Quantitation of fish culture wastes by biological (nutritional) and chemical (limnological) methods; the development of high nutrient dense (HND) diets, in *Nutritional Strategies and Aquaculture Waste* (eds. C.B. Cowey and C.Y. Cho), University of Guelph, Guelph, Canada, pp. 37-50.
Cho, C.Y and Bureau, D.P. (1997) Reduction of waste output from salmonid aquaculture through feeds and feeding. *Progressive Fish Culturist*, **59** 155-160.
Cornel, G.E. and Whoriskey, F.G. (1993) The effects of rainbow trout (*Oncorhynchus mykiss*) cage culture on the water quality, zooplankton, benthos and sediments of Lac du Passage, Quebec. *Aquaculture*, **109** 101-117.
Costa-Pierce, B.A. (1998) Constraints to the sustainability of cage aquaculture for resettlement from hydropower dams in Asia: an Indonesian case study. *Journal of Environment and Development*, **7** 333-363.
Costa-Pierce, B.A., Moreau, J. and Pullin, R.S.V. (1993) New introductions of common carp (*Cyprinus carpio* L.) and their impact on indigenous species in sub-Saharan Africa. *Discovery and Innovation*, **5** 211-221.

Cripps, S.J. and Kelly, L.A. (1996) Waste water treatment in aquaculture, in *Aquaculture and Water Resource Management* (eds. D.J. Baird, M.C.M. Beveridge, L.A. Kelly and J.F. Muir), Blackwell Science Ltd, pp. 166-201.

De Silva, S.S., Zhitang, Y. and Lin-Hu, X. (1991) A review of the status and practices of the reservoir fishery in Mainland China. *Aquaculture and Fisheries Management*, **22** 73-84.

Dillon, P.J. and Rigler, F.H. (1974) A test of a simple nutrient budget model predicting the phosphorus concentration in lake water. *Journal of the Fisheries Research Board of Canada*, **31** 1771-1778.

Djuangsih N., Siong, K. and Asaeda, T. (1999) Fish mortality in fish farming tropical reservoir in Indonesia, in *Book of Abstracts of ILEC '99 Sustainable Lake Management*, 8th Conference on the Conservation of Lakes, Copenhagen 17–21 May 1999, S11A-3.

Downing, J.A. and McCauley, E. (1992) The nitrogen:phosphorus relationship in lakes. *Limnology and Oceanography*, **37** 936-945.

Elberizon, I.R. and Kelly, L.A. (1998) Empirical measurements of parameters critical to modelling benthic impacts of freshwater salmonid cage aquaculture. *Aquaculture Research*, **29** 669-677.

Enell, M. and Löf, J. (1983) Changes in sediment phosphorus, iron and manganese dynamics caused by cage fish farming impact, in *11th Nordic Symposium on Sediments* (eds. T.R. Gulbrandsen and S. Sanni), Norsk Limnologforening, Oslo, pp. 80-89.

Faafeng, B.A. and Hessen, D.O. (1993) Nitrogen and phosphorus concentration and N:P ratios in Norwegian Lakes: perspectives on nutrient limitation. *Internationale Vereinigung für Theoretische und Angewandte Limnologie Verhandlungen*, **25** 465-469.

Folke, C., Kautsky, N. and Troell M. (1994) The costs of eutrophication from salmon farming: implications for policy. *Journal of Environmental Management*, **40** 173-182.

Folke, C., Kautsky, N. and Troell, M. (1997) Response to Black et al. *Journal of Environmental Management*, **50** 95-103.

Gavine, F.M., Phillips, M.J. and Murray, A. (1995) Influence of improved feed quality and food conversion ratio on phosphorus loadings from cage culture of rainbow trout, *Oncorhynchus mykiss* (Walbaum), in freshwater lakes. *Aquaculture Research*, **26** 483-495.

George, D.G and Davison, W. (1998) Managing the pH of an acid lake by adding phosphate fertiliser, in *Acidic Mining Lakes: Acid Mine Drainage, Limnology and Reclamation* (eds. W. Geller, H. Klapper and W. Salomons), Springer, Berlin, pp. 365-384.

Kaushik, S.J. (1995) Nutrient requirements, supply and utilization in the context of carp culture. *Aquaculture*, **129** 225-241.

Kautsky, N., Berg, H., Folke, C., Larsson, J. and Troell, M. (1997) Ecological footprint for assessment of resource use and development limitations in shrimp and tilapia aquaculture. *Aquaculture Research*, **28** 753-766.

Kelly, L.A. (1993) Release rates and biological availability of phosphorus released from sediment receiving aquaculture wastes. *Hydrobiologia*, **253** 367-372.

Kelly, L.A. (1995) Predicting the effect of cages on nutrient status of Scottish freshwater lochs using mass balance models. *Aquaculture Research*, **26** 469-477.

Kelly, L.A., Stellwagen, J. and Bergheim, A. (1996) Waste loadings from a freshwater Atlantic salmon farm in Scotland. *Water Resources Bulletin*, **32** 1-9.

Kestemont, P. (1995) Different systems of carp production and their impact on the environment. *Aquaculture*, **129** 347-372.

Lazur, A.M. (1995) Effects of continual water circulation on channel catfish production in cages. *Journal of Applied Aquaculture*, **5** 1-8.

Li, S. (1994) Fish culture in cages and pens, in *Freshwater Fish Culture in China: Principles and Practice. Developments in Aquaculture and Fisheries Science 28* (eds. S. Li and J. Mathias), Elsevier, Amsterdam, pp. 305-346.

Li, S. and Mathias, J. (eds.) (1994) *Freshwater Fish Culture in China: Principles and Practice. Developments in Aquaculture and Fisheries Science 28*, Elsevier, Amsterdam.

Lindbergh J.M. (1999) Private freshwater sites: are there enough? *Aquaculture Magazine*, **25** 36-45.
Maitland, P.S., Boon, P.J. and McClusky, D.S. (eds.) (1994) *The Freshwaters of Scotland: A National Resource of International Importance*. Wiley, Chichester.
Marsden, M.W., Fozzard, I.R., Clark, D., McLean, N. and Smith, M.R. (1995) Control of phosphorus inputs to a freshwater lake: a case study. *Aquaculture Research*, **26** 527-538.
Massik, Z. and Costello, M.J. (1995) Bioavailability of phosphorus in fish farm effluents to freshwater phytoplankton. *Aquaculture Research*, **26** 607-616.
Merican, Z.O. and Phillips M.J. (1985) Solid waste production from rainbow trout, *Salmo gairdneri* Richardson, cage culture. *Aquaculture and Fisheries Management*, **1** 55-59.
Middleton, N. (1999) *The Global Casino: An Introduction to Environmental Issues*. 2nd Edition, Arnold, London.
Moylette, M. (1998) An Environmental Appraisal of the Effects of an Atlantic Smolt Rearing Unit on Lough Fadda, Connemara, Co. Galway. Unpublished M.Sc. thesis, Department of Zoology, National University of Ireland, Galway.
Muir, J.F. (1996) A systems approach to aquaculture and environmental management, in *Aquaculture and Water Resource Management* (eds. D.J. Baird, M.C.M. Beveridge, L.A. Kelly and J.F. Muir), Blackwell Science Ltd, pp. 19-49.
NCC (1990) *Fish Farming and the Scottish Freshwater Environment*. Nature Conservancy Council, Edinburgh.
Pedrozo, F., Chillrud, S., Temporetti, P. and Diaz, M. (1993) Chemical composition and nutrient limitation in rivers and lakes of northern Patagonian Andes (39.5°–42°S; 71°W) (Rep. Argentina). *Internationale Vereinigung für Theoretische und Angewandte Limnologie Verhandlungen*, **25** 207-214.
Pillay, T.V.R. (1992) *Aquaculture and the Environment*. Fishing News Books, Oxford.
Remane, K. (ed.) (1997) *African Inland Fisheries, Aquaculture and the Environment*. Fishing News Books, Oxford.
Smith, P., Donlon, J., Coyne, R. and Cazabon, D.J. (1994a) Fate of oxytetracycline in a freshwater fish farm: influence of effluent treatment systems. *Aquaculture*, **120** 219-325.
Smith, P.R., Hiney, M. and Samuelsen, O.B. (1994b) Bacterial resistance to antimicrobial agents used in fish farming: a critical evaluation of method and meaning. *Annual Review of Fish Diseases*, **4** 273-313.
Starling, F., Beveridge, M, Lazzaro, X. and Baird, D. (1998) Silver carp biomass effects on the plankton community in Paranoá Reservoir (Brazil) and an assessment of its potential for improving water quality in lacustrine environments. *International Review of Hydrobiology*, **83** 499-508.
Starling, F., Beveridge, M., Lazzaro, X. and Baird, D. (1999) Silver carp culture and the control of algal blooms in tropical man-made lakes, in *Book of Abstracts of ILEC '99 Sustainable Lake Management*, 8th Conference on the Conservation of Lakes, Copenhagen 17-21 May 1999, S11B-3.
SNIFFER (1998) *Collection and Treatment of Waste Chemotherapeutants and the Use of Enclosed-cage Systems in Salmon Aquaculture*. SNIFFER Report SR 97(05), SEPA Head Office, Stirling, 50 pp.
Tacon, A.G.J., Phillips, M.J. and Barg, U.C. (1995) Aquaculture feeds and the environment: the Asian experience. *Water, Science and Technology*, **31** 41-59.
Troell, M. and Berg, H. (1997) Cage fish farming in the tropical Lake Kariba, Zimbabwe: impact and biogeochemical changes in sediment. *Aquaculture Research*, **28** 527-544.
UNEP (1994) *The Pollution of Lakes and Reservoirs*. UNEP Environment Library, No. 12. UNEP, Nairobi.
Vines, C.R., Braid, M.R., Raymond, C.B. and Sanders, W.S. (1996) A collapsible cage design for aquatic culture. *Journal of Applied Aquaculture*, **6** 73-79.
Wackernagel, M. and Rees, W. (1996) *Our Ecological Footprint: Reducing Human Impact on the Earth*. New Society Publishers, Gabriola Island, Canada.
Welch, E.B. (1992) *Ecological Effects of Wastewater*. Chapman and Hall, London.

Weston, D.P. (1996) Environmental considerations in the use of antibacterial drugs in aquaculture, in *Aquaculture and Water Resource Management* (eds. D.J. Baird, M.C.M. Beveridge, L.A. Kelly and J.F. Muir), Blackwell Science Ltd, pp. 140-165.

Yokom, S., Axler, R., McDonald, M. and Wilcox, D. (1997) Recovery of a mine pit lake from aquacultural phosphorus enrichment: model predictions and mechanisms. *Ecological Engineering*, **8** 195-218.

3 Ecological effects of shellfish cultivation
M.J. Kaiser

3.1 Introduction

In terrestrial systems, the development of intensive husbandry has concentrated mainly on four types of herbivorous mammal (cattle, pigs, sheep and goats) and four types of bird (chickens, turkeys, geese and ducks). This contrasts sharply with the marine environment, where we have seen the development of husbandry techniques for hundreds of fish, mollusc, crustacean and, latterly, echinoderm species. Of these, the bivalve molluscs are the most efficient in terms of their ability to convert consumed food (phytoplankton) into body tissues. Furthermore, cultivated bivalves are filter-feeders that require no additional input of feed. In contrast, fish and crustacea feed higher up the food chain and require feed with a high protein content. The overall energy conversion efficiency of non-molluscan marine species is relatively low (i.e. we often end up feeding fish to fish and crustacea in aquaculture systems). Finfish comprise 52% of global aquaculture production, whereas shellfish (molluscs and crustacea) comprise 24%. Nevertheless, the value of crustacean landings per unit weight far exceeds that of molluscs or finfish (Figure 3.1). Consequently, there are great financial incentives to invest in crustacean aquaculture systems, as this is a potentially lucrative business.

The cultivation of crustacea is largely based on penaeid shrimp (89% by weight of freshwater and marine global production) that were valued at approximately US$ 5 billion per year at first sale in 1993 (FAO, 1993). While molluscan species are less valuable than crustacea per unit weight, they require relatively little investment in infrastructure, feeding and husbandry. Nearly all cultivated molluscan species are bivalve molluscs, of which mussels contribute most in terms of production, followed by oysters, clams and scallops (Figure 3.1). Most of these species can be collected as spat in the wild and can be on-grown to market size with the minimum of technology, although there is an increasing tendency towards the hatchery production of spat. Consequently, these species are suitable for cultivation by a relatively unskilled workforce and can yield high profits due to the low level of initial investment required. As in other areas of the aquaculture industry, the lure of high financial returns and government subsidies has led to the uncontrolled expansion of crustacean and molluscan cultivation in some areas. This in turn has led to problems with over-exploitation of juveniles and seed, the spread of pathogens and alien species, destruction of coastal environments, pollution and outbreaks of disease. This chapter highlights some

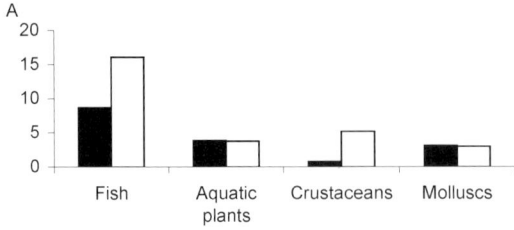
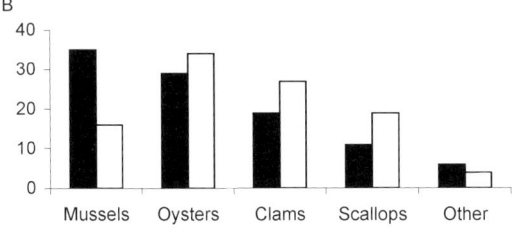
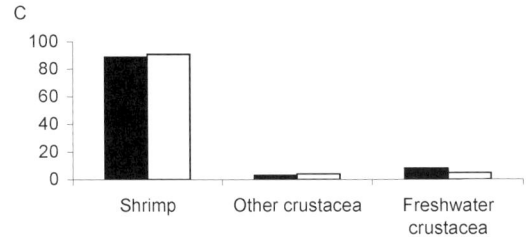

Figure 3.1 A) World aquaculture production by each species grouping in terms of the tonnage produced (10^6 mt, ■) and the value of the product (billions US$, □). B) Percentage of world molluscan aquaculture production (■) and value (□) based on weight produced by species groupings, 1991. C) Percentage of world crustacean aquaculture production (■) and value (□) based on weight produced by species groupings, 1985–1991. (Source FAO, 1993).

of the environmental problems that are associated with shellfish cultivation, with the greatest emphasis placed on the cultivation of crustacea and bivalve molluscs as these comprise the main area of current interest.

3.2 Cultivation of molluscs

As gastropods account for less than 4% of total molluscan production (Figure 3.1), this section focuses largely on the ecological effects associated with the cultivation of bivalve species. Bivalve mariculture is mainly restricted to coastal areas within several kilometres of the shoreline. The choice of cultivation sites is based on the prevalence of environmental factors that give conditions that are suitable for the successful collection and cultivation of commercially-exploited species. For example, cultivation sites must have a sufficient exchange of water to provide enough food and oxygen for adequate growth of the bivalves. Other criteria can be equally important when making site selection decisions; these may include considerations of water quality and contaminant load, the occurrence of disease organisms and conflicts with other coastal users (Laing and Spencer, 1997). Cultivation sites located in close proximity to rivers loaded with untreated sewage may yield high growth rates, but the bivalves produced are unlikely to be fit for human consumption as they will be loaded with pathogens from the sewage effluents. In Europe, coastal waters have been classified according to their faecal coliform counts. Bivalves cultivated in pristine coastal waters do not require depuration (purification) prior to consumption. In slightly contaminated waters, the bivalves have to be depurated for a specified period prior to consumption. At increased levels of bacterial contamination, bivalves must be harvested and relaid prior to depuration. At extreme levels of faecal contamination, the harvesting of bivalves for human consumption is prohibited.

The cultivation of bivalves can be a method of alleviating adverse environmental impacts arising from other activities in the coastal zone. For example, intensive fish farming has undesirable environmental impacts, particularly as the effluents are highly nutrient enriched, promoting the development of microalgal populations, some of which are toxic. It has been proposed that integrated fish/bivalve mariculture systems could ameliorate problems associated with algal blooms, as the bivalves would reduce algal densities and nutrients, which are effectively removed when the bivalve product is harvested (Folke and Kautsky, 1989; Shpigel et al., 1993). It has also been suggested that mussel culture can provide a means of removing excess particulate-bound nutrients from eutrophic systems (Haamer, 1996), or restoring enclosed water masses that are polluted (Russell et al., 1983). While these mussels are unfit for human consumption, they represent a cost-effective and self-perpetuating means of maintaining water quality.

There is an increasing awareness of the environmental effects that may result from the various stages of bivalve cultivation processes. Most notably, adverse effects have been associated with mussel and oyster farms on the continent of Europe (Tenore et al., 1982, Castel et al., 1989). Environmental problems have occurred at sites where hydrographical conditions were unsuitable for high-density cultivation of bivalves and the carrying capacity of the local marine environment was exceeded (Castel et al., 1989). So far, the majority of studies that have addressed the environmental impacts of bivalve cultivation have been largely concerned with the on-growing phase of cultivation. However, commercial cultivation of bivalves involves three distinct processes: seed collection; seed nursery and on-growing; and harvesting. The following section focuses on the environmental consequences of each of these phases in the cultivation process in turn.

3.3 Seed collection

3.3.1 Subtidal dredging

The adults of many species of bivalves (e.g. clams, *Arctica islandica*, *Mercenaria mercenaria*, *Ensis* spp. and scallops, *Pecten maximus*, *Aequipecten opercularis*) are harvested subtidally with dredges, yet very few species are harvested in this manner as spat or juveniles. One such species is the blue mussel *Mytilus edulis* which is dredged as seed for relaying and on-growing. There is increasing environmental concern regarding the ecological effects of dredging activities on non-target fauna and seabed habitats (Jennings and Kaiser, 1998). In the majority of experimental studies that have examined the incidental effects of shellfish dredging, the target species considered have been scallops or clams (Caddy, 1973; Peterson *et al.*, 1987; Thrush *et al.*, 1995; Currie and Parry, 1996; Hill *et al.*, 1996). These bivalves live as separate individuals, either just below the surface of or within the seabed, rather than in large beds as formed by oysters or mussels. Studies of the ecological impact of heavy commercial scallop dredges on benthic communities have demonstrated that large reductions in species numbers and abundance occur immediately after dredging and that, in some cases, recovery of the fauna and habitat can take many years (Collie *et al.*, 1997).

Whereas adult scallops and clams may be found in relatively low densities (<1 m^{-2}), seed mussels form dense aggregations (mussel seed beds) in discrete areas of the seabed (Dankers and Zuidema, 1995). Therefore, the extent of physical disturbance caused while dredging for mussel seed will be confined to relatively small areas of the seabed. In the UK, licences are issued to collect seed once a layer of 'mussel mud' has built up beneath the mussel seed bed. 'Mussel mud' is composed of the accumulated dead mussel shells, silt and pseudofaeces that build up beneath the living layer of mussels (Davies *et al.*, 1980). At this point, the mussels detach their byssus from the substratum and the whole bed becomes unstable and may be dispersed by wave action. Once the mussels have detached their byssus, fishers are able to skim-off the mussel seed, while leaving the substratum relatively unaffected. Mussel seed are vulnerable to heavy predation by starfish and crabs and it is not uncommon for entire seed beds to experience 100% mortality (Straaten, 1965; Seed, 1993). The Baird and Dutch dredges used to collect the seed mussels are employed singly and are relatively light, compared with gangs of up to 20 Newhaven dredges, which are fished in pairs by commercial scallop boats working the English Channel and Irish Sea (Dare, 1974; Kaiser *et al.*, 1996). Thus, it is unlikely that the environmental changes that may occur as a result of mechanical mussel seed harvesting will cause long-term environmental changes to the substratum and, furthermore, they will be restricted to the areas in which the seed beds have formed. In addition, dredging activities for mussel seed are seasonal, which allows a period of recovery both for the benthic fauna and seabed habitat. If properly managed, the benefits of harvesting a resource that may be lost to natural perturbations and predators probably outweigh the limited negative effects associated with the harvesting activity.

Of more concern than the incidental effects of harvesting is the over-exploitation of mussel seed resources. In extensive fisheries, such as the Wadden Sea in The Netherlands, the depletion of the seed mussel stocks has the greatest effect on trophic interactions within this ecosystem. Dankers and de Vlas (1992) estimated that from 1984 to 1990 30–130 $\times$ 10^6 kg of mussel seed were harvested from an estimated subtidal stock of 165 $\times$ 10^6 kg. One consequence of the heavy exploitation of these seed beds is that they are not permitted to develop into mature adult mussel beds. In 1990 and 1991, the entire intertidal mussel stock was removed by mussel seed dredging in the Wadden Sea. This resulted in increased mortalities of eider duck and reduced breeding success for oystercatchers, which depended on the mussels as a food source. Dankers and Zuidema (1995) suggested that these ecological effects would persist for many years, until new mussel beds developed and matured.

3.3.2 Intertidal collection

Juvenile mussels are gathered from intertidal rocky shores by scraping them by hand from the substratum. An estimated 1000 tonnes per annum is removed to support farming activities in Bantry Bay and the Kenmare estuary; this is the main source of seed for long-line mussel cultivation in southwestern Ireland (G. Burnell, personal communication). Although there is no published information on the direct impact of this harvest, there may be some effects associated with trampling across the shore (Brosnan and Crumrine, 1994), the disturbance of foraging birds and the removal of an important winter food source (Goss-Custard and Verboven, 1993). In addition, the removal of large patches of mussels from the middle to lower shore will presumably modify local community structure by altering small-scale hydrography, trophic interactions and the exchange of organic matter. Removal of mussel patches will also liberate space and, thereby, allow barnacles and algae to settle in the areas previously occupied by the mussels.

3.3.3 Use of collectors and cultch

Spat collectors are made from a wide variety of natural and man-made materials and are generally laid out on the shore or suspended in the water from ropes (Figure 3.2). Spat collectors are usually fibrous (e.g. frayed ropes or loose bundles of netting inside a net bag). The deployment of spat collectors has

Figure 3.2 Mussel seed collected on ropes in the Menai Strait, North Wales.

few environmental effects, apart from the removal of the spat of both target (bivalves) and non-target species (sea-squirts, starfishes, etc.). An alternative to spat collectors is the use of shell material, known as cultch, to encourage spatfalls. The laying down of cultch may alter the substratum significantly, for example, laying broken shell onto mudflats. The environmental implications of using cultch are discussed later in this chapter (Section 3.4.1).

3.3.4 Hatchery seed

As in finfish and crustacean hatcheries, antibiotics are used to some extent in bivalve hatcheries. So far, there is little information on the environmental effects of using antimicrobial products in bivalve hatcheries. Nevertheless, the problems associated with the excessive use of antibiotics in fish farming are also likely to occur in other aquaculture sectors. Drug resistance in fish pathogens is now well established, as are the effects of antibiotics on microfaunal communities beneath fish cages (Anonymous, 1996). While artificial feeds loaded with antibiotics reduce infection rates in intensive cultivation systems, these substances can affect microbial communities in the immediate vicinity of farm effluents leading to a reduction in microbial diversity and the possible development of drug resistance in potentially harmful species. The International Council for the Exploration of the Sea Working Group on the Environmental Interactions of Mariculture listed some of the concerns associated with the use of antimicrobials in fish feeds, these include:

- Development of drug resistance in fish pathogens
- Spread of drug-resistant plasmids to human pathogens
- Transfer of resistant pathogens from fish farming to humans
- Presence of antimicrobials in wild fish
- Impact of antimicrobials in the sediment beneath cages on the rates of microbial processes, the composition of the bacterial population and the relative size of the resistant subpopulation of bacteria

However, only a few bivalve hatcheries supply a much larger on-growing industry (for example only two hatcheries supply the entire UK seed requirements), therefore any environmental effects are likely to be highly localised.

3.3.5 Introductions of alien species

Among the many environmental concerns associated with aquaculture, the introduction of alien species is one of the most difficult to control. Introduced species may compete with native species for the same resources, and they may carry pests, predators and diseases to which native species are more vulnerable. Whenever bivalve seed are transported from one country to another, there is always a risk that alien species will be transported among them. It is now well established that alien marine organisms have become established after having been unintentionally introduced with imports of bivalve mollusc seed (Utting and Spencer, 1994; Eno, 1996; Minchin, 1996; Eno *et al.*, 1997).

One example of the introduction of a competitive species is the slipper limpet *Crepidula fornicata*. Slipper limpets can reach a high level of biomass and compete for food and space and can alter the sediment habitat through their feeding and excretory activities. Other introduced species include bivalve pests, such as the American whelk tingle *Urosalpinx cinerea* and diseases, such as *Bonamia*, which infects the blood cells of flat oysters, causing high mortalities under conditions of intensive cultivation. Indeed, the introduction of *Bonamia* into UK waters caused the collapse of the native oyster industry due to high mortalities. In the European Union, the spread of serious diseases of bivalves is restricted through regulations on movements of shellfish (EU Directive 95/70/EC). This legislation controls the movements of shellfish to prevent the introduction and spread of disease agents to EU countries, while encouraging free trade between member states. The International Council for the Exploration of the Sea (ICES) has produced a Code of Practice, entitled 'The Introductions and Transfers of Marine Organisms 1994' (Anon., 1995). The most recent version of the Code addresses three challenges that face aquaculture today: firstly, inadvertent co-introductions of harmful organisms associated with the target species, as occurred recently in Pacific oyster shipments from France to Ireland (Holmes and Minchin, 1995); secondly, the ecological and environmental impacts of introduced and transferred species; thirdly, the genetic impact of introduced and transferred species on indigenous stocks. The genetic impacts of transferring bivalve stocks from one area to another are yet to be addressed.

Introductions of algae, including toxic dinoflagellates, blooms of which can have a significant impact on commercial bivalve mollusc culture, have generally been attributed to the transportation of resting cysts in ships' ballast water (Hallegraeff and Bolch, 1991). However, normal trading, involving transport of shellfish stocks from one area to another followed by relaying or storage in open basins, can provide another mechanism of transfer. The faeces and digestive tracts of bivalves can be packed with viable dinoflagellate cells or can contain resting cysts (Scarratt et al., 1993). Viable cysts may also be found in the mud and sand retained with dredged mussels. These cysts may then be released into coastal waters at a new location. In the Netherlands, recirculating storage systems are used to quarantine bivalves as a precaution against such introductions (Dijkema, 1995). Invasive alien seaweeds, including *Sargassum muticum*, *Undaria pinnatifida* and *Laminaria japonica*, are also thought to have been introduced into European waters through transport of the sporophyte stage in oyster juveniles or as small plants attached to bivalve shells (Rueness, 1989).

3.4 On-growing

One of the main attractions of bivalve species for aquaculture is that they are on-grown to market size in the natural environment, without the requirement for supplemental or artificial feeds. Nevertheless, on-growing usually requires the introduction of structures into the marine environment, on or from which the bivalves are either supported or suspended. The introduction of such structures has an immediate effect on local hydrography and provides a new substratum upon which other epibiota can settle and grow. In addition, the introduction of high densities of cultivated organisms increases local oxygen demand and elevates the input of organic matter into the immediate environment. At high stocking densities of bivalves, the larval settlement of other benthic species may be reduced as their larvae are filtered and digested or become entrained within pseudofaeces (Baldwin et al., 1995).

Notwithstanding the above-documented effects on the marine ecosystem through commercial cultivation of stocks of bivalve molluscs, it should be remembered that natural beds of these animals were once extensive in certain areas (e.g. Chesapeake Bay, USA) and had an important functional role in the local ecosystem. Overharvesting, disease and, to some extent, pollution have in some places destroyed these native populations and the introduction of commercial cultivation may be no more than re-establishment of the status quo. The literature on the role of bivalve molluscs in estuarine ecosystems shows that they are an essential part of healthy estuaries around the world, where they fulfil an important role in the retention of phosphorus and nitrogen (e.g. Dame et al., 1989; Gottlieb and Schweighofer, 1996; Lenihan, 1999).

3.4.1 Intertidal cultivation

In intertidal systems, cultivation occurs essentially in two dimensions and, therefore, is limited by the available area of the intertidal zone that is suitable for adequate bivalve growth. In Arcachon Bay, France, 7% (10 km^2) of the total intertidal area within the bay is occupied by areas of cultivated oyster beds (Castel *et al.*, 1989). Oysters are on-grown either by relaying them directly onto the substratum, or by growing them in net bags (poches) suspended above the substratum on trestles. The trestles and poches often become fouled with green algae (Figure 3.3), which will increase levels of organic enrichment when the algae die back in autumn and winter. Castel *et al.* (1989) found that the presence of densely-stocked oyster parks elevated organic carbon levels in the local sediments, which increased oxygen demand and produced anoxic conditions in the underlying sediments. As a result, meiofauna increased in abundance by a factor of 3–4, while macrofaunal abundance decreased by nearly 50%. Nugues *et al.* (1996), examined environmental changes at a relatively small oyster farm in the River Exe, UK, and found that the abundance of macrofauna beneath trestles decreased by a half. Water currents were significantly reduced in close proximity to oyster trestles, which in turn doubled sedimentation rate and increased the organic content of the underlying sediments, leading to a reduction in the depth of the oxygenated layer of sediment (Nugues *et al.*, 1996). Nevertheless, the changes observed in the benthic fauna were restricted to the area immediately beneath the trestles. Therefore, at low stocking densities, the effects of oyster cultivation are relatively benign and highly localised. It is not surprising, however, that environmental effects are exacerbated as the carrying capacity of enclosed systems is approached and the extent of cultivated areas is increased (Castel *et al.*, 1989).

In some areas of Europe and North America, oysters are cultivated directly on the ground on a variety of substrata, including mud, sand and gravel (Simenstad

Figure 3.3 Oyster poches in a southern English estuary covered in a thick growth of *Enteromorpha* spp.

and Fresh, 1995). In North America, preparation of ground cultivation plots can involve severe levels of disturbance. In some areas, such as Willapa Bay, Washington, the insecticide, carbaryl, is sprayed on intertidal areas to kill populations of burrowing shrimp (*Neotrypaea californiensis* and *Upogebia pugettensis*), which destabilise the sediment and smother oysters through their burrowing activities. Spraying is carried out every 6 years and is strictly regulated due to its controversial nature (Simenstad and Fresh, 1995). In addition to the application of chemicals, oyster grounds are harrowed to level the ground prior to cultivation, and raked and dredged to distribute and thin oysters during on-growing (Simenstad and Fresh, 1995). Removal of burrowing fauna and harrowing are likely to induce habitat and community changes similar to those attributed to dredge harvesting techniques discussed in (Section 3.5). Intertidal cultivation often requires some form of habitat modification in the form of the addition of gravel or gravel and crushed shell to the substratum, the placement of protective plastic netting over seed bivalves, or laying the bivalves directly onto the sediment in poches. Not surprisingly, such habitat modifications lead to alterations in the local environment and, consequently, to faunal composition. The ecological significance of such modifications will be directly related to the extent of the area affected.

Simenstad and Fresh (1995) reported that the application of gravel to intertidal sediments resulted in a shift from a polychaete to a bivalve and nemertean dominated community, but emphasised that changes are likely to be site-specific. Such shifts in community composition could have repercussions at other trophic levels, for example, changes in the abundance of certain harpacticoid copepod populations, which are important prey for juvenile salmon and flatfish species (Simenstad and Fresh, 1995). In the UK, parliamentary law necessitates the use of protective netting in Manila clam, *Tapes philippinarum*, cultivation to prevent escape of this introduced species into the wild (Spencer *et al.*, 1997). Spencer and co-workers (1996, 1997) found that the application of plastic netting to an estuarine silty sand substratum led to an immediate increase in sedimentation rate over cultivated plots, which elevated the organic content of the sediment. Within 6 months of the application of the netting, the cultivated plots were dominated by opportunistic spionid worms. During the following 24 months, the spionids were replaced by a high abundance of larger deposit-feeding worm species. The plastic netting also became fouled with *Enteromorpha* spp., which in turn attracted grazing littorinid snails. Although the environment was significantly altered during the cultivation period, it was effectively restored to a condition comparable with surrounding control areas within 12 months of harvesting the clam plots (Table 3.1; Spencer *et al.*, 1998).

Hard-shelled clams, on-grown in plastic bags placed directly on the sediment, had no detectable effect on the benthic invertebrate population found in the sediment between the bags (Mojica and Nelson, 1993). Although Mojica and Nelson (1993) also found no differences in the infauna sampled directly beneath

Table 3.1 Changes in the abundance (per $0.176 \, m^2$) of the main infaunal species at a Manila clam cultivation site in the River Exe, Devon, UK

		Abundance				
		November 1994			November 1995	
		Controls	Treatments	Post-harvest	Controls	Treatments
Nephtys hombergii	Predatory polychaete	22.5 ± 1.3	20.0 ± 2.8	1.3 ± 0.9	22.3 ± 1.4	20.5 ± 1.6
Scoloplos armiger	Deposit-feeding polychaete	10.5 ± 1.8	3.3 ± 1.5	0.0 ± 0.0	4.2 ± 1.5	2.5 ± 0.8
Pygospio elegans	Deposit-feeding polychaete	13.3 ± 3.0	18.3 ± 0.5	0.3 ± 0.3	21.0 ± 3.2	17.2 ± 4.9
Cirratulidae	Deposit-feeding polychaetes	21.3 ± 7.8	86.2 ± 22.6	1.7 ± 1.2	25.5 ± 4.3	24.0 ± 3.0
Heteromastus filiformis	Deposit-feeding polychaete	3.2 ± 1.7	32.8 ± 6.4	1.3 ± 1.4	1.5 ± 0.6	2.7 ± 0.7
Melinna palmata	Suspension-feeding polychaete	3.8 ± 1.2	17.5 ± 5.8	3.7 ± 0.9	8.7 ± 2.8	11.2 ± 2.5
Ampharete acutifrons	Suspension-feeding polychaete	1.7 ± 0.8	111.5 ± 28.8	0.7 ± 0.7	0.3 ± 0.2	0.2 ± 0.2
Tubificoides benedii	Deposit-feeding polychaete	7.2 ± 2.6	89.5 ± 31.8	3.0 ± 1.0	12.0 ± 5.6	16.3 ± 5.7
Cerastoderma edule	Bivalve	10.2 ± 1.4	7.8 ± 1.2	1.0 ± 0.6	20.2 ± 2.5	15.7 ± 3.3
Macoma balthica	Bivalve	2.5 ± 1.1	2.5 ± 0.9	0.0 ± 0.0	1.0 ± 0.6	2.2 ± 1.4

Prior to harvesting in November 1994, the density of deposit feeding polychaetes was much higher in the cultivation plots than in adjacent control plots. Immediately after harvesting using suction pumps, the density of all infaunal species was greatly reduced. Twelve months later, there were no significant differences in the benthic community found in the control areas and the previously cultivated clam plots (data from Spencer *et al.*, 1998).

clam poches, they expressed caution about this result due to the low number of their samples. The extent of the intertidal given over to clam cultivation was relatively small, thus it is not surprising that Mojica and Nelson (1993) did not detect an ecological effect in their study.

Bivalves that form reef structures or beds can significantly alter seabed topography and, therefore, small-scale hydrography where they occur. Relaid mussels lead to the development of 'mussel mud' beneath the mussel bed, as the filtration and feeding activities of the mussels increase the rate of sedimentation. These deposits are composed of dead shells, silt and pseudofaeces, which persist in excess of 18 months after the mussels have been removed. The cohesive nature of the 'mussel mud' is degraded by a combination of bacterial activity and wave erosion (Davies *et al.*, 1980).

In summary, both the addition of gravel or shell substrata, the formation of 'mussel mud' and the use of protective netting induces localised change in the sedimentary environment and composition of the benthic community. Protective netting and poches are easily removed, and accumulated sediments are rapidly reworked by tidal currents, waves and bioturbatory activity (Spencer *et al.*, 1998). However, the addition of gravel and shell material effectively creates a new habitat, leading to more persistent changes in the composition of the local community.

3.4.2 Suspended cultivation

In suspended cultivation systems, bivalves are either attached directly to ropes or placed in net bags attached to ropes suspended in the water column from floats or rafts. The bivalves provide a surface on which many epibiotic organisms attach and grow. Thus, a large biomass of biota develops in suspended culture systems and this, in turn, has a major effect on phytoplanktonic, benthic and hydrographic processes in close proximity to the cultivation site. As with other bivalves, mussels provide a complex surface area on which epifaunal communities consisting of over 100 species can develop (Tenore and Gonzalez, 1976). As organisms die and fall off the ropes, they can provide a significant food resource for species found in the vicinity of the culture site. Small portunid crabs *Pisidia longicornis* were found to be abundant among fallen mussels beneath rafts in the Spanish rias. These, in turn, were fed upon opportunisitically by several fish species that normally consume polychaete worms (Lopez-Jamar *et al.*, 1984). *P. longicornis* are so abundant at cultivation sites that their larvae dominate (90% of the biomass) the zooplankton community that would normally be characterised by copepods (Alvarez-Ossorio, 1977). Mussels excrete high levels of ammonia (Tenore and Gonzalez, 1976), which promotes increased levels of productivity in algae attached to mussel lines. It is calculated that this is equivalent to algal production in local intertidal systems (Lapointe *et al.*, 1981). So great is the productivity associated with mussel lines in the Spanish rias, that Tenore *et al.* (1982) speculated that inshore fisheries were potentially enhanced by the bedload transport of organic-rich sediment into coastal areas.

Cultivation sites that are well flushed by tidal currents, as in the Spanish rias, do not encourage the accumulation of pseudofaeces beneath mussel rafts, which can lead to the development of anoxic conditions (Rodhouse and Roden, 1987). The relatively beneficial effects that occur in the Spanish rias contrast sharply with the effects observed by Dahlbäck and Gunnarsson (1981) in Sweden. They demonstrated organic sedimentation rates of 2.4–3.1 g organic carbon $m^{-2} day^{-1}$ beneath mussel longlines, which was twice as much as found in adjacent uncultivated areas. This excessive organic enrichment led to the development of anoxic sediment conditions. Mats of bacteria, *Beggiatoa* spp., then developed beneath the longlines at their study site. In this situation, the benthic infauna had low diversity and biomass, which is a well-documented response to excess organic enrichment (Pearson and Rosenberg, 1978). Similarly, the productivity of densely-stocked Japanese oyster grounds was detrimentally affected by the generation of large quantities of pseudofaeces and high filtration rates (Ito and Imai, 1955; Kusuki, 1977). Pseudofaeces production was so great beneath oyster cultivation rafts that it was at least equivalent to natural sources of sedimentation (Mariojouls and Kusuki, 1987).

3.4.3 Stock enhancement

In areas where wild stocks have been depleted, attempts have been made to restore populations to a viable stock size. In order to assess the success of hatchery release programmes, it is necessary to have an effective and workable means of tracking, observing and identifying the released organisms at a later date. The hatchery release of molluscs is complicated by considerations of the minimum density required to permit effective fertilization at spawning. Not only must the density at which the animals are released be considered but also local hydrography, which affects larval dispersal. Stocks of pink and green abalone, *Haliotis corrugata* and *H. fulgens*, increased dramatically after nearly 4500 mature adult animals were transplanted to the coastal waters of California, demonstrating that such releases can have a beneficial effect if properly planned (Bohnsack, 1996).

Reseeding and stock enhancement programmes are more likely to be financially viable if the species in question have a high financial or resource value. This is certainly the case with species such as pearl oysters and lobsters. Pearl oysters, as their name suggests, are harvested and cultured for the pearls they produce. Two species are commercially significant: the black-lipped pearl oyster *Pinctada margaritifera* that produces black pearls; and the gold- or silver-lipped pearl oyster *P. maxima*. Although natural pearls are rare in these oysters, both produce large, cultured pearls. Pearl oysters have suffered high levels of over-fishing and a whole suite of management techniques have been applied to maintain a sustainable population. Among these, reseeding depleted areas with spat collected from the wild and from hatcheries has proved successful, as long as the animals are permitted to reach maturity before they are havested (Sims, 1993). If the oysters are allowed to mature and spawn before they are harvested, then culture may be unnecessary in the future.

3.5 Harvesting

Subtidally-grown bivalve species are usually harvested with towed dredges or suction pumps, the effects of which were discussed in Section 3.3.1 and have been well reviewed previously (Dayton *et al.*, 1995; Jennings and Kaiser, 1998). The harvesting of trestle-grown and suspended bivalves has little, if any, effect as they are removed without interfering directly with the environment. The main exception to this might be the disturbance of feeding birds by the physical presence of humans harvesting bivalves. In contrast, the harvesting of intertidal species that are cultivated directly on or in the substratum necessitates the use of mechanical extraction to remove them from the substratum.

Physical disturbance of intertidal sediments by invasive, commercial, bivalve-harvesting activities is of concern to fisheries managers because of the direct effects on populations of target species and issues of non-catch mortality. In addition, interference with the feeding behaviour of wading birds (Goss-Custard and Verboven, 1993; Shepherd and Clark, 1994), habitat degradation and alteration of the infaunal invertebrate community structure are of concern to managers of the environment. The environmental effects of harvesting natural populations of intertidal and shallow sublittoral bivalves has received considerable attention because of the scale and intensity of operation, particularly with respect to intertidal tractor and suction dredging for cockles *Cerastoderma edule* (Cotter *et al.*, 1997; Hall and Harding, 1997). It is surprising that the environmental effects of cultivated bivalve harvesting have been little studied so far, especially as clam beds often occupy large areas of the intertidal zone (>1 ha), with individual commercial plots that usually measure $40\,m^2$ or more.

The most obvious visible effect of suction dredging or mechanical raking on the sediment is the creation of pits or trenches that may take between days to months to fill in, depending on sediment type and location (Dyrynda and Lewis, 1995; Hall and Harding, 1997). These trenches may encourage larval settlement by providing an environment subject to lower current velocities (Snelgrove and Butman, 1994). However, Thrush and co-workers (1996) reported that defaunated sediments became destabilized, leading to faunal emigration which greatly delayed recolonisation.

Recolonisation rate is likely to differ between habitat-types, depending on a combination of factors, including sediment stability and exposure to wave action and currents. In addition, the scale of disturbance will have important implications for recolonisation rate, depending on whether this occurs through active/passive movement of adults or through larval recruitment (Hall and Harding, 1997). Most studies of recolonisation rate have been performed on scales of $<1\,m^2$, which do not equate to the scale of commercial harvesting practices. However, more recent studies have been designed at a scale appropriate to examination of the effects of disturbance associated with commercial harvesting activities. In studying tractor dredging in the Solway Firth, Hall and Harding (1997) found that the benthic community within dredged plots was indistinguishable only 3 months after harvesting, regardless of the scale of disturbance, which ranged from 225 to $2025\,m^2$. They emphasised that the sum total of disturbed areas on a sandflat subjected to commercial harvesting would exceed that examined in their study, but that it would be patchily distributed and unlikely to extend the recovery trajectory much further. The rapid recovery in the Solway Firth was attributed to large-scale sediment movements that obliterated their treatment effects (Hall and Harding, 1997). In a similar study undertaken at a commercial Manila clam farm at Whitstable on the southeast coast of England, Kaiser et al. (1996) found that the infaunal community was restored within 7 months after suction harvesting clam parks (each park, 40 m^2). This site had an underlying sediment of cohesive mud/clay, overlayed with a veneer of coarser sediments. Suction dredging removed this veneer, leaving the mud/clay exposed. Tube dwelling polychaetes, such as *Lanice conchilega* and *Euclymene lumbricoides*, burrowed down into the mud/clay fraction and were less adversely affected than more mobile species, such as the bivalve, *Macoma balthica*, and the polychaete, *Scoloplos armiger*, which were found in the coarser overlying sediment. It is likely that these fauna would only recolonise harvested areas once this veneer of coarse sediment had been restored. Wave action was probably the main agent of sediment restoration at this site, which is exposed to prevailing northeasterly and easterly winds. Spencer and co-workers (1998) conducted a similar experiment in the River Exe, Devon, UK. Their site was much more sheltered than those studied by Hall and Harding (1997) and Kaiser et al. (1996), and characterised by fine muddy sand. Although sediment structure and profile was restored 3 months after suction harvesting, the benthic community was not fully restored until 9–12 months after harvesting had occurred (Spencer et al., 1998).

In general, soft sediments recover relatively quickly after physical disturbance (Collie et al., 2000); however, disturbance of key habitat structures, such as hard substrata or plants, particularly seagrasses, are likely to have much more severe effects. Seagrass beds are highly-specialised habitats, acting as nursery grounds for juveniles of many species, and they are important for the productivity of coastal areas. Harvesting activities in seagrass beds removes shoots and rhizomes, and can cause reduction in light penetration if harvesting leads to sediment resuspension (Short and Wyllie-Echeverria, 1996). Fishermen gathering molluscs in the intertidal zone may cause disturbance as they walk across mud flats or rocky shores. Recent studies by Chandrasekara and Frid (1996) and Fletcher and Frid (1996) have emphasised how repeated trampling can induce localised changes in invertebrate and plant communities on tideflats and rocky shores. As a result, they suggested the use of restricted walkways to minimise this disturbance.

3.6 Crustacean farming

Shrimp make up approximately 89% of world landings of crustacea by weight, with only 8% composed of other freshwater crustacea (e.g. crayfishes) and only 3% composed of other marine crustacea. As a result, most of this section concerns the environmental effects of shrimp cultivation. This subject is reviewed further by Choo in Chapter 4 of this volume. World landings of shrimp (including prawns) from capture fisheries reached a plateau of approximately 2 million tonnes in the 1980s. Worldwide production of farmed shrimp rose from 84,000 to 891,000 tonnes between 1982 and 1994 (FAO, 1996). Although shrimp farming has become a multibillion dollar business worldwide, its sustainability is currently adversely affected by widespread crop failures and associated negative environmental impacts that result from the farming practices. As a result, production has been unpredictable, and fluctuated between 789,000 and 891,000 tonnes from 1992 to 1994. Disease appears to be the major cause of recent declines in production. Initial signs of major disease problems occurred in 1988, when Taiwan's production dropped dramatically, shortly followed by a similar occurrence in China. Much of Asia's shrimp production is now seriously affected by rapidly spreading viral epidemics. So far, there are at least 20 identified shrimp viruses, of which the white-spot virus in Asia and Taura Syndrome (TS) virus in the Americas are the most threatening.

3.6.1 Disease outbreaks

Taiwan was among the leaders in shrimp production, based mainly on the output of small family-owned intensive farms that produced 115,000 tonnes of *Penaeus monodon* in 1987. However, a combination of industrial pollution, bacterial and viral diseases and the recirculation of pond effluents between farms led to a 66% decrease in production in 1988 (Chamberlain, 1997). Despite attempts to revive the industry by cultivating alternative *Penaeus* species, shrimp production collapsed to 25,000 tonnes in 1994 and most Taiwanese shrimp farmers have now moved into marine fish cultivation (Rosenberry, 1994). This collapse in the Taiwanese shrimp industry left a gap in the world market that was quickly filled by production from China, which rose at a rate of 80% per year and reached 199,000 tonnes in 1988. However, as in Taiwan, production became erratic and dropped to 88,000 tonnes in 1993. This decline was attributed to deterioration in water quality linked to industrial, agricultural and domestic pollution, and organic pollution from adjacent shrimp farms. The situation was further aggravated by outbreaks of disease and the occurrence of red tides. While the latter and pollution were at first blamed for increased death rates, it was later discovered that white-spot virus was mainly responsible, causing 100% mortality 2–3 days after infection (Huang *et al.*, 1995). This virus spread rapidly from China to other Asian countries, causing mass mortality in a wide range of shrimp species (Nakano *et al.*, 1994).

Although Thailand has a long history of shrimp cultivation, it was not until 1985 that intensive cultivation techniques were introduced from Taiwan. Most of these shrimp farms were small, family-owned and -operated concerns that occupied less than 2 ha. The shrimp are cultivated in ponds, which exchange water with the adjacent estuary or ocean to maintain water quality. However, the rapid spread of shrimp farms led to a situation in which wastewater from adjacent farms was recirculated among neighbouring shrimp operations. The resulting decrease in water quality in such circumstances leads to elevated levels of stress for the shrimp and causes a decline in growth rate and food conversion ratio, and an increase in disease transmission. These problems were exemplified by the development of over 5000 ha of intensive ponds in the northern Gulf of Thailand. This is an excellent example of poor site location. The gulf receives the run-off from four river systems, which create extensive mudflats that extend for up to 15 km at low water. As the tide inundates these mudflats, sediment is resuspended and transported into the shrimp ponds. The extreme sediment load lowered water quality to such an extent that shrimp farming was abandoned in this area (Chamberlain, 1997).

As in China, shrimp farming in Thailand was also affected by outbreaks of disease. In some farms, yellow-head disease caused total mortality within 5 days. The pathogen responsible for causing this disease was identified as a cytoplasmic virus that is carried by brackish water shrimp, *Palaemon styliferus* and *Acetes* sp. (Wongteerasupaya *et al.*, 1995). However, in seawater this virus is only viable for 72 h outside the host. The Thai Department of Fisheries was able to instigate a management programme as a result of these studies. Shrimp farmers were instructed to erect filters on their seawater intakes to prevent the entrance of the host shrimp into their ponds. The use of trash fish feed that often contains the host shrimps was prohibited. Incoming water had to be held in reservoirs for 72 h prior to use and, in the event of infection occurring, neighbouring farms had to be warned to prevent them pumping contaminated water for a period of 72 h. These measures contributed to a decline in this particular disease problem, although this may also have coincided with changes in host susceptibility to the viral pathogen. White-spot disease has also occurred in Thailand and was particularly prevalent from 1994 to 1996. Measures to control this disease have included: viral screening at water intakes; disinfection of pond water with chlorine or formalin; and pumping of wastewater offshore and clean oceanic water ashore. Nevertheless, despite these expensive efforts to limit the effect of the disease, production of shrimp in Thailand dropped by 25–40% in 1996, although the long-term impact of white-spot virus is unclear (Chamberlain, 1997).

Similar problems have occurred in the shrimp cultivation industry in the western hemisphere. Ecuador is the largest producer of shrimp in this region and production peaked at 113,000 tonnes in 1992. However, extended drought caused increases in salinity and nutrient concentrations in the Guayas River estuary, which supported a large number of shrimp farms. These environmental conditions favoured the proliferation of *Vibrio* spp., which then infected the shrimp causing them to swim erratically at the surface of the ponds. This shrimp behaviour attracted large numbers of gulls that fed on the disorientated shrimp, a phenomenon that gave its name to the syndrome 'Gaviota' which is the Spanish word for seagull. These outbreaks of disease were treated with antibiotics or by encouraging the growth of harmless bacteria that outcompeted the harmful *Vibrio* spp., but these measures had mixed results. Fortunately, El Niño rains lowered the salinity in the affected estuaries, diluted nutrient concentrations and eliminated the disease in 1993 (Chamberlain, 1997). Other areas in Ecuador were affected by TS virus, which caused 80–90% mortality in some pond systems (Chamberlain, 1994). Nevertheless, despite the continuing presence of the TS virus, high market prices have continued to drive Ecuador's shrimp production to a peak of 115,000 tonnes (Rosenberry, 1996).

3.6.2 Environmental issues

Water exchange between shrimp farms and their surrounding environment is a standard practice to avoid excessive build-up of waste products and increased eutrophication within ponds, and to maintain healthy planktonic blooms. Extensive cultivation systems require a daily water exchange rate of up to 5%, whereas intensive systems require an exchange rate of up to 30% (Clifford, 1985). Calculations of the nutrient budgets for such systems have indicated that more than 76% of the nitrogen and more than 87% of the phosphorous input is retained within the pond water and sediments (Robertson and Philips, 1995). These nutrients, bound-up in the water and sediment within the pond, are discharged into the surrounding ecosystem, whenever the pond water is exchanged, harvesting occurs or when the ponds are dredged out. However, the receiving body of water will have a limited capacity to assimilate concentrated and persistent pulses of highly nutrient-enriched waters and sediments. At some point, the critical load of nutrients will be attained beyond which the water quality will begin to deteriorate, with negative consequences for the local flora and fauna.

Mangroves are an obvious sink for such nutrient outputs, and Robertson and Philips (1995) calculated that approximately 3 ha of mangrove was required to assimilate the nutrient load generated by a 1 ha semi-intensive shrimp farm. This rose to 22 ha of mangrove for 1 ha of intensive shrimp pond. While the outputs from isolated shrimp farms do not tend to these critical levels, a successful farming operation often attracts the development of other farms in close proximity. As the number of farms in a restricted area increases, so the water quality begins to deteriorate as the concentration of nutrient discharges increases (Figure 3.4). This inevitably leads to elevated stress levels in the shrimp and increased susceptibility to infection by pathogens. Proactive management can help to prevent these situations occurring. For example, in Canada, the capacity of proposed salmon farm sites to assimilate nutrients is examined prior to consent for planning, while within enclosed bodies of water, salmon farms are required to be spaced at least 3 km apart to minimise environmental impacts (Black and Truscott, 1994).

3.6.3 Mangrove destruction

Mangroves are an important habitat, acting as nursery areas for estuarine fishes and invertebrates, and provide a feeding and breeding ground for birds and

Figure 3.4 High densities of shrimp farms in Taiwan have led to a significant degradation in water quality, thereby elevating stress levels experienced by farmed shrimp and making them more vulnerable to devastating outbreaks of disease (photograph courtesy of D.A. Jones).

mammals. They provide protection against storm activity, prevent soil erosion and provide an important source of income for poor coastal communities (Bailey, 1988). In the initial stages of the shrimp industry, mangroves were considered to be areas of low commercial value that were ideal for development for shrimp farming. Furthermore, mangroves are the natural habitat for many of the cultivated shrimp species, so that they seemed to be the ideal location for shrimp farms (Fegan, 1996). However, mangroves are actually very poor sites for shrimp farms as their acid sulphate soils have a pH 3–4 when dried out. On a global scale, shrimp farming has been responsible for less than 5% of mangrove destruction so far. However, on a localised scale, the impact of shrimp farming may be far more severe (Phillips *et al.*, 1993). Aquaculture pond construction projects have destroyed 20% of the mangrove forests in some areas of Ecuador (Snedaker *et al.*, 1986), while in Indonesia the majority of the 300,000 ha of mangrove forest cleared so far is currently used for shrimp cultivation.

Most shrimp-producing countries now recognise the ecological importance of mangroves and have legislation to protect them from development. However, as with the enforcement of fisheries regulations, many of these countries do not have the resources to monitor and police such legislation (Bailey, 1988). While there may be pressure on governments to turn a blind eye to small-scale shrimp farms within areas of protected mangrove, there is increasing pressure from environmental lobby groups in developed countries that advocate banning imports of shrimp farmed at the expense of mangrove forests (Woodhouse, 1996). Nevertheless, the high revenues earned from shrimp cultivation are likely to maintain the pressure for development of some areas of mangroves. Recent attempts to model the social and biological benefits of mangrove conservation have indicated that only 12% of the mangrove outside Thailand's mangrove conservation areas should be given over to shrimp farming, and that 61% of these mangroves should be maintained in their natural state (Pongthanapanich, 1996).

Shrimp farming has a number of other impacts on the environment, such as the intrusion of saltwater into neighbouring agricultural areas, and may even pollute groundwater supplies. Many chemical treatments used to control outbreaks of disease are occasionally applied in excess and pollute surrounding waters, or alternatively they may build up within sediments to high concentrations (e.g. copper compounds). The collection of postlarvae and reproductive adults from the wild for the aquaculture industry supports a large artisanal fishery but also threatens wild populations of shrimp. As virtually all shrimp cultivation is supported by the collection of wild juveniles, excessive collection of shrimp may also have effects on other biota, as shrimp are major predators of larval fishes and benthos in coastal waters.

3.6.4 Stock enhancement

Lobsters command a high unit value at market and their populations are easily overfished, factors that make them very suitable for stock enhancement programmes. Habitat suitability is a critical factor that determines the number of lobsters that can be supported in a given area of seabed. In particular, lobsters require shelter in the form of burrows or crevices throughout their life-history. One method of increasing the available habitat is to introduce artificial reefs into the marine environment (Jensen *et al.*, 1994). When released into a suitable habitat, juvenile lobsters *Homarus gammarus* have been shown to contribute to the commercial fishery (Bannister *et al.*, 1994). However, juvenile lobsters are costly to rear as they must be maintained in individual compartments to restrict their cannibalistic tendencies. One of the major obstacles to lobster release programmes has been the issue of ownership rights to areas of the seabed. As lobsters are highly mobile species, it is almost impossible to prevent them from migrating out of a stocked area. Therefore, few fishermen have been willing to invest in stock enhancement programmes, as they have been unable to ensure that they will have sole access to their investment once it is released onto the seabed. So far, the only way around such problems has been the investment of public money to fund stock enhancement programmes, such that all fishermen have equal access to the resource.

3.7 Summary

The contribution of aquaculture to protein production from the sea is increasingly important. Technological advances are improving product quality and the efficiency with which marine organisms are bred and grown, and as long as there is sufficient financial incentive there is no doubt that these areas will continue to improve. Nevertheless, it is quite clear that there are many ethical and environmental problems that need to be addressed. The financial success of shrimp farming has led to a gold rush mentality, with little environmental planning. The disastrous consequences of environmental mismanagement have severely hampered this sector of the aquaculture industry from reaching its full potential and sound a warning shot to other sectors of the industry. It is widely accepted that these environmental concerns need to be addressed in parallel with the technological obstacles that currently hinder the expansion of the aquaculture industry. These problems are not insurmountable but will require the input of considerable capital investment to overcome them and for the time being are probably restricted to developed countries.

References

Alvarez-Ossorio, M. (1977) Un estudia de la Ria de Muros en Noviembre de 1975. *Bolletin Instituto Espanol Oceanografia*, **2** 1-223.
Anonymous (1996) *Report of the Working Group on Environmental Interaction of Mariculture*. International Council for the Exploration of the Sea CM 1196/F: 5, 238 pp.
Anonymous (1995) ICES Code of practice on the introductions and transfers of marine organisms 1994. International Council for the Exploration of the Sea, Copenhagen, Denmark, 12 pp.
Bailey, C. (1988) The social consequences of tropical shrimp mariculture development. *Ocean and Shoreline Management*, **11** 31-44.
Baldwin, B., Borichewski, J. and Lutz, R.A. (1995) Predation on larvae: filtration by adult bivalves directly and indirectly kills bivalve larvae. 23rd Benthic Ecology Meeting, Rutgers State University, New Brunswick, NJ, USA, 17–19 March 1995.
Bannister, R.C.A., Addison, J.T. and Lovewell, S.R. (1994) Growth, movement, recapture rate and survival of hatchery-reared lobsters (*Homarus gammarus* Linnaeus, 1758) released into the wild on the English east coast. *Crustaceana*, **67** 156-172.
Black, E.A. and Truscott, J. (1994) Strategies for regulation of aquaculture site selection in coastal areas. *Journal of Applied Ichthyology*, **10** 294-306.
Bohnsack, J.A. (1996) Maintenance and recovery of reef fishery productivity, in *Reef Fisheries* (eds. N.V.C. Polunin and C.M. Roberts), Chapman and Hall, London, pp. 283-313.
Brosnan, D.M. and Crumrine, L.L. (1994) Effects of human trampling on marine rocky shore communities. *Journal of Experimental Marine Biology and Ecology*, **177** 79-97.
Caddy, J.F. (1973) Underwater observations on tracks of dredges and trawls and some effects of dredging on a scallop ground. *Journal of the Fisheries Research Board of Canada*, **30** 173-180.
Castel, J., Labourg, P.-J., Escaravage, V., Auby, I. and Garcia, M. (1989) Influence of seagrass beds and oyster parks on the abundance and biomass patterns of meio- and macrobenthos in tidal flats. *Estuarine, Coastal and Shelf Science*, **28** 71-85.
Chamberlain, G.W. (1994) Taura syndrome and China collapse caused by new shrimp viruses. *World Aquaculture*, **25** 22-25.
Chamberlain, G.W. (1997) Sustainability of world shrimp farming, in *Global Trends: Fisheries Management* (eds. E.K. Pikitch, D.D. Huppert and M.P. Sissenwine), American Fisheries Society Symposium 20, American Fisheries Society, Bethesda, Maryland, pp. 195-212.
Chandrasekara, W.U. and Frid, C.L.J. (1996) Effects of human trampling on tidalflat infauna. *Aquatic Conservation: Marine and Freshwater Ecosystems*, **6** 299-312.
Clifford, H.C. (1985) Semi-intensive shrimp farming, in *Texas Shrimp Farming Manual* (eds. G.W. Chamberlain, M.G. Haby and R.J. Miget). Publication of the Texas Agricultural Extension Service, College Station, Texas, pp. IV-15-IV-42.

Collie, J.S., Escanero, G.A. and Valentine, P.C. (1997) Effects of bottom fishing on the benthic megafauna of Georges Bank. *Marine Ecology Progress Series*, **155** 159-172.

Collie, J.S., Hall, S.J., Kaiser, M.J. and Poiner, I.R. (2000) A quantitative analysis of fishing impacts on shelf-sea benthos. *Journal of Animal Ecology*, **69** 785-798.

Cotter, A., Walker, P., Coates, P., Cook, W. and Dare, P. (1997) Trial of a tractor dredger for cockles in Burry Inlet, South Wales. *ICES Journal of Marine Science*, **54** 72-83.

Currie, D.R. and Parry, G.D. (1996) Effects of scallop dredging on a soft sediment community: a large-scale experimental study. *Marine Ecology Progress Series*, **134** 131-150.

Dalhbäck, B. and Gunnarson, L. (1981) Sedimentation and sulfate reduction under a mussel culture. *Marine Biology*, **63** 269-275.

Dame, R., Spurrier, J. and Wolaver, T. (1989) Carbon, nitrogen and phosphorus processing by an oyster reef. *Marine Ecology Progress Series*, **54** 249-256.

Dankers, N. and de Vlas, J. (1992) Multifunctioneel beheer in de Waddenzee-Integratie van natuurbeheer en schelpdiervisserij. Netherlands Institute for Sea Research, Texel, The Netherlands.

Dankers, N. and Zuidema, D. (1995) The role of the mussel (*Mytilus edulis* L.) and mussel culture in the Dutch Wadden Sea. *Estuaries*, **18** 71-80.

Dare, P. (1974) Damage caused to mussels (*Mytilus edulis* L.) by dredging and mechanized sorting. *Journal du Conseil Internationale de l'Exploration de la Mer*, **35** 296-299.

Davies, G., Dare, P.J. and Edwards, D.B. (1980) Fenced enclosures for the protection of seed mussels (*Mytilus edulis* L.) from predation by shore crabs (*Carcinus maenas* L.). Ministry of Agriculture, Fisheries and Food, Fisheries Research Technical Report No. 56. 14 pp.

Dayton, P.K., Thrush, S.F., Agardy, M.T. and Hofman, R.J. (1995) Environmental effects of marine fishing. *Aquatic Conservation: Marine and Freshwater Ecosystems*, **5** 205-232.

Dijkema, R. (1995) Large-scale recirculation systems for storage of imported bivalves as a means to counteract introduction of cysts of toxic dinoflagellates in the coastal waters of the Netherlands, in *Shellfish Depuration (Purification Des Coquillages)* (eds. R. Poggi and J.Y. Le-Gall). Plouzane France IFREMER, Centre de Brest, pp. 355-367.

Dyrynda, P. and Lewis, K. (1995) Ecological studies within the Crymlyn Burrows SSSI (Swansea bay, Wales): impacts of mechanised cockle harvesting. Marine Environment Research Group, University of Wales, 15 pp.

Eno, N.C. (1996) Non-native marine species in British waters: effects and controls. *Aquatic Conservation: Marine and Freshwater Ecosystems*, **6** 215-228.

Eno, N.C., Clark, R.A. and Sanderson, W.G. (1997) Non-native marine species in British waters: a review and directory. Joint Nature Conservation Committee, Peterborough, UK, 152 pp.

Fegan, D.F. (1996) Sustainable shrimp farming in Asia: vision or pipedream? *Asian Aquaculture*, **2** 22-24.

FAO (Food and Agricultural Organisation of the United Nations) (1993) Aquaculture production 1985–1991. FAO Fisheries Circular 815, Revision 5.

FAO (Food and Agriculture Organisation of the United Nations) (1996) Time series on aquaculture-quantities and values. FAO Fishery Information, Data and Statistics Unit (FIDI), Aquacult-PC, Release 8494/96, April 1996.

Fletcher, H. and Frid, C.L.J. (1996) Impact and management of visitor pressure on rocky intertidal algal communities. *Aquatic Conservation: Marine and Freshwater Ecosystems*, **6** 287-298.

Folke, C. and Kautsky, N. (1989) The role of ecosystems for a sustainable development of aquaculture. *Ambio*, **18** 234-243.

Goss-Custard, J.D. and Verboven, N. (1993) Disturbance and feeding shorebirds on the Exe estuary. *Wader Study Group Bulletin*, **68** 59-66.

Gottlieb, S.J. and Schweighofer, M.E. (1996) Oysters and the Chesapeake Bay ecosystem: a case for exotic species introduction to improve environmental quality? *Estuaries*, **19** 639-650.

Haamer, J. (1996) Improving water quality in a eutrophied fjord system with mussel farming. *Ambio*, **25** 356-362.

Hall, S.J. and Harding, M.J.C. (1997) Physical disturbance and marine benthic communities: the effects of mechanical harvesting of cockles on non-target benthic infauna. *Journal of Applied Ecology*, **34** 497-517.

Hallegraeff, G.M. and Bolch, C.J. (1991) Transport of toxic dinoflagellate cysts via ships' ballast. *Marine Pollution Bulletin*, **22** 27-30.

Hill, A.S., Brand, A.R., Wilson, U.A.W., Veale, L.O. and Hawkins, S.J. (1996) Estimation of by-catch composition and the numbers of by-catch animals killed annually on Manx scallop grounds. in *Aquatic Predators and Their Prey* (eds. S.P.R. Greenstreet and M.L. Tasker), Fishing News Books, Blackwell Science, Oxford, pp. 111-115.

Holmes, J.M.C. and Minchin, D. (1995) Two exotic copepods imported into Ireland with the Pacific oyster, *Crassosrea gigas* (Thunberg). *Irish Naturalist's Journal*, **25** 17-20.

Huang, J., Xiaoliang, S., Jia, Y. and Conghai, Y. (1995) Baculoviral hypodermal and hematopoietic necrosis: study on the pathogen and pathology of the explosive epidemic disease of shrimp. *Marine Fisheries Research*, **16** 7.

Ito, S. and Imai, T. (1955) Ecology of oyster bed. I. On the decline of productivity due to repeated culture. *Tokoku Journal of Agricultural Research*, **5** 251-268.

Jennings, S. and Kaiser, M.J. (1998) The effects of fishing on marine ecosystems. *Advances in Marine Biology*, **34** 201-352.

Jensen, A.C., Collins, K.J., Free, E.K. and Bannister, R.C.A. (1994) Lobster (*Homarus gammarus*) movement on an artificial reef: the potential use of artificial reefs for stock enhancement. *Crustaceana*, **67** 198-211.

Kaiser, M.J., Edwards, D.B. and Spencer, B.E. (1996) A study of the effects of commercial clam cultivation and harvesting on benthic infauna. *Aquatic Living Resources*, **9** 57-63.

Kusuki, Y. (1977) Fundamental studies on the deterioration of oyster grazing grounds. II. Organic content of faecal materials. *Bulletin of the Japanese Society of Science and Fisheries*, **43** 167-171.

Laing, I. and Spencer, B. (1997) Bivalve cultivation: criteria for selecting a site. The Centre for Environment, Fisheries and Aquaculture Science, Lowestoft.

Lapointe, B., Niell, F. and Fuentes, J. (1981) Community structure, succession and production of seaweeds associated with mussel-rafts in the Ria de Arosa, N.W. Spain. *Marine Ecology Progress Series*, **5** 243-253.

Lenihan, H.S. (1999) Physical-biological coupling on oyster reefs: how habitat structure influences individual performance. *Ecological Monographs*, **69** 251-275.

Lopez-Jamar, E., Iglesias, J. and Otero, J. (1984) Contribution of infauna and mussel-raft epifauna to demersal fish diets. *Marine Ecology Progress Series*, **15** 13-28.

Ludyanskiy, M.L., McDonald, D. and MacNeill, D. (1993) Impact of the zebra mussel, a bivalve invader. *Bioscience*, **43** 533-544.

Mariojouls, C. and Kusuki, Y. (1987) Appréciation des quantités de biodépôts émis par les huîtres en élevage suspendu dans la baie d'Hiroshima. *Haliotis*, **16** 221-231.

Minchin, D. (1996) Management of the introduction and transfer of marine molluscs. *Aquatic Conservation: Marine and Freshwater Ecosystems*, **6** 229-244.

Mojica, R. and Nelson, W. (1993) Environmental effects of hard clam (*Mercenaria mercenaria*) aquaculture in the Indian River Lagoon, Florida. *Aquaculture*, **113** 313-329.

Nakano, H., Koube, H., Umezawa, S., Momoyama, K., Hiraoka, M., Inouye, K. and Oseko, N. (1994) Mass mortalities of cultured kuruma shrimp, *Penaeus japonicus*, in Japan in 1993: epizootiological survey and infection trials. *Fish Pathology*, **29** 135-139.

Nugues, M.M., Kaiser, M.J., Spencer, B.E. and Edwards, D.B. (1996) Benthic community changes associated with intertidal oyster cultivation. *Aquaculture Research*, **27** 913-924.

Pearson, T. and Rosenberg, R. (1978) Macrobenthic succession in relation to organic enrichment and pollution of the marine environment. *Oceanography and Marine Biology: An Annual Review*, **16** 229-311.

Peterson, C.H., Summerson, H. and Fegley, S. (1987) Ecological consequences of mechanical harvesting of clams. *Fisheries Bulletin*, **85** 281-298.

Phillips, M.J., Lin, C.K., and Beveridge, M.C.M. (1993) Shrimp culture and the environment: lessons from the world's most rapidly expanding warm water aquaculture sector. Environment and Aquaculture in Developing Countries. *ICLARM Conference Proceedings*, 31, Manila, Philippines.

Pongthanapanich, T. (1996) Economic study suggests management guidelines for mangroves to derive optimal economic and social benefits. *Aquaculture Asia*, **1** 16-17.

Robertson, A.I. and Philips, M.J. (1995) Mangroves as filters of shrimp ponds effluent: predictions and biogeochemical research needs. *Hydrobiologia*, **295** 311-321.

Rodhouse, P. and Roden, C. (1987) Carbon budget for a coastal inlet in relation to intensive cultivation of suspension-feeding bivalve molluscs. *Marine Ecology Progress Series*, **36** 225-236.

Rosenberry, R. (1994) World shrimp farming 1994. Annual Report. Shrimp News International, San Diego, California, USA.

Rosenberry, R. (1996) World shrimp farming 1996. Annual Report. Shrimp News International. San Diego, California, USA.

Rueness, J. (1989) *Sargassum muticum* and other introduced Japanese macroalgae: biological pollution of European coasts. *Marine Pollution Bulletin*, **4** 173-176.

Russell, G., Hawkins, S.J., Evans, L.C., Jones, H.D. and Holmes, G.D. (1983) Restoration of a disused dock basin as a habitat for marine benthos and fish. *Journal of Applied Ecology*, **22** 43-58.

Scarratt, A.M., Scarratt, D.J. and Scarratt, M.G. (1993) Survival of live *Alexandrium tamarense* cells in mussel and scallop spat under simulated transfer conditions. *Journal of Shellfish Research*, **12** 383-388.

Seed, R. (1993) Invertebrate predators and their role in structuring coastal and estuarine populations of filter-feeding bivalves, in *Bivalve Filter Feeders in Coastal and Estuarine Ecosystem Processes* (ed. R. Dame), Springer-Verlag, Berlin, pp. 149-195.

Shepherd, M. and Clark, M. (1994) The effect of commercial cockling on the numbers of wintering waterfowl on the Solway estuary. British Trust for Ornithology.

Short, F.T. and Wyllie-Echeverria, S. (1996) Natural and human-induced disturbance of seagrasses. *Environmental Conservation*, **23** 17-28.

Shpigel, M., Neori, A., Popper, D.M. and Gordin, H. (1993) A proposed model for 'environmentally clean' land-based culture of fish, bivalves and seaweeds. *Aquaculture*, **117** 115-128.

Simenstad, C. and Fresh, K. (1995) Influence of intertidal aquaculture on benthic communities in Pacific northwest estuaries: scales of disturbance. *Estuaries*, **18** 43-70.

Sims, N.A. (1993) Pearl oysters, in *Nearshore Marine Resources of the South Pacific: Information for Fisheries Development and Management* (eds. A. Wrigth and L. Hill) Institute of Pacific Studies. Suva Forum Fisheries Agency, Honiara International Centre for Ocean Development, Canada, pp. 414-423.

Snedaker, S.C., Dickinson, J.C., Bown, M.S. and Lahmann, E.J. (1986) Shrimp pond siting and management alternatives in mangrove ecosystems in Ecuador. Final report. Office of the Science Advisor, U.S. Agency for International Development.

Snelgrove, P. and Butman, C. (1994) Animal-sediment relationships revisited: cause versus effect. *Oceanography and Marine Biology: An Annual Review*, **32** 111-177.

Spencer, B.E., Kaiser, M.J. and Edwards, D.B. (1996) The effects of Manila clam cultivation on an intertidal benthic community: the early cultivation phase. *Aquaculture Research*, **27** 261-276.

Spencer, B.E., Kaiser, M.J. and Edwards, D.B. (1997) Ecological effects of intertidal Manila clam cultivation: observations at the end of the cultivation phase. *Journal of Applied Ecology*, **34** 444-452.

Spencer, B.E., Kaiser, M.J. and Edwards, D.B. (1998) Intertidal clam harvesting: benthic community change and recovery. *Aquaculture Research*, **29** 429-437.

Straaten, L.v. (1965) De bodem der Waddenzee. Thieme, Zutphen.
Tenore, K. and Gonzalez, N. (1976) Food chain patterns in the Ria de Arosa, Spain, an area of intense mussel aquaculture, in *Population dynamics* (eds. G. Persoone and E. Jaspers), Ostend, Belgium, Vol. 2, pp. 601-609.
Tenore, K., Boyer, L., Cal, R., Corral, J., Garcia, F., Gonzalez, M., Gonzalez, E., Hanson, R., Iglesias, J. and Krom, M. (1982) Coastal upwelling in the Rias Bajas, NW Spain: constrasting the benthic regimes of the Rias de Arosa and Muros. *Journal of Marine Research*, **40** 701-772.
Thrush, S.F., Hewitt, J.E., Cummings, V.J. and Dayton, P.K. (1995) The impact of habitat disturbance by scallop dredging on marine benthic communities: what can be predicted from the results of experiments? *Marine Ecology Progress Series*, **129** 141-150.
Thrush, S.F., Whitlatch, R.B., Pridmore, R.D., Hewitt, J.E., Cummings, V.J. and Wilkinson, M.R. (1996) Scale-dependent recolonization: the role of sediment stability in a dynamic sandflat habitat. *Ecology*, **77** 2472-2487.
Utting, S.D. and Spencer, B.E. (1994) Introductions of marine bivalve molluscs into the United Kingdom for commercial culture—case histories. *ICES Marine Science Symposium*, **194** 84-91.
Wongteerasupaya, C., Sriurairatana, S., Vickers, J.E., Akrajamorn, A., Boonsaeng, V., Panyim, S., Tassanakajon, A., Withyachumnarnkul, B. and Flegel, T.W. (1995) Yellow-head virus of *Penaeus monodon* is an RNA virus. *Diseases of Aquatic Organisms*, **22** 45-50.
Woodhouse, C. (1996) Farms avoid new U.S. turtle curb on shrimp imports. *Fish Farming International*, **23** 24.

4 Environmental effects of warm water culture in ponds/lagoons

P.-S. Choo

4.1 Introduction

The origin of aquaculture has been associated with the Chinese, whose fish culture practices date back as early as 1100 BC. The common carp *Cyprinus carpio* was believed to be the first fish cultured in captivity and China remains the world's top producer of Chinese and common carps. In 1997, China produced a total of 2,619,738 tonnes of carp, thus accounting for 84.5% of the world's total production of 3,099,570 tonnes (FAO, 1999).

Brackish-water fish culture also has its origin in Asia. It is believed to have originated in Indonesia during the fifteenth century AD, when milkfish *Chanos chanos* were cultured in coastal ponds known as tambaks. By the eighteenth century, an estimated 32,389 ha were converted to tambaks for fish culture (Pillay, 1990).

The impact of aquaculture on the environment was already a subject of debate in the early nineteenth century. Francis (1865) produced a lengthy defence of the removal of trout spawn for culture, which some considered a serious loss to the natural fishery in England. However, aquaculture as practised in premodern times was usually on an extensive scale, which is generally considered to be environmentally friendly. It is only in recent years that farms have come to be operated on a semi-intensive or intensive scale and certain forms of aquaculture, especially shrimp culture in the tropics, have become highly controversial; many have questioned their sustainability and impacts on the environment. Apart from the culture system, location also influences the effects on the environment. Inland aquaculture is considered less hazardous, since many of the freshwater species cultured are herbivores or omnivores, which can adapt to environmentally benign polyculture or integrated farming systems, compared to the high-value, brackish-water species, which are often cultured in ponds sited in highly-sensitive habitats, such as mangroves.

This chapter examines the environmental effects of some of the warm-water culture systems in ponds/lagoons in the tropics. It discusses some of the more controversial aspects regarding their likely environmental impacts. These have become global issues, taking on a new meaning with the onset of trade globalisation and the associated issues involving trade barriers.

4.2 Culture systems and their environmental effects

In the tropics, aquatic organisms cultured in ponds are predominantly fresh water fish that are either herbivores or omnivores. In brackish water, penaeid shrimps are the main species under pond culture, followed by some herbivorous or omnivorous fish, like the mullet, milkfish and tilapia, which is a euryhaline species that can be cultured in either fresh or brackish water. Species of fish like the sea bass *Lates calcarifer*, grouper *Epinephelus* sp., and red snapper *Lutianus argentimaculatus*, are usually cultured in cages. The culture of mullet, milkfish and tilapia in brackish-water ponds becomes controversial if the ponds are sited in mangrove areas and are constructed without following proper regulations or guidelines. Although mullets and milkfish are able to breed in captivity, wild fry are still collected for grow-out in ponds.

The effects of aquaculture on the aquatic environment can either be positive or negative and have been extensively reviewed (Chua *et al.*, 1989; Pillay, 1992; Pullin *et al.*, 1993; Barg and Phillips, 1997; Phillips and Macintosh, 1997). Aquaculture, like any human activity, will have some effects on the environment. As long as these activities permit natural adjustments in the environment, it is recognised that their impacts will be minimal. In well-managed farms, the water quality of influents and effluents may not be significantly different (Pillay, 1992). Boyd (1985), for example, found that there was no accumulation of nitrogen and organic matter in the sediments of well-managed channel catfish ponds. Negative effects are normally associated with the intensive culture of carnivorous species and, in the tropics, of shrimps, especially when farmed on an intensive or superintensive scale.

Generally, environmental effects associated with extensive culture systems are considered minimal. Indeed aquaculture ponds can be designed to play a positive role in soil and water conservation programmes, by slowing down the force of erosion of run-off water and reducing downstream flooding (US Congress, 1988, cited in Harrison, 1994). Water storage in ponds can also help to irrigate vegetable farms with nutrient-rich water (ALCOM, 1988; Bailey and Skladany, 1990, both references cited in Harrison, 1994; Edwards, 1993), and is considered a good way to utilise marginal land (Pillai, 1988, cited in Harrison, 1994).

The practice of semi-intensive and intensive culture systems is a recent phenomenon, much later than the intensification of other forms of agricultural practice. The development of intensive aquaculture practices coincided with the increased awareness of the environment and the concerns over its unsustainable exploitation, which culminated in the Rio Conference in 1992. In recent years, aquaculture activities have been under great pressure from both government and non-government sectors to conform to either regulations or guidelines to ensure their sustainability. Many international agencies have also focused their attention on issues pertaining to aquaculture sustainability. Amongst these initiatives are programmes formulated by the Food and Agricultural Organisation (FAO) on the Code of Conduct for responsible Fisheries, with Article 9 of the Code devoted to aquaculture (FAO, 1995), and the development of technical guidelines for responsible aquaculture (FAO, 1997). The commitment to sustainable aquaculture was agreed upon in the World Food Summit held in 1996 (World Food Summit, 1996).

Adverse effects associated with aquaculture, especially intensive aquaculture, include: habitat destruction; discharge of effluents containing high concentrations of organic matter; and the contamination of the aquatic environment and organisms with chemicals. Common-user conflict and the introduction of exotics, which may alter the diversity of the natural flora and fauna, and feral organisms from culture systems are also contentious issues.

4.2.1 Freshwater culture systems

Freshwater bodies in the tropics, mainly ponds and reservoirs, are used to produce finfish, such as carps, tilapia and catfish. Most tropical freshwater fish are produced from ponds, with cages and pens accounting for about 10%, and tanks and raceways accounting for about 5% of production (Beveridge and Phillips, 1993). Tropical inland aquaculture systems are mostly semi-intensive, targeted at producing low-value herbivorous or omnivorous finfish that are fed a relatively cheap source of low-protein, plant-based diets.

In 1995, Chinese and Indian carps contributed to the greatest share of the total world aquaculture production, accounting for 45.6% (Rana, 1997). For many years, cyprinids ranked as the top species cultured (Table 4.1). They are amenable to polyculture and are either herbivores or omnivores with feeding

Table 4.1 World production of major species of freshwater fish (tonnes)

	World production (tonnes)					
	1988	1990	1992	1994	1996	1997
Hypophthalmichthys molitrix	1586816	1519219	1633850	2217292	2877035	3146410
Ctenopharyngodon idellus	608817	1046677	1251353	1817773	2437590	2661611
Cyprinus carpio	1092968	1139290	1146848	1536640	2034337	2237422
Osteichthys	442423	691154	914346	1133979	1247413	1407223
Aristichthyes nobilis	724728	676797	794696	1076680	1418579	1550995
Carassius carassius	121803	215579	257198	389328	692979	862148
Oreochromis niloticus	128339	231193	327297	424842	622858	741867
Labeo rohita	216318	243428	356153	432198	563968	598962
Catla catla	207000	234003	301306	353351	476806	491756
Crrihinus mrigala	141224	158857	295016	346072	462535	472556

Statistics compiled from FAO Aquaculture Statistics 1988–1997 (FAO, 1999).

habits that are met with diets that are low in protein or fishmeal. They are thus perfect candidates for sustainable practices and, moreover, have a good market and fetch reasonable prices in Asia.

One of the issues involving inland aquaculture in the tropics is conflict over land use. Most farms are sited on arable land, although marginal land may be used and, in some countries like Indonesia and the Philippines, conflicts arise when productive rice fields are converted to fishponds (Beveridge and Phillips, 1993). Surface waters are most commonly used for culture, although ground water is utilised in some instances. In Taiwan, the use of groundwater has led to salination and ground subsidence (Chien *et al.*, 1988). In the early 1990s in Malaysia, the pumping of groundwater at a massive 2000 ha aquaculture farm located in a peat swamp reportedly dried up wells in villages in the vicinity of the farm and also caused beach erosion. Apart from eels *Anguilla* sp., seeds of most freshwater fishes are usually obtained from hatcheries and there are few reports of excessive overfishing of wild stocks (Beveridge and Phillips, 1993).

4.2.1.1 Extensive culture
Initially culture systems were more of a semi-culture, fattening type operation, where the methods employed complimented the natural conditions of the environment. In early China, pisciculture originated with the erection of posts with faggots attached to them; these posts were planted in suitable places along rivers for fish to spawn. The eggs collected were then hatched and placed back into the river or transplanted to other rivers, or to artificial enclosures, such as ponds. Such practices may even have a positive effect on the environment, since they provide more spawning and breeding sites for fish. Fish under extensive culture systems rely only on natural food and do not require artificial feeding or fertiliser inputs. Such culture systems are, however, no longer suitable for the modern world. As the human population increases and land becomes more scarce and expensive, the low productivity from such cultures will no longer be economically viable and will not be able to produce sufficient food for the growing population.

4.2.1.2 Polyculture
Apart from the development of monoculture, the Chinese were also credited with the development of polyculture and integrated fish farming systems. They understood the basic principles of ecology and were able to fully utilise the whole three-dimensional niche structure of the pond to produce a mixed crop of fish. Polyculture is believed to have its origin during the Tang Dynasty (618–904 AD).

4.2.1.3 Integrated culture
The basic principles involved in integrated agriculture-aquaculture (IAA) farming are the utilisation of the synergistic effects of interrelated farm activities,

including the full utilisation of farm wastes (Pillay, 1990). IAA is based on the concept that there is no waste; waste is only a misplaced resource that can become a valuable material for another product (FAO, 1977). Positive effects from integrated farming include the recycling of nutrients and organic wastes. Feeding fish with wastewater also helps to reduce the level of excess nutrients getting into the aquatic environment and mud removed from the pond bottom can be used to fertilise crops.

The fertilisation of fishponds with organic manure comprising human or animal wastes has been widely practised in Asia and is still being practised in some rural areas. Most of these systems use fresh or only partially-treated wastes and the fish produced from such systems may be hazardous to health. The human health aspects of IAA and wastewater-fed aquaculture have been widely debated and reviewed (Feachem *et al.*, 1983; Strauch, 1987; Naegal, 1990; Edwards, 1992); in most cases, these reviews have been rich in conjecture but lacking in data (Pullin, 1998).

Although some studies show that bacterial contamination of fish cultured with wastewater can be reduced (Mara and Cairncross, 1989; Mara *et al.*, 1993), fears remain regarding the introduction of new human influenza viruses and increased incidence of diseases, such as malaria and schistosomiasis. Edwards (1990) suggested that wastewater-fed culture may be more sociologically acceptable if used to produce high-protein animal feed, rather than to produce fish meant for direct human consumption.

4.2.1.4 Semi-intensive and intensive culture
Semi-intensive (requiring some feed and fertiliser inputs) and intensive (relying heavily on formulated artificial feeds) culture of freshwater fish is less commonly practised compared to polyculture or IAA farming. Freshwater fish do not require feeds that are high in protein and fishmeal; very often plant-based protein from local sources can be used in the formulation of the feed. Uneaten feed ranges 1–30%, depending on the culture system used, type of feed and management (Beveridge *et al.*, 1991). The quantities of uneaten food and waste products vary with species and type of food and are influenced by body size and season (Beveridge and Phillips, 1993). Tropical inland pond culture is characterised by low water exchange rate, even in fairly intensive culture systems (Beveridge and Phillips, 1993), and the water quality of the day-to-day effluent is generally reasonable (Boyd, 1978, cited from Beveridge and Phillips, 1993). During cleaning and harvesting, however, the levels of biological oxygen demand (BOD), chemical oxygen demand (COD), total phosphorus and total ammoniacal nitrogen can be significantly higher than in the receiving streams. In the few tropical countries practising intensive pond-based aquaculture, the effluents are either drained into adjacent ponds or used for irrigation, but effluents from intensive tilapia farms in Central America and Caribbean effluents are known to discharge directly into relatively pristine rivers (Beveridge and Phillips, 1993).

Edwards (1993) recommended the use of the semi-intensive culture systems over intensive systems, since the latter are 26–44 times more polluting than the former in terms of nitrogen, and 12–15 times in terms of phosphorus (expressed as a percentage of the nutrients required to produce 1 kg of fish). Moreover, intensive culture systems do not generate a net gain in protein and are profitable only when high-value carnivorous species are produced.

Chemicals and drugs are not often used in inland aquaculture practised in the tropics, except for lime and fertilisers. Antibiotics are not often used, although they may be used occasionally for broodstock treatment (Beveridge and Phillips, 1993). Disease problems or epidemics, like those associated with shrimp farming, are rarely encountered.

4.2.2 Shrimp farming

Globally, the cultured shrimp subsector grew at an annual percentage rate of 16.8 between 1984 and 1995, compared to only 2.6 from capture fisheries (Rana, 1997). Cultured shrimps are predominantly produced from Asia, with Thailand as the world's largest producer. In 1997, Thailand accounted for about 23%, or 214,000 tonnes, of the penaeid shrimps produced (FAO, 1999). Table 4.2 shows the world's top ten producers of farmed shrimps from 1991 to 1997. In 1997, the world's penaeid shrimp production totalled 942,000 tonnes, with the tiger prawn *Penaeus monodon* occupying top position at a production of 490,195 tonnes (or 52% of the total).

4.2.2.1 Habitat destruction

One of the most serious environmental concerns in relation to shrimp farming is the loss of natural habitats, such as the mangroves and other wetland

Table 4.2 The World's top producers of farmed penaeid shrimps from 1991–1997 (ktonnes)

	Annual production (ktonnes)						
	1991	1992	1993	1994	1995	1996	1997
Thailand	162	185	224	264	260	223	214
Indonesia	116	120	116	107	121	125	131
Ecuador	105	113	83	89	106	99	133
India	36	40	72	91	97	95	55
Vietnam	34	36	40	43	49	62	76
Bangladesh	20	21	29	29	34	49	57
China	220	207	88	64	78	89	103
Philippines	51	79	96	93	91	78	42
Taiwan	23	17	12	9	12	13	6
Other	88	72	88	101	104	116	125
Total	832	890	848	890	952	949	942

Statistics obtained from Aquaculture Production Statistics 1988–1997 (FAO, 1999).

ecosystems. The conversion of mangrove land into prawn ponds has also transformed a common property and multiuse resource into a privately owned, single-use resource (Lacanilao, 1989; Bailey and Skladany, 1990, cited from Harrison, 1994).

Mangrove habitats are known to be important spawning, breeding and nursery grounds for many species of fish and shrimp. They play a significant role in sustaining the fisheries resources through the tidal flushing of detritus and nutrients that form the food base for microorganisms, which in turn support the coastal and near-shore fisheries. Prawn resources, in particular, have been closely correlated to the presence of mangroves (Sasekumar and Chong, 1987; Chong, 1996). Mangroves play a role in maintaining the water quality in coastal areas and act as a buffer against soil erosion. Studies have shown that unprotected coastlines have an erosion rate about 20 times greater than shoreline protected with mangroves (Salleh and Chan, 1988). Mangrove areas are important foraging and stopover sites for local and migratory birds, and many species of terrestrial animals reside permanently or temporarily in the forests. These areas also contribute to the livelihood of the coastal communities, which are dependent on wood products that are harvested from the mangrove forests, and artisanal fishermen, who are dependent on fisheries resources for a living.

Although shrimp farms have been sited in mangrove swamps, many other coastal activities have also utilised such areas. In Malaysia, about 30% of the total mangrove areas (641,172 ha) have been cleared for various purposes. Of the 30%, less than 1% has been developed for shrimp farming, and yet this industry has borne the brunt of the criticism with respect to mangrove destruction. Concerned about the increasing destruction of mangrove areas, the Malaysian government has imposed a moratorium on the clearing of new areas for development since mid-1996. Unfortunately, in some other Southeast Asian countries, a far greater percentage of the mangrove areas has been cleared for shrimp farming. In Thailand, the total mangrove area has decreased from an estimated 372,448 ha in 1961 to 168,000 ha in 1993; of the area cleared, 64% has been converted to shrimp ponds (Yashiro, 1997). In the Philippines, over a period of nearly 70 years (1920–1988), more than 310,275 ha of mangroves were cleared from an estimated 450,000 ha available in 1920, and a large proportion (210,457 ha) of the mangrove has been converted to brackish-water ponds (Zamora, 1989). In Indonesia, about 5% of the mangrove forests, or 200,000 ha, has been converted to shrimp farms, and this conversion has led to significant denudation in Java, Sulawesi and Sumatra (FAO-NACA, 1994, cited from Phillips, 1997).

In 1969, Ecuador lost 42,000 ha out of an estimated 204,000 ha of mangrove forest cover, primarily to the shrimp mariculture industry, which has devastated huge areas in the Manabi province (51%) and in Rio Chone, which has lost 80% of its mangroves (Bodero and Robadue, 1995). The widespread loss of mangrove areas led to the formulation of the Mangrove Governance Policies in the mid-1980s, with the objective of centralising enforcement of a virtual prohibition on mangrove use.

Many countries in the Association of South East Asian Nations (ASEAN), such as Indonesia, Malaysia, the Philippines and Thailand, advocate a mangrove buffer zone between the sea and the ponds. The mangrove belt acts as a sink and can improve the water quality to the shrimp farms, as well as protecting the ponds from erosion.

4.2.2.2 Fishing pressure on postlarvae and gravid females
In some countries, such as Bangladesh, Ecuador and the Philippines, some of the postlarvae for stocking grow-out ponds are still collected from the wild, which may adversely affect the recruitment of fry in their natural fisheries. The collection of wild-caught fry is the backbone of the shrimp culture industry in Ecuador, and wild-caught shrimp is also the cheapest source of seed shrimp in the country (Olsen and Coello, 1995). This heavy exploitation has led to the suggestion that the protection and management of the wild shrimp stock be given priority (Olsen and Figueroa, 1986). In other countries, such as Thailand, Taiwan and Malaysia, there is heavy fishing pressure on gravid females and, in the 1980s and early 1990s, huge quantities were exported to Taiwan from Malaysia. Each spawner fetched a price of US$150–300 in Taiwan. In recent years, owing to the shortage of spawners for home industry, the export of wild-caught tiger prawn spawners from Malaysia has been banned. There is still heavy reliance on wild-caught spawners in shrimp farming, since the technique (usually by ablation of the eyestalk) to induce spawning has not yielded consistent results with the production of high-quality fry. Moreover, the production of broodstock in captivity is costly (Kongkeo, 1995) and farmers, therefore, still favour the purchase of wild spawners.

4.2.2.3 Pond effluents
Much attention has also been given to the effluents from shrimp farms that have been perceived to be highly polluting. The main problem associated with newly-excavated extensive shrimp ponds is the acidic nature of the effluents, which may have a pH in the range 2.7–3.9 (Cholik and Poernomo, 1987; Phillips *et al.*, 1993).

The use of fertilisers to stimulate plankton bloom during the first two months of culture, and artificial feeds, coupled with the faeces produced by the shrimps, may cause the build-up of organic matter and result in hypernutrification as well as eutrophication. Intensive culture systems will contribute to a higher organic load than extensive or semi-intensive culture systems. The high organic matter in the effluents will increase the suspended solids and nutrient levels, decrease the dissolved oxygen level and increase the BOD in coastal waters. This, in turn, may cause the bottom sediments to become anoxic, leading to changes in the benthic community. Hypernutrification may cause a change in the composition of plankton species, while eutrophication may cause toxic phytoplankton blooms. Incidences of increased occurrence of harmful algal bloom (HAB), linked to increased eutrophication, are reported to be on the rise worldwide. Occurrences of HAB in aquaculture areas are also reported to be on the increase. In Malaysia in 1983, red tides were reported in the coastal areas in the vicinity of shrimp farms, leading to mass mortality of prawns in the ponds (Khoo, 1985). The number of red tide incidences in China has also increased over the last two decades and the outbreaks have coincided with the expansion of mariculture, coastal development and international maritime traffic (Qi et al., 1992). The rapid expansion of mariculture, specifically the alteration of coastal wetlands for shrimp farming, is suspected by some to be the prime cause of red tide events (Qi et al., 1992). However, not all red tide occurrences are linked to eutrophication; many instances of red tides were reported in areas with minimal anthropogenic activities and pollution problems (Maclean, 1984, 1989, 1993; Taylor, 1990; Choo, 1994a).

Although effluents from intensive shrimp ponds have been associated with high organic load, recent studies have shown that the level of pollution from shrimp farms is a few orders less than the discharges from some other agricultural activities (Table 4.3). Studies in China showed that shrimp farming contributed only 1.2–5.1% of total COD load in two coastal provinces (FAO-NACA, 1995, cited from Phillips, 1997). In Thailand, intensive shrimp farming contributed only 0.5% of nitrogen and 0.1% of phosphorus to the total loading from major rivers of the central region (Phillips, 1997).

Although the volume of effluents discharged from shrimp ponds is greater than those from other industries, recent advances in shrimp farming have seen

Table 4.3 Biological oxygen demand (BOD) and ammoniacal nitrogen from various agricultural discharges

Parameter	Discharge	Concentration (mg/l)	Source
BOD	PW	1900–21600	Ho et al., 1984
	POME	10250–43750	Chew and Yeoh, 1987
	LCRE	4849	Wong, 1980
	DC	1.8± 0.9	Tookwinas, 1998
Ammoniacal nitrogen	PW	75–950	Ho et al., 1984
	POME	4–77	Chew and Yeoh, 1987
	LCRE	466	Wong, 1980
	DC	0.4	Tookwinas, 1998
	Shrimp ponds	0.12	Tookwinas, 1998
	Shrimp ponds	0.003–0.4 (Av. 0.11)	Choo and Tanaka, 2000
	Shrimp ponds (harvesting)	1.1–8.4	Choo and Tanaka, 2000

Abbreviations: PW, raw waste from pig pens; POME, untreated palm oil mill effluent; LCRE, raw latex concentrate rubber effluent; DC, drainage canal of shrimp ponds.

the development of the semi-closed or closed systems in the grow-out phase, with minimal water exchange. In Thailand, semi-closed or closed systems are usually used in areas that are prone to disease problems. Water from the sea is pumped directly into reservoirs, where it is treated before being channelled into ponds used for grow-out purposes. In other Southeast Asian countries, such as Malaysia, semi-closed or closed systems have yet to become widespread.

In more traditional practices, no water exchange is usually carried out in the first two months of culture. In the third month, as much as 10% of water is changed daily, while the fourth month may require daily changes of 20–30% of the pond volume. Recent culture practices in Thailand show a shift towards closed-culture systems where there is no need for water exchange from external sources for the duration of the growing period (Kongkeo, 1995; see also Waller, Chapter 5 of this volume, for a discussion of recirculating systems). Such practices are commonly carried out in disease-prone or polluted areas.

4.2.2.4 The use of chemicals and antibiotics
Chemicals commonly used in shrimp farming include tea seed cake *Camellia* sp., lime, calcium/sodium hypochlorite, malachite green, formalin, copper sulphate, benzalkonium chloride (BKC), glutaraldehyde, zeolite and povidone chloride (Yashiro, 1997; Choo, 1998a). Apart from lime, very small concentrations of the other chemicals are used, therefore their impacts on the environment will probably be minimal. Although a considerable amount of lime is used (2–5 tonnes applied per hectare of pond) for pond preparation, to neutralise acid sulphate conditions and also as a disinfectant, its impact on the environment is considered minimal (Pillay, 1992; Beveridge and Phillips, 1993).

Amongst the drugs used in shrimp culture, antibiotics pose the greatest concern (Choo, 1998a). The development of antibiotic-resistant bacterial strains and the presence of antibiotic residues in cultured species as well as non-target groups are perceived as the two main threats. In the early 1990s, antibiotic residues in cultured shrimps were commonly reported and shrimp consignments from Thailand and Indonesia have been rejected by Japan because of antibiotic residues (Yashiro, 1997). Shrimps from Thailand were rejected by Japan in the early 1990s because of the presence of oxolinic acid (OXA) (Saitanu *et al.*, 1994). Studies carried out by Saitanu *et al.* (1994) reported high incidence of antibiotic (oxytetracycline and oxolinic acid) residues in shrimps sampled between October 1990 and October 1991. Since the mid-1990s, shrimp exporting countries in Southeast Asia are more aware of problems caused by antibiotic tainting and most of these countries have since initiated monitoring programmes to ensure the compliance with food safety regulations from importing countries. In Thailand, regulatory laboratories are established in every province to monitor for antibiotic residues in shrimps before harvesting and sale to processing plants (Kongkeo, 1995).

The use of antibiotics in fish culture in many Southeast Asian countries is not regulated and only guidelines exist for their use (Choo, 1999). Antibiotics used in shrimp farms include oxytetracycline, oxolinic acid, nitrofurans and chloramphenicol (Yashiro, 1997; Choo, 1998a). In Malaysia, nitrofurans and chloramphenicol are not permitted to be present in food or foodstuff. Nitrofurans are not permitted for use in the farming of livestock and fish, but chloramphenicol is allowed for use under prescription in Malaysia; their use, however, is still permitted in some Southeast Asian countries, such as Thailand.

There is a lack of data, especially in the tropics, concerning the impacts of antibiotics on human health and on the environment, and more data-driven relationships are needed to establish proven rather than perceived impacts. Impacts of antibiotic use on the aquatic environment are based mainly on work carried out in developed countries in temperate zones. Results of studies carried out in temperate countries may not necessarily apply in the tropics, since the pharmacokinetic behaviour of antibiotics varies with pH, salinity, temperature and light. Generally, antibiotics with long half-lives will persist in the environment and will have greater adverse impacts compared to those with shorter half-lives.

Amongst the most well-studied antibiotics in temperate countries is oxytetracycline (OTC), which is one of the most commonly used antibiotics for fish farming. Lunestad (1991), based on studies carried out in the temperate zone, and Choo (1994b), based on studies carried out in the tropics, showed that OTC has a short half-life in water. Coupled with dilution of farm effluents from the aquatic environment and the loss of antibacterial activity in seawater, the danger posed by OTC in marine waters is probably negligible. However, OTC has been reported to persist in sediments on the sea floor. It has been found in bottom sediments of fish farms, in concentrations capable of causing antimicrobial effects, for up to 12 weeks after its use (Jacobsen and Berglind, 1988). Samuelsen (1989) found that the half-life of OTC was 32 days under low sedimentation rate. Antibiotics in sediments may alter the bacterial flora and have adverse effects on sediment processes.

Nitrofurans are a group of antimicrobials that possess either carcinogenic or mutagenic properties. Amongst the nitrofurans, whose use is banned in many countries but still permitted in some, furazolidone (FZ) ranks as one of the most commonly used. Choo (1998b) found that FZ does not persist in water, having a half-life of 170 h in freshwater (average pH 7.4, average temperature 26.7°C) and 135 h in seawater (average pH 8.2, average temperature 26.7°C). FZ was also found to degrade rapidly in sediments, having a half-life of 18 h at 4°C (Samuelsen et al., 1991).

Laboratory studies have shown that oxolinic acid (OXA) has a long half-life under tropical conditions and remained at initial concentrations for up to 45 days in both fresh- and seawater experiments (Choo, 1999). A preliminary study conducted by Lunestad (1991) also showed that OXA has a long half-life and remained stable for at least two months, even when illuminated. OXA, however, does not persist in sediments. Björklund et al. (1991) found that OXA in sediments lost its antibacterial effects within 10 days, compared to 77 days for OTC.

Antibiotic tainting of non-target groups in the vicinity of fish farms has fuelled concerns over its use. The residues may threaten human health by being acutely or cumulatively allergenic, toxic, teratogenic, mutagenic or carcinogenic (Anon, 1999). Detectable concentrations of OTC were found in blue mussels 80 m from a farm using this antibiotic (Moster, 1986). OTC was also found in wild fish caught in the vicinity of two farms (Björklund et al., 1990). OXA residues were found in wild fish in the vicinity of fish cages where medication was used; however, 12 days after the cessation of medication, no OXA residues were detected in the wild fish (Lunestad, 1991). Coyne et al. (1997) reported the transient presence of OTC residues in blue mussels in the vicinity of a farm during and after a therapeutic treatment, but OTC was not detected in mussels 20 m away from the farm. They suggested that the only possible potential risk represented by the mussels would be from those harvested in the vicinity during or immediately after a period of therapy. All the studies were, however, conducted in temperate countries, and the results may not necessarily be the same in the tropics.

Long-term exposure to low concentrations of antibiotics is known to enhance development of resistant strains of bacteria, which can influence treatment therapy for fish and humans, and alter the environment of fish farms. In Japan, it was found that 111 out of 139 strains of *Vibrio anguillarum* in ayu *Placoglossus altivelis* were resistant to tetracyclines (Aoki et al., 1985), and strains of *Aeromonas salmonicida* that were resistant to OXA have also been reported (Aoki et al., 1983; Hastings and McKay, 1987). The persistence of antibiotics in the marine environment increases the risk of building up resistant bacterial strains, as well as the risk of transferring plasmids coding for resistance into human pathogens. Although this risk exists, some consider this possibility to be very remote. Smith et al. (1994) contended that the risk of the transfer of resistant plasmids from fish to humans is very low in temperate countries, where most of the studies were conducted. However, they cautioned that more information is needed from developing countries in the tropics before the same conclusion can be made. Although a link can be demonstrated between the use of antibiotics in food animals, the development of resistant microorganisms in those animals and the zoonotic spread to humans, the incidence of the spread of human disease through this route is historically very low (Anon, 1999). Moreover, the abuse of antibiotics in human medicine is far more prevalent and is the principal cause of bacterial resistance (Lieberman and Wooten, 1998; Anon, 1999).

4.3 Culture of exotic species

The culture of exotic species in fresh, brackish and marine waters has been widely practised worldwide and, for some species, like the common carp, the practice originated in historical times. Data compiled by FAO in 1994 show that aquaculture is the main route of introduction in 38.7% of the database records and, globally, 9.7% of aquaculture production comes from introduced species (Garibaldi, 1996). There have been concerns over adverse environmental impacts, including threats to biodiversity with the introduction of exotics, and many national and international organisations have responded to these concerns by formulating regulations, guidelines and codes of practice for their introduction. The Database on Introduction of Aquatic Species (DIAS), created by the Food and Agriculture Organization of the United Nations (FAO), has so far recorded a total of 654 aquatic species belonging to over 140 families that have been introduced throughout the world (Bartley, 1999). Among the objectives of the DIAS are the determinations of the ecological and socioeconomic effects of introduced species. It is not easy to determine the ecological impacts from the introduction of exotic species, but overall the introductions have been considered beneficial (Bartley, 1999). Pillay (1992) and Bartley (1999), however, warned against complacency, since very few of the introductions have been adequately studied and little documentation exists on the ecological impacts of such introductions. Ecological impacts include effects of the interaction of the introduced species on the local fauna and flora, transmission of hitherto unknown diseases and alteration of habitats.

4.3.1 Interaction with local flora and fauna

Predation by introduced carnivorous fish on prey from the natural habitat is often regarded as having the greatest ecological impact (Moyle and Light, 1996). Pacific salmon and rainbow trout introduced into Australia, Chile and New Zealand were reported to have displaced gallaxiids (McDowell, 1990; Arthington and Bluhdorn, 1996, cited in Bartley, 1999). The introduction of *Oreochromis mossambicus* in Lake Butu, the Philippines, has almost resulted in the extinction of *Mistichthys luzonensis* stocks (Baluyut, 1983).

Certain introduced species become a pest in the new environment. The Louisiana red crayfish *Procambarus clarkii* became a pest when it was introduced to Japan, eating rice crops and undermining rice field dykes (Penn, 1954, cited from Pillay, 1992). The tilapia *O. mossambicus* has been introduced to many countries, and is considered a pest and a threat to native diversity (Ang *et al.*, 1989; De Silva, 1989; Juliano *et al.*, 1989; Pullin *et al.*, 1997). In Sri Lanka, however, the introduction of this species of tilapia has not brought about adverse impacts on biodiversity, but instead has made a considerable contribution to the supply of animal protein, increased rice yield, controlled malaria and recycled aquatic vegetation (de Zylva, 1999). Likewise, the introduction of some other fish species as considered beneficial by some but a disaster by others. The introduction of the Nile perch *Lates niloticus* into Lake Victoria in Africa caused massive species extinction of haplochromine fish (Gophen *et al.*, 1995), but its introduction in Lake Kioga was considered a success (Eccles, 1985).

4.3.2 Disease transmission

The spread of diseases through introduced species is a serious concern. Disease agents introduced with exotic species or strains may be more pathogenic in their new environment, where they may spread to atypical hosts or encounter more favourable conditions (Langdon, 1990). Whirling disease in rainbow trout is caused by a myxosporean that is non-pathogenic in brown trout, and *Penaeus vannamei* is a carrier of infectious hypodermal and haematopoietic necrosis virus (IHHN) that can devastate *P. stylirostris* (Bartley, 1999). The introduction of east Pacific and South American penaeid shrimp species to Taiwan is believed to be responsible for the widespread disease problems in penaeid shrimps experienced in that country in the late 1980s (Pillay, 1992; Bartley, 1999). The introduction of the grass carp from the Far East to Europe has been reported to be responsible for the spread of the cestode *Bothriocephalus acheilognathi* to that continent (Ivasik *et al.*, 1969). Fish parasites, such as the fish louse *Argulus* and the anchorworm *Lernaea cyprinacea* were brought into Australia through fish importation (Roberts, 1978, cited from Arthington, 1989; Williams, 1980).

4.3.3 Habitat alteration

Some introduced species, such as the common carp and the grass carp, have the ability to alter the aquatic environment. In Australia, carps were reported to graze extensively on the aquatic plant *Potamogeton pectinatus* (Fletcher *et al.*, 1985). The Asian clam *Corbicula fluminea* has spread in large numbers after their introduction to the United States and has altered the flow and substrate in streams and lakes (CRS, 1999, cited from Bartley, 1999). Seaweeds and seagrasses found in water used for the shipment of introduced oysters were reported to have transformed thousands of square kilometres of mudflats of the Pacific coast (Posey, 1988).

4.4 Feral animals

Farmed species may occasionally escape from the holding facilities because of accident or natural catastrophe. Escape of exotic species may have major consequences for the environment. The escapees may predate on other local species and successfully compete with them to the point of elimination. Introduced fish may also breed with local species and produce hybrids that may be sterile or inviable.

The loss of genetic variability in farmed fish has been widely documented, but studies from the tropics are scarce. Interbreeding of farmed fish with wild fish, leading to alteration in the genetic make-up of wild populations, has become an important issue. In temperate countries, the escape of salmon from cages has given rise to concern. Farmed salmon have been observed to be less fit for survival in the wild and the dilution of gene pool in the wild stock may make the wild fish less adaptable. However, most studies have merely documented genetic change and not actual change in populations or in fitness (Campton, 1995, cited from Bartley, 1999). See Kapuscinski and Brister, Chapter 6, for a further discussion of genetic impacts of aquaculture.

4.5 Mitigation measures

Although there may be some adverse environmental impacts arising from aquaculture activities, in general these impacts are not very significant and the effects could be remedied by proper farming technology and management measures. In the tropics, the culture system that has caused the most concern is shrimp farming. With proper planning and farming technology, this activity could be made sustainable (Pillay, 1992; Choo, 1996, 1997; Phillips, 1997). Greater awareness and training should also be given to stakeholders and government officials approving aquaculture projects, so that they have a fuller understanding of sustainability and will be better committed to ensuring that the projects implemented will have minimal adverse environmental impacts. Many countries in the tropics are in the process of formulating guidelines and taking steps to make aquaculture activities more environmentally friendly.

One useful management tool would be to conduct an environmental impact assessment (EIA) study before granting approval for the construction of farms. In many countries in the tropics, an EIA study is required before a project is approved. In Malaysia, development projects involving a land area of 50 ha or more require an EIA for prescribed activities as described in the Environmental Quality Act 1974 (Amendment 1985). Land-based aquaculture is considered to be one of the prescribed activities under the EIA Order 1987. However, only projects involving the clearing of mangroves of 50 ha or more are required to submit an EIA. In its present form, there is much abuse by the stakeholders in Malaysia, as they very often develop farms in phases of less than 50 ha each time, although the eventual farm size may total many times more than the mandatory size where an EIA is required. Clearly this calls for tightening of the loophole to prevent abuse by stakeholders.

Siting of farms in the mangrove should also be regulated and strictly enforced. In Malaysia, the National Mangrove Committee (NATMANCOM, 1986) recommended a 100 m wide buffer zone along the coast between the pond site and the mean high water level of the sea. In Sulawesi, Indonesia, the Land Rehabilitation and Soil Conservation Services established a 200 m wide buffer to protect the coastal mangroves (Nurkin, 1994). India has implemented a buffer zone of 500 m landward of the high water mark (Jones, 1997). The Supreme Court of India delivered an order in December 1996, directing all semi-intensive and intensive farms along the coasts within the 500 m buffer zone to close by March 1997. In the Philippines, the Forestry Reform Code requires farms in mangrove areas to retain a buffer zone, but this law is seldom enforced (Primavera, 1995).

Guidelines and Codes of Practice to promote good aquacultural practices can also help to minimise the adverse environmental impacts from aquaculture. Issues related to the siting of farms, the judicious use of chemicals and therapeutants, as well as the proper way to manage pond effluents could all be addressed under the Codes of Practice. Malaysia has already formulated a workplan on sustainable aquaculture development and is now actively promoting greater awareness of the workplan to stakeholders (Choy, 1999). In Thailand, policies to ensure the sustainable development of shrimp farms have been formulated (Tookwinas and Kittiwanich, 1996) and the Pollution Control Department has set standards for effluents from shrimp ponds; these standards are awaiting legal approval from the Pollution Control Committee (Suksomjit *et al.*, 1999).

Appropriate feeds and feeding husbandry are important ways of minimising environmental effects from aquaculture practices. Overfeeding should be avoided and the use of pelleted feeds that are less polluting than trash fish should be encouraged. These should eventually replace all use of trash fish. In Southeast Asia, formulated feeds developed for shrimps are adequate for growth but much research is needed to make them less polluting. Feeding strategies for the farmed species should also be improved. Information pertaining to meal timing, meal size, feeding level and optimal ratios of nutrients is lacking for species commonly farmed in the tropics (Chou, 1997). Feed instability and wastage are generally high. In an intensive shrimp culture pond, 17% of the feed is converted into animal tissue, 15% (on dry matter basis) lost through leaching and unconsumed feeds and 20% evacuated in faeces (Primavera, 1994). Most fish diets in the tropics contain phosphorus in concentrations far above the required level (Chou, 1997). There is a need to minimise dietary phosphorus to keep pollution of the aquatic environment as low as possible (Kibria *et al.*, 1996). Digestibility of feed ingredients should be better studied, so that faeces discharged can be minimised and a high energy and nutrient concentration achieved in the diet (Chou, 1997).

Wastes, which may become a problem during harvesting and pond cleaning, need to be properly managed. There are several ways of managing pond effluents. Harvesting of catfish in farms in America does not necessarily require ponds to be drained, thereby avoiding the discharge of effluents into the environment (Tucker, 1985). In Thailand, all shrimp farms of more than 8 ha in size are required to have sedimentation ponds covering not less than 10% of pond area for waste treatment (Tookwinas and Kittiwanich, 1996). Effluents can also be treated with other treatment facilities. Sand filtration, microstraining and air flotation can be utilised, but simple sedimentation has proved to be more cost-effective in commercial farms (Pillay, 1992).

Studies to determine the carrying capacity of the culture site help to avoid pollution, especially self-pollution, which may bring about poor water quality and disease problems. Predictive and hydrological modelling can be used as tools to forecast the environmental effects of aquaculture. Models, however, have to be used with care, and predictions should be followed up with appropriate monitoring (Pillay, 1992; Beveridge and Phillips, 1993).

Initially, the fastest way to ensure sustainability in the farms may be through the regulatory and enforcement approach. However, for the 'control and command' approach to work, a considerable amount of money and manpower must be invested. In future, aquaculture farms, especially the bigger ones, must be more proactive and must follow the approach taken by the industrial sector. They should voluntarily seek accreditation through the implementation of environmental programmes, like the ISO 14000 series, to ensure the sustainability of the aquatic environment, as well as the Hazard Analysis Critical Control Point (HACCP) system in farm management and production to ensure food safety.

References

ALCOM (1988) Report of the Technical Consultation on Aquaculture in Rural Development, Lusaka, Zambia, 27–30 October 1987, FAO/SIDA, Rome.

Ang, K.J., Gopinath, R. and Chua, T.E. (1989) The status of introduced fish species in Malaysia, in *Exotic Aquatic Organisms in Asia* (ed. S.S. De Silva). Proceedings of the Workshop on Introduction of Exotic Aquatic Organisms in Asia, Asian Fisheries Society Special Publication No. 3, Asian Fisheries Society, Manila, Philippines, pp. 71-82.

Anon. (1999) *The Use of Drugs in Food Animals: Benefits and Risks*. National Academy Press, Washington, D.C.

Aoki, T., Kitao, T., Iemura, N., Mitoma, Y. and Nomura, T. (1983) The susceptibility of *Aeromonas salmonicida* strains isolated in cultured and wild salmonids to various chemotherapeutics. *Bulletin of the Japanese Society of Scientific Fisheries*, **49** 17-92.

Aoki, T., Kanazawi, T. and Kitao, T. (1985) Epidemiological surveillance of drug-resistant *Vibrio anguillarum* strains. *Fish Pathology*, **20** 199-205.

Arthington, A.H. (1989) Impacts of introduced and translocated freshwater fishes in Australia, in *Exotic Aquatic Organisms in Asia* (ed. S.S. De Silva). Proceedings of the Workshop on Introduction of Exotic Aquatic Organisms in Asia, Asian Fisheries Society Special Publication No. 3, Asian Fisheries Society, Manila, Philippines, pp. 7-20.

Arthington, A.H. and Bluhdorn, D.R. (1996) The effects of species interactions resulting from aquaculture operations, in *Aquaculture and Water Resource Management* (eds. D.J. Baird, M.C.M. Beveridge, L.A. Kelly and J.F. Muir), Blackwell Science Ltd, Oxford, UK, pp. 114-139.

Bailey, C. and Skladany, M. (1990) Sociology and Aquaculture: the Political Economy of Aquaculture Development in the Third World. ICA Communicae, Auburn University International Center For Aquaculture, Communicae, **12** 6-8.

Baluyut, E. (1983) A review of inland water capture fisheries in Southeast Asia with special reference to fish stocking. *FAO Fisheries Report,* **288** 13-57.

Barg, U. and Phillips, M.J. (1997) Environment and sustainability, in *Review of the State of World Aquaculture.* FAO Inland Water Resources and Aquaculture Service, Fishery Resources Division, FAO Fisheries Circular, No. 886, Rev. 1, Rome, FAO, pp. 55-66.

Bartley, D.M. (1999) The introduction of exotic species through mariculture and its impact on the environment. Paper presented in the Conference on Mariculture and the Environment—Towards the New Millennium, 24–25 August 1999, Maritime Institute of Malaysia (MIMA), Kuala Lumpur.

Beveridge, M.C.M., Phillips, M.J. and Clarke, R.M. (1991) A quantitative and qualititative assessment of wastes from aquatic animal production, in *Advances in World Aquaculture,* vol. 3 (eds. D.E. Brune and J.R. Tomasso), World Aquaculture Society, pp. 506-533.

Beveridge, M.C.M. and Phillips, M.J. (1993) Environmental impact of tropical inland aquaculture, in *Environment and Aquaculture in Developing Countries* (eds. R.S.V. Pullin, H. Rosenthal and J.L. Maclean), ICLARM Conference Proceedings 31, pp. 213-236.

Björklund, H., Bondestam, J. and Bylund, G. (1990) Residues of oxytetracycline in wild fish and sediments from fish farms. *Aquaculture,* **86** 359-367.

Björklund, H., Räbergh, C.M.I. and Bylund, G. (1991) Residues of oxolinic acid and oxytetracycline in fish sediments from fish farms. *Aquaculture,* **97** 85-96.

Bodero, A. and Robadue, D. (1995) Strategies for managing mangrove ecosystems, in *Eight Years in Ecuador: The Road to Integrated Coastal Management* (ed. D. Robadue Jr.), Coastal Resources Center, University of Rhode Island, pp. 43-69.

Boyd, C.E. (1978) Effluent from catfish ponds during fish harvest. *Journal of Environmental Quality,* **7** 59-62.

Boyd, C.E. (1985) Chemical budget for channel catfish ponds. *Transactions of the American Fisheries Society,* **114** 291-298.

Campton, D.E. (1995) Genetic effects of hatchery fish on wild populations of Pacific salmon and steelhead: what do we really know? in *Uses and Effects of Cultured Fishes in Aquatic Ecosystems* (eds. H.L. Schramm and R.G. Piper), American Fisheries Society Symposium 15, Bethseda, MD, pp. 337-353.

Chew, T.Y. and Yeoh, B.G. (1987) Palm oil mill effluent as a source of bioenergy. *Proceedings of the National Symposium on Oil Palm By-products for Agro-based Industry,* pp. 155-168.

Chien, Y.H., Liao, I.C. and Yang, C.M. (1988) The evolution of prawn grow-out systems and their management in Taiwan, in *Aquaculture Engineering Technologies for the Future* (ed. K. Murray), Institute of Chemical Engineers Symposium Series No. 111, Hemisphere Publishing Corporation, New York, pp. 143-168.

Cholik, F. and Poernomo, A. (1986) Development of aquaculture in mangrove areas and its relationship to the mangrove ecosystem, in Papers contributed to the Workshop on Strategies for Management of Fisheries and Aquaculture in Mangrove Ecosystems, 23–25 June 1986 (eds. R.H. Mepham and T. Petr), Bangkok, Thailand, pp. 93-104.

Chong, V.C. (1996) The prawn-mangrove connection—fact or fallacy? in *Sustainable Utilization of Coastal Ecosystems* (eds. M. Suzuki, M.S. Hayase and S. Kawahara), Proceedings of the Seminar on Sustainable Utilization of Coastal Ecosystems for Agriculture, Forestry and Fisheries in Developing Regions, JIRCAS Working Report No. 4, Japanese International Research Center for Agricultural Sciences (JIRCAS), Ministry of Agriculture, Forestry and Fisheries, pp. 3-20.

Choo, P.S. (1994a) A review on red tide occurrences in Malaysia. Risalah Perikanan No. 60, Department of Fisheries, Ministry of Agriculture, Malaysia, 9 pp.

Choo, P.S. (1994b) Degradation of oxytetracycline hydrochloride in fresh- and seawater. *Asian Fisheries Science,* **7** 195-200.

Choo, P.S. (1996) Aquaculture development in the mangrove, in *Sustainable Utilization of Coastal Ecosystems* (eds. M. Suzuki, S. Hayase and S. Kawahara), Proceedings of the Seminar on Sustainable Utilization of Coastal Ecosystems for Agriculture, Forestry and Fisheries in Developing

Regions, JIRCAS Working Report No. 4, Japanese International Research Center for Agricultural Sciences (JIRCAS), Ministry of Agriculture, Forestry and Fisheries, pp. 63-72.

Choo, P.S. (1997) The utilisation of mangrove areas for aquaculture: can it be sustainable? in *Proceedings of the Seminar on Wise Use of Wetlands* (eds. H. Jamil and Y.F. Chew), Wetlands International-Asia Pacific, Kuala Lumpur, Malaysia, pp. 32-38.

Choo, P.S. (1998a) Effluents from penaeid prawn ponds and their possible impacts on the aquatic environment, in *Proceedings of the National Conference on Chemistry and Technological Development in Environmental Management* (eds. P.E. Lim, S.C. Eng, K.Y. Liew and M.A. Mohd. Nawi), Universiti Sains Malaysia, Penang, pp. 65-78.

Choo, P.S. (1998b) Degradation of furazolidone in fresh- and seawater. *Asian Fisheries Science*, **11** 295-301.

Choo, P.S. (1999) Environmental and regulatory issues pertaining to the use of antibiotics in mariculture. Paper presented in the Conference on Mariculture and the Environment—Towards the New Millennium, 24–25 August 1999, Maritime Institute of Malaysia (MIMA), Kuala Lumpur.

Choo, P.S. and Tanaka, K. (2000) Nutrient levels in ponds during the grow-out and harvest phase of *Penaeus monodon* under semi-intensive or intensive culture. Japanese International Research Center for Agricultural Sciences (JIRCAS), Journal No. 8 13-20.

Chou, R. (1997) Feeds and feeding in mariculture—towards sustainable development of the industry. Paper presented in the Seminar on Sustainable Development of Mariculture Industry in Malaysia, 30–31 July 1997, Maritime Institute of Malaysia (MIMA), Kuala Lumpur.

Choy, S.K. (1999) Workplan on sustainable aquaculture development in Malaysia. Paper presented in the Conference on Mariculture and the Environment—Towards the New Millennium, 24–25 August 1999, Maritime Institute of Malaysia (MIMA), Kuala Lumpur.

Chua, T.E., Paw, J.N. and Tech, E. (1989) The environmental impact of aquaculture and the effects of pollution on coastal aquaculture development in Southeast Asia. *Marine Pollution Bulletin*, **20** 335-343.

Coyne, R., Maura, H. and Smith, P. (1997) Transient presence of oxytetracycline in blue mussels (*Mytilus edulis*) following its therapeutic use at a marine Atlantic salmon farm. *Aquaculture*, **149** 175-181.

CRS (1999) Harmful non-native species: issues for Congress. CRS Report for Congress. Congressional Research Service. Order No. RL30123.

De Silva, S.S. (1989) Exotics: a global perspective with special reference to finfish introductions to Asia, in *Exotic Aquatic Organisms in Asia* (ed. S.S. De Silva), Asian Fisheries Society Special Publication No. 3, Asian Fisheries Society, Manila, Philippines.

De Zylva, E.R.A. (1999) The introduction of exotic fish in Sri Lanka with special reference to tilapia. *NAGA, The ICLARM Quarterly*, **22** 4-8.

Eccles, D.H. (1985) Lake flies and sardines—a cautionary note. *Biological Conservation*, **33** 305-333.

Edwards, P. (1990) An alternative excreta-reuse strategy for aquaculture: the production of high-protein animal feed, in *Wastewater-fed Aquaculture* (eds. P. Edwards and R.S.V. Pullin), Environmental Sanitation Information Center, Asian Institute of Technology, Bangkok, pp. 209-221.

Edwards, P. (1992) Reuse of human waste in aquaculture, a technical review. Water and Sanitation Report No. 2, UNDP-World Bank Water and Sanitation Information Center, Asian Institute of Technology, Bangkok.

Edwards, P. (1993) Environmental issues in integrated agriculture-aquaculture and wastewater-fed fish culture systems, in *Environment and Aquaculture in Developing Countries* (eds. R.S.V. Pullin, H. Rosenthal and J.L. Maclean), ICLARM Conference Proceedings 31, pp. 139-170.

FAO (1977) China: recycling of organic wastes in agriculture. *FAO Soils Bulletin*, **40**.

FAO (1995) Code of conduct for responsible fisheries. Food and Agricultural Organization of the United Nations, Rome.

FAO (1997) Aquaculture development. Technical Guidelines for Responsible Fisheries No. 5. Food and Agricultural Organization of the United Nations, Rome.

FAO (1999) Aquaculture Production Statistics 1988–1997. FAO Fisheries Circular No. 815, Revision 11, Fishery Information, Data and Statistics Unit, FAO Fisheries Department, Rome.

FAO-NACA. (1994) Regional study and workshop on the environmental assessment and management of aquaculture development. Food and Agricultural Organization of the United Nations and Network of Aquaculture Centres in Asia-Pacific, Bangkok, Thailand.

FAO-NACA. (1995) Regional study and workshop on the environmental assessment and management of aquaculture development. Food and Agricultural Organization of the United Nations and Network of Aquaculture Centres in Asia-Pacific, Bangkok, Thailand.

Feachem, R.G., Bradley, D.J., Garelick, H. and Mara, D.D. (1983) Sanitation and Disease. Health Aspects of Excreta and Wastewater Management, Vol. 3. World Bank Studies in Water Supply and Sanitation, John Wiley and Sons, Chichester, U.K.

Fletcher, A.R., Morrison, A.K. and Hume, D.J. (1985) Effects of carp, *Cyprinus carpio* L., on communities of aquatic vegetation and turbidity of waterbodies in the Lower Goulburn River Basin. *Australian Journal of Marine and Freshwater Research*, **36** 311-327.

Francis, F. (1865) *Fish-Culture: A Practical Guide to the Modern System of Breeding and Rearing Fish*, Second Edition, Routledge, Warne and Routledge, London.

Garibaldi, L. (1996) List of animal species used in aquaculture. FAO Fisheries Circular No. 914, Food and Agricultural Organization, Rome.

Gophen, M., Ochumba, P.B.O. and Kaufman, L.S. (1995) Some aspects of perturbation in the structure and biodiversity of the ecosystem of Lake Victoria (East Africa). *Aquatic Living Resources*, **8** 27-41.

Harrison, E. (1994) Aquaculture in Africa: Socio-Economic Dimensions, in *Recent Advances in Aquaculture. V* (eds. J.F. Muir and R.J. Roberts), Blackwell Science Ltd, London.

Hastings, T.S. and McKay, A. (1987) Resistance of *Aeromonas salmonicida* to oxolinic acid. *Aquaculture*, **61** 165-171.

Ho, Y.C., Chee, Y.S., Chong, C.N., Choy, C.K., Yeoh, B.G. and Yew, K.K. (1984) Environmental, social and economic considerations in the management of swine wastes in Malaysia. Proceedings of the Regional Seminar on the Technical Utilization and Management of Agricultural Wastes.

Ivasik, V.M., Kulakovskaya, O.P. and Vorona, N.I. (1969) Parasite exchange of herbivorous fish species and carps in ponds of the western Ukraine. *Hydrobiological Journal*, **5** 68-71.

Jacobsen, P. and Berglind, L. (1988) Persistence of oxytetracycline in sediments from fish farms. *Aquaculture*, **70** 365-370.

Jones, P. (1997) India's coasts saved from shrimp peril. *Marine Pollution Bulletin*, **34** 146.

Juliano, R.O., Guerrero III, R. and Ronquillo, I. (1989) The introduction of exotic aquatic species in the Philippines, in *Exotic Aquatic Organisms in Asia* (ed. S.S. De Silva), Asian Fisheries Society Special Publication No. 3, Asian Fisheries Society, Manila, Philippines, pp. 83-90.

Khoo, E.W. (1985) Occurrences of 'red tide' along Johore Straits, Malaysia, resulted in heavy mortality of shrimp. *World Mariculture Society Newsletter*, **16** 4.

Kibria, G., Nugegoda, D., Lam, P. and Fairclough, R. (1996) Aspects of phosphorous pollution from aquaculture. *NAGA, The ICLARM Quarterly*, July 1996, pp. 20-24.

Kongkeo, H. (1995) How Thailand made it to the top. *INFOFISH International*, **1/95** 25-31.

Lacanilao, F. (1989) In crisis: our coastal ecosystems. *Wallaceana*, **58** 6-8.

Langdon, J.S. (1990) Disease risks of fish introductions and translocations, in *Introduced and Translocated Fishes and their Ecological Effects* (ed. D.A. Pollard), Australian Society for Fish Biology Workshop, Proceedings No. 8, Bureau of Rural Resources, Canberra, pp. 69-82.

Lieberman, P.B. and Wootan, M.G. (1998) Protecting the Crown Jewels of Medicine: A Strategic Plan to Preserve the Effectiveness of Antibiotics. Report prepared by the Center for Science in the Public Interest, USA (downloaded from http://www.cspinet.org/reports/abiotic.htm).

Lunestad, B.T. (1991) Fate and effects of antibacterial agents in aquatic environment, in *Chemotherapy in Aquaculture: From Theory to Reality* (eds. J. Alderman and C. Michel), Office International Des Epizooties, Paris, pp. 151-161.

Maclean, J.L. (1984) Indo-Pacific toxic red-tide occurrences, 1972–1984, in *Toxic Red Tides and Shellfish Toxicity in Southeast Asia* (eds. A. White, M. Anraku and K.K. Hooi), Southeast Asian Fisheries Development Center, Singapore, and International Development Research Centre, Ottawa, pp. 92-93.

Maclean, J.L. (1989) Indo-Pacific Red Tides, 1985–1988. *Marine Pollution Bulletin*, **20** 304-310.

Maclean, J.L. (1993) Developing-country aquaculture and harmful algal blooms, in *Environment and Aquaculture in Developing Countries* (eds. R.S.V. Pullin, H. Rosenthal and J.L. Maclean), ICLARM Conference Proceedings 31, pp. 252-284.

Mara, D. and Cairncross, S. (1989) Guidelines for the safe use of wastewater and excreta in agriculture and aquaculture. World Health Organisation, Geneva.

Mara, D.D., Edwards, P., Clark, D. and Mills, S.W. (1993) A rational approach to the design of wastewater-fed fishponds. *Water Research*, **27** 1797-1799.

McDowell, R.M. (1990) Filling the gaps—the introduction of exotic fishes into New Zealand, in *Introduced and Translocated Fishes and Their Ecological Effects* (ed. D.A. Pollard), Australian Society for Fish Biology Workshop, Proceedings No. 8, Bureau of Rural Resources, Canada, pp. 69-82.

Moster, H. (1986) Bruk av antibiotika fiskeoppdrett (The use of antibiotics in Norwegian fish farming). Sogn og Fjordane Distrikshfgskole, 5800 Sogndal, Norway (in Norwegian).

Moyle, P.B. and Light, T.L. (1996) Biological invasions of freshwater: empirical rules and assembly theory. *Biological Conservation*, **78** 149-161.

Naegel, L.C.A. (1990) A review of public health problems associated with the integration of animal husbandry and aquaculture, with emphasis on Southeast Asia. *Biological Wastes*, **31** 69-83.

NATMANCOM. (1986) Guidelines on the use of the mangrove ecosystem for brackish water aquaculture in Malaysia. The Working Group to the Malaysian National Mangrove Committee and the National Council for Scientific Research and Development, Ministry of Science, Technology and the Environment, Kuala Lumpur, Malaysia.

Nurkin, B. (1994) Degradation of mangrove forests in South Sulawesi, Indonesia, in *Ecology and Conservation of Southeast Asian Marine and Freshwater Environments Including Wetlands* (eds. A. Sasekumar, N. Marshall and D.J. Macintosh), *Hydrobiologia*, **285** 271-276.

Olsen, S. and Figuera, E. (1986) An integrated strategy to promote a sustainable shrimp mariculture industry for Ecuador. Coastal Resources Center, University of Rhode Island, Narrangansett, R.I.

Olsen, S. and Coello, S. (1995) Managing mariculture development, in *Eight Years in Ecuador: the Road to Integrated Coastal Management* (ed. and translated by D. Robadue, Jr.), Coastal Resources Center, University of Rhode Island, Narragansett, R.I.

Penn, G.H. (1954) Introductions of American crayfishes into foreign lands. *Ecology*, **35** 296.

Phillips, M.J. (1997) Mariculture and the environment—strategies for sustainability. Paper prepared for the Seminar on Sustainable Development of Mariculture Industry in Malaysia, 30–31 July 1997, Maritime Institute of Malaysia (MIMA), Kuala Lumpur.

Phillips, M.J. and Macintosh, D.J. (1997) Aquaculture and the environment: challenges and opportunities, in *Sustainable Aquaculture* (ed. by K.P.P. Nambiar and T. Singh), Proceedings of INFOFISH–AQUATECH 1996 International Conference on Aquaculture, Kuala Lumpur, Malaysia, 25–27 September 1996, INFOFISH.

Phillips, M.J., Lin, C.K. and Beveridge, M.C.M. (1993) Shrimp culture and the environment: lessons from the world's most rapidly expanding warmwater aquaculture sector, in *Environment and Aquaculture in Developing Countries* (eds. R.S.V. Pullin, H. Rosenthal and J.L. Maclean), ICLARM Conference Proceedings 36, International Center for Living Aquatic Resource Management, Manila; Deutsche Gessellschaft für Technische Zusammenarbeit (GTZ) GmbH, Eschborn, pp. 171-196.

Pillai, T.G. (1988) Aquaculture development and planning: village level aquaculture development in Africa, in *Proceedings of the Commonwealth Consultative Workshop on Village Level Aquaculture*

Development in Africa, Freetown, Sierra Leone, 14–20 February 1985 (ed. by H. King and K. Ibrahim), Commonwealth Secretariat, London, pp. 85-92.

Pillay, T.V.R. (1990) Aquaculture: Principles and Practices, Fishing News Books, London.

Pillay, T.V.R. (1992) Aquaculture and the Environment, Fishing News Books, London.

Posey, M.H. (1988) Community changes associated with the spread of an introduced seagrass, *Zostrea japonica. Ecology*, **69** 974-983.

Primavera, J.H. (1994) Environmental and socioeconomic effects of shrimp farming: the Philippines experience. *INFOFISH International*, **1/94** 44-49.

Primavera, J.H. (1995) Mangroves and brackish-water pond culture in the Philippines. *Hydrobiologia*, **295** 303-309.

Pullin, R.S.V. (1998) Aquaculture, integrated resources management and the environment, in *Proceedings of a Workshop on Integrated Fish Farming*, 11–15 October 1994, Wuxi, China (ed. J.A. Mathias, A.T. Charles and H. Baotong), CRC Press, Boca Raton, New York.

Pullin, R.S.V., Rosenthal, H. and Maclean, J.L. (eds) (1993) Environment and aquaculture in developing countries. ICLARM Conference Proceedings 31, Manila, Philippines.

Pullin, R.S.V., Palomares, M.L., Casal, C.V., Dey, M.M. and Pauly, D. (1997) Environmental impacts of tilapia. ICLARM Contribution No. 1350.

Qi, Y., Hong, Y., Qian, H. and Lu, S. (1992) Problems caused by harmful algal blooms in China. BMTC-IOC-POLARMAR International Workshop on Training Requirements in the Field of Eutrophication in Semi-Enclosed Seas and Harmful Algal Blooms, Bremerhaven, 29 September–3 October 1992. Intergovernmental Oceanographic Commission Workshop Report No. 94, UNESCO.

Rana, K.J. (1997) Aquatic environments and use of species groups, in *Review of the State of World Aquaculture*, FAO Fisheries Circular No. 886, Rev. 1, FAO, Rome.

Roberts, R.J. (1978) Fish Pathology. Ballier Tindall, London.

Saitanu, K., Amornsin, A., Kondo, F. and Tsai, C.-E. (1994) Antibiotic residues in tiger shrimp (*Penaeus monodon*). *Asian Fisheries Science*, **7** 47-52.

Salleh, M.N. and Chan, H.T. (1988) Mangrove forests in Peninsular Malaysia: an unappreciated resource. Paper presented at the ENSEARCH seminar on Marine Environment: Challenges and Opportunities, 31 March–2 April 1988, Kuala Lumpur.

Samuelsen, O.B. (1989) Degradation of oxytetracycline in seawater at two different temperatures and light intensities, and the persistence of oxytetracycline in the sediment from a fish farm. *Aquaculture*, **83** 7-16.

Samuelsen, O.B., Solheim, E. and Lunestad, B.T. (1991) Fate and microbiological effects of furazolidone in a marine aquaculture sediment. *Science of the Total Environment*, **108** 275-283.

Sasekumar, A. and Chong, V.C. (1987) Mangroves and prawns: further perspectives, in *Proceedings of the 10th Annual Seminar of the Malaysian Society of Marine Sciences*, pp. 10-22.

Smith, P., Hiney, M.P. and Samuelsen, O.B. (1994) Bacterial resistance to antimicrobial agents used in fish farming: a critical evaluation of method and meaning. *Annual Review of Fish Diseases*, **4** 273-313.

Strauch, D. (1987) Animal production and environmental health. Elsevier, Amsterdam.

Suksomjit, M., Prarakkamo, P., Chongprasith, P., Sukasam, W. and Srinetr, V. (1999) Enforcement of shrimp culture in Thailand through setting up a new effluent standard, in *ASEAN Marine Environmental Management: Towards Sustainable Development and Integrated Management of the Marine Environment in ASEAN*. Proceedings of the Fourth ASEAN-Canada Technical Conference on Marine Science, 26–30 October 1998, Langkawi, Malaysia (eds. I. Watson, G. Vigers, K.S. Ong, C. McPherson, N. Millson, A. Tang and D. Gass), EVS Environment Consultants, North Vancouver, and Department of Fisheries, Malaysia.

Taylor, F.J.R. (1990) Red tides, brown tides and other harmful algal blooms: the view into the 1990s, in *Toxic Marine Phytoplankton* (eds. E. Granéli, B. Sundström, L. Edler and D.M. Anderson), Elsevier, New York.

Tookwinas, S. (1998) The environmental impact of intensive marine shrimp farming effluents and carrying capacity estimation at Kung Krabaen Bay, Eastern Thailand. *Asian Fisheries Science*, **11** 303-316.

Tookwinas, S. and Kittiwanich, J. (1996) Overview of marine shrimp culture in Thailand. Paper presented in the Seminar on Coastal Aquaculture, Golok River Basin Malaysia-Thailand Joint Committee, 11.6.96, Penang, Malaysia.

Tucker, C.S. (ed.) (1985) Channel catfish culture. Elsevier, Amsterdam.

United States Congress (1988) Assessment, Enhancing Agriculture in Africa: A Role for US Development Assistance. Office of Technology, US Congress, Washington, DC.

Williams, W.D. (1980) Australian Freshwater Life, Macmillan, Sydney.

World Food Summit (1996) World Food Summit Plan of Action, in *Rome Declaration on World Food Security and World Food Summit Plan of Action*; World Food Summit, 13–17 November 1996, Rome, Italy, Food and Agricultural Organisation, Rome.

Wong, K.K. (1980) Application of ponding systems in the treatment of palm oil mill and rubber mill effluents. *Pertanika*, **3** 133-141.

Yashiro, R.H. (1997) Sustainable development of shrimp culture: a Thai perspective. Paper presented in the Seminar on Sustainable Development of Mariculture Industry in Malaysia 30–31 July 1997, Maritime Institute of Malaysia (MIMA), Kuala Lumpur.

Zamora, P.M. (1989) Philippine mangroves: their depletion, conversion and decreasing productivity. *Wallaceana*, **58** 1-5.

5 Tank culture and recirculating systems
U. Waller

5.1 Introduction: the benefits of closed and semi-closed systems

The spectrum of containment of aquaculture species ranges from sea ranching (where fish and shellfish are released into the environment, utilising naturally occurring food prior to recapture) to total-no-loss recirculating systems (where all the environmental requirements of the culture species are met from within the system). Much of the focus of environmental concern is on intensive culture with low containment, for example, marine cage culture, where high energy wastes are lost directly to the environment. In this chapter, land-based systems, where containment and treatment are more easily accomplished, are described and the positive advantages of several of these systems highlighted in terms of the benefits to fish performance, to the environment and thus to sustainable development of aquaculture.

Tank and raceway aquaculture systems are almost always situated on land and connected to a water supply and disposal. Tanks are square, rectangular or circular structures, while raceways are extended along the axis of flow. They are made of concrete, fibreglass and other suitable materials. In some cases, tanks or raceways may also be constructed as floating systems providing holding space at the water level (Martin and Heard, 1987; Yoo et al., 1995).

Tanks and raceways are, in principle, artificial ponds. However, the most obvious distinction is the lack of an internal biology, which is an important feature of natural ponds. For this reason, tanks and raceways need to have a continuous flow of water. The water supply can be via open (flow-through) or closed (recirculating) systems. While flow-through systems require large quantities of water that are discharged after passing through the culture vessels, recirculating systems recycle more than 90% of the water volume of the system (EIFAC, 1986). In recirculating facilities, a wide variety of organisms can be maintained and species having particular demands can be cultured, because size, shape, and water quality can be adjusted. Even in flow-through systems, post-treatment of effluent water is possible, and this will mitigate the environmental impact.

The majority of freshwater and marine fish aquaculture is carried out in ponds and net cages. Carps, for example, make up 35% of world aquaculture production and these are mainly reared in earthen ponds. Around 7% of world production comprises diadromous and marine fish, which are commonly reared in net cages along the coast but are also produced in industrial land-based tank or raceway systems (Steffens, 1988; Kestemont, 1995), although the contribution to world production is still minor. This may change in the near future as further regulation is imposed on aquaculture ventures. The worldwide expansion of finfish aquaculture (Figure 5.1) will call for improved concepts that sidestep limitations imminent in open-pond and net-cage culture facilities. The annual increment of production is still ~10%, and the lower growth of recent years may reflect the competition for the means of production, that is, production space and especially water. Tanks and raceways are suitable for the culture of a vast number of animals, plants and algae and, therefore, may add to the diversification of production. For both reasons, these culture systems will increasingly contribute to world aquaculture, particularly to finfish production. This chapter concentrates on finfish and attempts to bring together the available information concerning new ideas and concepts aimed at an improvement in conventional technologies.

A major step from traditional finfish aquaculture towards industrial production was achieved through the development of tanks or raceway systems that allowed increased rearing densities and, at the same time, reduced land usage and water consumption. In traditional fish farms, tanks and raceways are often used in combination with ponds or net cages. These tanks and raceways may be comparably small and probably applied only to certain life stages, but there is an obvious trend in finfish farming towards large culture vessels to save investment and labour costs. The capital expenditure and the running costs (energy and maintenance) for such installations are still comparatively high. The operation is demanding and requires sound knowledge and skills. The risk of failure is high, especially if the biological and engineering concepts do not fit the species.

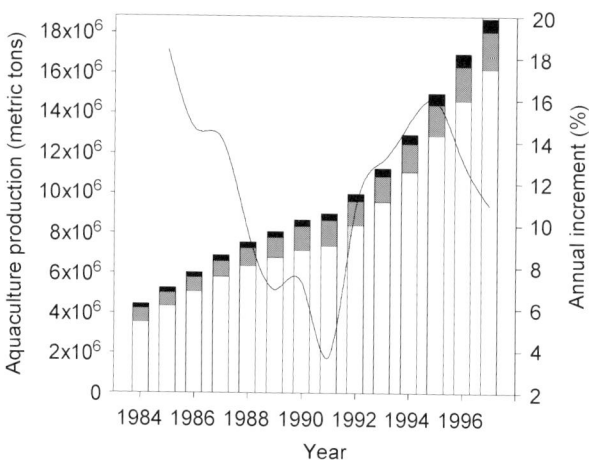

Figure 5.1 Projections for the development of finfish aquaculture with data from FAO (1999). Key: ▭ freshwater fish; ▨ diadromous fish; ▮ marine fish; — year vs annual increment.

Tanks and raceways serve a wide range of applications in aquaculture. They are used for incubation, hatching and weaning of early stages in the life history of aquatic organisms and for the rearing of juveniles and adults or the maintenance of parental stocks. Aquaculture in tanks and raceways can embrace the entire life cycle of a species or only certain stages of life history.

These systems usually allow control of the rearing environment. For instance, the rearing temperature can be controlled and growth may be much faster; optimum growth temperatures can be maintained throughout the rearing period. Without temperature control, fish will grow satisfactory only during seasons with suitable temperatures (Ewing *et al.*, 1998a) and growth will be retarded during the rest of the year. In a controlled-temperature regime, the whole rearing process can be shifted towards a harvest in the off-season, supporting the economy of such aquaculture operations.

A distinctive feature of open installations, such as ponds or net cages, is the immediate contact with the surrounding environment through water flow or through infiltration or leakage. This has raised concerns about the environmental impacts that may harm ecosystems in the receiving water bodies (Rosenthal *et al.*, 1993; Ackefors and Enell, 1994; Rennert, 1994; Brown and King, 1995; Hennessy *et al.*, 1996). In this respect, tanks and raceways are different. They are usually situated at a distance from natural waters, so that the waste water can be treated and unwanted effluents avoided.

As land-based systems allow the separation of farmed organisms and natural populations, possible impacts of escaped animals on natural stocks can be precluded. This has the potential to allow culture of domesticated strains or even transgenic fish that may otherwise threaten natural ecosystems (Devlin *et al.*, 1999). When tank or raceway rearing units are operated as recirculating systems, it is in fact possible to farm aquatic animals totally independent of the surrounding environment. Non-indigenous species can be reared at almost any location, and it is possible to guard against unwanted transfers through careful removal of the organisms from the drain (for further discussion of containment, see Kapuscinski and Brister, Chapter 6 of this volume). This will be important in future as markets demand new products of global origin. Thus, it is necessary to continue development of new or improved culture facilities in which husbandry conditions can be adjusted to the needs of the species.

The shortage of suitable fresh, brackish and marine water is already causing problems for aquaculture in many areas and it is to be expected that this will become even more severe in the near future. A major problem that aquaculture plants are faced with is the increasing levels of general environmental pollution. Lack of oxygen, the input of nutrients, the discharge of toxic substances from industry, as well as toxic algal blooms are particularly feared. Open systems are run at a particularly high operational risk, as environmental disasters may lead to high losses. Tank and raceway systems can be operated with a controlled water flow and, therefore, safety can be maintained even at critical locations. The risks, however, very often stem from dense aquaculture operations in the near neighbourhood; especially the transfer of parasites and disease that may be facilitated by neighbouring fish farms (Jarp and Karlsen, 1997). The deterioration of seabeds and water quality by excess release of nutrients may have several backlash effects (e.g. release of toxic hydrogen sulphide from sediments, hypertrophication of enclosed bays) on nearby cage operations (Lumb, 1989; Rosenthal et al., 1995). Land-based aquaculture installations, especially recirculating systems, get around these limitations and will play an important role for the future development of a sustainable aquaculture.

5.2 General design of rearing units

Several authors have thoroughly reviewed the design and characteristics of tanks and raceways (Wheaton, 1977; Cripps and Poxton, 1992; Timmons et al., 1998) and it is not intended to duplicate their work in this chapter. For this reason, only the characteristic features of the different culture vessels are examined for the most common types of rearing units. Flow pattern and profiles for different types of vessels and hydrodynamic engineering have been discussed in detail by Ross et al. (1995) and Ross and Watten (1998).

Tanks for fish culture are of various shapes, sizes and colours. According to their function, they can be rather small (having a diameter of <1 m) or can reach diameters of >6 m. The water depth varies from <1 m to 1–2 m. Giant tanks, having diameters of >30 m and water depth of >5 m, are under development. The form of tanks may be circular or rectangular, but rectangular tanks usually have rounded corners so that circular flow pattern may develop. The corner area, however, is a dead space that does not contribute to the holding volume available to the livestock (EIFAC, 1986). This is very often neglected during planning and may result in unsuitable stocking densities during operation.

The hydrodynamic properties of circular tanks, as compiled from reports by Wheaton (1977), Rosenthal (1981) and Timmons et al. (1998), are outlined in Figure 5.2. The inlet by means of jets is usually mounted at the water surface, distributing the water tangentially. Additional pipes with orifices may be mounted vertically to improve the flow profile. The outlet is in the tank centre. Additional outlets may be mounted at the water level. Two distinct flow patterns can be distinguished: the primary circular flow driven by the nozzles at the surface; and the secondary flow accounting for the self-cleaning properties at the bottom of the tank (Timmons et al., 1998) (Figure 5.2). The current profile can be subdivided into three zones with different hydrodynamic properties. Zone A, that is, the outside circumference of the tank, is characterised by a high water velocity and turbulent flow profile (Wheaton, 1977; Timmons et al., 1998). This part of the water column is well mixed and provides homogeneous living space. Zone B embodies slower water velocities and poorer mixing of

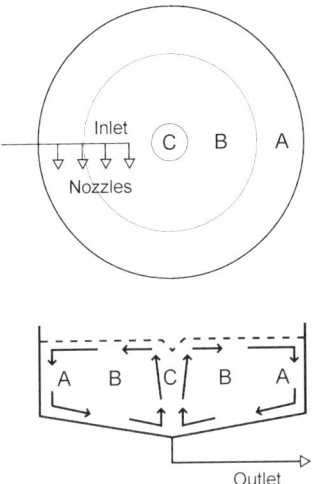

Figure 5.2 Schematic representation of the hydrodynamic properties of circular tanks. The top panel shows the view from above, and the lower panel the side view of a circular tank. (Drawings after Rosenthal, 1981; Timmons et al., 1998; and Wheaton, 1977).

the water column (Timmons *et al.*, 1998). Zone C is the drain zone, with an outward flow at the surface and a bottom flow towards the drain (self-cleaning); it is characterised by relatively poor water quality (Rosenthal, 1982), and is not likely to provide suitable living conditions. Basically, this area has to be excluded from the calculations of stocking density as fish will not voluntarily swim into such water bodies. Fish kept in circular tanks usually swim in zones A and B (Figure 5.2) (Rosenthal, 1982). The fish are usually oriented towards the water flow, as rheotaxis is a typical behavioural pattern in fish (Arnold, 1974; Montgomery *et al.*, 1997; Veselov *et al.*, 1998). Thus, the flow profile and velocity in the tank forces fish to swim at certain speeds.

Raceways are shallow longitudinal tanks. According to the application, the dimension of raceways varies markedly. Raceways for the fattening of juvenile, subadult and adult fish may have a length of >50 m, a width of 2–6 m and a water depth of 1–2 m. Installations for rearing larvae and fingerlings are much smaller and may have a length, width and depth of 6, 1 and 0.5 m, respectively. For certain species, such as flatfish, shallow raceways for early larval rearing may have a water depth of only a few millimetres (Klokseth and Øiestad, 1999), and similar designs for larger fish are under development (Øiestad, 1999). An obvious advantage of a raceway design is the possibility of separating different groups of fish by grids facilitating handling, sorting and harvesting.

Raceways may be operated in parallel or in series (Figure 5.3). The latter design requires less water, while parallel raceways need large waterflows.

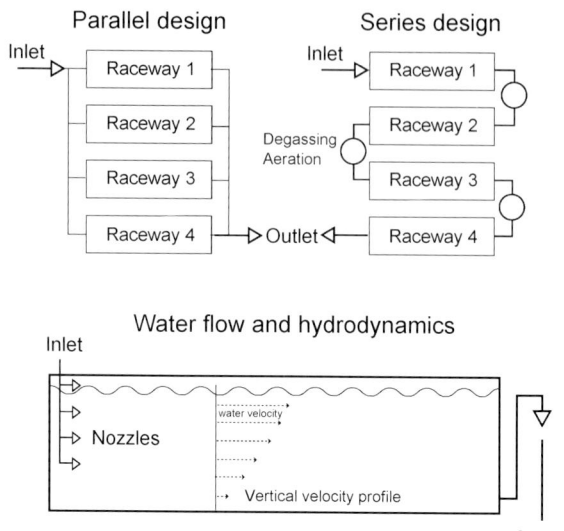

Figure 5.3 The set-up and hydrodynamics of plug-flow raceway installations. Vertical velocity profile after Wheaton (1977).

Raceways installed in series are faced with a downstream accumulation of metabolites, increase of carbon dioxide and depletion of dissolved oxygen levels. Precautionary measures, such as aeration (Figure 5.3), can mitigate some of these problems, but metabolites can only be reduced by exchanging the water, which may require large waterflows leading to inappropriate water velocities within the rearing unit.

The hydrodynamic properties of raceways are outlined in Figure 5.3. Water velocity is higher at the surface and low at the bottom, due to frictional forces. A series of nozzles mounted vertically may support a more homogeneous velocity profile, while a single inlet will lead to a very steep decrease of water velocity towards the bottom (Wheaton, 1977). Shallow raceways will, to a certain degree, improve the vertical flow pattern. Because the self-cleaning of raceways depends on the strength of the bottom flow, a low current may lead to an inappropriate removal of suspended solids. Self-cleaning properties can be enhanced through baffles, which increase the bottom flow by reducing the cross-sectional area (Boersen and Westers, 1986), but these may have an impact on fish distribution, growth and health as reported by Kindschi *et al.* (1991), Barnes *et al.* (1996) and Ewing *et al.* (1998b). Another possible measure is the installation of crosswise installed particle traps. These traps are in the shape of small canals running all the way through the raceway (Wheaton, 1977). The outlet is usually mounted at one side and several traps may be connected by a main pipe directing the water towards the drainage system.

Circular tanks have certain drawbacks linked to their hydrodynamic properties that are very much related to stocking density, fish distribution and water inlet design (Rosenthal, 1981). However, to a great extent, this handicap can be overcome by proper design and operation. The circular flow allows fish to swim in a virtually endless space, resulting in less frequent turns of individual fish (Ross and Watten, 1998), which may reduce restless behaviour. The plug-flow in raceway systems is a major limitation because fish are not able to move as freely as is possible in circular tanks. To overcome this disadvantage, raceways may be constructed in the form of an oval tank (Figure 5.4). The water flow is induced by airlift-pumps (Loyless and Malone, 1998) that simultaneously aerate the water. Such tanks get around the limitations of plug-flow raceways. The influent water is quickly mixed and, over the entire tank area, a homogeneous water column develops (Western and Pratt, 1977). Water velocity can be adjusted to the ability of the species and, therefore, optimised for growth.

Tanks and raceways are made of a range of materials, such as glass fibre, thermoplastic or concrete. Concrete tanks are suited for freshwater applications but in seawater aquaculture they may not be acceptable due to corrosion. To prevent the growth of algal or bacterial films, which can lead to deterioration of water quality, the interior surface of concrete tanks or raceways may be plastic coated to minimise porosity; the coating also improves resistance to saltwater and minimises fin lesions through abrasion (Bosakowski and Wagner, 1995). Thermoplastic is a recyclable material that is non-toxic and can be easily formed into different shapes. Nevertheless, the surface is rough and is easily damaged during cleaning. This material is often used for small units providing temporary holding space. Glass fibre is a solid material, having a smooth surface. Because the surface coating is robust and glass fibre reinforced plastic tanks are durable, they are very common. However, the raw materials used are highly toxic and may release hazardous substances (e.g. organic solvents, glass fibres) into the water over a prolonged period, depending on the manufacturing process; this particularly endangers the survival of early life-history stages. Tanks may also be made from a robust frame or jointed panels, and a plastic liner made from vinyl or polyethylene, or a coated fabric. These tanks are problematic in some ways, although this construction allows the transportation in parts of tanks of any size. The lining material may wrinkle and waste particles and bacteria

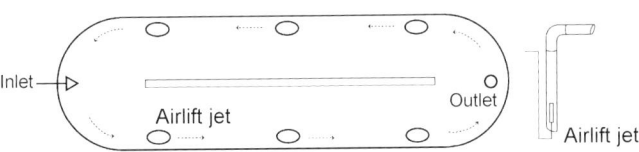

Figure 5.4 Oval tank design with airlift-jets to induce water flow.

may accumulate in pockets, leading to sanitary problems by reducing the self-cleaning efficiency. The material is easily cut or damaged and may not be as durable as other materials.

5.3 Environmental control

The particular advantage of land-based tank or raceway systems is that external factors can be controlled and adjusted to the requirements of the species. Besides better growth, food conversion efficiency can be maximised, at the same time reducing the total load of dissolved nutrients and particulate matter in the effluent water. Careful management of land-based systems allows not only the optimisation of production but also the reduction of the environmental burden. The following section considers some key factors that must be considered carefully during planning and operation.

5.3.1 Light regime

Tanks and raceways may be located indoors or outdoors. Outdoor tanks exposed to ambient light may support the natural behaviour of the species under culture. Several investigations into the influence of photoperiod on fish have shown that behaviour and metabolic expenditure fluctuate with a predictable pattern. In turbot *Scophthalmus maximus* a bimodal pattern with an increase at dawn and dusk was found (Waller, 1992), obviously coinciding with the feeding instinct. In coho salmon *Oncorhynchus kisutch* a steep increase was observed during the night with a minor rise in the morning (Waller *et al.*, 2000), which tallies with the voluntary food uptake of Pacific salmon (Brett, 1971). Hossain *et al.* (1999) found that the feeding activity of the catfish *Clarias gariepinus* was strictly related to the photoperiod. Similar results were reported by Kadri *et al.* (1997) for Atlantic salmon *Salmo salar* Azzaydi *et al.* (1998) concluded from their experiments with sea bass *Dicentrarchus labrax*, that the feeding schedule of farmed fish should reflect the voluntary feeding activity. Therefore, in outdoor culture facilities it is necessary to adjust the feeding schedule to the season in order to support foraging behaviour. However, following the natural seasonal growth pattern may reduce overall growth in animals kept under a natural photoperiod (Jobling, 1987; Hansen *et al.*, 1992). If the rearing units are indoors, the light level and diurnal pattern can be set. Under such conditions, growth rate of particular fish species can be enhanced by prolonged light periods; manipulation of light toward a long photoperiod stimulates the growth of several fish species (Sigholt *et al.*, 1989; Saunders and Harmon, 1990; Silva-Garcia, 1996; Skilbrei *et al.*, 1997). However, with respect to the feeding instinct, the photoperiod must still support the feeding behaviour of the species under culture. Furthermore, in certain fish species, it is possible to control aggressive behaviour (Kushnirov and Degani, 1991; Nicieza and Metcalfe, 1997). Another important aspect is the possibility of shifting the spawning time through modification of the photoperiod (Carrillo *et al.*, 1989). This allows the extension of the spawning season and, as a result, the period available for restocking of systems.

5.3.2 Temperature regime

Holding temperature determines growth and overall performance of poikilothermic aquatic organisms. The final preferendum for temperature (Richards *et al.*, 1977; McCauley and Casselman, 1981), that is the temperature in which fish would congregate, is an indicator of the optimum growth temperatures (Jobling, 1981). The term 'final preferendum' implies that fish behave in temperature gradients and avoid undesirable conditions. The most favourable temperature conditions must be maintained in order to maximise growth and optimise production efficiency (Jobling, 1988). In outdoor facilities with an open water supply, water temperature will vary throughout a year and possibly during a day, depending on solar radiation and the flow rate. Appropriate temperature control can be attained by adjusting water exchange rates. However, if water flow rate is increased, inappropriate current regimes may develop, as discussed in Section 5.3.3 of this chapter.

Diel temperature variations may not affect growth, even if temperature changes will quickly act on feeding activity (Alanara, 1996). In fact, Berg *et al.* (1990) found that diel temperature variations slightly increased growth rate in Atlantic salmon, compared to a control group maintained at a median temperature. However, optimum holding temperatures are a prerequisite for successful farming of fish and other aquatic animals. This may require protection and insulation of outdoor installations. In unprotected outdoor facilities, ultraviolet radiation may also harm the animals (Blazer *et al.*, 1997; Walters and Ward, 1998), therefore provision of cover has benefits in addition to moderating temperature fluctuations. In indoor installations, temperature control is much easier to achieve. Solar radiation can be used by means of collecting panels, and heat exchangers allow the recovery of heat from the effluent water or from industrial waste heat. The ambient air, as well as natural water, may be used as a heat source or sink, at least during certain periods of the year.

5.3.3 Swimming activity

Swimming activity represents a large and variable component of a fish's energy budget (Boisclair and Sirois, 1993) and excess activity in fish may compromise the amount of energy available for digestion and growth processes (Calow, 1985). The swimming performance of fish depends on size and, during the rearing process, it is necessary to adjust water velocity in order to optimise growth. Investigations by several authors have emphasised a strong relationship between the swimming exercise of fish and growth (weight gain). Greer Walker and Emerson (1978) found that maximum growth in trout *Oncorhynchus mykiss* is associated with a swimming speed of 1 bodylength^{-1}. In brown trout, *Salmo trutta*, best growth is obtained at a 1.5 bodylength^{-1} swimming speed. In *Salmo salar*, a swimming speed in the range 1.3–1.5 bodylength^{-1} results in highest growth rates (Davison and Goldspink, 1977; Joergensen and Jobling, 1993; Joergensen *et al.*, 1996). Arctic charr, *Salvelinus alpinus*, growth is significantly improved with increasing swimming speed; maximum growth rate is observed at around 1.75 body lengths^{-1} swimming speed (Christiansen and Jobling, 1990). However, under restricted feeding, exercised fish may have a reduced growth rate (Mehner and Wieser, 1994; Forster and Hiroshi, 1996) as less energy will be allocated to growth. Thus, a net economic gain is expected only if fish are fed at satiation levels. In view of these findings, swimming metabolism is a key factor in fish culture and should be considered carefully during planning and operation. Water velocity must range within the acceptable scope (Beamish, 1978), which may vary with temperature and fish size. Periodic observations of swimming behaviour are necessary to maintain best possible husbandry conditions.

Swimming activity may be strongly rhythmic and may fluctuate considerably during the day. In seabass *Dicentrarchus labrax*, daylight swimming activity was up to 10 times higher compared with nighttime (Bégout Anras *et al.*, 1997). Corresponding results are reported by Waller *et al.* (2000) for coho salmon *O. kisutch* maintained in net pens. Swimming behaviour is very often ignored in system design, even though it is widely accepted that this supports fish welfare in the culture environment. In tank and raceway systems, the hydrodynamic design can be modified to very nearly match the spontaneous behaviour (activity pattern) of the species, in order to significantly improve the husbandry conditions. For obvious reasons this is difficult or impossible in net cages or ponds.

The maintenance of fish in a tank environment that induces steady swimming activity is beneficial in several ways. One important aspect is the reduction of agonistic behaviour, which may cause significant losses due to harmful injuries and secondary infections. Reducing aggression will also be beneficial by avoiding stress (Cutts *et al.*, 1998). Pronounced dominant and aggressive behaviour is known in salmonid fish but the frequency of aggressive encounters can be lessened by forcing fish to swim (Totland *et al.*, 1987). Besides the possible detrimental effect of aggressive behaviour, energy pathways may also be altered. Reallocation of metabolic energy to avoidance behaviour as a response to social stress may significantly reduce growth performance. Christiansen and Jobling (1990) have shown that growth of Arctic charr *Salvelinus alpinus* is affected by high levels of aggressive interaction.

Swimming exercise also has a supporting effect on development of muscle fibres and, as a result, on product quality. By forcing fish to swim, white muscles develop an increased number of hypertrophied fibres (Totland *et al.*, 1987), improving the quality of fish as food for human consumption (Johnston, 1999). This aspect is of paramount importance for future aquaculture development, as public concern regarding product quality is increasing.

5.3.4 Rearing density

Rearing density has a pronounced effect on growth (Vijayan and Leatherland, 1988; Holm *et al.*, 1990; Suresh and Lin, 1992; Björnsson, 1994) and health (Wagner *et al.*, 1997). In circular tanks, fish are distributed more uniformly compared to raceways, where they congregate upstream or downstream. As a consequence, the risk of injuries or disease transfer may be higher in raceways, where contact with walls or other fish will be more frequent (Ross *et al.*, 1995; Ross and Watten, 1998), compared to circular tanks, which provide virtually infinite space.

In territorial fish, such as tilapia, group size determines behaviour, swimming activity and, as a result, energy allocation (Antoniou, 1989). In larger fish groups, agonistic, territorial behaviour may be reduced (Ballarin and Haller, 1982). Thus, agonistic behaviour may be diminished by increasing the group size, as long as the competition for space and resources will not reverse the beneficial effect. A well-known behavioural pattern in many fish species in circular tanks is the establishment of 'fish mills' that mitigate agonistic behaviour. Although Atlantic salmon may be territorial in natural habitats, under farm conditions they preferentially show pelagic behaviour, that is, they tend to be positioned in the water column (Mork *et al.*, 1999) and will not defend territories. These fish are typically organised in fish mills that may break down during feeding and during external disturbance.

Feed must be distributed over the tank or raceway surface to allow all fish to have access to it. Agonistic behaviour in fish is often directed towards smaller individuals (Efthimiou *et al.*, 1994). A strong relationship between nutrition, growth and aggressive behaviour is known for fish (Olsen and Ringo, 1999). If food is distributed from a point source, dominant fish will take a greater part of the meal (Alanara, 1996; Adams *et al.*, 1998) and monopolise food (Gélineau *et al.*, 1998). Circumferential water currents in circular tanks result in a better spreading of food over tank surfaces, allowing fish to access the food more easily; greater homogeneity in feeding will improve overall growth (Joergensen *et al.*, 1996). High stocking densities act against this, hindering fish access to food, even at excess feeding rates (Refstie and Kittelsen, 1976; Refstie, 1977).

Spreading of food will improve the current regime, as crowding of fish around the feeding spot may interfere with the hydrodynamics within the tank or raceway and interrupt the self-cleaning of rearing units. The installation of feeders, or the feeding position, must fit the foraging strategy of the particular species. Davenport *et al.* (1990) found that halibut *Hippoglossus hippoglossus* feed in midwater, while lemon sole *Microstomus kitt* acquire food at the tank bottom. To satisfy halibut, feed must be distributed such that fish can swim several body lengths before taking food items. Observations of foraging patterns are of great importance in optimising feeding and energy distribution. For pelagic species, pelleted food having different specific weights and sinking speeds are available, allowing improved ingestion and reduced losses. The food quality and feeding procedure has to be adjusted with regard to size, feeding behaviour and food demand of the species. Periodic grading of fish is necessary to establish groups of similar size and feeding requirements.

5.4 Water supply

Environmental burdens from aquaculture installations are mainly linked to the pollution of water with dissolved and particulate matter. Aquaculture operations themselves may be threatened by the poor quality of natural waters. Appropriate design and engineering of water supply and discharge make it possible to circumvent these problems, which is an obvious advantage over conventional open installations, such as net-cage farms and ponds. However, at present it appears that water treatment is often neglected for economic reasons. This, however, will change in future as the shortage of water due to pollution or depression of water tables will provoke strict regulations for water usage and discharge. In several countries, certain regulations and laws are already in use (Rosenthal *et al.*, 1993).

The water effluents of aquaculture installations can carry numerous different compounds. In the first place, nutrients contribute to the eutrophication and hypertrophication of adjacent waters. Liming materials and coagulants are widely used in traditional pond farming. In addition, various substances are used in modern aquaculture, including disinfectants, antibiotics, bactericides, fungicides, parasiticides, insecticides, algicides, herbicides, piscicides, molluscicides and hormones (Bergheim and Asgard, 1996; Boyd and Massaut, 1999). In many cases, the risk to the environment and food safety is small or absent (Boyd and Massaut, 1999). However, the latter group of substances, in particular, undoubtedly threaten human health and well-being and impact ecosystem functions in receiving water bodies. In view of this, it is reasonable to demand that water effluents are treated. However, an important first step is to reduce drug use and treatment by improving husbandry conditions. It is widely accepted that the outbreak of diseases in fish farms is linked to stress (Rigos *et al.*, 1998), which is often a result of inappropriate holding conditions. From this, it must be concluded that the first step towards environmentally acceptable and sustainable aquaculture is to maintain living conditions that fit the demands of the species. In doing so, the use of excess drugs and hazardous chemicals should not be necessary—at least not on a regular basis.

The water supply to tank and raceway systems can be open (through-flow) or closed (recirculating system). Figure 5.5 outlines the general set-up. In open systems, the water is pumped from natural waters, passed through the tanks and discharged into adjacent water bodies. In such installations, the incoming water often needs to be treated before entering the holding tanks. This may include settling of suspended solids, filtering to remove plankton and debris, heating or cooling and/or degassing or aeration. Particulate matter is known to impact fish health by direct physical damage or secondary infections. Particles may carry bacteria and fungi and, therefore, act as carriers for pathogens. Heating may shift total gas pressure towards critical values, and removal of dissolved gases and subsequent controlled aeration to restore dissolved oxygen is essential. The water effluent to tank and raceway aquaculture systems must be clarified to remove suspended matter and nutrients and to restore dissolved oxygen. Post-treatment is the key to the mitigation of environmental impacts. This can be achieved in public utilities as well as in small integrated units as part of the aquaculture installation.

Nowadays, an increasing number of tanks and raceways are operated as recirculating systems, with a water exchange rate of less than 10% of the system volume per day. In such systems, a water treatment system restores the water quality according to the needs of the species. Fundamental concepts of recirculation technology have been detailed by Rosenthal (1981, 1993) and Lucchetti and Gray (1988). Such systems are complicated and, even if it can be said that the basic technologies are available, improvements directed at better control of system function are still needed. The make-up water is pumped from

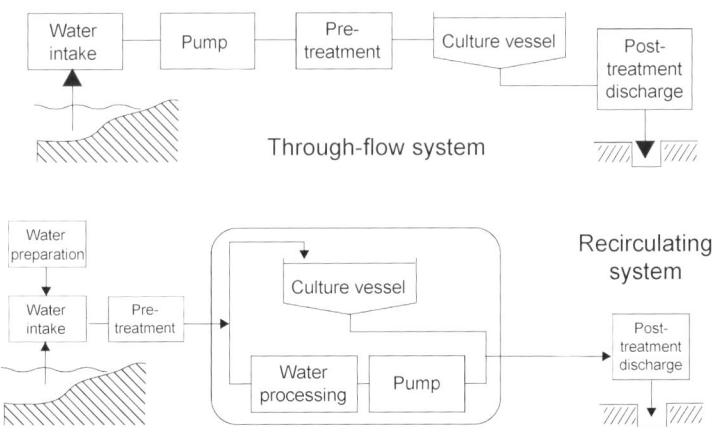

Figure 5.5 Schematic representation of the general set-up of a flow-through and a recirculating system. The culture vessel is representative of the whole number of tanks or raceways that may be linked to the system. Tanks are usually installed parallel, while raceways may also be connected in series.

natural waters; in marine systems, it may also be prepared from artificial sea salt mixtures. If water is delivered from natural waters, a pretreatment may be necessary, as described above. The effluent water is released into the natural environment after post-treatment.

5.4.1 Through-flow systems

The scheme in Figure 5.5 for through-flow installations is simple in view of the available technology. In Figure 5.6, a more sophisticated and desirable set-up is outlined. The water intake from natural water bodies can be by means of pumps or by using tidal flow. Engineering details have been summarized by Huguenin and Colt (1989) and Rosenthal *et al.* (1995). Pretreatment is by mechanical filtration to remove particles and organisms, including hazardous plankton, from the water. Common practice and filtration technology was presented in detail by Wheaton (1977) and Huguenin and Colt (1989). The water is stored in a head tank in order to maintain a constant flow rate (constant pressure head) and to buffer fluctuations or failures in the primary water supply. Subsequently, the water is heated or chilled and total gas pressure is controlled by degassing and aeration. Both processes are interdependent and critical gas pressures (oversaturation) may develop if the water is heated and immediately supplied to the tanks (Colt, 1984). Disinfection by UV-light is usually employed to reduce germ counts in the water influent to the holding tanks (Wheaton, 1977;

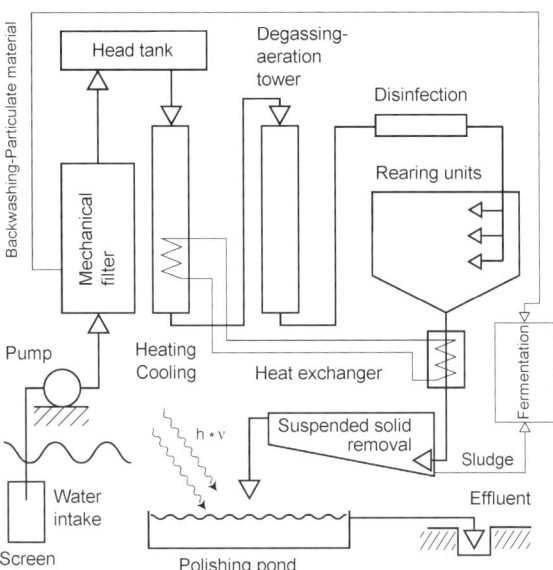

Figure 5.6 Schematic representation of an open water supply to land-based aquaculture installations.

Liltveld and Cripps, 1999). Alternatively, ozone or, to a lesser extent, chemicals like chlorine, iodine or formalin are used. These compounds are toxic (Stoskopf, 1993) and not easily removed from water. The technical application of ozone, which is increasingly used in aquaculture, together with its biological effects, was reviewed by Rosenthal and Wilson (1987).

After the water has passed through the rearing units, the effluent water should be treated (the post-treatment) to remove suspended solids, thus diminishing the organic load. This post-treatment step lowers the biochemical (BOD) and chemical oxygen demand (COD) in the waste water and is a prerequisite in reducing the impact of discharged water on neighbouring ecosystems. The sludge removed consists mostly of organic matter and is highly putrescible. The sludge from pretreatment and post-treatment can be supplied to an anaerobic digestor to produce biogas (methane), and eventually to feed back heat or electric energy into the system. The anaerobic digestion is basically a three-step process that is highly dependent on temperature and pH (Metcalf and Eddy, 1979). Reliable technologies for treatment of manure from livestock breeding are available that can be applied to aquaculture effluents.

Nutrients in the effluent water can be recovered in plant or algal tissue in a polishing pond or in hydroponics. A well-known technique for freshwater fish ponds is the so-called 'plant filter', that is, artificial gravel beds with plants like *Phragmites* sp., *Typha* sp. and *Juncus* sp. Investigations by Langer *et al.* (1996) showed a minimum of 30% removal of nutrients through plant growth and/or bacterial activity. Sansana-yuth *et al.* (1996) studied the possibility of artificial wetland for the treatment of effluent waters from shrimp farms, on an experimental scale. The results showed that suspended solids, BOD, total organic carbon (C), total nitrogen (N) and total phosphorus (P) can be significantly reduced (30–>90%). In an experiment by Lefebvre *et al.* (1996), the feasibility of using fish farm effluents was evaluated as a source of inorganic nutrients for mass production of marine diatoms. Various reports on the post-treatment of aquaculture effluents are available for freshwater (Seawright *et al.*, 1998) and marine (Cohen and Neori, 1991; Neori *et al.*, 1991; Krom *et al.*, 1995) systems. Further steps toward an integrated management of fish farm effluents have been presented by Shpigel *et al.* (1993) and Neori *et al.* (2000). The combination of fish, seaweed and invertebrate animals reduces nutrient discharge and increases marketable yield by providing a secondary and tertiary crop based on the nutrient effluents of the fish tanks.

An estimation of necessary waterflow into culture systems depends on the species under culture and their physiological state. The waterflow must be sufficient to control dissolved gas levels and to remove toxic metabolites. Dissolved oxygen may also be restored by means of aeration or oxygenation. Because oxygenation supersaturates the inflowing water, this process must be carefully controlled. The target production can only be achieved when optimum conditions can be maintained throughout the rearing period. Huguenin and Colt

(1989) demonstrated a simple mathematical formulation for the quantification of waterflow and carrying capacity in through-flow systems. A more refined approach by means of numerical modelling requires a sound knowledge of metabolic processes of the target species. For Atlantic salmon, numerical expressions describing oxygen uptake and nitrogen excretion are available from Bergheim *et al.* (1991), Fivelstad and Smith (1991) and Fivelstad *et al.* (1991). However, if several cohorts are maintained in the same system, the situation becomes more complicated. Figure 5.7 shows the oxygen consumption derived from metabolic data for a sea bass *Dicentrarchus labrax* farming system with three sequential stockings per year. After an initial phase with a steadily increasing oxygen consumption, the uptake rates fluctuate depending on cohort growth and harvest. The same patterns apply to food ingestion, excretion rates and all other interdependent metabolic processes. Theoretical considerations of this and algorithms were provided by Watten (1992). However, in view of the potential fluctuations resulting from changes in spontaneous activity (Figure 5.8), food ingestion and/or aggressive behaviour, a continuous control of waterflow is indispensable and may also help to reduce energy costs during periods with low biomass subsequent to the harvest of a cohort.

5.4.2 *Recirculation systems*

Recirculating devices are used to achieve complete environmental control in rearing systems. Attempts to maintain appropriate rearing conditions are manifold and have not been without drawbacks. Water quality control in recirculation systems is attained through water exchange and water treatment, whereby the necessary water exchange depends on the efficiency of water treatment

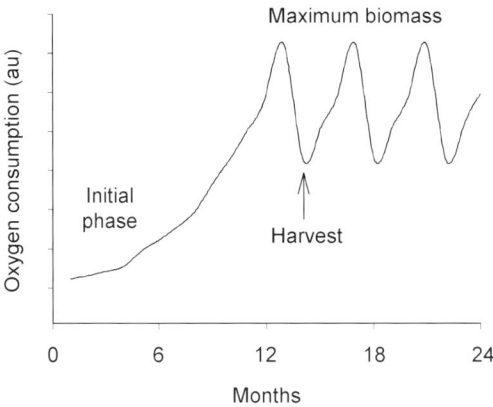

Figure 5.7 Oxygen consumption in arbitrary units (au) for a sea bass *Dicentrarchus labrax* farming system with three sequential stockings per year. Each cohort is harvested after a 390 day rearing period.

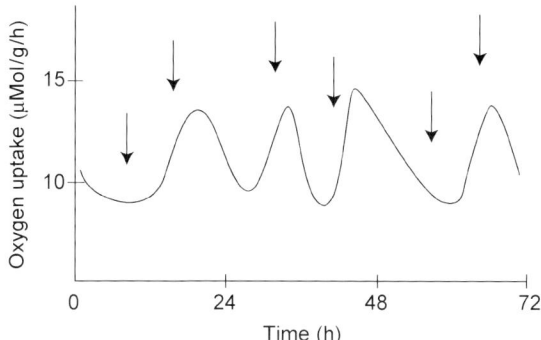

Figure 5.8 The oxygen consumption of sea bass *Dicentrarchus labrax* during three subsequent days obtained from rearing experiments (Thetmeyer and Waller, unpublished data). The arrows indicate feeding of fish (ration = 1% bodyweight per day).

components. Therefore, the design of recirculating facilities is important, not only to ensure optimum holding conditions but also to reduce the operational costs that are still comparatively high and presently hamper development. Due to the intensity of production in such systems, the ability to control the rearing process is a prerequisite and the design must carefully reflect the requirements of the species.

Figure 5.9 presents a schematic drawing of a recirculating system consisting of primary and secondary water treatment. While the primary treatment is to maintain water quality in the rearing units, the secondary treatment is to purify the water discharged from the primary loop. After the secondary treatment, the water is fed back into the primary recirculation or discharged. This set-up allows recycling of nearly 100% of the system's water and makes it possible to operate such systems almost anywhere. However, system control is much more complicated compared to through-flow and other open aquaculture systems. At present, most of the commercially-operated recirculating systems contain only a subset of the components outlined in Figure 5.9 and, consequently, water demand is still high and often exceeds 10% of the system volume per day. On the other hand, secondary treatment is important in view of environmental issues because the effluent from a recirculating aquaculture system may have a serious impact when discharged. An effective 'downstream' treatment is necessary for these systems to meet certain discharge standards.

The primary loop (Figure 5.9) consists of the holding tanks, a separation for suspended solids, biological filters, degassing and oxygenation units and temperature control. The water from the rearing units is clarified in two steps, according to the particle size (Chen *et al.*, 1993a; Patterson *et al.*, 1999). The large particles that accumulate at the tank bottom may be processed separately from the smaller particles floating in the water column. While large particles

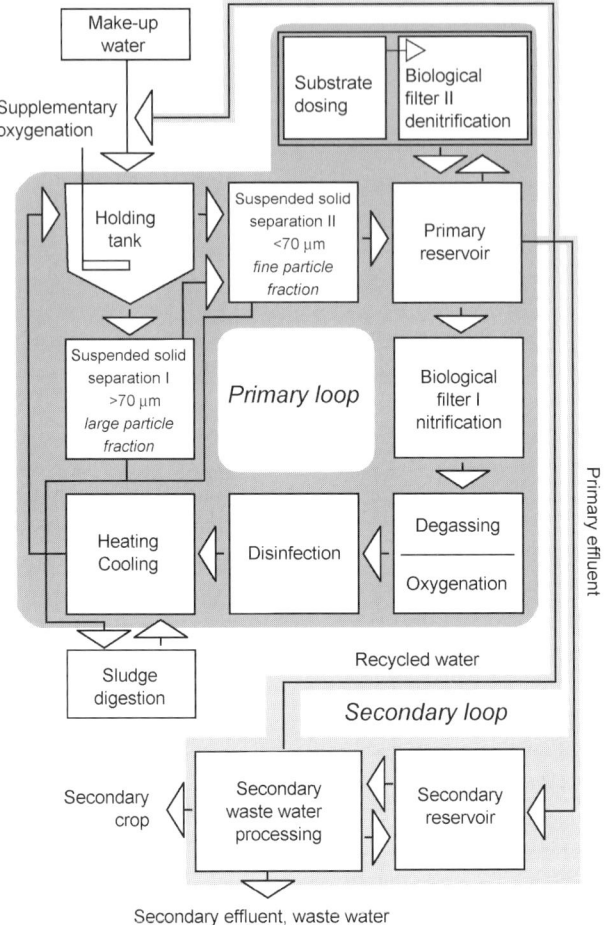

Figure 5.9 Schematic representation of a recirculation system consisting of a primary and secondary treatment. At present most commercially used recirculation systems contain only a subset of the outlined components and, consequently, may be operated with a comparably high water demand.

($>70\,\mu m$) can be removed by means of coarse screens, rotary microscreens, settling tanks, swirl separators or hydrocyclones (Wheaton, 1977; Chen et al., 1994; Twarowska et al., 1997), small particles ($<70\,\mu m$) often remain in the water, affecting the health of the cultured animals. Small suspended solids damage the gill tissues of fish and may facilitate the outbreak of bacterial diseases (Bullock et al., 1994) and induce stress (Redding et al., 1987). They carry nitrogenous and phosphorous compounds that may mineralise and build up critical concentrations in the recirculated water. Furthermore, the biofilter surfaces may be blocked and the nitrification process, detoxifying the nitrogenous metabolic waste, may be diminished. It is, therefore, necessary to separate the fine particle fraction from the recycled water. This is very often disregarded during planning and the effects associated with the fine particle fraction cause significant losses during operation. Solid separation is a two-step process in recirculating systems (Figure 5.9), as separators for large particles will not remove the fine fraction and may increase the share of smaller particles (<20 µm), as emphasised by Langer and co-workers (1996).

Small-particle filters are relatively complex elements of recirculating systems and use air bubbles as the filter medium, whereby two different principles for freshwater and seawater are available. In freshwater, flotation tanks are employed. Flotation tanks are usually of circular shape and have a large surface area. The water effluent to the rearing tanks or primary suspended solid separation is supplied at the bottom, after being pressurised and supersaturated with air. The lower pressure in the flotation tank results in a 'boiling out' of gas as very fine bubbles. The bubbles attach to particles and ascend to the surface, where the waste is periodically removed by an automatic wiper. Flotation filters are the state of the art and appropriate products are available from different manufacturers. However, to attain highest removal efficiencies it is important to generate the micro-sized bubbles continuously and to control and maintain the components correctly.

In seawater, foam fractionators (Sander, 1967; Otte and Rosenthal, 1979) are applicable. The basic principle is the build up of foams from surface active compounds (e.g. proteins) in the water that collect small particles (Chen *et al.*, 1993b). This will also partly remove soluble proteins from the culture water (Chen *et al.*, 1993c) and, consequently, reduce the build up of nitrogenous waste through mineralisation. The foam containing the particles is skimmed at the water/atmosphere interface and passed to the sludge discharge or a sludge digestor (Figure 5.9), together with the waste from the large particle separation. The basic principles of particle skimming were described by Spotte (1979), and the application to aquaculture recirculation systems was discussed by Wheaton *et al.* (1979) and Timmons (1994). The design needs to include an automated sludge removal device and water flow and water head control systems in order to minimise water consumption and avoid losses. These are often not included and this causes poor removal performance and critical operation. The foam-building process can be enhanced by injection of ozone gas to the inflowing water, which will generally improve water quality in recirculation systems (Sutterlin *et al.*, 1984; Kobayashi *et al.*, 1993). Ozone is a highly toxic gas (Crecelius, 1979; Wedemeyer *et al.*, 1979; Lenwood *et al.*, 1981), therefore the dosage control is of utmost importance and is usually achieved by continuous measurement of the redox potential (Sander, 1998). Another possible measure is the post-treatment of ozonated water with activated carbon (Ozawa *et al.*, 1991). The application of ozone in aquaculture and the possible effects on livestock have been discussed by Rosenthal (1980).

In recirculating aquaculture systems, the control of nitrogenous excretory products is of utmost importance. Aquatic organisms excrete ammonia (NH_3), which is toxic in small concentrations. In recirculating systems, ammonia is broken down into nitrite (NO_2) and nitrate (NO_3), with nitrate being the least toxic compound. A comprehensive summary of nitrogen toxicity and threshold values has been published by Colt and Armstrong (1979). In addition to physiological changes, a high level of ammonia may directly impact growth (Guillen et al., 1993; Wajsbrot et al., 1993), possibly as the result of decreased food ingestion (Person Le Ruyet et al., 1997). Inappropriate nitrite and nitrate levels may diminish the oxygen carrying capacity of fish haemoglobin (Colt and Armstrong, 1979; Williams et al., 1993) and, in view of the energetic pathways in fish (Calow, 1985), reduced growth performance may result (Tomasso, 1993). Even if some of the effects may be reversible (Schorre et al., 1995), all nitrogenous metabolites threaten the well-being of farmed animals, and threshold levels are to be considered thoroughly during planning and operation. The nitrogenous waste is usually removed by means of biological filtration, which embraces two processes: the transformation of toxic nitrogenous waste (ammonia and nitrite) into the least toxic nitrogous compound (nitrate); and the removal of this compound from the water (denitrification) (Wheaton 1977). However, the reduction of nitrogenous compounds from the water may also be attained by means of 'plant filters' (Naegel, 1977; Rennert, 1992).

Nitrification is usually attained in submerged biofilters, trickling filters, rotating biodrums, bead filters or fluidized bed filters. A thorough summary of the different filter types, with advice on the estimation of size and flow, was provided by Wheaton and co-workers (1994). The removal efficiency of different filter types has been investigated by several authors (Nijhof, 1995; Greiner and Timmons, 1998; Kamstra et al., 1998; Lekang and Kleppe, 2000). The ozonation of the water prior to nitrification may be beneficial, because ozonation, in combination with foam separation, may reduce the overall organic load in the water and favour growth of nitrifying (Hagopian and Riley, 1998) over heterotrophic bacteria (Paller and Lewis, 1988).

Denitrification is commonly carried out in submerged anaerobic biofilters, with injection of an organic substrate to promote bacterial growth. Methanol is usually used as the carbon source and simple equations for dosage are available from Wheaton (1977). The most crucial point of the denitrification process is the biomass production (Rheinheimer et al., 1988), which results in a strong bacterial matrix on the biosubstrate in submerged filters. The plugging or, at an earlier state, the channelling of the filterbed will decrease removal efficiency and finally bring the process to a standstill. Fluidised filters bed with an agitated filter bead avoid these limitations and are now widely used (Aboutboul et al., 1995; Arbiv and van Rijn, 1995). The denitrification is usually employed as a bypass system with only little waterflow (Figure 5.9).

Degassing is mainly to strip the carbon dioxide from the water that is produced by the fish in the system. If carbon dioxide is not controlled, the low water exchange in recirculation systems may lead to an accumulation in the system water. The increase in carbon dioxide concentration shifts the acid/base equilibrium and may impact fish health (Fivelstad *et al.*, 1998) and biofilter function. Therefore, carbon dioxide has to be stripped from the water (Grace and Piedrahita, 1994) before it gets reoxygenated to restore oxygen levels (Watten, 1994). In parallel to the stripping of carbon dioxide, the acid/base equilibrium may be controlled by alkalinity supplements (Bisogni and Timmons, 1994). Dissolved oxygen may be supplied under pressure to oversaturate the water, in order to supply ample oxygen to the culture tanks. Moreover, supplementary oxygen supply in every culture tank is vital to supply oxygen during periods of elevated demand (Figure 5.9).

The disinfection (Figure 5.9) of water can be by means of UV irradiation (Wheaton, 1977; Liltved and Cripps, 1999) or ozone treatment (Wedemeyer and Nelson, 1977; Colberg and Lingg, 1978; Wedemeyer *et al.*, 1978). In freshwater systems, ozone contact towers are installed, while in seawater systems with foam fractionation the disinfection is in parallel with particle removal. An effective removal of fine particles, including bacteria, may considerably lessen the need to disinfect the water and may alleviate the danger of introducing unwanted toxic substances or secondary products of the disinfecting process.

The secondary loop of recirculating aquaculture systems (Figure 5.9) is analogous with the post-treatment of effluents from land-based farms. Seaweeds as well as phytoplankton can be reared in order to reduce nutrient load, and it is desirable that at least a part of the water is recycled to be used again in the primary loop. This step is important in order to reduce the water demand and the amount of discharge water. Naegel (1977) and Rennert (1992) have demonstrated the feasibility of such systems. Some important aspects are the light and nutrient requirements of plants and algae (Chapman, 1980; Becker, 1993), the selection of substrates, and the waterflow pattern and turbulence in the second loop rearing units. Another possible method of secondary treatment may be the utilisation of natural wetlands, which are highly significant ecosystems. Limnic, brackish and marine wetlands are highly productive and may have the potential to process excess nutrients in their food chain without any adverse effects on species assemblage, biological pathways and overall system function. However, this requires precautionary investigations and modelling in order to estimate the carrying capacity and to prevent these system from being destroyed by excess particle or nutrient load. Moreover, the technology of the second loop treatment appears to be still in its infancy and comprehensive developments are required in order to successfully introduce these components in commercial aquaculture operations.

References

Aboutboul, Y., Arviv, R. and van Rijn, J. (1995) Anaerobic treatment of intensive fish culture effluents: volatile fatty acid mediated denitrification. *Aquaculture*, **133** 21-32.

Ackefors, H. and Enell, M. (1994) The release of nutrients and organic matter from aquaculture systems in Nordic countries. *Journal of Applied Ichthyology*, **10** 225-241.

Adams, C.E., Huntingford, F.A., Turnbull, J.F. and Beattie, C. (1998) Alternative competitive strategies and the cost of food acquisition in juvenile Atlantic salmon (*Salmo salar*). *Aquaculture*, **167** 17-26.

Alanara, A. (1996) The use of self-feeders in rainbow trout (*Oncorhynchus mykiss*) production. *Aquaculture*, **145** 1-20.

Antoniou, E. (1989) Der Einfluß der Gruppengröße auf den Stoffwechsel und das Verhalten von Fischen. Diplomarbeit, Kiel, Institut für Meereskunde.

Arbiv, R. and van Rijn, J. (1995) Performance of a treatment system for inorganic nitrogen removal in intensive aquaculture systems. *Aquacultural Engineering*, **14** 189-203.

Arnold, G.P. (1974) Rheotropism in fishes. *Biological Reviews*, **49** 515-576.

Azzaydi, M., Madrid, J.A., Zamora, S., Sanchez-Vasquez, F.J. and Martinez, F.J. (1998) Effect of three feeding strategies (automatic, *ad libitum* demand-feeding and time-restricted demand-feeding) on feeding rhythms and growth in European sea bass (*Dicentrarchus labrax* L.). *Aquaculture*, **163** 285-296.

Ballarin, J.D. and Haller, R.D. (1982) The intensive culture of tilapia in tanks, raceways and cages, in *Recent Advances in Aquaculture* (eds. J.F. Muir and R.J. Roberts), Croom Helm, pp. 216-265.

Barnes, M.E., Sayler, W.A. and Cordes, R.J. (1996) Baffle usage in covered raceways. *The Progressive Fish-Culturist*, **58** 286-288.

Beamish F.W.H. (1978) Swimming capacity, in *Fish Physiology VII* (eds. W.S. Hoar and D.J. Randall), Academic Press, pp. 101-187.

Becker, E.W. (1993) Microalgae: biotechnology and microbiology. Cambridge University Press, 293 pp.

Bégout Anras, M.L., Lagardere, J.P. and Lafaye, J.Y. (1997) Diel activity rhythms of seabass tracked in a natural environment: group effects on swimming pattern and amplitudes. *Canadian Journal of Fisheries and Aquatic Science*, **54** 162-168.

Berg, K., Finstad, B., Grande, G. and Wathne, E. (1990) Growth of Atlantic salmon (*Salmo salar* L.) in a variable diel temperature regime. *Aquaculture*, **90** 261-266.

Bergheim, A. and Aasgaard, T. (1996) Waste production from aquaculture, in *Aquaculture and Water Resource Management* (eds. D.J. Baird, L.A. Beveridge, L.A. Kelly and J.F. Muir), Blackwell Science pp. 51-80.

Bergheim, A., Seymour, E.A., Sanni, S. and Tyvold, T. (1991) Measurements of oxygen consumption and ammonia excretion of Atlantic salmon (*Salmo salar* L.) in commercial-scale, single-pass freshwater and seawater land-based culture systems. *Aquacultural Engineering*, **10** 251-267.

Bisogni Jr., J.J. and Timmons, M.B. (1994) Control of pH in cycle aquaculture systems, in *Aquaculture Water Reuse Systems: Engineering Design and Management* (eds. M.B. Timmons and T.M. Losordo), Elsevier, pp. 235-245.

Björnsson, B. (1994) Effects on stocking density on growth rate of halibut (*Hippoglossus hippoglossus* L.) reared in large circular tank for three years. *Aquaculture*, **123** 259-270.

Blazer, V.S., Fabacher, D.L., Little, E.E., Ewing, M.S. and Kocan, K.M. (1997) Effects of ultraviolet-B radiation on fish: histological comparison of a UVB-sensitive and a UVB-tolerant species. *Journal of Aquatic Animal Health*, **9** 132-143.

Boersen, G. and Westers, H. (1986) Waste solids control in hatchery raceways. *The Progressive Fish-Culturist*, **48** 151-154.

Boisclair, D. and Sirois, P. (1993) Testing assumptions of fish bioenergetics models by direct estimation of growth, consumption, and activity rates. *Transactions of the American Fisheries Society*, **122** 784-796.

Bosakowski, T. and Wagner, E.J. (1995) Experimental use of cobble substrates in concrete raceways for improving fin condition of cutthroat (*Oncorhynchus clarki*) and rainbow trout (*O. mykiss*). *Aquaculture*, **130** 159-165.

Boyd, C.E. and Massaut, L. (1999) Risk associated with the use of chemicals in pond aquaculture. *Aquacultural Engineering*, **20** 113-132.

Brett, J.R. (1971) Energetic responses of salmon to temperature: a study of some thermal relations in the physiology and freshwater ecology of sockeye salmon (*Oncorhynchus nerka*). *American Zoologist*, **11** 99-113.

Brown, C.A. and King, J.M. (1995) The effects of trout-farm effluents on benthic invertebrate community structure in rivers in the south-western Cape, South Africa. *South African Journal of Aquatic Science*, **21** 3-21.

Bullock, G., Herman, R., Heinen, J., Noble, A., Weber, A. and Hankins, J. (1994) Observations on the occurrence of bacterial gill disease and amoeba gill infestation in rainbow trout cultured in a water recirculation system. *Journal of Aquatic Animal Health*, **6** 310-317.

Calow, P. (1985) Adaptive aspects of energy allocation, in *Fish Energetics* (eds. P. Tytler and P. Calow), Croom Helm, pp. 13-31.

Carillo, M., Bromage, N., Zanuy, S., Serrano, R. and Prat, F. (1989) The effect of modifications in photoperiod on spawning time, ovarian development and egg quality in the sea bass (*Dicentrarchus labrax* L.). *Aquaculture*, **81** 351-365.

Chapman, V.J. (1980) Seaweeds and their uses. Chapman and Hall, 287 pp.

Chen, S., Timmons, M.B., Aneshansley, D.J. and Bisogni Jr., J.J. (1993a) Suspended solid characteristics from recirculating aquacultural systems and design implications. *Aquaculture*, **112** 143-155.

Chen, S., Timmons, M.B., Bisogni Jr., J.J. and Aneshansley, D.J. (1993b) Suspended-solids removal by foam fractionation. *The Progressive Fish-Culturist*, **55** 69-75.

Chen, S., Timmons, M.B., Bisogni Jr., J.J. and Aneshansley, D.J. (1993c) Protein and its removal by foam fractionation. *The Progressive Fish-Culturist*, **55** 76-82.

Chen, S., Stechney, D. and Malone, R.F. (1994) Suspended solid control in recirculating aquaculture systems, in *Aquaculture Water Reuse Systems: Engineering Design and Management* (eds. M.B. Timmons and T.M. Losordo), Elsevier, pp. 61-100.

Christiansen, J.S. and Jobling, M. (1990) The behaviour and the relationship between food intake and growth of juvenile Arctic charr, *Salvelinus alpinus* L., subjected to sustained exercise. *Canadian Journal of Zoology*, **68** 2185-2191.

Cohen, I. and Neori, A. (1991) *Ulva lactuca* biofilters for marine fishpond effluents. I. Ammonia uptake kinetics and nitrogen content. *Botanica Marina*, **34** 475-482.

Colberg, P.J. and Lingg, A.J. (1978) Effect of ozonation on microbial fish pathogens, ammonia, nitrate, nitrite and BOD in simulated reuse hatchery water. *Journal of the Fisheries Research Board of Canada*, **35** 1290-1296.

Colt, J. (1984) Computation of dissolved gas concentrations in water as functions of temperature, salinity and pressure. Special publication, American Fisheries Society, **14**, 154 pp.

Colt, J. and Armstrong, D. (1979) Nitrogen toxicity to fish, crustaceans and molluscs. Department of civil engineering, University of California, 30 pp.

Crecelius, E.A. (1979) Measurement of oxidants in ozonised seawater and some biological reactions. *Journal of the Fisheries Research Board of Canada*, **36** 1006-1008.

Cripps, S.J. and Poxton, M.G. (1992) A review of the design and performance of tanks relevant to flatfish culture. *Aquacultural Engineering*, **11** 71-91.

Cutts, C.J., Metcalfe, N.B. and Taylor, A.C. (1998) Aggression and growth depression in juvenile Atlantic salmon: the consequences of individual variation in standard metabolic rate. *Journal of Fish Biology*, **52** 1026-1037.

Davenport, J., Kjørsvik, E. and Haug, T. (1990) Appetite, gut transit, oxygen uptake and nitrogen excretion in captive Atlantic halibut, *Hippoglossus hippoglossus* L., and lemon sole, *Microstomus kitt* (Walbaum). *Aquaculture*, **90** 267-277.

Davison, W. and Goldspink, G. (1977) The effect of prolonged exercise on the lateral musculature of the brown trout (*Salmo trutta*). *Journal of Experimental Biology*, **70** 1-12.

Devlin, R.H., Johnsson, J.I., Smailus, D.E., Biagi, C.A., Jönsson, E. and Björnsson, B.T. (1999) Increased ability to compete for food by growth hormone-transgenic coho salmon, *Oncorhynchus kisutch* (Walbaum). *Aquaculture Research*, **30** 479-482.

Efthimiou, S., Divanach, P. and Rosenthal, H. (1994) Growth, food conversion and agonistic behaviour in common dentex (*Dentex dentex*). *Aquatic Living Resources*, **7** 267-275.

EIFAC Technical Paper 49 (1986) Flow-through and recirculation systems. Report of the working group on terminology, format and units of measurement, FAO, 100 pp.

Ewing, R.D., Sheanan, J.E., Lewis, M.A. and Palmisano, A.N. (1998a) Effects of rearing density and raceway conformation on growth, food conversion, and survival of juvenile spring chinook salmon. *The Progressive Fish-Culturist*, **60** 167-178.

Ewing, R.D., Lewis, M.A., Sheahan, J.E. and Ewing, S.K. (1998b) Evaluation of inventory procedures for hatchery fish. III. Nonrandom distributions of chinook salmon in raceways. *The Progressive Fish-Culturist*, **60** 159-166.

Fivelstad, S. and Smith, M.J. (1991) The oxygen consumption rate of Atlantic salmon (*Salmo salar* L.) reared in a single pass land-based seawater system. *Aquacultural Engineering*, **10** 227-235.

Fivelstad, S., Bergheim, A. and Tyvold, T. (1991) Studies of limiting factors governing the waterflow requirement for Atlantic salmon (*Salmo salar* L.) in land-based seawater systems. *Aquacultural Engineering*, **10** 237-249.

Fivelstad, S., Haavik, H., Løvik, G. and Olsen, A.B. (1998) Sublethal effects and safe levels of carbon dioxide in seawater for Atlantic salmon postsmolts (*Salmo salar* L.): ion regulation and growth. *Aquaculture*, **160** 305-316.

Forster, I.P. and Hiroshi, O. (1996) Growth and whole-body lipid content of juvenile red sea bream reared under different conditions of exercise training and dietary lipid. *Fisheries Science*, **62** 404-409.

Gélineau, A., Corazze, G. and Boujard, T. (1998) Effects of restricted rations, time-restricted access and reward level on voluntary food intake, growth and growth heterogeneity of rainbow trout (*Oncorhynchus mykiss*) fed on demand with self-feeders. *Aquaculture*, **167** 247-258.

Grace, G.R. and Piedrahita, R.H. (1994) Carbon dioxide control, in *Aquaculture Water Reuse Systems: Engineering Design and Management* (eds. M.B. Timmons and T.M. Losordo), Elsevier, pp. 209-234.

Greer Walker, M. and Emerson, L. (1978) Sustained swimming speeds and myotomal muscle function in the trout, *Salmo gairdneri*. *Journal of Fish Biology*, **13** 475-481.

Greiner, A.D. and Timmons, M.B. (1998) Evaluation of the nitrification rates of microbead and trickling filters in an intensive recirculating tilapia production facility. *Aquacultural Engineering*, **18** 189-200.

Guillen, J.L., Endo, M., Turnbull, J.F., Hiroshi, K., Richards, R.H. and Aoki, T. (1993) Depressed growth rate and damage to cartilage of red sea bream associated with exposure to ammonia. *Bulletin of the Japanese Society of Scientific Fisheries*, **59** 1231-1234.

Hagopian, D.S. and Riley, J.G. (1998) A closer look at the bacteria of nitrification. *Aquacultural Engineering*, **18** 223-244.

Hansen, T., Stefansson, S. and Taranger, G.L. (1992) Growth and sexual maturation in Atlantic salmon, *Salmo salar* L., reared in sea cages at two different light regimes. *Aquaculture and Fisheries Management*, **23** 275-280.

Hennessy, M.M., Wilson, L., Struthers, W. and Kelly, L.A. (1996) Waste loadings from two freshwater Atlantic salmon juvenile farms in Scotland. *Water, Air, Soil Pollution*, **86** 235-249.

Holm, J.C., Refstie, T. and Bo, S. (1990) The effect of fish density and feeding regimes on individual growth rate and mortality in rainbow trout (*Oncorhynchus mykiss*). *Aquaculture*, **89** 225-232.

Hossain, M.A.R., Batty, R.S., Haylor, G.S. and Beveridge, M.C.M. (1999) Diel rhythm of feeding activity in African catfish, *Clarias gariepinus* (Burchell 1822). *Aquaculture Research*, **30** 901-905.

Huguenin, J.E. and Colt, J. (1989) Design and operating guide for aquaculture seawater systems. Elsevier Science Publisher B.V., Amsterdam, 264 pp.

Jarp, J. and Karlsen, E. (1997) Infectious salmon anaemia (ISA) risk factors in sea-cultured Atlantic salmon, *Salmo salar*. *Diseases of Aquatic Organisms*, **28** 79-86.

Jobling, M. (1981) Temperature tolerance and final preferendum—rapid methods for the assessment of optimum growth temperatures. *Journal of Fish Biology* **19** 439-455.

Jobling, M. (1987) Growth of Arctic charr (*Salvelinus alpinus* L.) under conditions of constant light and temperature. *Aquaculture*, **60** 243-249.

Jobling, M. (1988) A review of the physiological and nutritional energetics of cod, *Gadus morhua* L., with particular reference to growth under farmed conditions. *Aquaculture*, **70** 1-19.

Joergensen, E.H. and Jobling, M. (1993) The effects of exercise on growth, food utilization and osmoregulatory capacity of juvenile Atlantic salmon, *Salmo salar*. *Aquaculture*, **116** 233-246.

Joergensen, E.H., Baardvik, B.M., Eliassen, R. and Jobling, M. (1996) Food acquisition and growth of juvenile Atlantic salmon (*Salmo salar*) in relation to spatial distribution of food. *Aquaculture*, **143** 277-289.

Johnston, I.A. (1999) Muscle development and growth: potential implications for flesh quality in fish. *Aquaculture*, **177** 99-115.

Kadri, S., Metcalfe, N.B., Huntingford, F.A. and Thorpe, J.E. (1997) Daily feeding rhythms in Atlantic salmon. 2. Size-related variation in feeding patterns of post-smolts under constant environmental conditions. *Journal of Fish Biology*, **50** 273-279.

Kamstra, A., van der Heul, J.W. and Nijhof, M. (1998) Performance and optimisation of trickling filters on eel farms. *Aquacultural Engineering*, **17** 175-192.

Kestemont, P. (1995) Different systems of carp production and their impacts on the environment. *Aquaculture*, **129** 347-372.

Kindschi, G., Thompson, R.G. and Mendoza, A.P. (1991) Use of raceway baffles in rainbow trout culture. *The Progressive Fish-Culturist*, **53** 97-101.

Klokseth, V. and Øiestad, V. (1999) Forced settlement of metamorphising halibut (*Hippoglossus hippoglossus* L.) in shallow raceways: growth pattern, survival and behaviour. *Aquaculture*, **176** 117-133.

Kobayashi, T., Yotsumoto, H., Ozawa, T. and Kawahara, H. (1993) Closed circulatory system for mariculture using ozone. *Ozone Science and Engineering*, **15** 311-330.

Krom, M.D., Ellner, S., van Pijn, J. and Neori, A. (1995) Nitrogen and phosphorous cycling and transformations in a prototype 'non-polluting' integrated mariculture system, Eilat, Israel. *Marine Ecology Progress Series*, **118** 25-36.

Kushnirov, D. and Degani, G. (1991) Growth performance of European eel (*Anguilla anguilla*) under controlled photocycle and shelter availability. *Aquaculture Engineering*, **10** 219-226.

Langer, J., Efthimiou, S., Rosenthal, H. and Bronzi, P. (1996) Drum filter performance in a recirculating eel culture unit. *Journal of Applied Ichthyology*, **12** 61-65.

Lefebvre, S., Hussenot, J. and Brossard, N. (1996) Water treatment of land-based fish farm effluents by outdoor culture of marine diatoms. *Journal of Applied Phycology*, **8** 193-200.

Lekang, O.-I. and Kleppe, H. (2000) Efficiency of nitrification in trickling filters using different filter media. *Aquacultural Engineering*, **21** 181-199.

Lenwood Jr., W.H., Burton, D.T. and Richardson, L.B. (1981) Comparison of ozone and chlorine toxicity to the developmental stages of stripped bass, *Morone saxatilis*. *Canadian Journal of Fisheries and Aquatic Sciences*, **38** 752-757.

Liltveld, H. and Cripps, S.J. (1999) Removal of particle-associated bacteria by prefiltration and ultraviolet irradiation. *Aquaculture Research*, **30** 445-450.

Loyless, J.C. and Malone, R.F. (1998) Evaluation of air-lift pump capabilities for water delivery, aeration, and degasification for application to recirculating aquaculture systems. *Aquacultural Engineering*, **18** 117-133.

Lucchetti, G.L. and Gary, G.A. (1988) Water reuse systems: a review of principal components. *The Progressive Fish-Culturist*, **50** 1-6.

Lumb, C.M. (1989) Self-pollution by Scottish salmon farms? *Marine Pollution Bulletin*, **20** 375-379.

Martin, R.M. and Heard, W.R. (1987) Floating vertical raceways to culture salmon (*Oncorhynchus* spp.). *Aquaculture*, **61** 295-302.

McCauley, R.W. and Casselman, J.M. (1981) The final preferendum as an index of the temperature for optimum growth in fish. *Proceedings of the World Symposium on Aquaculture in Heated Effluents and Recirculation Systems*, **2** 81-93.

Mehner, T. and Wieser, W. (1994) Effects of temperature on allocation of metabolic energy in perch (*Perca fluviatilis*) fed submaximal rations. *Journal of Fish Biology*, **45** 1079-1086.

Metcalf, L. and Eddy, H.P. (1979) Wastewater Engineering: Treatment, Disposal and Reuse. McGraw Hill.

Montgomery, J.C., Baker, C.F. and Carton, A.G. (1997) The lateral line can mediate rheotaxis in fish. *Nature*, **389** 360-364.

Mork, O.I., Bjerkeng, B. and Rye, M. (1999) Aggressive interactions in pure and mixed groups of juvenile farmed and hatchery-reared wild Atlantic salmon, *Salmo salar* L. in relation to tank substrate. *Aquaculture Research*, **30** 571-578.

Naegel, L.C.A. (1977) Combined production of fish and plants in recirculating water. *Aquaculture*, **10** 17-24.

Neori, A., Cohen, I. and Gordin, H. (1991) *Ulva lactuca* biofilters for marine fishpond effluents. II. Growth rate, yield and C:N ratio. *Botanica Marina*, **34** 483-489.

Neori, A., Shipgel, M. and Ben-Ezra, D. (2000) A sustainable integrated system for culture of fish, seaweed and abalone. *Aquaculture*, **186** 279-291.

Nicieza, A.G. and Metcalfe, N.B. (1997) Effects of light level and growth history on attack distances of visually foraging juvenile salmon in experimental tanks. *Journal of Fish Biology*, **51** 643-649.

Nijhof, M. (1995) Bacterial stratification and hydraulic loading effects in a plug-flow model for nitrifying trickling filters applied in recirculating fish culture systems. *Aquaculture*, **134** 49-64.

Øiestad, V. (1999) Shallow raceways as a compact, resource-maximizing farming procedure for marine fish species. *Aquaculture Research*, **30** 831-840.

Olsen, R.E. and Ringo, E. (1999) Dominance hierarchy formation in Artic charr, *Salvelinus alpinus* (L.): nutrient digestibility of subordinate and dominant fish. *Aquaculture Research*, **30** 667-671.

Otte, G. and Rosenthal, H. (1979) Management of a closed brackish water system for high-density fish culture by biological and chemical water treatment. *Aquaculture*, **18** 169-181.

Ozawa, T., Yotsumoto, H., Sasaki, T. and Nakayama, S. (1991) Ozonation of seawater—applicability of ozone for recycled hatchery cultivation. *Ozone Science and Engineering*, **13** 697-710.

Paller, M.H. and Lewis, W.M. (1988) Use of ozone and fluidized-bed biofilters for increased ammonia removal and fish loading rates. *The Progressive Fish-Culturist*, **50** 141-147.

Patterson, R.N., Watts, K.C. and Timmons, M.B. (1999) The power law in particle size analysis for aquaculture facilities. *Aquacultural Engineering*, **19** 259-273.

Person Le Ruyet, J., Galland, R., Lee Roux, A. and Chartois, H. (1997) Chronic ammonia toxicity in juvenile turbot (*Scophthalmus maximus*). *Aquaculture*, **154** 153-169.

Redding, J.M., Schreck, C.B., Everest, F.H. (1987) Physiological effects on coho salmon and steelhead of exposure to suspended solids. *Transactions of the American Fisheries Society*, **116** 737-744.

Refstie, T. (1977) Effect of density on growth and survival of rainbow trout. *Aquaculture*, **11** 329-334.

Refstie, T. and Kittelsen, A. (1976) Effect of density on growth and survival of artificially reared Atlantic salmon. *Aquaculture*, **8** 319-326.

Rheinheimer, G., Hegemann, W., Raff, J. and Sekoulov, I. (1988) Stickstoffkreislauf im Wasser. R. Oldenbourg Verlag, 394 pp.

Rennert, B. (1992) Simple recirculation systems and the possibility of combined fish and vegetable production, in *Progress in Aquaculture Research* (eds. B. Moav, V. Hilge and H. Rosenthal), Special Publication of the European Aquaculture Society, pp. 91-97.

Rennert, B. (1994) Water pollution by a land-based trout farm. *Journal of Applied Ichthyology*, **10** 373-378.

Richards, F.P., Reynolds, W.W. and McCauley, R.W. (1977) Temperature preference studies in environmental impact assessments: an overview with procedural recommendations. *Journal of the Fisheries Research Board of Canada*, **34** 728-761.

Rigos, G., Grigorakis, K., Nengas, I., Christophilogiannis, P., Yiganisi, M., Koutsodimou, M., Andriopoulou, A. and Alexis, M. (1998) Stress-related pathology seems a significant obstacle for the intensive farming of common dentex, *Dentex dentex* (Linneaus 1758). *Bulletin of the European Association of Fish Pathologists*, **18** 15-18.

Rosenthal, H. (1980) Ozonation and sterilization. *Symposium on New Developments in the Utilization of Heated Effluents and of Recirculation Systems for Intensive Aquaculture*, Stavanger, Norway 28-30 May 1980.

Rosenthal, H. (1981) Recirculation systems in Western Europe. *Proceedings of the World Symposium on Aquaculture in Heated Effluent and Recirculation Systems*, Berlin, Heinemann, pp. 305-315.

Rosenthal, H. (1993) The history of recycling technology: a lesson learned from the past experience? in *Fish Farming Technology* (eds. H. Reinertsen, L.A. Dahle, L. Jørgensen, and K. Tvinnereim). A.A. Balkema, pp. 341-349.

Rosenthal, H. and Wilson, J.S. (1987) An updated bibliography (1845-1986) on ozone, its biological effects and technical applications. Canadian Technical Report of Fisheries and Aquatic Science, **1542** 249 pp.

Rosenthal, H., Hoffman, R., Jörgensen, L., Krüner, G., Peters, G., Schlotfeldt, H.-J. and Schomann, H. (1982) Water management in circular tanks of a commercial intensive culture unit and its effects on water quality and fish conditions. ICES Mariculture Committee, F:22.

Rosenthal, H., Hilge, V. and Kamstra, A. (1993) Workshop on fish farm effluents and their control in EC countries. Report. Department of fishery biology, Institute for Marine Science at the Christian-Albrechts-University of Kiel, 205 pp.

Rosenthal, H., Allen, J.H., Helm, M.M. and McInerney-Northcott, M. (1995) Aquaculture technology: its application, development and transfer, in *Cold-water Aquaculture in Atlantic Canada* (ed. A. Boghen). The Canadian Institute for Research on Regional Development, Moncton, Canada, pp. 393-450.

Ross, R.M. and Watten, B.J. (1998) Importance of rearing-unit design and stocking density to the behaviour, growth and metabolism of lake trout (*Salvelinus namaycush*). *Aquacultural Engineering*, **19** 41-56.

Ross, R.M., Watten, B.J., Krise, W.F., DiLauro, M.N. and Soderberg, R.W. (1995) Influence of tank design and hydraulic loading on the behaviour, growth and metabolism of rainbow trout (*Oncorhynchus mykiss*). *Aquacultural Engineering*, **14** 29-47.

Sander, E. (1967) Skimmers in the marine aquarium. *Petfish Monthly*, **2** 49-51.

Sander, M. (1998) Aquarientechnik in Süß- und Seewasser. Stuttgart, Ulmer, 256 pp.

Sansana-yuth, P., Phadungchep, A., Ngammontha, S., Ngdngam, S., Sukasem, P., Hoshino, H., Ttabucanon, M.S. (1996) Shrimp pond effluent: pollution problems and treatment by constructed wetlands, in *Water Quality International '96* (eds. D. Bally, T. Asano, R. Bhamidimarri, K.K. Chin, W.O.K. Grabow, E.R. Hall, S. Ohgaki, D. Orhon, A. Milburn, C.D. Purdon and P.T. Nagle). *Water Science and Technology*, **34** 93-98.

Saunders, R.L. and Harmon, P.R. (1990) Influence of photoperiod on growth of juvenile Atlantic salmon and development of salinity tolerance during winter-spring. *Transactions of the American Fisheries Society*, **119** 689-697.

Schorre, E.J., Simco, B.A. and Davis, K.B. (1995) Response of blue catfish and channel catfish to environmental nitrite. *Journal of Aquatic Animal Health*, **7** 304-311.

Seawright, D.E., Stickney, R.R. and Walker, R.B. (1998) Nutrient dynamics in integrated aquaculture-hydroponics systems. *Aquaculture*, **160** 215-237.

Shpigel, M., Neori, A., Popper, D.M. and Gordin, H. (1993) A proposed model for 'environmentally clean' land-based culture of fish, bivalves and seaweeds. *Aquaculture*, **117** 115-128.

Sigholt, T., Järvi, T. and Loftus, R. (1989) The effect of constant 12-hour light and simulated natural light on growth, cardiac-somatic index and smolting in the Atlantic salmon (*Salmo salar*). *Aquaculture*, **82** 127-136.

Silva-Garcia, A.J. (1996) Growth of juvenile gilthead seabream (*Sparus aurata* L.) reared under different photoperiod regimes. *Israeli Journal of Aquaculture Bamidgeh*, **48** 84-93.

Skilbrei, O.T., Hansen, T., Stefansson, S.O. (1997) Effects of decreases in photoperiod on growth and bimodality in Atlantic salmon, *Salmo salar* L. *Aquaculture Research*, **28** 43-49.

Spotte, S. (1979) Seawater aquariums: the captive environment. New York, John Wiley and Sons, 413 pp.

Steffens, W. (1988) Fischproduktion in Rinnenanlagen. Fischerei-Forschung Rostock, **26** 54-58.

Stoskopf, M.K. (1993) Fish medicine. W.B. Saunders Co., 882 pp.

Suresh, A.V. and Lin, C.K. (1992) Effect of stocking density on water quality and production of red tilapia in a recirculated water system. *Aquacultural Engineering*, **11** 1-22.

Sutterlin, A.M., Coutourier, C.Y. and Devereaux, T. (1984) A recirculating system using ozone for the culture of Atlantic salmon. *The Progressive Fish-Culturist*, **46** 239-244.

Timmons, M.B. (1994) Use of foam fractionators in aquaculture, in *Aquaculture Water Reuse Systems: Engineering Design and Management* (eds. M.B. Timmons and T.M. Losordo), Elsevier, pp. 247-279.

Timmons, M.B., Summerfelt, S.T. and Vinci, B.J. (1998) Review of circular tank technology and management. *Aquacultural Engineering*, **18** 51-69.

Tomasso, J.R. (1993) Toxicity of nitrogenous wastes to aquaculture animals. *Journal of Aquatic Animal Health*, **5** 64-72.

Totland, G.K., Kryvi, H., Jodestol, K.A., Christiansen, E.N., Tangeras, A. and Slinde, E. (1987) Growth and composition of the swimming muscle of adult Atlantic salmon (*Salmo salar* L.) during long-term sustained swimming. *Aquaculture*, **66** 299-313.

Twawarowska, J.G., Westermann, P.W. and Losordo, T.M. (1997) Water treatment and waste characterization evaluation of an intensive recirculating fish production system. *Aquacultural Engineering*, **16** 133-147.

Veselov, A.E., Kazakov, R.V., Sysoyeva, M.I. and Bahmet, I.N. (1998) Ontogenesis of rheotactic and optomotor responses of juvenile Atlantic salmon. *Aquaculture*, **168** 17-26.

Vijayan, M.M. and Leatherland, J.F. (1988) Effects of stocking density on the growth and stress-response in brook charr, *Salvelinus fontinalis*. *Aquaculture*, **75** 159-170.

Wagner, E.J., Jeppsen, T., Arndt, R., Routledge, M.D. and Bradwisch, Q. (1997) Effects of rearing density upon cutthroat trout hematology, hatchery performance, fin erosion, and general health and condition. *The Progressive Fish-Culturist*, **59** 173-187.

Wajsbrot, N., Gasith, A., Diamant, A., Popper, D.M. (1993) Chronic toxicity of ammonia to juvenile gilthead seabream, *Sparus aurata*, and related histopathological effects. *Journal of Fish Biology*, **42** 321-328.

Waller, U. (1992) Factors influencing routine oxygen consumption in turbot, *Scophthalmus maximus*. *Journal of Applied Ichthyology*, **8** 62-71.

Waller, U., Black, E., Burt, D., Groot, C. and Rosenthal, H. (2000) The reaction of young coho, *Oncorhynchus kisutch*, to declining oxygen levels during long-term exposure. *Journal of Applied Ichthyology*, **16** 14-19.

Walters, C.J. and Ward, B. (1998) Is solar radiation responsible for declines in marine survival rates of anadromous salmonids that rear in small streams? *Canadian Journal of Fisheries and Aquatic Sciences*, **55** 2533-2538.

Watten, B.J. (1992) Modelling the effects of sequential rearing on the potential production of controlled environment fish-culture systems. *Aquacultural Engineering*, **11** 33-46.

Watten, B.J. (1994) Aeration and oxygenation, in *Aquaculture Water Reuse Systems: Engineering Design and Management* (eds. M.B. Timmons and T.M. Losordo), Elsevier, pp. 173-208.

Wedemeyer, G.A. and Nelson, N.C. (1977) Survival of two bacterial fish pathogens (*Aeromonas salmonicida* and the enteric redmouth bacterium) in ozonated, chlorinated and untreated waters. *Journal of the Fisheries Research Board of Canada*, **34** 429-432.

Wedemeyer, G.A., Nelson, N.C. and Smith, C.A. (1978) Survival of the salmonid viruses infectious hematopoietic necrosis (IHNV) and infectious pancreatic necrosis (IPNV) in ozonated, chlorinated and untreated waters. *Journal of the Fisheries Research Board of Canada*, **35** 875-879.

Wedemeyer, G.A., Nelson, N.C. and Yasutake, W.T. (1979) Physiological and biochemical aspects of ozone toxicity to rainbow trout (*Salmo gairdneri*). *Journal of the Fisheries Research Board of Canada*, **36** 605-614.

Western, H. and Pratt, K.M. (1977) Rational design for intensive salmonid culture, based on metabolic characteristics. *The Progressive Fish-Culturist*, **39** 157-165.

Wheaton, F.W. (1977) *Aquacultural Engineering*, New York, John Wiley and Sons, 708 pp.

Wheaton, F.W., Lawson, T.B. and Lomax, K.M. (1979) Foam fractionation applied to aquaculture systems. *Proceedings of the World Mariculture Society*, **10** 795-808.

Wheaton, F.W., Hochheimer, J., Kaiser, G.E., Malone, E.F., Krones, M.J., Libey, G.S. and Easter, C.C. (1994) Nitrification filter design methods, in *Aquaculture Water Reuse Systems: Engineering Design and Management* (eds. M.B. Timmons and T.M. Losordo), Elsevier, pp. 127-171.

Williams, E.M., Glass, M.L. and Heisler, N. (1993) Effects of nitrite-induced methaemoglobinaemia on oxygen affinity of carp blood. *Environmental Biology of Fishes*, **37** 407-413.

Yoo, K.H., Masser, M.P. and Hawcroft, B.A. (1995) An in-pond raceway system incorporating removal of fish wastes. *Aquacultural Engineering*, **14** 175-187.

6 Genetic impacts of aquaculture
A.R. Kapuscinski and D.J. Brister

6.1 Introduction

Aquaculture, like many human activities in the environment, has the potential to alter the genetic make-up and diversity within and between wild populations of aquatic organisms. Why should the genetic impacts of aquaculture be a major environmental concern? The answer lies in the fact that genetic variation is the foundation of biological diversity. Although genetics clearly does not encompass all important aspects of biological diversity, it lies at the root of the structural, compositional, and functional dimensions of biological diversity (Noss, 1990). The maintenance of adequate levels of genetic variation, both within and between populations, is essential for their long-term sustainability and evolutionary potential. This fundamental importance of genetic diversity applies to all populations and species, whether they reproduce without human intervention in natural environments, are managed by humans for only part of their life-cycle (e.g. sea-ranching of salmon), are propagated and reared entirely in confined aquaculture systems, or lie elsewhere along this continuum of human intervention in the life cycle.

It is important to protect populations *in situ* because they harbor coevolved gene complexes capable of continually responding to evolutionary forces on the planet. *Ex situ* conservation methods (e.g. germplasm banks, genome maps, genomics databases) are helpful and necessary in triage situations (Harvey, 1999; Ryder *et al.*, 2000) but they are clearly insufficient. In the case of germplasm banks, they capture only a static snapshot of extant genotypes with the further limitation, in many aquatic species, of preserving only the paternal complement in sperm banks due to a lack of feasible cryopreservation methods for ova or zygotes (Harvey, 1999). In the case of genomics databases, the genetic make-up of an organism is reduced to the linear genetic code. Missing is the ability to reproduce such genetic structures and processes as coadapted gene complexes, tissue-specific and phenological differences in gene expression and pleiotropy that act in concert to generate phenotypic variation of living organisms within and between living populations. Like other forms of production, aquaculture depends on this critical role of genetic variation to sustain productivity, prevent inbreeding depression and keep the door open for new products and increased yields.

Aquaculture has come into its own as a major economic force of development and food production precisely at the time when widespread genetic erosion due to human actions has become a major global problem. As a leading technology of the twenty first century, the aquaculture sector should aim to avoid exacerbating this problem. Making the genetic conservation of wild aquatic populations a primary goal of sustainable aquaculture would be an act of enlightened self-interest and of responsible global citizenship.

This chapter presents a systematic assessment of whether or not escapees from a particular aquaculture operation in a particular site pose one or more genetic hazards to wild populations in accessible aquatic environments. It is derived from a more broadly-based assessment tool, the *Environmental Assessment Tool for Cage Aquaculture in the Great Lakes, Version 1.1* (Brister and Kapuscinski, 2000). Much of the scientific literature cited and documents referred to are primarily relevant for the Laurentian Great Lakes of North America, although the scientific principles and rationales presented can be easily adapted to other geographic regions. To assist readers unfamiliar with genetics, a glossary of genetic terms is provided at the end of the chapter.

6.2 Genetically-engineered organisms

The human ability to genetically modify organisms has expanded greatly with the advent of novel techniques of genetic engineering. A genetically-engineered organism (GEO) is one that has been constructed by isolating nucleic acid molecules (molecules that encode genetic information) from one organism, and introducing these molecules into another organism, in a manner that makes them part of the permanent genetic make-up of the recipient, that is, capable of being inherited by offspring (Scientists' Working Group on Biosafety, 1998). This definition also includes those organisms constructed by the transfer of subcellular organelles from one cell to another, followed by the regeneration of an adult organism from the genetically-altered cell, so long as the alteration can be transmitted to offspring.

In the case of aquatic organisms, interspecific hybridization and chromosome manipulations are so novel that they also warrant careful biosafety assessment (Agricultural Biotechnology Research Advisory Committee, 1995; Scientists' Working Group on Biosafety, 1998). Furthermore, many interspecific hybrids and chromosomal-manipulated finfish, shellfish or plants are derived from parental populations that are close to the wild-type, so these genetically-engineered offspring will be ecologically competent if they escape into the wild (Kapuscinski and Hallerman, 1994).

6.3 Assessing biosafety

The *Manual for Assessing Ecological and Human Health Effects of Genetically Engineered Organisms* is appropriate for assessing commercial-scale aquaculture of genetically engineered animals or plants (Scientists' Working Group on Biosafety, 1998). It is an expanded version of the USDA's *Performance Standards for Safely Conducting Research with Genetically Modified Fish and Shellfish* (Agricultural Biotechnology Research Advisory Committee, 1995; available at: *www.nbiap.vt.edu/perfstands/psmain.html*). The manual leads the user through a set of flowcharts, with each user following a case-specific pathway. The manual offers procedures for identifying potential hazards associated with the release of GEOs created from aquatic plants, finfish and shellfish. Where a specific hazard is identified, recommendations are made for minimizing the perceived risk (i.e. minimizing the likelihood that a potential hazard will actually occur).

The scientific community has barely begun to conduct the appropriate studies to test for ecological risks of aquatic GEOs. Risk assessment tests need to address two broad issues: What is the ability and probability of a transgene to spread from escaped GEOs into a natural population through outbreeding of the GEO? and What is the potential for ecological disruptions, for instance, excessive predation on a prey species or competitive displacement of a wild population, due to altered traits of organisms bearing the transgenes? In addressing both issues, one needs to search for altered traits of the GEO that could affect the outcome. For instance, large size at sexual maturity is known to give a mating advantage to males or females in many fish species. If growth-enhanced transgenic fish are larger than non-transgenics at sexual maturity, they would have a mating advantage that could increase the spread of transgenes into a wild population (discussed in further detail below).

The *Manual for Assessing Ecological and Human Health Effects of Genetically Engineered Organisms* (Scientists' Working Group on Biosafety, 1998) directs the user to first assess the potential for transgene spread and, depending on the outcome, then proceed to assess the potential for ecological disruptions. The user assesses the risk of transgene spread by taking a case-specific pathway. In certain cases, the user goes on to assess the potential for ecological disruptions. This order of priority makes sense because conclusions about the potential spread of the transgene into wild populations will affect the range of situations for which one needs to assess ecological disruptions.

One should go on to assess the risk of ecological disruptions when any of three scenarios might apply:

1) the escaped GEOs could survive and interbreed with wild relatives in the accessible ecosystems and the transgene could spread through the wild population;
2) the escaped GEOs could survive and reproduce among themselves and establish a new population in an accessible ecosystem that lacks wild relatives; and
3) the escaped GEOs cannot reproduce in the wild (e.g. rendered sterile via triploidy induction in fish) but could survive long enough in the wild

to prey on, compete with or otherwise displace wild organisms in the ecosystem.

The first and second scenarios are of concern for frequent leakage of relatively small numbers of escapee (e.g. small wear and tear holes in netting of farm cages), as well as infrequent but potentially very large numbers of escapees (e.g. storm damage destroying entire net cages). The third scenario is primarily of concern for infrequent, potentially large numbers of aquaculture escapees, particularly if these recur often enough so that a new wave of escapees tends to replace the earlier wave as it dies off.

The few existing scientific publications that might aid in ecological risk assessment of transgenic fish, although welcome in light of scanty support for such studies (Kapuscinski and Hallerman, 1994), have important shortcomings. They have not estimated the probability of the transgene spreading in wild populations (except for the studies discussed below). Devlin *et al.* (1999) found that dramatically faster growing transgenic coho salmon *Oncorhynchus kisutch* had extraordinarily high plasma growth hormone (GH) levels and consumed 2.9 times more feed pellets than the non-transgenic controls in tanks. The elevated GH levels apparently increased feeding motivation or appetite, raising the possibility that escaped GH transgenic fish could compete successfully with wild fish for food. This study confirmed that genetic engineering usually changes non-target traits (feeding motivation, appetite) in addition to changing the target trait (growth rate), thus supporting the need to search for unintended trait changes when assessing the risk/safety of a GEO. This study was not designed to determine whether changes in other behavioral traits, such as increased predation exposure due to increased foraging for natural prey, could counteract the higher feeding motivation of the transgenic fish. A second study examining critical swimming speed in tanks suggested that this same transgenic strain might have an inferior swimming ability (Farell *et al.*, 1997). We are left, however, not knowing if swimming ability would offset any feeding-related competitive advantage were these transgenic salmon to escape into natural ecosystems. It is also unclear whether swimming ability and food competition are the most crucial traits to measure in order to assess the ecological impacts of these fish.

Stevens *et al.* (1998) found that a line of growth-enhanced transgenic Atlantic salmon had higher oxygen uptake (indicating higher metabolic rate) but similar critical swimming speed to similarly sized non-transgenic controls. The company that has developed these transgenic fish also attests that these fish have better food conversion than controls and produce GH in their tissues year round (Entis, 1997, 1999; Yoon, 2000; A/F Protein unpublished brochures). These isolated bits of information, while potentially useful for demonstrating the desirability of these fish for aquaculture, do not provide the data needed to estimate the probabilities of transgenes spreading from escapees into wild populations and of ecological disruption.

We need a more effective and systematic means of testing aquatic GEOs for possible ecological risk or safety. One step in this direction is the methodology of Muir and Howard (1999, 2000) for assessing the risk of transgene spread to wild relatives (scenario 1 discussed above). Their approach focuses on estimating the overall fitness of a GEO by collecting data at critical 'check points' in its life history (Prout, 1971a,b; Muir and Howard, 2000). The first step is to conduct controlled experiments to test the transgenic organisms for changes in six fitness components: viability (survival to sexual maturity), longevity, age at sexual maturation, fecundity (clutch or spawn size), male fertility, and mating success of both females and males. Then, the fitness component data are integrated to predict gene flow from escapees to wild relatives. Integration of the fitness component data requires the use of simulation models (or multiple generation experiments in simplified, confined ecosystems) to estimate the joint effects of all altered fitness components on transgene spread and population size in the wild population. This methodology improves the chances of identifying one type of ecological disruption: major decreases or increases in wild population size resulting directly from a trade-off between the altered fitness components of transgenic individuals. Such information can then assist further assessment for other undesired ecological disruptions; for instance, changes in predation that might hurt biodiversity or economically important species, or competitive displacement of native species by transgenic organisms that display invasive or pest characteristics.

Muir and Howard (1999), for instance, showed how critically important it is to examine interactions among different fitness components that can be changed by one transgene. They examined genetically engineered Japanese medaka *Oryzias latipes* containing extra GH genes. The transgenic medaka grew faster, reached sexual maturity earlier, and had lower viability than non-engineered controls. Among medaka, as is true in many fish species, larger males have a substantial mating advantage over smaller males. Computer simulations combining the data on mating advantage and lower viability led to a troubling result, called the Trojan gene effect. A transgene introduced into a wild population by interbreeding with a small number of transgenic fish spread quickly as a result of enhanced mating advantage; however, the reduced viability of offspring drove the mixed population to half its size in less than six generations and to extinction in about 40 generations.

If the Trojan gene effect held true in a real situation, particularly whenever the wild population was already depleted, the local extinction of a wild population could have cascading negative effects on the biological community. It is possible that researchers will eventually identify biological factors that prevent the Trojan gene effect from happening in nature (and researchers are presently designing experiments to test the Trojan gene effect on fish populations in confined ecosystems). Meanwhile, taking a precautionary approach to any proposed aquaculture of a GEO would involve first requiring laboratory testing for a mating advantage and changes in other fitness components in the aquatic GEO. In the absence of such key information, the *Manual for Assessing Ecological and Human Health Effects of Genetically Engineered Organisms* (Scientists' Working Group on Biosafety, 1998) recommends to 'consider disallowing the release' or to implement multiple types of barriers to the escape of culture organisms; the latter will probably require relocation of cage aquaculture operations to land-based systems.

In summary, lack of a systematic biosafety assessment of the genetically engineered organisms proposed for aquaculture poses a hazard to aquatic biological communities. Although few empirical risk assessments have been conducted on genetically engineered aquatic organisms, a number of studies indicate possible ecological risks. Modern evolution and ecology further point to the complex ways in which genetically-engineered organisms could harm aquatic communities (Kapuscinski and Hallerman, 1991; Kapuscinski *et al.*, 1999).

6.4 Mitigating risks

If one or more hazards are identified, then the feasibility of implementing risk reduction measures must be determined. A guiding principle is to apply a mix of different types of confinement measures, where each type has a fundamentally different vulnerability to failure. By mixing confinement measures with different vulnerabilities, one increases the chances that failure of one barrier will not breach all the barriers to escape of GEOs from the aquaculture operation. Physical barriers induce 100% mortality through such physical alterations as imposing lethal water temperatures or pH to water flowing out of fish tanks or ponds before the effluent is discharged to the environment. Mechanical barriers are devices, such as screens, that hold back any life stage of the GEO from leaving the aquaculture facility. Biological barriers, such as induced sterilization, are those that prevent any possibility of the GEO reproducing or surviving in the natural environment.

Within-lake aquaculture systems, such as cage aquaculture, pose a major challenge when it comes to trying to mix types of barriers. Physical barriers are not an option for cage farming of salmon because there is no 'end of the pipe' effluent that can be so treated. Mechanical barriers are highly vulnerable to breaching in net-cage farming. Materials such as extra predator barrier nets and rigid netting can help but cannot alone prevent large escapes of GEOs due to storm damage, predator damage, or wear and tear. Floating enclosed bags, a new technology, may work well in waters of the Great Lakes, North America, where potentially damaging physical force of tides is not an issue, but these bags need to be thoroughly tested for their ability to prevent the escape of fish while providing cost-effective rearing conditions (Dodd, 2000).

The exclusive farming of monosex, triploid fish that are functionally sterile is a feasible biological barrier for cage culture of some transgenic fish species, such as salmon and trout (Solar and Donaldson, 1991; Donaldson et al., 1996; Cotter et al., 2000). However, sole reliance on biological barriers in net-cage farms would violate the risk management principle of applying multiple barrier types. Furthermore, biological barriers to reproduction are unknown for some aquaculture species. For a freshwater alga, there is no feasible way to make it sterile to prevent either sexual or asexual reproduction if some plants were to release propagules into aquaculture effluents or escape the culture facility. Sterilization of farmed, genetically engineered algae, therefore, is not an option for helping to reduce establishment of a self-propagating population or to reduce gene flow to locally present wild relatives.

Induction of triploidy is widely accepted as the most effective method for producing sterile fish for aquaculture (Tave, 1993; Benfey, 1999). Triploidy induction disrupts gonadal development to some extent. Typically, gonadal development is more fully disrupted in females than in males *m* in general, ovarian growth is greatly retarded, whereas testes grow to near normal size. Triploid males often produce viable sperm but in greatly reduced numbers and with aneuploid chromosome numbers and other abnormalities. In most, though not all species, fertilization of eggs with milt from triploid males produces progeny that die at embryonic or larval stages. Typically, triploid females do not produce mature oocytes, although several studies that continued beyond the normal first time of sexual maturation in diploids did report the occasional production of mature oocytes by triploid females. In summary, the production of all-female lines of triploids in fish and shellfish (Thorgaard and Allen, 1992; Benfey, 1999) is the best way to maximize disruption of gonadal development as a biological barrier to reproduction of aquacultural escapees. The commercial culture of all-female lines is now widespread in chinook salmon farming in British Columbia and rainbow trout farming in North America, Europe and Japan. Monosex triploid trout are also widely grown and monosex triploid Atlantic salmon are grown commercially in Tasmania (reviewed by Donaldson and Devlin, 1996).

Methods of triploidy induction have been well described (for reviews, see Thorgaard, 1995; Benfey, 1999). Triploidy has been induced in numerous aquaculture species, such as channel catfish, African catfish, various trout species, various salmon species, common carp, grass carp, various tilapia species, yellow perch, red sea bream, and various loach species (Benfey, 1999). The methods for production of all-female lines of fish vary depending on whether the species has an XY or a WZ sex-determining system. These have also been well described and have been used successfully on a broad variety of aquacultural species (reviewed by Tave, 1993, pp. 268-277).

Donaldson et al. (1996, figure 5) summarized the production cycle for integrating triploidy induction into a monosex line, with additional detail provided by Donaldson and Devlin (1996) for salmon, trout and other species with an

XY sex-determining system. Applying this production cycle to transgenic fish involves initially developing an all-female line of transgenic fish, then fertilizing transgenic eggs with milt from the sex-reversed females and inducing triploidy on the newly fertilized eggs. Triploidy induction must occur every time the all-female transgenic line is bred to produce offspring for grow-out. Under experienced hands, one can expect rates of successful triploidy in the 90th percentile in large-scale production but this will vary with fish strain, egg quality, age of spawners and induction conditions.

The critical risk management issue is whether to screen every individual destined for grow-out for the all-female triploid condition or only a subsample of each production lot. Screening for the all-female condition only needs to occur once in the development process. The most common screening method is progeny testing, although male-specific DNA probes provide a faster alternative in chinook salmon and perhaps, someday, in other species (Devlin et al., 1994; Donaldson et al., 1996; Clifton and Rodriguez, 1997). Screening for triploidy must occur in every generation of production fish.

Individual screening has long been required for large-scale stocking of grass carp in Florida (Griffin, 1991; Wattendorf and Phillippy, 1996). The most effective screening method involves particle size analysis of fish blood samples using a Coulter counter and channelyzer (Wattendorf, 1986; Harrell and Van Heukelem, 1998). Estimated labor and supply costs in 1986 were US$0.08–0.20 per screened fish (Wattendorf, 1986). It should be possible to maintain or lower this cost at year 2000 prices by economies of scale and the application of computer automation technology. In any event, the cost of individual screening is a small fraction of the current market price of salmon smolts, trout fingerlings, or other early life stages purchased by grow-out farmers.

It is hypothetically possible to induce sterility in fish through gene transfer that aims to disrupt the production of a key enzyme or hormone involved in gonadal development. Some fish research in this direction is at a very early stage of development (e.g. Alestrom et al., 1992). The feasibility of this approach has not yet been proved. Induction of sterility solely by gene transfer might not be a good option because of vulnerabilities known to be inherent to gene transfer. Expression of the transgene responsible for sterility induction could be turned off at any time through methylation, something that genetic engineers do not know how to prevent. The transgene could also undergo rearrangement in the founders or descendants, thus possibly disrupting the expression needed to induce sterility.

If cage culture operations ever produce transgenic fish, the most secure biological barrier would be to raise transgenics that are exclusively all-female, triploid fish and to provide individual confirmation of triploidy. However, these biological measures alone would not fit well with the principle of multiple barrier types. Land-based farming of transgenic salmon fits this principle much better because it allows use of effective mechanical and physical barriers in addition to sterilization of production fish. The diversity and number of barriers may need to be higher in flow-through systems than in recirculating aquaculture systems. The risk of fish escaping is typically lowest in recirculating systems; this is because no more than 10% of the rearing water is discharged daily and many upstream components of the system (such as solids removal) also act as mechanical barriers to fish escape.

It may be wise to monitor for permanent sterility in triploids; reversion to the diploid and fertile condition was recently discovered in a group of triploid oysters, much to the surprise of shellfish biologists (Blankenship, 1994). So far, no one has reported reversion in fish.

6.5 Interbreeding with natural populations

At issue are naturally reproducing populations of the same species as the culture species or a closely-related species with which the aquaculture escapees can interbreed. The natural populations of concern may be indigenous to the Great Lakes or naturalized descendants of an introduced species that has become socioeconomically important (see example of genetically distinct steelhead trout populations discussed in section 6.5.3). Many aquaculture operations raise organisms from non-local broodstock sources. In most of these cases, organisms escaping from the aquaculture operation will be capable of surviving to reproduce and interbreed with natural populations in surrounding waters.

It is important to assess whether the aquaculture escapees could cross with any closely-related species in the accessible ecosystem. Interspecific hybridization among aquatic species is quite common, particularly among fish (Hubbs, 1955; Lagler *et al.*, 1977; Turner, 1984; Collares-Pereira, 1987), often yielding fertile hybrids that can back-cross to wild populations of either parental species. Interspecific hybrids and their back-crossed descendants may occur naturally but usually at low frequencies; walleye containing introgressed sauger genes, for example, have been found in waters draining into Georgian Bay, Lake Huron (Billington and Hebert, 1988). Frequent or large-scale escapes of fertile hybrids or either parental species from aquaculture operations can substantially increase these frequencies. This then poses the genetic hazard of losing a taxonomically distinct population of a native species. For instance, walleye × sauger hybrids have become a popular culture organism (Held and Malison, 1996). A wild population of walleye could lose its taxonomic and genetic distinctness if large numbers of walleye × sauger hybrids escaping from an aquaculture operation successfully out-crossed with the wild walleye.

It is in the long-term interest of parties interested in aquaculture or capture fisheries to prevent losses of taxonomically distinct populations in the wild. Taxonomically distinct, wild populations are an irreplaceable reservoir of genes (live gene bank) harboring coadapted gene and chromosomal complexes that aquaculture breeders can tap to improve economically important traits, such as disease resistance. Introgressive hybridization would disrupt these gene complexes as well as dilute rare alleles that could be crucially important for aquacultural performance traits. Furthermore, if one half of a hybrid cross comes from an introgressed rather than a pure parental species, the offspring will not show hybrid vigor for the target performance traits, thus undermining the very purpose of making interspecific hybrids in aquaculture. Indeed, Billington (1996) found saugeye genes in some aquaculture broodstocks presumed to be pure walleye. The loss of coadapted gene and chromosomal complexes and of rare alleles also threatens the long-term sustainability of capture fisheries, for reasons explained in greater detail below.

6.5.1 Panmictic populations versus genetically distinct populations

In cases where wild relatives belong to one panmictic population (probably a rarity in the Great Lakes), interbreeding with aquaculture escapees poses the genetic hazard of reducing the fitness, and thus the productivity, of wild populations due to outbreeding depression. In cases where the wild relatives have a number of genetically distinct populations in the Great Lakes, their interbreeding with aquaculture escapees poses two hazards: firstly, outbreeding depression that might reduce the short-term fitness and productivity of the wild fish; and, secondly, homogenization of the genetic differences between populations that might reduce the long-term sustainability of wild populations. Evidence of adverse effects of interbreeding between fish coming from genetically divergent sources has grown in recent years. For instance, see reviews by Kapuscinski and Jacobson (1987), Krueger and May (1991), Heggberget *et al.* (1993), Busack and Currens (1995), Leary *et al.* (1995), Allendorf and Waples (1996), Lynch (1996), National Research Council (1996), Reisenbichler (1997), Gross (1998), Youngson and Verspoor (1998) and Miller and Kapuscinski (2000).

Although information concerning the genetic population structure is missing for many important species in the Great Lakes, substantial information exists for some species. Genetic data may be Great Lakes basin-wide or only lake-wide. For instance, there are data on population structure of lake trout (Ihssen *et al.*, 1988; Krueger *et al.*, 1989; Krueger and Ihssen, 1995), walleye (Billington and Hebert, 1988; Ward *et al.*, 1989; Todd, 1990; Billington *et al.*, 1992; Stepien, 1995), steelhead trout (Krueger *et al.*, 1994; O'Connell *et al.*, 1997), brook trout (Danzmann *et al.*, 1991, 1998; Angers *et al.*, 1995), and Northern pike (Senanan and Kapuscinski, 2000). Because new genetic studies are underway all the time, users need to actively seek out the most current information. This involves searching the scientific literature as well as consulting with practicing fisheries geneticists in the region to find out about unpublished results from the most recent studies (e.g. yellow perch population genetic analysis is currently underway for Lake Michigan).

Genetic analyses of population structure should examine variation in at least one type of nuclear genetic marker that is polymorphic for the species in question. For example, protein electrophoresis is inadequate for assessing population structure in Northern pike *Esox lucius* because studies have shown virtually no variability in these genetic markers (Healy and Mulcahy, 1980; Seeb *et al.*, 1987). Instead, one should use microsatellite DNA, a nuclear genetic marker that has much higher levels of variation and has been used to delineate distinct populations (Miller and Kapuscinski, 1996; Senanan and Kapuscinski, 2000). Likewise, proteins and mitochondrial DNA markers exhibit low variability in yellow perch and genetic studies with such markers have found very little population structure across broad geographic regions (Todd and Hatcher, 1993; Billington, 1996). However, the existence of distinct breeding populations within single lakes has been proposed based on tagging studies, comparative growth and behavior studies, and patterns of egg mass deposition (Aalto and Newsome, 1990). Studies are presently underway to develop higher resolution nuclear DNA markers to search for genetic population structure in yellow perch (Miller and Kapuscinski, unpublished data). It is desirable to confirm population structure results by looking for concurrence between results from two or more types of genetic markers (nuclear or mitochondrial).

The objective is to prevent decline in the short-term fitness and productivity and long-term sustainability of wild populations that could be wrought by interbreeding with aquacultural escapees. Genetic diversity is "part of the fabric of a biological resource" (National Research Council, 1996, p. 146). The productivity of the resource, Great Lakes fish populations in this case, cannot be separated from its genetic basis. Escapees that survive and spread to the breeding grounds of a naturally reproducing population could interbreed with the wild organisms. If this happens on a large enough scale, genetic differences between the aquacultural and wild population are eroded, making all the populations simultaneously more vulnerable to environmental change (e.g. pathogens, contaminants, changes in water quality or temperature regimes). An additional outcome can be reduced fitness of the introgressed wild population resulting from outbreeding depression or maladaptive genes from the partially domesticated aquacultural broodstocks. For a review of the genetic basis for fitness and outbreeding depression in wild fish populations, see Busack and Currens (1995) and Campton (1995).

6.5.2 *Increased vulnerability to environmental change due to loss of genetic differences between populations*

Genetic differences between naturally reproducing populations of a species provide an evolutionary 'bet-hedging' strategy analogous to the adage: "don't put all your eggs in one basket". The 'eggs' are the different alleles (total genetic variation) harbored within each species. The 'basket' is each distinct population. As initially distinct populations become genetically homogenized, they develop the same vulnerability to stressful environmental conditions. The National Research Council (1996) expressed the critical importance of conserving between-population genetic differences as follows:

> 'Consider the extreme where no differences exist between local populations. In that case, a species consists of many copies of the same genetic population and is extremely vulnerable to environmental change. For example, a new disease might be introduced to which most individuals are genetically susceptible; the disease would jeopardize all populations and therefore the entire species. However, in the usual case, where genetic differences do exist between local populations, it is likely that some populations would have a higher frequency of genetically resistant individuals and thus would be relatively unaffected by the disease.'

A graphic example of the extreme case was the widespread crash in yields of genetically-uniform corn crops across North America in the 1970s due to rapid spread of corn blight disease. Following the precautionary principle, it is desirable to prevent erosion of any existing between-population genetic differences in naturally reproducing populations of fish and other aquatic species in the Great Lakes.

6.5.3 Decreased production and fitness of wild populations due to outbreeding depression

Outbreeding depression is a loss of fitness in the offspring produced as a result of interbreeding between two groups because the parents are too distantly related (Templeton, 1986). Local adaptation in naturally reproducing populations increases the probability that farmed fish × wild fish matings will yield outbreeding depression in the offspring. Outbreeding depression may result from the loss of local adaptation (i.e. through introduction of maladaptive genes) or a disruption in coadapted gene complexes that have evolved through many generations of natural selection (Shields, 1993). Reductions in fitness due to loss of local adaptation may occur as early as the first generation of outbred progeny (F_1). Reductions in fitness because of a disruption of coadapted gene complexes are more likely to occur in the next generation (F_2). For instance, Gharrett and Smoker (1991) documented severe outbreeding depression in F_2 hybrids between even- and odd-year pink salmon from the same stream in Alaska. The reduction in fitness could not be due to loss of local adaptation because both populations are native to the same stream. Instead, the appearance of outbreeding depression in the F_2, but not the F_1 generation, was probably due to breakdown of coadapted gene or chromosomal complexes (Allendorf and Waples, 1996).

If a substantial proportion of wild fish secure matings with escaped farmed fish, outbreeding depression could cause a decline in abundance of the wild population, posing a variety of ecological and socioeconomic concerns. Reznick *et al.* (1997) found adaptive evolution of guppies to a new wild environment in only seven generations (a mere 4 years for this species). It is thus reasonable to assume that populations of fish and other aquatic organisms in the Great Lakes have persisted in their local environments over enough generations that they have evolved local adaptation.

For example, two studies suggest that local adaptation is important in walleye, a native and economically important species of the Great Lakes. Fox (1993) compared the embryo hatching success of two populations of walleye from two neighboring rivers in Georgian Bay, Ontario. The rivers were 30 km apart and hatching success of both stocks was compared in both rivers. The native population showed significantly higher hatching rates than the non-native population in both rivers. Jennings *et al.* (1996) found that walleye recruitment to the spawning grounds had a heritable component. Walleye progeny from a river spawning population and a reef spawning population were stocked into an Iowa reservoir containing both river and reef spawning habitat. Upon reaching sexual maturity, the stocked walleye preferred the spawning habitat of their parental populations.

The effects of interbreeding and introgression between genetically divergent populations on the fitness and performance of fish in the wild have not been extensively studied (Campton, 1995; Leary *et al.*, 1995). The published data show that interbreeding between genetically different populations and introgression seldom improve performance of fish in natural environments (reviewed by Krueger and May, 1991; Waples, 1991, 1995; Leary *et al.*, 1995). In a recent study of genetic impacts of a non-indigenous hatchery stock of brown trout on two indigenous populations, Skaala *et al.* (1996) found that survival was nearly three times higher in wild trout than in hybrids of wild and introduced trout. McGinnity *et al.* (1997) compared the performance of wild, farmed and hybrid Atlantic salmon progeny in a natural spawning stream. The progeny of farmed salmon had significantly lower survival to the smolt stage than wild salmon, but they grew fastest and competitively displaced the smaller native fish downstream. A related study showed that progeny of farmed fish in this stream and other sites successfully migrated to the sea, homed to their river of escape, and interbred with wild salmon (Clifford *et al.*, 1998). Such introgression is likely to reduce the fitness and productivity of wild populations.

Negus (1999) examined the effects of interbreeding between two genetically distinct populations of *Oncorhynchus mykiss* from Lake Superior, a long-naturalized population of steelhead trout and a hatchery-propagated 'kamloops' strain of rainbow trout. Embryo survival to hatching and the fright response behavior of fry were compared across progeny of four crosses: pure steelhead crosses, pure kamloops crosses, and the two reciprocal hybrid crosses (steelhead × kamloops and kamloops × steelhead). Survival to hatching was greatest in the pure steelhead cross. Pure steelhead fry displayed a greater fright response than pure kamloops fry when startled by movements over their tanks. Survival to hatching and fry fright response of hybrids was intermediate to both pure crosses but more closely resembled the maternal source. These results confirm a genetic basis for traits affecting survival and productivity of fish in the wild. They also suggest that interbreeding between a partly domesticated strain (kamloops) and a naturalized strain (steelhead) could reduce the naturalized strain's short-term fitness in the wild. It is reasonable to expect similar fitness reductions in wild populations if partly domesticated strains of rainbow trout escaped from cage culture operations and hybridized with naturalized steelhead trout in the Great Lakes.

Some of the best evidence for outbreeding depression comes from studies comparing the post-stocking performance and introgression between genetically distinct populations of largemouth bass. Long-term studies documented genetic and physiological differences between Northern largemouth bass *Micropterus salmoides salmoides* and Florida largemouth bass *Micropterus s. floridanus*. The non-native stocks exhibited poorer fitness and performance traits than the native stock (Philipp, 1991; Philipp and Whitt, 1991). Because these comparisons involved stocks that were very distant geographically, follow-up studies compared two much geographically closer stocks, a northern Illinois and a southern Illinois largemouth bass population (Philipp and Claussen, 1995). The Northern Illinois stock demonstrated better survival, reproductive success and growth than did the Southern Illinois stock in northern Illinois and the reverse was true in southern Illinois. This result strongly supports the existence of local adaptation and, consequently, outbreeding depression if non-native fish interbreed with a locally adapted population.

Outbreeding between genetically distinct populations is most likely to yield hybrids with improved fitness in the wild (outbreeding enhancement) when hybridization alleviates inbreeding depression that existed within one or both populations (Waples, 1995). However, inbreeding depression is unlikely in most naturally reproducing populations of aquatic species in the Great Lakes. Ferguson *et al.* (1988) did find some evidence for superior fitness of first-generation hybrids between two non-inbred populations of cutthroat trout. The superior fitness of hybrids often disappears in subsequent generations when the hybrids back-cross to a parental population (Gharrett and Smoker, 1991). Non-native populations of organisms escaping from aquaculture operations would, therefore, pose a genetic risk to the wild population in the second and subsequent generations, even if offspring in the first hybrid generation exhibited superior fitness.

*6.5.4   Escapees from domesticated aquacultural stocks
         increase the hazard of outbreeding depression*

Most performance traits of aquacultural organisms are partly controlled by genes, and are thus partly heritable (reviewed in Tave, 1993). Compared to wild-type ancestors, the aquacultural organisms will genetically adapt to the new natural selection forces in the aquaculture environment, even when farmers do not actively practice selective breeding. As the organisms become domesticated by genetic adaptation to the aquaculture environment, their adaptation to natural environments declines. This does not mean, however, that aquaculture escapees will be so maladapted to the wild that natural selection will weed them out before they can cross with wild relatives and possibly trigger outbreeding depression (see further discussion below).

Domestication and the commensurate maladaptation to the wild can happen in a fairly small number of generations. Fleming and Einum (1997) documented differences in numerous morphological, behavioral and physiological traits between a seventh-generation farm strain and its wild founder population of Atlantic salmon. These changes were adaptive responses to the farm environment but most are maladaptive to the natural environment. Another study confirmed that innate predator avoidance ability can be negatively altered through short-term domestication (Berejikian, 1995). Hatchery steelhead fry, whose parents were between one and seven generations removed from the wild population of the Quinault River, Washington, survived predation significantly less than fry raised from fertilized eggs of wild Quinault River steelhead adults.

A growing number of studies reveal large differences in aggressive behavior between domesticated finfish and wild counterparts. Heritable changes in aggression in wild offspring of matings between aquaculture escapees and wild fish could make them less fit through various ecological mechanisms. Depending on the life history of the species and its interactions with other species in the wild, either increased or decreased aggression could reduce fitness in the wild. The precautionary approach to sustaining wild populations of aquatic organisms, therefore, is to avoid genetic changes in aggression caused by human actions.

Numerous studies have shown increased aggression in offspring of domesticated broodstocks, for example, in brook trout (Vincent, 1960; Moyle, 1969) and Atlantic salmon (Einum and Fleming, 1997). Increased aggression (or increased competitive ability) has also been found in hatchery fish, including brown trout (Johnsson et al., 1996), and hatchery coho salmon and cutthroat trout (Swain and Riddell, 1990, 1991; Mesa, 1991; Ruzzante, 1991, 1992, 1994; Holtby and Swain, 1992). The reasons for differences in aggressiveness between hatchery and wild fish could be unintentional artificial selection (imposed when broodstock are chosen) or natural selection to the more domestic hatchery environment (reviewed by Jonsson, 1997). For all these salmonine species, increased aggression in wild offspring of hatchery × wild matings would make them more vulnerable to predators (Johnsson and Abrahams, 1991).

Some analysts have argued that maladaptation of escaped farmed fish ensures that their genes would be quickly purged from wild populations by natural selection. Unfortunately, virtually no aquacultural broodstocks have become so intensively domesticated as to assure a high death rate in the wild and, thus, rapid purging of maladaptive genes. Furthermore, the ability of natural selection to purge wild populations of maladaptive traits will be severely hindered whenever there is year-after-year escape and interbreeding of farmed fish with wild fish. Frequent and relatively large escapes of partially-domesticated organisms that successfully interbreed with wild organisms would lead to a chronic reduction (genetic load) in the wild population's fitness and productivity. The decline in the wild population's well being will be in proportion to the frequency of individuals in the mixed population that carry genes from the domesticated farmed fish. Estimation of this frequency is a key step towards quantifying the possible genetic load. Although natural selection is expected to remove maladaptive genes from a population, the number of generations required for the process to be completed can be very large (Hartl, 1988).

6.5.5 Estimating the probability of interbreeding

The risk of increased vulnerability to environmental change is the product of the probability of interbreeding between escapees and wild fish × the probability of loss of genetic differences between the aquacultural and wild stocks. The risk of decreased production is the product of the probability of interbreeding between escapees and wild fish × the probability of outbreeding depression.

Factors to consider in estimating the probability of interbreeding include, but are not limited to:

- *Entry potential*—frequency of farmed fish escaping at different seasons; travel distance to all areas harboring wild fish with which they can mate; probability of surviving in transit to these areas.
- *Introgression potential*—probability of surviving to reproductive stage; degree of similarity in reproductive development, timing of spawning and mating behaviors between aquacultural and wild fish; fecundity and gamete viability of aquacultural escapees.

The probability of loss of genetic differences increases as the genetic distance between the aquacultural and wild populations increases. Estimation of this probability requires knowledge of the genetic population structure of wild populations with which the aquacultural escapees could interbreed, as well as the genetic distance of these populations from the aquacultural stock. In the absence of population genetic data, it may be possible to identify major groupings of genetically-divergent populations based on knowledge of adult fish movements between spawning grounds, geographical distances and geographical barriers to gene flow. However, geographical distance does not always parallel genetic distance. For example, chinook salmon *Oncorhynchus tshawytscha* in California's Klamath River appear to be descended from a lineage quite distinct from that of chinook in the adjacent coastal populations (Utter *et al.*, 1989;

Bartley and Gall, 1990). In another example, Tessier et al. (1997) discovered greater genetic differences between land-locked Atlantic salmon *Salmo salar* populations from two tributaries of a single river than between them and a population from a neighboring river. Thus, any attempt to delineate genetically different groups solely on the basis of geographical proximity should expect surprises (i.e. large error terms).

The probability of outbreeding depression will increase as the number of generations of domestication of the farmed broodstock increases and the genetic distance between the farmed and wild populations increases. Laboratory or adequately confined field experiments conducted to test directly for outbreeding depression between the populations at issue can greatly assist in estimation of this probability.

In the total absence of outbreeding data, one might turn to a cruder estimation of the risk of outbreeding depression. This involves assessing the degree of similarity (or difference) in life-history patterns and ecology of originating environments between the aquaculture production stock (including its founding source) and the wild populations (Miller and Kapuscinski, 2000). As the degree of similarity increases, the potential for outbreeding depression because of introduction of maladaptive genes from the aquaculture stock should decrease. Similarity in life-history patterns partly reflects similarity in genetic make-up for these evolutionarily important traits (Ricker, 1972) and increases the chances that the life-history patterns of outbred individuals will remain locally adaptive. Similarity in ecology of originating environment is indicative of similarity in evolutionary history, also increasing the chances that outbred individuals will remain locally adapted. This, albeit crude, approach fits with principles of evolution but is unproven as a risk estimation technique.

6.5.6 Interactions with naturalized populations

It is important to consider distinct populations of native species as well as naturalized populations of introduced species that have become socioeconomically important. For instance, assessment of a proposed rainbow trout cage culture operation in Lake Superior should include consideration of naturalized genetically-distinct populations of steelhead trout *Oncorhynchus mykiss*. This migratory form of rainbow trout was introduced to Lake Superior approximately 100 years ago through hatchery stockings and quickly became established in several parts of the Lake Superior basin. Genetic analyses of fish collected along the North Shore of Lake Superior showed that these fish have evolved into genetically-distinct populations that breed in different tributary streams (Krueger et al., 1994).

In the 20 generations of natural reproduction since introduction, steelhead trout populations have had adequate opportunity to evolve local adaptation to Lake Superior streams. Local adaptation in naturalized steelhead trout would increase the probability that farmed rainbow trout × wild steelhead trout matings in streams generate hybrid offspring with reduced fitness. These steelhead populations form the basis of a recreational fishery but have recently experienced declines in abundance. The declines have heightened the concerns of anglers and focused attention on gaining the information needed to successfully rehabilitate naturalized steelhead populations.

The current policy of the Minnesota Department of Natural Resources (DNR) is to protect the genetic differences among these naturalized steelhead populations (Schreiner, 1992, 1995). Thus, the DNR would be interested in preventing introgressive hybridization caused by farmed rainbow trout escaping into the wild and mating with wild steelhead. (Note that ongoing field research in two Lake Superior streams is measuring the fitness of hybrids compared to pure steelhead trout; Miller and Kapuscinski, unpublished data).

6.6 Conclusion

The number of possible interactions between escaped farm-raised organisms (especially those transgenic organisms that end up displaying dramatically novel but ecologically competent traits) and wild aquatic organisms is daunting. Many of these interactions would not even be considered until detected firsthand. Rather than becoming paralyzed by this uncertainty and human limits to predicting the behavior of complex living systems, an adaptive approach should be taken that involves active and ongoing learning about the expected and unexpected genetic consequences of aquaculture operations.

If the genetic risk assessment of an aquaculture operation identifies one or more hazards and the responsible organizations have accepted the associated risks, the approach of adaptive biosafety assessment and management (Figure 6.1) calls for treating each aquaculture project as a large experiment and putting in place effective and on-going monitoring for the specific hazard(s) and possible environmental effects (Kapuscinski et al., 1999). When monitoring reveals new information regarding genetic risks or safety, oversight bodies should re-evaluate the aquaculture project and perhaps seek revisions in specific practices. Adaptive biosafety assessment and management recognizes: 1) that we have a limited basis of well-documented experience with genetic and ecological effects of escaped farmed fish; 2) that we need to expect surprises in how complex living systems will respond; and, 3) that even the best guide to genetic risk assessment may have significant weaknesses or blind spots that only long-term monitoring will reveal.

An adaptive approach necessitates a systematic, technically and financially feasible monitoring plan. Baseline biological measurements should be taken to allow valid comparison of changes against preoperation conditions. Threshold limits should be identified and agreed upon before the start of production,

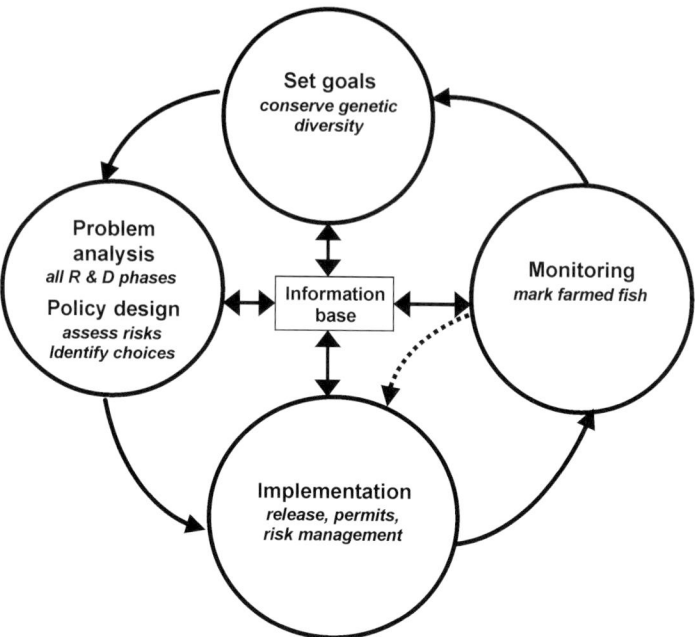

Figure 6.1 The interconnected phases of adaptive biosafety assessment and management. These phases should be applied iteratively and across multiple spatial scales from local to regional, with adequate provisions for information exchange among people implementing biosafety policies at different levels. Abbreviation: R & D, research and development.

thereby reducing the need for emergency measures. With adaptive management approaches in place, we can attempt to minimize the genetic risks to wild populations and conserve biological integrity as the aquaculture sector proceeds into new and relatively uncharted methods of genetic selection and manipulation.

Acknowledgments

This project was supported in part by the Great Lakes Fishery Commission, a grant to ISEES from the Graduate School, University of Minnesota, the MacArthur Interdisciplinary Program on Global Change, Sustainability and Justice, the Minnesota Agricultural Experiment Station and Minnesota Sea Grant, Department of Commerce under grant NOAA-NA86-RG0033, J.R. 472. The US Government is authorized to reproduce and distribute reprints for government purposes, notwithstanding any copyright notation that may appear hereon. This is article 004410002 of the Minnesota Agricultural Experiment Station Scientific Journal Article Series.

Glossary

Fitness—in population and evolutionary biology, the success in terms of survival and reproduction of an individual organism, a population or a species, relative to other individuals, populations or species; the number of offspring that survive to reproduce.

Hazard—an act or phenomenon that has the potential to produce harm or other undesirable consequences to humans or what they value (e.g. a fish species, biodiversity, an entire ecosystem). Hazards may come from physical phenomena (such as floods, fire), chemicals (pesticides, antimicrobial agents), organisms (introduced species, pathogens), commercial products, or human behavior.

Nuclear marker—information about nuclear genes (in contrast to genes found in animal mitochondria or plant chloroplasts); includes proteins, which are encoded by nuclear genes, chromosomal structures (such as chromosome banding patterns), RNA or DNA.

Outbreeding depression—a reduction in fitness due to mating of genetically divergent individuals. Like inbreeding depression, outbreeding depression can result from loss of local adaptation, or breakdown of coadapted genes or chromosomes at different loci. Reductions in fitness due to loss of local adaptation may occur in the F_1 generation, whereas reductions due to breakdown of coadapted gene complexes are more likely to occur in the F_2 generation because F_1 hybrids retain an entire chromosomal array from each parent (Allendorf and Waples, 1996).

Panmictic—refers to a population in which mating is completely random (as opposed to assortative mating between certain adults in the population).

Polymorphic—having two or more forms (alleles) of a gene.

Population—a local (geographically defined) group of conspecific organisms sharing a common gene pool; also called deme.

Propagule—asexual portions of an organism that are capable of dispersal and formation of a new individual.

Risk—an estimate of the probability or likelihood of occurrence of an identified hazard.

Strain—an intraspecific group of organisms possessing only one or a few distinctive traits, usually genetically homozygous (pure-breeding) for those traits and maintained as an artificial breeding group by humans for domestication (e.g. in agriculture or aquaculture) or experimentation.

Transgenic—refers to organisms whose genetic composition has been altered to include specific genes from other organisms of the same or different species by methods other than those used in traditional breeding; this is typically accomplished through recombinant DNA or cloning methods.

Some definitions come from or are adapted from: King and Stansfield (1990); Stern and Fineberg (1996); Scientists' Working Group on Biosafety (1998).

References

Aalto, S.K. and Newsome, G.E. (1990) Additional evidence supporting demic behavior of a yellow perch (*Perca flavescens*) population. *Canadian Journal of Fisheries and Aquatic Sciences*, **47** 1959-1962.

Agricultural Biotechnology Research Advisory Committee, Working Group on Aquatic Biotechnology and Environmental Safety. (1995) *Performance Standards for Safely Conducting Research with Genetically Modified Fish and Shellfish. Parts I & II*. United States Department of Agriculture, Office of Agricultural Biotechnology, Documents No. 95-04 and 95-05. Washington, D.C.

Alestrom, P., Kisen, G., Klungland, H. and Anderson, O. (1992) Fish gonadotropin-releasing hormone gene and molecular approaches for control of sexual maturation: development of a transgenic fish model. *Molecular Marine Biology and Biotechnology*, **1** 376-379.

Allendorf, F.W. and Waples, R.S. (1996) Conservation and genetics of salmonid fishes, in *Conservation Genetics: Case Histories from Nature* (eds. J.C. Avise and J.L. Hamrick), Chapman and Hall, New York, pp. 238-280.

Angers, B., Bernatchez, L., Angers, A. and Desgroseillers, L. (1995) Specific microsatellite loci for brook charr reveal strong population subdivision on a microgeographic scale. *Journal of Fish Biology*, **47** (Suppl. A) 177-185.

Bartley, D.M. and Gall, G.A.E. (1990) Genetic structure of chinook and gene flow in chinook salmon populations of California. *Transactions of the American Fisheries Society*, **119** 55-71.

Benfey, T.J. (1999) The physiology and behavior of triploid fishes. *Reviews in Fisheries Science*, **7** 39-67.

Berejikian, B.A. (1995) The effects of hatchery and wild ancestry and experience on the relative ability of steelhead trout fry (*Oncorhynchus mykiss*) to avoid a benthic predator. *Canadian Journal of Fisheries and Aquatic Sciences*, **52** 2476-2482.

Billington, N. (1996) Geographical distribution of mitochondrial DNA (mtDNA) variation in walleye, sauger and yellow perch. *Annales Zoologici Fennici*, **33** 699-706.

Billington, N. and Hebert, P.D.N. (1988) Mitochondrial DNA variation in Great Lakes walleye (*Stizostedion vitreum*) populations. *Canadian Journal of Fisheries and Aquatic Sciences*, **45** 643-654.

Billington, N., Barette, R.J. and Ward, R.D. (1992) Management implications of mitochondrial DNA variation in walleye stocks. *North American Journal of Fisheries Management*, **12** 276-284.

Blankenship, K. (1994) Experiment with Japanese oysters ends abruptly: oysters thought to be sterile found capable of reproducing. *Bay Journal*, **4** 1-4. Alliance for Chesapeake Bay, Baltimore, MD.

Brister, D.J. and Kapuscinski, A.R. (2000) *Environmental Assessment Tool for Cage Aquaculture in the Great Lakes, Version 1.1.* Prepared for the Great Lakes Fishery Commission by the University of Minnesota Institute for Social, Economic and Ecological Sustainability. St Paul, Minnesota. (Available on the Internet at: http:www.glfc.org).

Busack, C.A. and Currens, K.P. (1995) Genetic risks and hazards in hatchery operations: fundamental concepts and issues. *American Fisheries Society Symposium*, **15** 71-80.

Campton, D.E. (1995) Genetic effects of hatchery fish on wild populations of Pacific salmon and steelhead: what do we really know? *American Fisheries Society Symposium*, **15** 337-353.

Clifford, S.L., McGinnity, P. and Ferguson, A. (1998) Genetic changes in an Atlantic salmon population resulting from escaped juvenile farm salmon. *Journal of Fish Biology*, **52** 118-127.

Clifton, D.R. and Rodriguez, R.J. (1997) Characterization and application of a quantitative DNA marker that discriminates sex in chinook salmon (*Oncorhynchus tshawytscha*). *Canadian Journal of Fisheries and Aquatic Sciences*, **54** 2647-2652.

Collares-Pereira, M.J. (1987) The evolutionary role of hybridization: the example of natural Iberian fish populations, in *Selection, Hybridization and Genetic Engineering in Aquaculture* (ed. K. Tiews), Verlag, H. Heenemann GmbH, Berlin, Germany, pp. 83-92.

Cotter, D., O'Donovan, V., Maoiléidigh, N.O., Rogan, G., Roche, N. and Wilkins, N.P. (2000) An evaluation of the use of triploid Atlantic salmon (*Salmo salar* L.) in minimizing the impact of escaped farmed salmon on wild populations. *Aquaculture*, **186** 61-75.

Danzmann, R.G., Ihssen, P.E. and Hebert, P.D.N. (1991) Genetic discrimination of wild and hatchery populations of brook charr, *Salvelinus fontinalis* (Mitchill), in Ontario using mitochondrial DNA analysis. *Journal of Fish Biology*, **39** (Suppl. A) 69-78.

Danzmann, R.G., Morgan II, R.P., Jones, M.W., Bernatchez, L. and Ihssen, P. (1998) A major sextet of mitochondrial DNA phylogenetic assemblages extant in eastern North American brook trout (*Salvelinus fontinalis*): distribution and postglacial dispersal patterns. *Canadian Journal of Zoology*, **76** 1300-1318.

Devlin, R. Kelly, H., McNeil, B., Solar, I.I. and Donaldson, E.M. (1994) A rapid PCR-based test for Y-chromosomal DNA allows simple production of all-female strains of chinook salmon. *Aquaculture*, **128** 211-220.

Devlin, R.H., Johnsson, J.I., Smailus, D.E., Biagi, C.A., Jönsson, E. and Björnsson, B.T. (1999) Increased ability to compete for food by growth hormone-transgenic coho salmon, *Oncorhynchus kisutch* (Walbaum). *Aquaculture Research*, **30** 479-482.

Dodd, Q. (2000) Closed containment promise and pitfalls. Aquaculture and the Environment (special supplement). *Northern Aquaculture*, **6** 13-14.

Donaldson, E.M. and Devlin, R.H. (1996) Uses of biotechnology to enhance production, in *Principles of Salmonid Culture* (eds. W. Pennell and B.A. Barton), Developments in Aquaculture and Fisheries Science No. 29, Elsevier Publishers, Amsterdam, pp. 969-1020.

Donaldson, E.M., Devlin, R.H., Piferrer, F. and Solar, I.I. (1996) Hormones and sex control in fish with particular emphasis on salmon. *Asian Fisheries Science*, **9** 1-8.

Einum, S. and Fleming, I. (1997) Genetic divergence and interactions in the wild among native, farmed and hybrid Atlantic salmon. *Journal of Fish Biology*, **50** 634-651.

Entis, E. (1997) Aquabiotech: a blue revolution? *World Aquaculture*, **28** 12-15.

Entis, E. (1999) Policy implications for commercialization of transgenic fish, in *Towards Policies for Conservation and Sustainable Use of Aquatic Genetic Resources* (eds. R.S.V. Pullin, D.M. Bartley and J. Kooiman), ICLARM Conference Proceedings, International Center for Living Aquatic Resources Management, Makati City, Philippines, pp. 35-42.

Farrell, A.P., Bennet, W. and Devlin, R.H. (1997) Growth enhanced transgenic salmon can be inferior swimmers. *Canadian Journal of Zoology*, **75** 335-337.

Ferguson, M.M., Danzman, R.G. and Allendorf, F.W. (1988) Developmental success of hybrids between two taxa of salmonid fishes with moderate structural gene divergence. *Canadian Journal of Zoology*, **66** 1389-1395.

Fleming, I.A. and Einum, S. (1997) Experimental tests of genetic divergence of farmed-from-wild Atlantic salmon due to domestication. *ICES Journal of Marine Science*, **54** 1051-1063.

Fox, M.G. (1993) Comparison of zygote survival of native and non-native walleye stocks in two Georgian Bay rivers. *Environmental Biology of Fishes*, **38** 379-383.

Gharrett, A.J. and Smoker, W.W. (1991) Two generations of hybrids between even- and odd-year pink salmon (*Oncorhynchus gorbuscha*): a test for outbreeding depression? *Canadian Journal of Fisheries and Aquatic Sciences*, **48** 1744-1749.

Griffin, B.R. (1991) The US Fish and Wildlife Service's Triploid Grass Carp Inspection Program. *Aquaculture Magazine*, **1** 188-189.

Gross, M. (1998) One species with two biologies: Atlantic salmon (*Salmo salar*) in the wild and in aquaculture. *Canadian Journal of Fisheries and Aquatic Sciences*, **55** (Suppl. 1) 131-144.

Harrell, R.M. and Van Heukelem, W. (1998) A comparison of triploid induction validation techniques. *The Progressive Fish Culturist*, **60** 221-226.

Hartl, D. (1988) *A Primer of Population Genetics*, 2nd edition, Sinauer Associates, Sunderland, MA.

Harvey, B. (1999) Fish genetic conservation in Canada and Brazil: field programs and policy development, in *Towards Policies for Conservation and Sustainable Use of Aquatic Genetic Resources* (eds. R.S.V. Pullin, D.M. Bartley and J. Kooiman), ICLARM Conference Proceedings, International Center for Living Aquatic Resources Management, Makati City, Philippines, pp. 17-22.

Healy, J.A. and Mulcahy, M.F. (1980) A biochemical genetic analysis of populations of northern pike, *Esox lucius* L., from Europe and North America. *Journal of Fish Biology*, **17** 317-324.

Heggberget, T.G., Johnsen, B., Hindar, K., Jonsson, B., Hansen, L., Hvidsten, N. and Jensen, A. (1993) Interactions between wild and cultured Atlantic salmon: a review of the Norwegian experience. *Fisheries Research*, **18** 123-146.

Held, J.A. and Malison, J.A. (1996) Culture of walleye to food size, in *Walleye Culture Manual* (ed. R.C. Summerfelt), NCRAC Culture Series 101, North Central Regional Aquaculture Center Publications Office, Iowa State University, Ames, pp. 231-232.

Holtby, L.B. and Swain, D.P. (1992) Through a glass, darkly: a response to Ruzzante's reappraisal of mirror image stimulation studies. *Canadian Journal of Fisheries and Aquatic Sciences*, **49** 1968-1969.

Hubbs, C.L. (1955) Hybridization between fish species in nature. *Systematic Zoology*, **4** 1-20.

Ihssen, P.E., Casselman, J.M., Martin, G.W. and Phillips, R.B. (1988) Biochemical genetic differentiation of lake trout (*Salvelinus namaycush*) stocks of the Great Lakes region. *Canadian Journal of Fisheries and Aquatic Sciences*, **45** 1018-1029.

Jennings, M.J., Claussen, J.E. and Philipp, D.P. (1996) Evidence for heritable preferences for spawning habitat between two walleye populations. *Transactions of the American Fisheries Society*, **125** 978-982.

Johnsson, J.I. and Abrahams, M.V. (1991) Interbreeding with domestic strain increases foraging under threat of predation in juvenile steelhead trout (*Oncorhynchus mykiss*): an experimental study. *Canadian Journal of Fisheries and Aquatic Sciences*, **48** 243-247.

Johnsson, J.I., Petersson, E., Jönsson, E., Björnsson, B.T. and Järvi, T. (1996) Domestication and growth hormone alter antipredation behavior and growth pattern in juvenile brown trout, *Salmo trutta*. *Canadian Journal of Fisheries and Aquatic Sciences*, **53** 1546-1554.

Jonsson, B. (1997) A review of ecological and behavioral interactions between cultured and wild Atlantic salmon. *ICES Journal of Marine Science*, **54** 1031-1039.

Kapuscinski, A.R. and Jacobson, L.D. (1987) *Genetic Guidelines for Fisheries Management*. Minnesota Sea Grant College Program, St Paul, Minnesota.

Kapuscinski, A.R. and Hallerman, E.M. (1991) Implications of introduction of transgenic fish into natural ecosystems. *Canadian Journal of Fisheries and Aquatic Sciences*, **48** (Suppl. 1) 99-107.

Kapuscinski, A.R. and Hallerman, E.M. (1994) *Benefits, Risks and Policy Implications: Biotechnology in Aquaculture*. Contract report for Office of Technology Assessment (US Congress). Aquaculture: Food and Renewable Resources from US Waters. 80 pp.

Kapuscinski, A.R., Nega, T. and Hallerman, E.M. (1999) Adaptive biosafety assessment and management regimes for aquatic, genetically-modified organisms in the environment, in *Towards Policies for Conservation and Sustainable Use of Aquatic Genetic Resources* (eds. R.S.V. Pullin, D.M. Bartley and J. Kooiman), ICLARM Conference Proceedings, International Center for Living Aquatic Resources Management, Makati City, Philippines, pp. 225-251.

King, R.C. and Stansfield, W.D. (1990) *A Dictionary of Genetics*. 4th edition, Oxford University Press, New York.

Krueger, C.C. and May, B. (1991) Ecological and genetic effects of salmonid introductions in North America. *Canadian Journal of Fisheries and Aquatic Sciences*, **48** (Suppl. 1) 66-77.

Krueger, C.C. and Ihssen, P.E. (1995) Review of genetics of lake trout in the Great Lakes: history, molecular genetics, physiology, strain comparisons and restoration management. *Journal of Great Lakes Research*, **21** (Suppl. 1) 348-363.

Krueger, C.C., Marsden, J.E., Kincaid, H.L. and May, B. (1989) Genetic differentiation among lake trout strains stocked in Lake Ontario. *Transactions of the American Fisheries Society*, **118** 317-330.

Krueger, C.C., Perkins, D.L., Everett, R.J., Schreiner, D.R. and May, B. (1994) Genetic variation in naturalized rainbow trout (*Oncorhynchus mykiss*) from Minnesota tributaries to Lake Superior. *Journal of Great Lakes Research*, **20** 299-316.

Lagler, K.F., Bardach, J.E. and Miller, R.R. (1977) *Ichthyology*. John Wiley and Sons, New York, NY.

Leary, R.F., Allendorf, F.W. and Sage, G.K. (1995) Hybridization and introgression between introduced and native fish. *American Fisheries Society Symposium*, **15** 91-101.

Lynch, M. (1996) A quantitative genetic perspective on conservation issues, in *Conservation Genetics: Case Histories from Nature* (eds. J.C. Avise and J.L. Hamrick), Chapman and Hall, New York, pp. 471-501.

McGinnity, P., Stone, C., Taggart, J.B., Cooke, D., Cotter, D., Hynes, R., McCamley, C., Cross, T. and Ferguson, A. (1997) Genetic impact of escaped farmed Atlantic salmon (*Salmo salar* L.) on native populations: use of DNA profiling to assess freshwater performance of wild, farmed and hybrid progeny in a natural river environment. *ICES Journal of Marine Science*, **54** 998-1008.

Mesa, M.G. (1991) Variation in feeding, aggression and position choice between hatchery and wild cutthroat trout in an artificial stream. *Transactions of the American Fisheries Society*, **120** 723-727.

Miller, L.M. and Kapuscinski, A.R. (1996) Microsatellite DNA markers reveal higher levels of genetic variation in northern pike. *Transactions of the American Fisheries Society*, **125** 971-977.

Miller, L.M. and Kapuscinski, A.R. (2000) Genetic guidelines for hatcheries used to rebuild fish populations, in *Population Genetics of Fishes* (ed. E.M. Hallerman), American Fisheries Society, Bethesda. *In press*.

Moyle, P.B. (1969) Comparative behavior of young brook trout of wild and hatchery origin. *Progressive Fish Culturist*, **31** 51-56.

Muir, W.M. and Howard, R.D. (1999) Possible ecological risks of transgenic organism release when transgenes affect mating success: sexual selection and the Trojan gene hypothesis. *Proceedings of the National Academy of Sciences*, **96** 13853-13856.

Muir, W.M. and Howard, R.D. (2000) Methods to assess ecological risks of transgenic fish releases, in *Genetically Engineered Organisms: Assessing Environmental and Human Health Effects* (eds. D.K. Letourneau and B.E. Burrows), CRC Press, *In press*.

National Research Council. (1996) *Upstream: Salmon and Society in the Pacific Northwest*. National Academy Press.

Negus, M.T. (1999) Survival traits of naturalized hatchery and hybrid strains of anadromous rainbow trout during egg and fry stages. *North American Journal of Fisheries Management*, **19** 930-941.

Noss, R.F. (1990) Indicators for monitoring biodiversity: a hierarchical approach. *Conservation Biology*, **4** 355-364.

O'Connel, M., Danzmann, R.G., Cornuet, J.-M., Wright, J.M. and Ferguson, M.M. (1997) Differentiation of rainbow trout populations in Lake Ontario and the evaluation of the stepwise mutation and infinite allele mutation models using microsatellite variability. *Canadian Journal of Fisheries and Aquatic Sciences*, **54** 1391-1399.

Philipp, D.P. (1991) Genetic implications of introducing Florida largemouth bass, *Micropterus salmoides floridanus*. *Canadian Journal of Fisheries and Aquatic Sciences*, **48** (Suppl. 1) 58-65.

Philipp, D.P. and Whitt, G.S. (1991) Survival and growth of northern, Florida and reciprocal F_1 hybrid largemouth bass in central Illinois. *Transactions of the American Fisheries Society*, **120** 56-64.

Philipp, D.P. and Claussen, J.E. (1995) Fitness and performance differences between two stocks of largemouth bass from different river drainages within Illinois. *American Fisheries Society Symposium*, **17** 236-243.

Prout, T. (1971a) The relation between fitness components and population prediction in Drosophila. I. The estimation of fitness components. *Genetics*, **68** 127-149.

Prout, T. (1971b) The relation between fitness components and population prediction in Drosophila. II. Population prediction. *Genetics*, **68** 151-167.

Reisenbichler, R.R. (1997) Genetic factors contributing to declines of anadromous salmonids in the Pacific Northwest, in *Pacific Salmon and Their Ecosystems: Status and Future Options* (eds. D.J. Stouder, P.A. Bisson and R.J. Naiman), Chapman and Hall, pp. 223-244.

Reznick, D.N., Shaw, F.H., Rodd, F.H. and Shaw, R.G. (1997) Evaluation of the rate of evolution in natural populations of guppies (*Poecilia reticulata*). *Science*, **275** 1934-1937.

Ricker, W.E. (1972) Hereditary and environmental factors affecting certain salmonid populations, in *The Stock Concept in Pacific Salmon* (eds. R.C. Simon and P.A. Larkin), H.R. MacMillan Lectures in Fisheries, University of British Columbia, Vancouver, pp. 19-160.

Ruzzante, D.E. (1991) Variation in agonistic behavior between hatchery and wild populations of fish: a comment on Swain and Riddell (1990). *Canadian Journal of Fisheries and Aquatic Sciences*, **48** 1966-1968.

Ruzzante, D.E. (1992) Mirror image stimulation, social hierarchies and population differences in agonistic behavior — a reappraisal. *Canadian Journal of Fisheries and Aquatic Sciences*, **49** 1966-1968.

Ruzzante, D.E. (1994) Domestication effects on aggressive and schooling behavior in fish. *Aquaculture*, **120** 1-24.

Ryder, O.A., McLaren, A., Brenner, S., Zhang, Y.-P. and Benirschke, K. (2000) DNA banks for endangered animal species. *Science*, **288** 275-277.

Schreiner, D.R. (ed.) (1992) *North Shore Steelhead Plan*. Minnesota Department of Natural Resources, St Paul, Minnesota.

Schreiner, D.R. (ed.) (1995) *Fisheries Management Plan for the Minnesota Waters of Lake Superior*. Section of Fisheries Special Publication No. 149. Minnesota Department of Natural Resources, St Paul, Minnesota.

Scientists' Working Group on Biosafety. (1998) *A Manual for Assessing Ecological and Human Health Effects of Genetically Engineered Organisms. Volumes 1 and 2*. Edmonds Institute, 20319-92 Ave. West, Edmonds, WA 98020, USA, (www.edmonds-institute.org).

Seeb, J.E., Seeb, L.W., Oates, D.W. and Utter, F.M. (1987) Genetic variation and postglacial dispersal of populations of northern pike (*Esox lucius*) in North America. *Canadian Journal of Fisheries and Aquatic Sciences*, **44** 556-561.

Senanan, W. and Kapuscinski, A.R. (2000) Genetic relationships among populations of northern pike (*Esox lucius*). *Canadian Journal of Fisheries and Aquatic Sciences*, **57** 1-14.

Shields, W.M. (1993) The natural and unnatural history of inbreeding and outbreeding, in *The Natural History of Inbreeding and Outbreeding* (ed. N.W. Thornhill), University of Chicago Press, Chicago, pp. 143-169.

Skaala, Ø., Jørstad, K.E. and Borgstrom, R. (1996) Genetic impact on two wild brown trout (*Salmo trutta*) populations after release of non-indigenous hatchery spawners. *Canadian Journal of Fisheries and Aquatic Sciences*, **53** 2027-2035.

Solar, I.I. and Donaldson, E.M. (1991) A comparison of the economic aspects of monosex chinook salmon production versus mixed sex stocks for aquaculture. *Bulletin of the Aquaculture Association of Canada*, **91** 28-30.

Stepien, C. (1995) Population genetic divergence and geographic patterns from DNA sequences: examples from marine and freshwater fishes. *American Fisheries Society Symposium*, **17** 263-287.

Stern, P. and Fineberg, H. (eds.) (1996) *Understanding Risk: Informing Decisions in a Democratic Society*. Committee on Risk Characterization, Commission on Behavioral and Social Sciences and Education, National Research Council. National Academy Press. Washington, D.C.

Stevens, E.D., Sutterlin, A. and Cook, T. (1998) Respiratory metabolism and swimming performance in growth hormone transgenic Atlantic salmon. *Canadian Journal of Fisheries and Aquatic Sciences*, **55** 2028-2035.

Swain, D.P. and Riddell, B.E. (1990) Variation in agonistic behavior between newly-emerged juveniles from hatchery and wild populations of coho salmon, *Oncorhynchus kisutch*. *Canadian Journal of Fisheries and Aquatic Sciences*, **47** 566-577.

Swain, D.P. and Riddell, B.E. (1991) Domestication and agonistic behavior in coho salmon: reply to Ruzzante. *Canadian Journal of Fisheries and Aquatic Sciences*, **48** 520-522.

Tave, D. (1993) *Genetics for Fish Hatchery Managers*, Van Nostrand Reinhold, New York.

Templeton, A.R. (1986) Coadaptation and outbreeding depression, in *Conservation Biology: the Science of Scarcity and Diversity* (ed. M. Soulé) Sinauer Associates, Sunderland, MA, pp. 105-116.

Tessier, N., Bernatchez, L. and Wright, J.M. (1997) Population structure and impact of supportive breeding inferred from mitochondrial and microsatellite DNA analyses in land-locked Atlantic salmon, *Salmo salar* L. *Molecular Ecology*, **6** 735-750.
Thorgaard, G.H. (1995) Biotechnology approaches to broodstock management, in *Broodstock Management and Egg and Larval Quality* (eds. N.R. Bromage and R.J. Roberts), Blackwell Science Ltd, Oxford, pp. 76-93.
Thorgaard, G.H. and Allen Jr, S.K. (1992) Environmental impacts of inbred, hybrid and polyploid aquatic species, in *Dispersal of Living Organisms into Aquatic Ecosystems* (eds. A. Rosenfield and R. Mann), Maryland Sea Grant College Program, College Park, Maryland, pp. 281-288.
Todd, T.N. (1990) *Genetic Differentiation of Walleye Stocks in Lake St. Clair and Western Lake Erie*. US Department of the Interior, Fish and Wildlife Service, Fish and Wildlife Technical Report 28, 1-19.
Todd, T.N. and Hatcher, C.O. (1993) Genetic variability and glacial origins of yellow perch (*Perca flavescens*) in North America. *Canadian Journal of Fisheries and Aquatic Sciences*, **50** 1828-1834.
Turner, B.J. (ed.) (1984) *Evolutionary Genetics of Fishes*. Plenum Press, New York.
Utter, F.M., Miller, G., Stahl, G. and Teel, D. (1989) Genetic population structure of chinook salmon, *Oncorhynchus tshawytscha*, in the Pacific Northwest. *US National Marine Fisheries Service Fishery Bulletin*, **87** 239-264.
Vincent, R.E. (1960) Some influences of domestication upon three stocks of brook trout (*Salvelinus fontinalis* Mitchell). *Transactions of the American Fisheries Society*, **89** 35-52.
Waples, R.S. (1991) Genetic interactions between hatchery and wild salmonids: lessons from the Pacific Northwest. *Canadian Journal of Fisheries and Aquatic Sciences*, **48** (Suppl. 1) 124-133.
Waples, R.S. (1995) Genetic effects of stock transfers of fish, in *Protection of Aquatic Biodiversity* (eds. D.P. Philipp, J.M. Epifanio, J.E. Marsden, J.E. Claussen and R.J. Wolotira), Proceedings of the World Fisheries Congress, Theme 3. Oxford and IBH Publishing Co. Pvt. Ltd., New Delhi, pp. 51-69.
Ward, R.D Billington, N. and Hebert, P.D.N. (1989) Comparison of allozyme and mitochondrial variation in populations of walleye, *Stizostedion vitreum*. *Canadian Journal of Fisheries and Aquatic Sciences*, **46** 2074-2084.
Wattendorf, R.J. (1986) Rapid identification of triploid grass carp with a Coulter counter and channelyzer. *The Progressive Fish Culturist*, **48** 125-132.
Wattendorf, R.J. and Phillippy, C. (1996) Administration of a state permitting program, in *Managing Aquatic Vegetation with Grass Carp, a Guide for Water Resource Managers* (ed. J.R. Cassani), American Fisheries Society, Bethesda, Maryland, pp. 130-176.
Yoon, C.K. (2000) Altered salmon lead the way to the dinner plate, but rules lag. *The New York Times*, May 1, 2000:A1, A20.
Youngson, A. and Verspoor, E. (1998) Interactions between wild and introduced Atlantic salmon (*Salmo salar*). *Canadian Journal of Fisheries and Aquatic Sciences*, **55** (Suppl. 1) 153-160.

7 Modelling impacts
W. Silvert and C.J. Cromey

7.1 Introduction

Modelling is an essential but poorly-understood component of our knowledge of the environmental effects of aquaculture. Essential, because everything we do, from data collection and monitoring to interpretation and site evaluation, is based on models; poorly-understood, because most scientists in the field have a distorted idea of what models are and what modelling is all about.

7.2 Why model?

The main goals of modelling in the context of aquaculture are as follows:

- Provision of information on environmental issues to regulatory agencies—this is essential for effective planning of development in the coastal zone, where defining the holding capacity and/or carrying capacity are of great importance
- Design of practical and appropriate monitoring strategies that allow assessment of regulatory thresholds and acceptable risk
- Provision of advice on coastal zone management, including socioeconomic factors, such as value of production, costs of input, employment implications and other macroeconomic issues related to the contribution of mariculture to regional scale economy and employment
- Modelling as a means to improve husbandry and optimize productivity, providing advice on site selection and management practices
- Modelling to test process understanding of a particular system

The first three of these goals address the broader issues of what society and the environment need. The fourth is sometimes neglected in discussions of the environmental impacts of aquaculture, but the impacts of farms on themselves is important and needs to be considered, even if only to gain the cooperation of the fish farmers without whom one would be working in a data vacuum.

The importance of modelling in the regulation of aquaculture is increasingly recognised as an essential aspect of the management process (SEPA, 1998). For example, the highest priority for future research to be proposed by a major workshop on fish farm effluents was, 'The development and validation of models for: a) predicting the scale of impacts associated with aquaculture waste; and

b) predicting the capacity of water bodies to assimilate anthropogenic inputs' (Rosenthal, 1994). These issues have been further discussed by Ervik *et al.* (1997), Gillibrand and Turrell (1997), Hansen *et al.* (1997) and Cromey *et al.* (2000).

7.3 What are models?

It is a common misconception that a model must be some sort of fancy and complicated mathematical construct that can only be understood by highly specialised experts. The reality is that anything which helps us understand how a system functions is a model of that system. It may of course be a highly mathematical simulation which is implemented on a powerful computer, but it might also be a 'back of the envelope' calculation or even a hunch based on experience.

For example, if Pat wants to set up a fish farm along the shoreline adjoining his property, he may look across the inlet at his friend Mike and think, "Mike has a good farm going, why can't I do the same?" In this case, he is using Mike's farm as a model for his own. It is of course a crude model, one which he would do well to refine before actually investing in a farm operation. He should make sure that the water depth, current regime, temperature range and other factors are really comparable before he accepts the idea that he can do as well as Mike. Even if he establishes that his site is a good as Mike's, there may be other factors operating at a different level which could invalidate the conclusions of his model; for example, it may be that the water body cannot assimilate the additional loading from a second fish farm. In retrospect, it was found that the model correctly predicted that Pat's farm could do as well as Mike's if his were the only farm and Mike's did not exist. However, it did not correctly predict that, with both farms in operation, both operators would go out of business!

There are several lessons to be learned from this simple example. Most of all, models do not need to be complex or mathematical. Complexity arises when a model is refined, but generally is not an inherent feature of models. Another fundamental point to keep in mind is that models do not have an independent existence of their own, but exist to help in the understanding of systems and to answer questions about them. The responsibility of developers and users of models is to ensure that the system has been described correctly and the right questions asked. For example, Pat's model dealt with the farm as a system and addressed the question of whether his planned farm would work as well as Mike's. It did not address the larger system of the entire water body and, since it failed to ask whether it could support twice the production of Mike's farm alone, it neglected to investigate the critical question.

If we think about how Pat should have tackled this problem, we also realise that a single model would not suffice to determine whether he should construct his own farm. Mike's farm may be a good model for addressing some of the questions, such as whether the water was deep enough and sufficiently well oxygenated, but a separate model was needed to determine the total assimilative capacity of the environment. This latter model would need to deal with quite another aspect of the system, different in scale and function.

All of the chapters in this book deal, to some degree, with modelling. The purpose of the present chapter is to focus attention on the best ways to develop and refine models, and to address some of the special issues that arise. The focus is on modelling the environmental impacts of finfish culture. These tend to be more significant than the impacts of (bivalve) shellfish culture because finfish farming involves the introduction of substantial amounts of nutrient that would not otherwise be present, whereas shellfish generally feed on plankton naturally present in the system. The density of shellfish is, therefore, limited by the natural productivity of the ecosystem, but much higher densities of finfish can be raised, at least in the short term. Because of this potential for heavier nutrient loading, finfish farming can lead to much more serious environmental degradation than is usually found with shellfish culture. Furthermore, the impacts of shellfish farms are usually a subset of those from finfish farms, for example, both produce faeces and soluble wastes, but excess feed is associated with finfish farming only.

7.4 What is a system?

The term 'system' is intimately connected with modelling but is so general that it tends to lead to confusion. We speak of modelling a system, but when we ask what the system is, the answer is usually that it is whatever we are modelling! It is hardly surprising that this sort of circular definition promotes despair.

Even so, the circular definition is not that far from the truth. We can of course define a system in the broadest possible terms, so that the system that we are modelling when we are dealing with aquaculture encompasses not only the farm under consideration but also the entire sociopolitical context, the feed industry, foreign competitors, global change, etc. This is clearly impossible (although some misguided souls have tried to model systems on this scale!). Systems must be defined on a scale that can be managed, which basically means on a scale that can be modelled. Consequently, the definition of a system as the thing we model is not as absurd as it sounds.

This idea is formalised in the context of 'hierarchical system theory', which deals with the way in which systems (formally, subsystems) are nested. In the context of aquaculture, there are a number of hierarchical levels that are easy to identify. Fish farms and the environment surrounding fish farms (i.e. the water body in which they are located) can be modelled. Moving in the other direction, individual fish, and their internal organs and disease factors can be modelled. If we look at the scientific specialities which deal with aquaculture issues, we find that they reflect this hierarchical structure—there are all kinds of specialists from biochemists, microbiologists and parasitologists to engineers and ecologists, but each group deals mostly with issues on its own scale of interest and there is a clear division of labour between specialities.

That is not to say that different levels function in total isolation from each other. Clearly, there must be communication between, say, microbiologists and engineers about the design of recirculation systems. However, the hierarchical nature of the system means that the engineer only needs to know a narrow range of details about microbes (size, resistance to ozonation, etc.), while the microbiologist is not likely to be concerned about the kinds of piping and concrete used, beyond issues such as leaching behaviour.

These considerations have important implications for modelling. Each speciality has its own set of models, and it seldom works to combine different hierarchical systems in a single model. The approach taken in this chapter reflects this—the outcome of a modelling project to study the impacts of an aquaculture development will rarely consist of a single massive model, but rather a number of very different models addressing different aspects of the issue, differentiated by scale, function and position in a hierarchy of related systems. This is true of almost all complex systems and, fortunately, they can usually be decomposed into weakly coupled subsystems that can be modelled quasi-independently. The modeller can take advantage of this by building a number of different models of these subsystems and using then to calculate different effects.

Although building multiple models may sound like more work, this is rarely true. The independent models are simpler, and often the total size of a set of such models (as measured, say, by lines of computer code) is much smaller than any single model would be.

There are, however, more fundamental reasons for constructing multiple models than merely simplifying the computational process. One of the most important is to isolate the effects of any parameter errors to a limited subset of models. This is true of any attempt to model complex systems, but with aquaculture modelling, the danger of biasing the results with erroneous data cannot be ignored.

7.5 Scale issues

One of the critical issues in developing a model is identifying the scale of the system it represents. Using an inappropriate scale for the model is not only inefficient, but usually unreliable. For example, if you want to model the long-term deposition from a fish farm over a period of months or years, it is probably pointless to use a model with hourly time steps. Silvert (1992) identified three major time and space scales for the impacts of finfish farms:

- Immediate neighbourhood—this is a very small scale in both space and time, illustrated by depletion of oxygen in and near the pens at slack tide
- Local effects—such as the deposition of carbon and other wastes on the seabed in the vicinity of a fish farm
- Regional impacts—usually due to the release of soluble nutrients and disease organisms into the water column

In marine systems, there is usually a correlation between space and time scales (Steele, 1974), but this is not always true for fish farm impacts. Some impacts, such as oxygen depletion, operate on small time and space scales, and there is a degree of correlation between the distance over which effluents spread and the time it takes to cover the impacted area. However, an obvious and important exception is benthic deposition, which is usually significant only in the immediate vicinity of a fish farm, but which can take months or even years to reach a steady-state (Silvert, 1994a).

The significance of this for the modeller is clear. Any attempt to build a single model to encompass all possible impacts of a fish farm is likely to prove futile. Even if such an ambitious effort were to prove successful, the time and effort that would be required to build and test such a model would probably be far greater than that necessary for a set of more appropriate models fitted to the various impacts of concern.

Of course, impacts on different scales can interact, so these models are not totally independent of each other. Part of the work of the modeller lies in identifying how the various subsystems interact, and translating this into connections between models. Knowing which variables to include is vital, but knowing which ones to omit can be just as important, since inclusion of irrelevant variables is inefficient and can degrade the performance of the model.

For example, rates of benthic deposition are highly variable, and it is not always clear how much of this variation is relevant to determining the impacts. There is certainly a diurnal component to the rate at which shellfish produce wastes, and for finfish there is a short interval during feeding when a spike of unconsumed feed reaches the bottom. This short-term variability is probably not significant and can be ignored by using daily, weekly or even monthly average inputs. However, the rates of assimilation of wastes vary with temperature and other environmental conditions, so if too long a time scale is used, important factors may be overlooked. Furthermore, if wastes are removed by short-term episodic events, such as storms or migratory scavengers, the use of a long time step may lead to erroneous results. Determination of an appropriate scale and time step can be a very complicated process, which requires a good understanding of how the system functions.

7.6 Modular approaches

One way to address the scale problem is to use a modular approach, in which each component of the total system (meaning the farm and its environment) is represented by a module and the linkages between these modules are carefully thought out and defined. In this way, one can model each component to the degree of detail that is necessary without adding to the complexity of the other component models.

For example, in the development of a set of models of fish farm impacts, a central module is one representing an individual fish (Silvert, 1994b). This module is very detailed, as it represents physiological processes, temperature dependencies, and other factors operative only on this very small scale. The outputs of this model include both predictions of growth, which enable the model to be used to simulate the entire lifetime development of the fish, and nutrient budgets, which represent the inputs for separate models of nutrient dynamics and dissipation, benthic deposition and water quality (Silvert, 1994c). This kind of model, in turn, can serve as input to models of assimilative processes, such as the long-term assimilation of carbon loading by the benthos investigated by Sowles *et al.* (1994). The end result is typically a network of models, where the lines connecting them represent only essential variables; for example, the model of an individual fish generates numerous outputs, including production of particulate carbon and soluble ammonia, with particulate carbon being the primary input to a model of deposition, while ammonia fluxes are passed on to the water quality model.

This modular approach is conceptually very similar to object-oriented modelling, in which an 'object' is basically an independent submodel that communicates with the main model through a well-defined set of 'public' variables. Object-oriented modelling can be a very effective way to model hierarchical biological systems (Silvert, 1993), but it has not been widely used in the field and is not discussed here in any detail.

Although different types of environmental impact can arise independently and can be described by separate models, there are some points in common. In particular, since most of the effluents come from the fish in the pens (exceptions are bloodwater and wastes associated with the operation of the farm, such as fuel slicks), a physiological model of the fish themselves lies at the core of most of these models. Because of this, each model can be envisioned as a collection of modules, or submodels, some of which are shared with other models.

In a paper by Silvert (1992), a simple modular scheme was laid out for a set of three models of fish farm impacts. This is illustrated in Figure 7.1. The basic module (FISH) describes a single fish and simulates the feeding, growth and effluent production as a function of size, age and water temperature. A second module (POINT) describes a farm as a collection of fish of different size and age and, by combining the outputs of the single-fish module, it is possible to describe the farm as a point source of effluents, particulates and oxygen

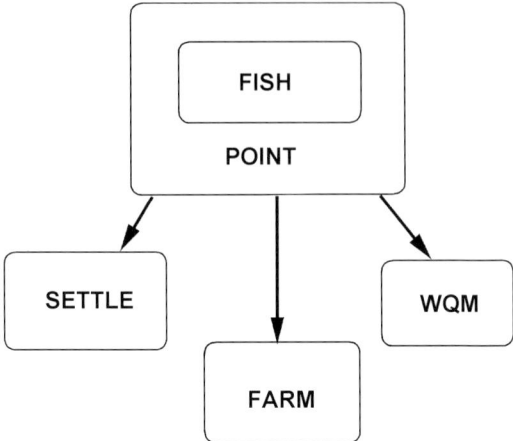

Figure 7.1 Nested set of modules for modelling impacts of a fish farm. Abbreviation: WQM, water quality module (see text for further explanation.)

demand. This module is used as input for three further modules, dealing with different variables on different space-time scales: oxygen depletion at the farm site (FARM); benthic carbon loading under the farm (SETTLE); and water quality over a region whose size depends on the physical mixing in the inlet (WQM). This approach can obviously be extended to cover broader issues. For example, the water quality model describes nutrient enrichment and biological oxygen demand (BOD), which can be used as inputs to models of primary production and other ecological effects.

It is important to maintain balance between the degree of detail in the modules, in order to avoid wasting time on details which will not contribute to the overall performance of the model. In particular, there are some very detailed models of fish growth that are used to optimise feeding schedules and other husbandry concerns but, since the major uncertainties in predicting impacts are in the transport and degradation of wastes, there is little advantage to using very precise and specific models rather than very general models that are much simpler (Telfer, 1995). Sensitivity analysis is a valuable tool in determining how much uncertainties in different parameter values and submodel structures contribute to the overall performance of the model, and it should be used in planning the allocation of effort to different aspects of the modelling process.

7.7 Modelling individual fish

Models of individual fish can range from the very general one described by Silvert (1994c) and Silvert and Sowles (1996) to extremely detailed physiological models of individual strains. While the latter can be used in optimising the day-to-day operation of fish farms, they are probably much too complex for modelling impacts and, since they require precise data on the number, size and genetic make-up of the fish, as well as detailed information on the type of feed used, they are of little value in long-term prediction of environmental impacts.

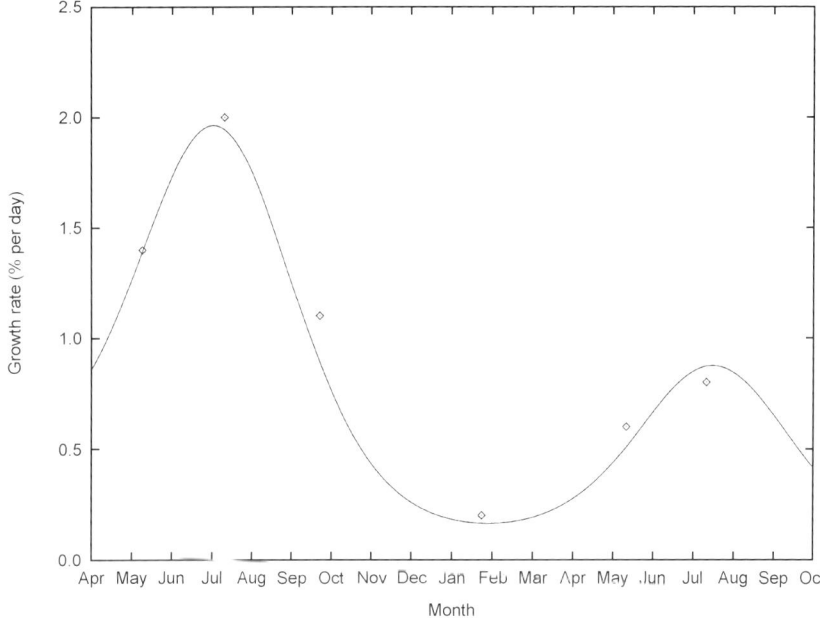

Figure 7.2 Simulated growth of salmon compared with data from the Bay of Fundy, Novia Scotia.

The model described by Silvert (1994c) has two basic components: a nutrient budget for the fish; and a component describing the outputs (faeces, waste feed, dissolved substances, etc). These depend on many factors—the type and composition of the feed, the size, age and, of course, the species of fish, as well as environmental factors. Not all of these have been adequately investigated. For example, based on data from the Bay of Fundy, Nova Scotia, the temperature dependence of the growth rate is given by e^{QT}, where Q is in the range 6–7, but this Q value is probably too high to be explained by temperature alone. However, since the data are based on annual cycles, part of the effect is undoubtedly due to changes in photoperiod during the year, with temperature serving as a proxy for the length of daylight (Silvert and Sowles, 1996). Clearly, a model like this could not be used at other latitudes without further testing and recalibration. Even so, this kind of very simple model can provide a completely satisfactory fit to the data, as shown in Figure 7.2, which compares the basic allometric form:

$$Growth \sim W^a exp(Q_{10}T) \qquad (1)$$

to data from the Bay of Fundy—despite the simplicity of the functional relationship, it is hard to imagine a better fit (Silvert, 1994b).

The nutrient budget for a fish is fairly straightforward; Figure 7.3 represents the one described by Silvert (1994c). The major problem in using such a budget is determining the appropriate parameter values. Changes in feed composition have a major effect (Talbot and Hole, 1994). Feed companies are constantly developing new formulations to optimise the efficiency of assimilation and growth, and since feed is generally the greatest single expense in a farming operation, these new types of feed tend to be rapidly adopted. Furthermore, since selection of a good feed can offer an important competitive advantage to a farmer, the type of feed used is often considered a commercial secret and may not be known to the modeller.

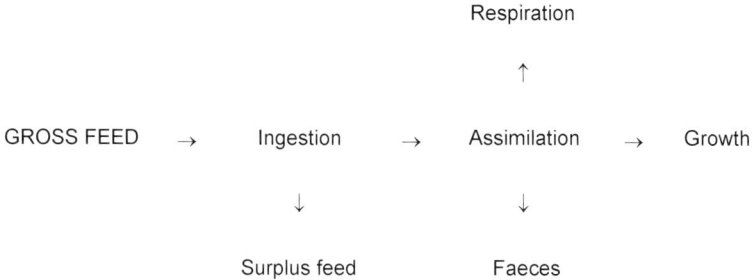

Figure 7.3 The nutrient budget for a fish as described by Silvert (1994c).

7.8 Physical transport modelling

A quick look at the aquaculture literature suggests that the modelling of physical transport is a well-developed field, so that good physical models should be readily available (e.g. Falconer and Hartnett, 1993). Certainly, there are some very good models available to describe water motion in estuaries and other inlets, so it might seem that the need for good physical oceanography is adequately fulfilled. Unfortunately, this is not the case as 3-dimensional hydrodynamic models are lacking for many environments in which fish farms are located, such as fjordic systems. More careful examination shows that the major efforts in this area deal with estimating the availability of planktonic food resources to shellfish and have little relevance to the modelling of impacts. Furthermore, many of the modelling studies directed at fish farm impacts are based on questionable assumptions, which may invalidate their results.

There are many different ways in which physical transport determines the environmental impacts of fish farms. There is of course the motion of water masses which distribute effluents throughout the water column. There are also processes which affect depositional processes, such as flocculation and resuspension.

Probably the most important single physical parameter needed for aquaculture impact modelling is the flushing rate for a water body. To calculate effluent levels, one uses a model which is a variant on the standard input-output model:

$$dC/dt = P - fC \qquad (2)$$

where, C is the effluent concentration, P is the production rate (i.e. the rate at which the farm releases the effluent), and f is the flushing rate. (This form of the model assumes that the effluent is not degraded or assimilated in the water column, but these effects are easy to include.) The parameter P is not difficult to calculate, given sufficient information about the number and size of fish in the farm, but the flushing parameter, f, is usually very difficult to obtain.

This is not as surprising as it may at first appear. For one thing, flushing is difficult to measure directly, so modelling the flushing process does not always receive the attention it requires. Large-scale tracer experiments (such as those using fluorescent dyes) are usually impractical and drogue observations do not reflect the flushing of the entire water column. Basically, this reflects a general principle that it is easier to measure the point properties of a system, namely those which can be associated with a specific location, than to measure the extensive properties of the entire system.

One of the greatest difficulties with flushing calculations, and one which has major significance for modellers, is the problem of re-entrant water masses. In a tidally-flushed estuary, the amount of water leaving the estuary on the ebb tide is the tidal prism, which is easily calculated. However, some of this water re-enters the estuary on the next flood tide, so it may take several tidal cycles before the volume of the tidal prism is permanently removed.

Calculation of the fraction of water that re-enters an estuary is very dependent on the overall geometry of the system. If the estuary in question is actually the inner part of a long inlet, then much of the water probably just sloshes back and forth between the inner and outer estuaries and takes a long time to be totally flushed. On the other hand, some estuaries empty through a narrow channel that forms a jet, but the water that enters on the flood tide comes from along the shore, so flushing is almost complete.

The benthic effects of fish farms are probably the most thoroughly investigated impacts, not necessarily because they are greater than water-column changes, but because they are localised and thus easier to measure and identify (Brown *et al.*, 1987). Even here, it is difficult to find a clear consensus on how to model these effects and interpret the results. Understanding the effects involves three steps:

1. Calculating the flux of carbon and other waste products to the bottom.
2. Understanding the dynamics of depositional materials once they are on the bottom.

3. Identifying the biological and chemical processes associated with the degradation of depositional materials (Harvey and Phillips, 1996).

There does not appear to be general agreement on how to carry out any of these steps!

The first of these steps might appear to be the simplest and most straightforward, since it involves identification of what the waste products are and how they settle. The particulate wastes are mainly faeces and unconsumed feed particles, both of which should be easy to measure. Having determined what the settling particles are, the rest should be straightforward Stokes Law settling, which can be mathematically complicated but not terribly sophisticated (e.g. Gowen et al., 1994). In the laboratory, empirical measurement of the settling velocities of feed and faecal particles can be used to determine general values for use in modelling (Chen et al., 1999a,b), but there are many complications if only particle size data are available. The particles are not inert spheres as assumed in a Stokes Law calculation, but rather dynamic organic particles with active bacterial surfaces. Reports from divers indicate that salmonid faeces consist of mucosoid strings, which tend to drift in the current and can foul nets and other structures (B. Hargrave, personal communication; B. O'Connor, mailing list posting). Tlusty et al. (2000) have found that different types of faeces can be produced by the same kind of fish, which they call 'cohesive' and 'granular'. These different types of faeces differ in structure, sinking rate, friability and carbon content. Feed particles from sloppy feeding rapidly break up in turbulent water and are subject to flocculation. Fouling organisms on the pen structures may ingest the particles, subsequently releasing their own faecal wastes.

An additional complication is that these intermediate processes can lead to episodic deposition and increases in oxygen demand (Findlay and Watling, 1997), which have very different benthic effects from the steady rain of nutrients which most models assume. Storms and other events involving strong currents can shake masses of material loose from pen structures in a matter of hours. Careless removal of fish pens can cause tonnes of attached organic material (epiphytes, mussels, etc.) to fall to the bottom. Even if the average rate of deposition is within a range that the benthic community can assimilate, these episodic events may cause serious damage.

Little is known about how the material behaves after it reaches the bottom. There are several mechanisms which can move sediments around, but given the biological activity of these sediments it is difficult to place a high degree of confidence in traditional models of sediment transport. For example, *Beggiatoa* mats vary greatly in texture and can be very soft and fluffy, while material that has been incorporated into polychaete tubes may be extremely difficult to activate. Bioturbation can also be a factor in disturbing sediments and making them subject to resuspension, in addition to the actual removal of material by bioturbating organisms (Angel *et al.*, 1998). These processes are very difficult to predict, although a semi-phenomenological model has been developed by Sowles *et al.* (1994).

There are practical limitations to our ability to model benthic impacts, arising in large part from the very localised nature of these effects. A farm located in a depositional pocket may experience far greater environmental effects than a larger farm positioned over an erosional bottom only a kilometre away. Furthermore, although the immediate effects in the 'footprint' of the farm may be easy to monitor, transport processes may move the depositional material to a relatively remote location. It is possible for an inlet to contain several farms, under all of which a healthy benthic community can be found, but at the same time for a seriously impacted area to develop at some distance from the farms due to mechanisms which focus the depositional material in that region. This can be seen as a small-scale analogue of the well-known problem of anoxia in the Baltic Sea, which is almost certainly due to nutrient inputs from land, even though the anoxic regions are generally not found close to the shore-based sources of the effluents.

7.9 Primary *vs* secondary effects

One of the hardest parts of modelling the environmental impacts of aquaculture is to identify and quantify secondary effects. For example, it is easy to calculate the rate at which nutrients are released into the water column and, with a simple flushing model, the change in nutrient levels can be determined (Aure and Stigebrandt, 1990). This is a primary effect, but nutrient levels in themselves are not very important. What really matters is the result of this nutrification—does it lead to enhanced primary productivity (e.g. eutrophication) and, in particular, harmful algal blooms? Similarly, it is relatively easy to calculate the benthic deposition rate of carbon, but it is much more difficult to determine the rate at which this carbon can be assimilated and to predict whether there will be deleterious changes in the benthic community.

There are very few cases in which the secondary effects of aquaculture have been quantified to the extent that we can claim any consensus, and there are many crucial issues which have not proceeded beyond the level of heated debate. For example, the question of whether fish farm effluents can increase the incidence of harmful algal blooms has received worldwide attention, but there are few regions where there is even general agreement within the scientific community. Moreover, whenever the scientists agree that effluents are not causing harm, there is certain to be disagreement from the environmental community as well as anyone who considers their welfare at risk.

Apart from the obvious political factors that go into the determination of secondary effects, there are very complex scientific issues that are extremely difficult to resolve. The question of the connection between effluents and harmful algal blooms (HABs) is part of the entire field of phytoplankton dynamics and cannot be answered without addressing the following issues:

1. What limits primary production in this region? If nutrients are limiting, then fish farm effluents might increase primary production, depending on the nature of the critical nutrients. If, on the other hand, primary production is light-limited, increased turbidity might reduce primary production (and presumably the risk of HABs, although the reduction on net productivity may itself be undesirable).
2. Going beyond total primary production, how will the effluents affect the structure of the algal community? Different combinations of nutrients, or different light regimes, may favour some species over others. To use a simple illustration, added silicate is likely to increase diatom production, but since most (but my no means all) HABs are caused by dinoflagellates, increased silicate is probably not as much of an issue as increased nitrate or phosphate. Of course, silicate is not a major component of fish feed, so this is not a great source of comfort!

These two points alone show that modelling secondary effects is, at best, a challenging task, and may prove impossible, at least in any practical timeframe dictated by the needs of regulatory decision-making. The only realistic alternative may be to set arbitrary levels by negotiation among the various 'stakeholders', with little solid scientific input to the process.

7.10 Water quality models

While almost all fish farm impacts arise from material discharged into the water (the effects of moorings on the seabed are a major exception), water quality usually refers to material that is passively transported with the water masses, either as dissolved substances or as fine particulates whose settling speed is negligible. Since water quality modelling involves mainly the motion of water masses, it entails a lot of physical oceanography (Wu et al., 1999). However, because most effluents are chemically and biologically active, transformation and decay processes have to be taken into account.

The basic water quality model is a differential equation of the form:

$$d(Concentration)/dt = (Input) - (Loss) \qquad (3)$$

representing the balance between input to the water column from a fish farm and the loss of material through both physical transport and degradation. An equation of this form holds for each effluent of interest.

Because effluents are not independent, these equations can be linked. Some effluents, such as ammonia, degrade rapidly. BOD declines more slowly, and some nutrients persist until they are taken up by primary production and other biological processes. However, if one is careful to avoid excessive detail, many of these complications can be avoided. For example, it is seldom worthwhile to separate ammonia from other nitrogen compounds, so a single equation for nitrogen can be used in place of three coupled equations for ammonia, nitrite and nitrate.

The inputs are easy to estimate, at least in principle. If one knows the number and size distribution of the fish, composition of the feed, and efficiency of feeding (usually expressed as the feed conversion ratio, FCR), it is possible to construct a nutrient budget that includes the amount of material wasted. In practice, it can be very difficult to obtain this information, especially without the full cooperation of the fish farmer. Even with full cooperation, many farmers seem to be unaware of some critical operational parameters (which may explain why so many of the operations fail).

The loss terms are much harder to determine. These include flushing and other physical processes, as well as the uptake and degradation of nutrients. Even the word 'loss' can be difficult to define precisely when looking at aquaculture impacts; for example, buried carbon can be considered lost to the system (Cranston, 1994), but it can re-enter the system after a storm event which disturbs the sediments. It can also re-enter the system during recovery of a heavily-impacted bottom with the reintroduction of bioturbating organisms. These factors are not likely to occur as much with water quality effects as with benthic impacts, but they illustrate the difficulty of defining clearly what effects one is dealing with.

It can also be extremely difficult to identify whether known effects should be viewed as positive or negative impacts. Nutrients released into the water column can stimulate primary production, but what form does this primary production take?

- Does it stimulate the growth of desirable algae, which are a food resource for shellfish, and enrich the food chain?
- Or, does it cause harmful algal blooms?

It can also provide nutrition for macroalgae, in which case the questions are:

- Does it promote the spread of salt marshes, which are highly productive habitats?
- Or, does it contribute to the spread of smelly sludge along the water line?

These are not questions that are easily answered, but they are typical of the issues that arise in any discussion of aquaculture impacts and they must be addressed.

7.11 Models of benthic loading

Benthic impacts of aquaculture have received the greatest attention in the modelling literature, for two main reasons:

- Benthic impacts are easy to identify and can be clearly associated with fish farms, especially when unconsumed feed or faecal pellets are present
- It is easy to model benthic deposition, but the models are elaborate enough to be satisfyingly impressive

However, the benthic impacts really have to be modelled in two stages, and the above considerations apply mainly to the first stage only, that is, the initial deposition of fish farm wastes on the bottom. The second stage, which is much more complicated and uncertain, gives rise to questions about what happens to carbon and other nutrients after the initial deposition on the bottom, and these questions involve both physical transport and biological degradation and removal.

7.11.1 Modelling primary deposition

Most models are based on settling velocities of waste particles, either measured or computed by Stokes Law, as described by Gowen and Bradbury (1987). The earlier models assumed uniform current and a bottom of uniform depth, but these restrictions were relaxed by Gowen *et al.* (1994). The most detailed model of particulate settling was described by Silvert and Sowles (1996), and can be summarised as follows:

A particle falling at speed, S, in current, V, falls at an angle, θ, from the vertical, given by:

$$tan\theta = V/S \qquad (4)$$

thus, if the current is uniform, and if the depth of the water is Z, then the particle falls at a horizontal distance, $D = Z \, tan \, \theta = ZV/S$, from where it started. This means that the 'footprint' of the cage, namely the region in which the particles fall, is displaced by this distance from directly under the cage. Furthermore, fish farms generate a mix of particles that fall at different speeds, so for different types of particle the points at which they hit the bottom are different. This is illustrated in Figure 7.4 showing the trajectories of faecal pellets and feed particles. The result is a superposition of different footprints, each corresponding to particles of different size and density, and thus of different settling rate (as shown in Figure 7.4).

If, however, the current varies over time, then the footprint moves around. If the cage is circular with radius, R, then the area of the cage is πR^2. However, the footprint moves around with changing current, and if the current is of constant speed, V, but randomly varying direction, then the averaged footprint is a larger

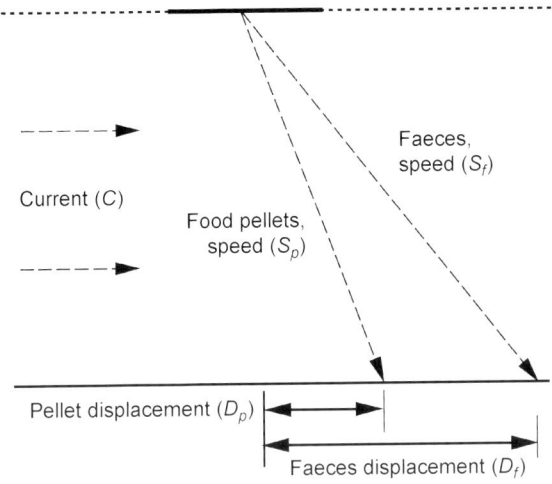

Figure 7.4 Deposition of particles which fall at different speeds.

circle of radius $R + D = R + ZV/S$, so the area is: $\pi(R + ZV/S)^2$. If the particulate flux from the cage is F (measured say in g carbon/m^2/day), then the flux to the bottom is diluted by the ratio of these areas, so the mean flux to the bottom is: $F' = F/(1 + ZV/RS)^2$. This is, of course, only an approximate result, and the flux to the bottom will be uneven, with the densest deposition under the central part of the cage.

Similar results can be obtained for different cage geometries and current regimes. For example, if we have a square cage of side, L, and a reciprocating current that changes back and forth parallel to one set of sides, then the depositional area is a rectangle of area, $L \cdot (L + 2D)$, so the depositional flux is: $F' = F/(1 + 2D/L) = F/(1 + 2ZV/SL)$. More complicated geometry leads to more complicated equations, but the principle is straightforward and some of the simpler calculations may be undertaken using a spreadsheet.

A more important consideration is that the current speed usually varies with depth, especially in fjordic systems, where current shear in the water column is a common feature. The depth-averaged value of the current should be used in this case (Gowen et al., 1994; Silvert and Sowles, 1996), but if the depth varies, this can be a complex quantity to calculate. This situation is illustrated in Figure 7.4, which shows a decrease in current speed with depth, and which also illustrates how variations in depth can lead to uneven deposition rates. The simple model described above and illustrated in Figure 7.4 predicts that the displacement of the depositional point should be proportional to the depth of the water under the pen; however, Figure 7.5 shows that this is not generally the case and, in deeper water, the particles fall almost vertically, especially into a trench geometry where the currents are attenuated.

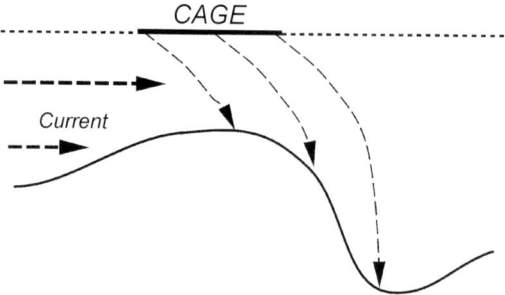

Figure 7.5 Deposition of particles over an uneven bottom, with a current that varies with depth.

There are several ways of dealing with this challenging mathematical problem. Gowen *et al.* (1994) and Silvert and Sowles (1996) described an algorithm to calculate deposition by working backwards from the bottom to the top of the water column (more precisely, to the bottom of the pen), while Hevia *et al.* (1996) used a graphical programming approach to map the deposition on the seabed. A simulation using particle-tracking algorithms can also be used to describe the trajectory of individual particles from sea surface to seabed (Panchang and Richardson, 1992; Cromey *et al.*, 2000). Advection and random walk as a representation of turbulence (Allen, 1982) are the main processes in these models, which may be driven by sophisticated hydrodynamics. Panchang *et al.* (1997) described a model for a coastal area, which included tidal/wind driven currents and waves, for determining dispersion from numerous fish farm discharges. Each of these approaches can be used to produce detailed charts of where particles of a given settling speed fall. As a mathematical challenge this is an interesting but solvable problem, which is undoubtedly why it has received so much attention in the literature.

The choice of depositional model is relatively unimportant, however, since none of the approaches just described address the critical issue of how fast the particles really do fall. A wide range of particle types contribute to benthic deposition, including intact feed particles, fragments of feed from disintegration in the water column and sloppy feeding, and different types of faecal matter. Furthermore, the properties of these particles may not remain fixed as they fall— feed particles may disintegrate and bacterial action can have effects over the time scale of settling. Small particles can stick together as they fall, a process known as flocculation, and this affects their settling speed. Waste particles from different types of discharge will be affected differently by flocculation. For example, fish farm wastes are generally much coarser with higher settling velocities than a typical sewage discharge, so the effect of flocculation on the overall settling characteristics of the fish farm discharge will be less important.

7.11.2 After the fall

Once particles have reached the bottom, a whole series of new processes begins, involving physical removal and transport, as well as many biological effects which can lead to the degradation and removal of nutrients.

Physical removal and transport of material away from a point source via resuspension often results in a reduction in the material available to the benthic community, and the degree of resuspension is dependent on a number of processes (Clarke and Elliot, 1998; Cromey et al., 1998). The fluctuation of near bed current speed, the degree of bed consolidation, biological activity and the degree of stickiness of freshly-deposited material all affect the potential for resuspension. Typically, resuspension models include parameters such as a critical threshold speed for resuspension, where erosion takes place at a certain rate when near bed current speed exceeds the critical threshold level. Despite the potential for resuspension being affected by all these factors, attempts have been made to measure thresholds for freshly deposited material, in laboratory experiments and *in situ*, arising from both natural sources and discharged matter. Typically, freshly-deposited material, so called fluff, has a much lower threshold than consolidated bed material, with threshold values for fluff close to $10\,cm\,s^{-1}$ being reported by numerous researchers (Burt and Turner, 1983; de Jonge and van den Bergs, 1987; Sanford et al., 1991; Washburn et al., 1991; Lund-Hansen et al., 1997). These resuspension models determine rates of erosion, transport, deposition and consolidation on a time-step basis. This approach may be computationally very long and there are advantages in summarising these processes in rate equations, such as Equations (5) and (6) below. This approach might allow better understanding of the overall fluxes in the system.

The biological pattern after the fall has been described by many authors, notably Pearson and Rosenberg (1987) and Weston (1990), and can be summarised as follows: firstly, there is a period of increased productivity due to the additional nutrient input. In most cases, this ultimately exceeds the assimilative capacity of the benthos, and there is then a decline in species richness as the community becomes dominated by organisms that can survive under highly-impacted conditions of low oxygen, high sulphide, etc. (typically *Capitella* sp.). Ultimately, even these vanish when the sediments become anoxic and only bacterial mats remain.

These processes do not always proceed in a smooth and continuous manner. There are many different kinds of event that can disturb the sediments. Angel et al. (1998) described several extreme bioturbation events that occurred during the recovery of a farm site, involving invasions by organisms not normally found on the site. However, Sowles et al. (1994) developed a model of continuously competing processes, which seems to describe adequately the degradation and recovery of a benthic ecosystem under conditions of heavy carbon loading. Their model consists of two differential equations, the first of which is a straightforward uptake-loss model—if C is the rate of carbon loading and A is the amount of accumulated carbon, then:

$$dA/dt = C - kA \tag{5}$$

where, k is a constant. This equation states that the rate of accumulation is the difference between two terms, one of which is the carbon input and the other a constant fractional loss. The second equation is similar, but has a less clear basis:

$$dD/dt = gA - rD \tag{6}$$

where, D is the degree of degradation and g and r are constants, describing the rate at which accumulated carbon degrades and the rate at which improvement occurs (the remediation rate). This is a difficult model to quantify, since the degree of degradation is largely subjective (it was determined by consulting a panel of expert divers, who had examined transects through the sites), and the degradation and remediation rates cannot be independently assessed. The output of this model is represented in Figure 7.6, which shows that although carbon begins to accumulate immediately, it can take considerably longer for deterioration of the benthic community to be observed.

The model just described is very highly speculative and should be viewed as an example of how the more difficult aspects of modelling fish farm impacts

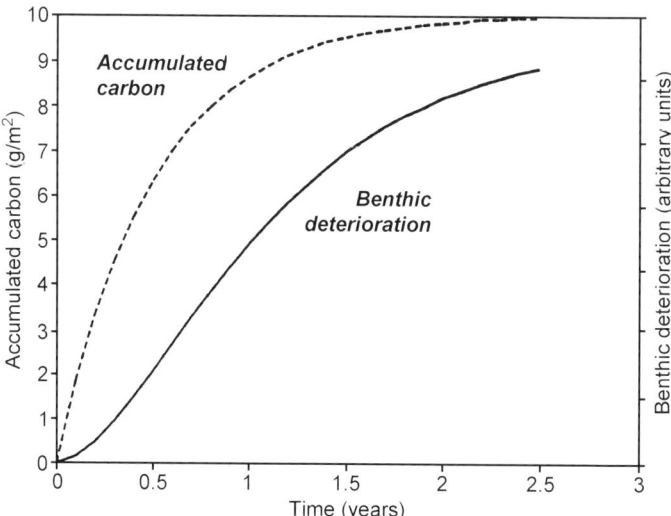

Figure 7.6 Hypothetical deterioration of the benthic community with time, under conditions of constant carbon loading.

can be attacked, rather than as a complete answer to the problem of predicting benthic degradation and deterioration. In fact, analysis of a site under recovery conditions by Angel *et al.* (1998) suggests that the recovery process might be biphasic, that is, there may be two or more remediation rates. They saw an initial rapid improvement, but there was not the kind of exponential approach to an unperturbed state that the above model predicts. This indicates that there may be two or more components to the degraded system, which have to be modelled separately. This is not surprising since there are many different processes going on, some geochemical, some bacteriological and some involving higher levels in the ecosystem (Pearson and Rosenberg, 1987). Moreover, of course, parameters obtained from curve-fitting in one location cannot be assumed to have the same values in other locations, without some experimental or theoretical justification. The recovery rates fitted to the data for the Gulf of Maine system by Sowles *et al.* (1994) are much slower than the initial rates measured in the Red Sea by Angel *et al.* (1998), as one would expect from the large temperature difference between the two bodies of water.

7.12 Disease transmission

One of the major ironies of modelling the environmental impacts of aquaculture is that some of the most important impacts, ones of great concern both to the industry and to regulatory bodies, are among the most poorly understood and most difficult to model reliably. These are the release of disease organisms and the pharmaceuticals used to treat them into the water column, the transmission of disease to other farms and to wild stocks, and the effects of antibiotics and other treatments on natural communities.

Little is known about the transmission of disease from farmed stocks to wild stocks, but the risk of contagion from one farm to another is of course a serious concern to the industry and must be considered an environmental problem, since any farm which serves as a source of contagion is releasing disease organisms into the natural environment.

Epidemiological models are among the most mathematically sophisticated encountered in ecology, and theoretical epidemiology is too complex a field to deal with here in any depth. A brief discussion of how one constructs epidemiological models follows.

To calculate the probability that a population will succumb to an epidemic, two major factors have to be evaluated, namely, exposure and susceptibility. In the context of aquaculture, the exposure that a wild population or another farm receives from an infected farm site depends on the source strength (the prevalence of disease at the infected site), the degree of dilution of infectious organisms, and the virulence of the disease vectors. In calculating contagion from one farm site to another, this involves a lot of physical oceanography, both to calculate the dispersion of the disease agents and to estimate how long it takes them to arrive, since virulence may decrease over time.

Although this sounds complicated, it is in the calculation of susceptibility that the really difficult modelling problems arise. Most epidemics involve continual feedback within the infected population, where individuals which resisted their original exposure are repeatedly subjected to contact with infected individuals and are eventually no longer able to resist the disease. This is one of the reasons why epidemics tend to originate in large crowded populations, although there are of course other factors, including the stress caused by overcrowding (Richardson and Panchang, 1992), that contribute.

Wild populations often receive higher exposure to diseases present in fish farms, since they can swim very close to the site and often feed in the immediate vicinity of fish farms but their susceptibility is much lower. For one thing their exposure is often transitory, and for another, infected individuals may not remain in contact with the rest of the population. The dense aggregations of farmed animals are unfortunately ideal breeding grounds for disease, and their susceptibility is far higher. For this reason, the spread of disease from one fish farm to another is usually a much more serious problem than the infection of wild populations (even if we ignore the commercial aspects of the problem).

The complexity of these issues makes it unlikely that modelling will be able to answer questions about disease propagation with sufficient precision to be useful. For example, one of the major decisions concerning regulatory policy for aquaculture is the minimum separation distance between farm sites—should it be 500 m, 2 km or 10 km? It is difficult to imagine how an epidemiological model could provide strong support for one value over another. Even in almost ideal situations, it is wise to be sceptical. Suppose, for example, that one is dealing with farms in a calm water body with negligible tides, where drogues spread at a leisurely 200 m/day and the most common disease organisms cannot survive for more than 3 or 4 days away from their host. A separation of 500 m is clearly too little, but it would seem that 1 km or more should be sufficient. However, it only takes one storm or other wind event to change the picture, and since disease propagation does not require continual exposure, a model based on the most common conditions is unlikely to give an adequate margin of safety.

7.13 Pharmaceuticals

Modelling the effects of antibiotics and other pharmaceuticals on the environment is, in some ways, even more difficult than modelling disease transmission. In principle, modelling the primary effects, namely the initial distribution of these and any other additives, is the same as modelling the distribution of nutrients. However, so little is known about how the various chemicals used in fish farming affect wild populations that it is almost impossible to estimate the secondary effects, that is, how they actually affect the environment.

But the major problem is lack of reliable information about what is being used. There is general agreement that many banned substances are widely used and information about the actual use of legal additives is almost impossible to obtain. In some cases, the environmental consequences provide the only evidence of what is being used; for example, in one case a large die-off of lobsters was attributed to the use of a chemical known to be highly toxic to crustaceans (including sea lice), but the fish farmers charged with using this chemical vigorously denied resorting to it.

7.14 Tools to improve modelling

There are numerous tools, in the sense of sets of data and software, that can help in the modelling process. Many of these are site-specific or, on the other hand, so general that they need not be mentioned here. Examples of very general tools are statistical packages, simulation programs and database management software.

One category of tools that is general but provides site-specific information is Geographical Information Systems (GISs), which are becoming increasingly widely used in modelling aquaculture impacts, and indeed in many other areas related to coastal zone management (Ali *et al.*, 1991; Ross *et al.*, 1993; Nath *et al.*, 2000). GISs offer a powerful graphical means to store, and more importantly to combine, geographical data. This is normally done by segregating the data into layers, which can be superimposed. For example, when a plan of a proposed aquaculture development is plotted on a master chart of an inlet, it is possible to superimpose layers showing environmental factors, such as depth and current, in order to calculate the dissipation of wastes. It is also possible to superimpose layers showing distributions of commercially important benthic species, which can be used to identify areas of greatest sensitivity to deposition. Other layers can be superimposed to show possible conflicts with transportation or recreational uses of the water body.

7.15 Providing management advice

The result of a modelling programme is generally to provide advice to management at some level. This may simply involve advising an individual farmer how best to avoid adverse impacts at one site, or it may address the general formulation of a broad aquaculture policy for a large jurisdiction. The outputs of the model must, therefore, be expressed in terms appropriate for the ultimate audience, commonly referred to as 'clients'. This can be difficult, since managers are rarely familiar with or comfortable with scientific jargon, and other clients may have little or no scientific background.

Furthermore, it is not always reasonable to assume that the clients will be sympathetic to the scientific work and, therefore, willing to make an extra effort to interpret and understand it. Farmers and pro-business politicians will probably see the modelling project as an effort to throw imaginary obstacles in the way of legitimate development, while environmental groups will be certain that you have ignored the most serious effects and are being paid to whitewash impacts. You cannot always expect to win in a situation like this, but you have to be prepared to put up a good fight and maintain scientific integrity.

7.15.1 Decision support systems

One approach to management advice that offers a practical means of implementing a marriage of modelling to decision-making is through the development of Decision Support Systems (DSSs). This normally refers to a computer program that can be used by a person without extensive technical training to access the kind of information that could normally only be provided by a scientific specialist (Silvert, 1994d). Typically, a DSS incorporates a GIS that automatically accesses relevant information, such as bathymetry and current data (Silvert, 1994e,f). There are several advantages to this kind of approach:

- By automatically accessing data, it avoids errors of transcription and data entry by unskilled personnel, especially when the data are highly technical and unlikely to be understood by the person using the DSS
- DSS programs can be used in areas where skilled scientists are not available for consultation—e.g. when information on anticipated impacts is needed for site-licensing purposes, a DSS can be used to predict these impacts at the time and place where the application is made
- Computer programs may be more acceptable than scientists!

This last point may seem a bit odd, but there is often resistance to seeking and accepting scientific advice and, by depersonalising the advisory process, it can sometimes become more effective. Aquaculture licensing, like other regulatory processes, does not exist in a political vacuum and extraneous factors can enter into the decision-making process. This is why managers usually like to make decisions in private, so scientific advisors are seldom present when a final decision is made. If the manager has a DSS on his desk, he can consult it in private in situations where he might be reluctant to have to defend his position to a living scientist.

Underlying the concept of the DSS is the unfortunate reality that the modelling of aquaculture impacts, and ecological modelling in general, has not had as great an influence on the management of ecological habitats as most scientists feel that it deserves. One of the major concerns of the modeller should be to present results as effectively as possible. A DSS is one possible way of doing this.

7.16 The three Ms

Modelling does not occur in a vacuum. It is a key part of an overall approach to environmental management, which can be characterised as the three Ms—modelling, monitoring and mitigation. Models that predict environmental impacts, no matter how good, have no credibility unless they are supported by empirical evidence. For predictive impact models, that means that there must be a programme of monitoring in place to determine whether the predicted impacts are actually occurring and whether the effects are greater or lesser than predicted. It also means that the modelling work should include the needs of monitoring and should, therefore, deal with impacts that are easily measured and that are part of a standard monitoring protocol.

Furthermore, even under the most carefully controlled circumstances, impacts may prove to be worse than anticipated and a degree of mitigation will be required. This also needs to be taken into account in the modelling process, which should provide some guidance regarding the factors that contributed to the impacts and what measures would be most effective and practical in minimising these impacts.

One way in which modelling can help in monitoring and mitigation is by identifying critical variables that are sensitive indices of degradation, even though they may not themselves be important. For example, some ecological indicators, like diversity and the presence of pollution-tolerant species, are important even though the organisms that contribute to these measures may not themselves have much ecological or economic importance.

Furthermore, while models are powerful tools for predicting impact levels, they are of little value without substantial contribution from field monitoring. A constant interchange of ideas between experts and modellers (and managers and other interested parties) is necessary: firstly, to improve the reliability of the prediction given by the model (i.e. model validation and calibration); secondly, to give applicable interpretation of the results by connecting effects to prediction; and, thirdly, to focus on the critical parameters and their value, which may have been identified by sensitivity analysis. This last point is especially important, because it is essential to keep in mind that a model by itself cannot be used to determine what is or is not acceptable. What the model can do is show what is likely to happen, and thus provide the best advice on possible outcomes of different management decisions.

7.17 Summary

Models of fish farm impacts fall into two categories—some are straightforward both to develop and to apply, while others are so complex and involve so many uncertainties that they are almost useless. It is easy to become so involved in the mathematical aspects of model development that one loses an adequate sense of perspective about what one is doing.

It is, therefore, essential to keep in mind the biological and environmental realities, even if they start to appear intractable. Some of the problems faced in modelling the impacts of aquaculture seem to range from the sublimely unpredictable—storms that suddenly shake debris from the pens and overturn the sediments—to the ridiculous (but necessary) effort to determine the size, texture and density of fish faeces.

All of this effort is wasted if it does not contribute to the processes of planning and regulating aquaculture. It is, therefore, essential to address the needs of decision-makers and to present results in a form that they find both acceptable and useful.

7.18 Additional resources

The modelling of aquaculture impacts is a rapidly changing field, so much of the most useful material has appeared in technical reports, conference proceedings and other ephemeral publications. These include several interesting studies from Nordic sources, such as an overview by Håkanson *et al.* (1988) and a sequel by Mäkinen *et al.* (1991), both funded by the Nordic Council of Ministers. A very informative workshop on 'Fish Farm Effluents and their Control in EC Countries' was held in Hamburg in 1992, and many of the papers were published in a special issue of the Journal of Applied Ichthyology (1994, 10). Hargrave (1994) collected some of the recent Canadian and American work in a technical report. Silvert and Hargrave (1995) organised an ICES workshop on Modelling Environmental Interactions of Mariculture, which has not been formally published, but many of the contributions are available on the Web at http://www.mar.dfo-mpo.gc.ca/science/mesd/he/eim. The proceedings of the ICES Symposium on Environmental Effects of Mariculture held in September 1999 will be an extremely valuable resource, and they are scheduled to appear in the April 2001 issue of the ICES Journal of Marine Science (Vol. 58, no.2).

A more ephemeral, although very valuable, source of information is to be found on mailing lists. The most relevant of these is probably the Environmental Impacts of Mariculture (EIM) list, although this is not very active. The aquaculture list, AQUA-L, covers a wider range of topics, but includes much of interest to anyone modelling impacts. These mailing lists have the added advantage that they are interactive, in that one can post a question and often receive useful replies in a matter of hours. Unfortunately, these postings are not really citable, other than as personal communications, which can be a problem in preparing publications.

Acknowledgements

Many colleagues within the Marine Environmental Science Division have helped over the years in developing an understanding of aquaculture impacts and how to model them, but special recognition is certainly due Dr Barry Hargrave for his great contributions. Dr James Stewart's vast knowledge in this field has proved invaluable, and Tim Milligan has helped all of us understand the sedimentary processes that determine the dynamics of benthic impacts. The authors wish to thank Michael Tlusty for providing a prepublication copy of his work on particulate wastes (Tlusty *et al.*, 2000). But, more than anyone else, thanks are due to Dr Harald Rosenthal for his constant interest, ideas and help. Research workers in this field have benefited widely from Dr Rosenthal's extraordinary depth of knowledge and generosity.

References

Ali, C.Q., Ross, L.G. and Beveridge, M.C.M. (1991) Microcomputer spreadsheets for the implementation of geographic information systems in aquaculture: a case study on carp in Pakistan. *Aquaculture*, **92** 199-205.

Allen, C.M. (1982) Numerical simulation of contaminant dispersion in estuarine flow. *Proceedings of the Royal Society London (A)*, **381** 179-194.

Angel, D., Krost, P. and Silvert, W. (1998) Describing benthic impacts of fish farming with fuzzy sets: theoretical background and analytical methods. *Journal of Applied Ichthyology*, **14** 1-8.

Aure, J. and Stigebrandt, A. (1990) Quantitative estimates of the eutrophication effects of fish farming on fjords. *Aquaculture*, **90** 35-156.

Brown, J.R., Gowen, R.J. and McLusky, D.S. (1987) The effect of salmon farming on the benthos of a Scottish sea loch. *Journal of Experimental Marine Biology and Ecology*, **109** 39-51.

Burt, T.N. and Turner, K.A. (1983) Deposition of sewage sludge on a rippled sand bed. Hydraulics Research Report IT248, Wallingford, Oxon, UK.

Chen, Y.S., Beveridge, M.C.M. and Telfer, T.C. (1999a) Physical characteristics of commercial pelleted Atlantic salmon feeds and consideration of implications for modelling of waste dispersion through sedimentation. *Aquaculture International*, **7** 89-100.

Chen, Y.S., Beveridge, M.C.M. and Telfer, T.C. (1999b) Settling rate characteristics and nutrient content of the faeces of Atlantic salmon, *Salmo salar* L., and the implications for modelling of solid waste dispersion. *Aquaculture Research*, **30** 395-398.

Clarke, S. and Elliot, A.J. (1998) Modelling suspended sediment concentrations in the Firth of Forth. *Estuarine Coastal and Shelf Science*, **47** 235-250.

Cranston, R. (1994) Dissolved ammonium and sulphate gradients in surficial sediment pore water as a measure of organic carbon burial rate, in *Modelling Benthic Impacts of Organic Enrichment from Marine Aquaculture* (ed. B.T. Hargrove), Bedford Institute of Oceanography, Dartmouth, NS Canada, Biological Science Branch, **1949** 93-120.

Cromey, C.J., Black, K.D., Edwards, A. and Jack, I.A. (1998) Modelling the deposition and biological effects of organic carbon from marine sewage discharges. *Estuarine Coastal and Shelf Science*, **47** 295-308.

Cromey, C.J., Nickell, T.D. and Black, K.D. (2000) DEPOMOD (v1.5) software: a model for predicting the effects of solids deposition to the benthos from mariculture. CCMS Dunstaffnage Marine Laboratory, P.O. Box 3, Oban, Argyll, UK, PA34 4AD.

de-Jonge, V.N. and van den Bergs, J. (1987) Experiments on the resuspension of estuarine sediments containing benthic diatoms. *Estuarine Coastal and Shelf Science*, **24** 725-740.

Ervik, A., Kupka-Hansen, P., Aure, J., Stigebrandt, A., Johannessen, P. and Jahnsen, T. (1997) Regulating the local environmental impact of intensive marine fish farming. I. The concept of the MOM system (Modelling-Ongrowing fish farms-Monitoring). *Aquaculture*, **158** 85-94.

Falconer, R.A. and Hartnett, M. (1993) Mathematical modelling of flow, pesticide and nutrient transport for fish-farm planning and management. *Ocean and Coastal Management*, **19** 37-57.

Findlay, R.H. and Watling, L. (1997) Prediction of benthic impact for salmon net-pens based on the balance of benthic oxygen supply and demand. *Marine Ecology Progress Series*, **155** 147-157.

Gillibrand, P.A. and Turrell, W.R. (1997) The use of simple models in the regulation of the impact of fish farms on water quality in Scottish sea lochs. *Aquaculture*, **159** 33-46.

Gowen, R.J. and Bradbury, N.B. (1987) The ecological impact of salmonid farming in coastal waters. *Oceanographic and Marine Biology Annual Review*, **25** 563-575.

Gowen, R.J., Smyth, D. and Silvert, W. (1994) Modelling the spatial distribution and loading of organic fish farm waste to the seabed, in *Modelling Benthic Impacts of Organic Enrichment from Marine Aquaculture* (ed. B.T. Hargrave). *Canadian Technical Report of Fisheries and Aquatic Sciences*, **1949** 19-30.

Håkanson, L., Ervik, A., Mäkinen, T. and Möller, B. (1988) *Basic Concepts Concerning Assessments of Environmental Effects of Marine Fish Farms*. Nordic Council of Ministers, Copenhagen, 103 pp.

Hansen, P.K., Ervik, A., Aure, J., Johannessen, P., Jahnsen, T., Stigebrandt, A. and Schaanning, M. (1997) MOM: (Modelling-Ongrowing Fish Farms-Monitoring). Concept and revised edition of monitoring programme. *Fisken og Havet*, **7** 51 pp.

Hargrave, B.T. (ed.) (1994) Modelling benthic impacts of organic enrichment from marine aquaculture. *Technical Report of Fisheries and Aquatic Sciences*, **1949** xi+125 pp.

Harvey, S.M. and Phillips, C.J. (1996) The degradation of fish farm waste in the marine environment, in *Aquaculture and Sea Lochs* (ed. K.D. Black), Scottish Association for Marine Science, P.O. Box 3, Oban, Argyll, PA34 4AD, 56-58.

Hevia, M., Rosenthal, H. and Gowen, R.J. (1996) Modelling benthic deposition under fish cages. *Journal of Applied Ichthyology*, **12** 71-74.

Lund-Hansen, L.C., Valeur, J., Pejrup, M. and Jensen, A. (1997) Sediment fluxes, resuspension and accumulation rates at two wind-exposed coastal sites and in a sheltered bay. *Estuarine Coastal and Shelf Science*, **44** 521-531.

Mäkinen, T., Henriksson, S.-H., Håkanson, L., Kupka Hansen, P., Laurén-Määttä, C., Persson, J., Uotila, J. and Wallin, M. (1991) *Marine Aquaculture and Environment*. Nordic Council of Ministers, Copenhagen. iv+126 pp.

Nath, S.S., Bolte, J.P., Ross, L.G. and Aguilar-Manjarrez, J. (2000) Applications of geographical information systems (GIS) for spatial decision support in aquaculture. *Aquaculture Engineering*, **23** 233-278.

Panchang, V. and Richardson, J. (1992) A review of mathematical models used in assessing environmental impacts of salmonid net-pen culture. *Journal of Shellfish Research*, **11** 204-205.

Panchang, V., Cheng, G. and Newell, C. (1997) Modelling hydrodynamics and aquaculture waste transport in coastal Maine. *Estuaries*, **20** 14-41.

Pearson, T.H. and Rosenberg, R. (1987) Macrobenthic succession in relation to organic enrichment and pollution of the marine environment. *Oceanographic and Marine Biology Annual Review*, **16** 229-311.

Richardson, J.E. and Panchang, V. (1992) Estimation of mussel seeding densities by mathematical modelling. *Journal of Shellfish Research*, **11** 205-206.

Rosenthal, H. (1994) Fish farm effluents and their control in EC countries: summary of a workshop. *Journal of Applied Ichthyology*, **10** 215-224.

Ross, L.G., Mendoza, E.A. and Beveridge, M.C.M. (1993) The application of geographical information systems to site selection for coastal aquaculture: an example based on salmonid cage culture. *Aquaculture*, **112** 165-178.

Sanford, L.P., Panageotou W. and Halka, J.P. (1991) Tidal resuspension of sediments in northern Chesapeake Bay. *Marine Geology*, **97** 87-103.

SEPA (1998). Regulation and monitoring of marine cage fish farming in Scotland: a manual of procedures. Scottish Environment Protection Agency, Erskine Court, Castle Business Park, Stirling, UK, FK9 4TR, 71 pp.

Silvert, W. (1992) Assessing environmental impacts of finfish aquaculture in marine waters. *Aquaculture*, **107** 67-71.

Silvert, W. (1993) Object-oriented ecosystem modelling. *Ecological Modelling*, **68** 91-118.

Silvert, W. (1994a) Modelling environmental aspects of mariculture: problems of scale and communication. *Fisken og Havet*, **13** 61-68.

Silvert, W. (1994b) Simulation models of finfish farms. *Journal of Applied Ichthyology*, **10** 349-352.

Silvert, W. (1994c) Modelling benthic deposition and impacts of organic matter loading, in *Modelling Benthic Impacts of Organic Enrichment from Marine Aquaculture* (ed. B.T. Hargrave). *Canadian Technical Report of Fisheries and Aquatic Sciences*, **1949** 1-18. xi+125 pp.

Silvert, W. (1994d) Putting management models on the manager's desktop. *Journal of Biological Systems*, **2** 519-527.

Silvert, W. (1994e) A decision support system for regulating finfish aquaculture. *Ecological Modelling*, **75/76** 609-615.

Silvert, W. (1994f) Decision support systems for aquaculture licensing. *Journal of Applied Ichthyology*, **10** 307-311.

Silvert, W. and Hargrave, B. (1995) Report on International Workshop on Modelling Environmental Interactions of Mariculture. *Proceedings of ICES Annual Science Conference*, C.M. 1995/F:6 (Session F).

Silvert, W. and Sowles, J.W. (1996) Modelling environmental impacts of marine finfish aquaculture. *Journal of Applied Ichthyology*, **12** 75-81.

Sowles, J.W., Churchill, L. and Silvert, W. (1994) The effect of benthic carbon loading on the degradation of bottom conditions under farm sites, in *Modelling Benthic Impacts of Organic Enrichment from Marine aquaculture* (ed. B.T. Hargrave). *Canadian Technical Report of Fisheries and Aquatic Sciences*, **1949** 31-46.

Steele, J.H. (1974) *The Structure of Marine Ecosystems*. Harvard University Press.

Talbot, C. and Hole, R. (1994) Fish diets and the control of eutrophication resulting from aquaculture. *Journal of Applied Ichthyology*, **10** 258-270.

Telfer, T. (1995) Modelling of environmental loading: a tool to help marine fish cage management. *Aquaculture News*, p. 17.

Tlusty, M., Snook, K. and Pepper, V.A. (2000) The potential for soluble and transport loss of particulate aquaculture wastes. *Aquaculture Research*, **31** 1-11.

Washburn, L., Jones, B.H., Bratkovich, A., Dickey, T.D. and Chen, M.S. (1991) Mixing, dispersion, and resuspension in vicinity of an ocean wastewater plume. *Journal of Hydraulic Engineering*, **118** 38-58.

Weston D.P. (1990) Quantitative examination of macrobenthic community changes along an organic enrichment gradient. *Marine Ecology progress Series*, **61** 233-244.

Wu, R.S.S., Shin, P.K.S., MacKay, D.W., Mollowney, M. and Johnson, D. (1999) Management and marine farming in the subtropical environment: a modelling approach. *Aquaculture*, **174** 279-298.

8 Aquaculture in the age of integrated coastal management (ICM)

D.W. Donnan

8.1 Introduction

With a few notable exceptions, marine aquaculture is a relatively recent phenomenon that has experienced the bulk of its growth over the last three decades. For example, Asia, the predominant aquacultural region of the world, has a long history of farming marine animal and plant species as a small-scale, labour-intensive activity. However, this situation has been transformed into a large-scale, food-producing, export-earning industry during the relatively recent past, essentially from the 1970s onwards (Chua and Graces, 1994). A similarly rapid expansion has been seen elsewhere in the world and, during the past 15 years, the global production of farmed fish and shellfish has more than doubled (FAO, 1999a,b).

Aquaculture is, however, only one facet of the numerous pressures currently being exerted on the world's coastal resources. An expanding coastal population, pollution, capture fisheries, the exploitation of minerals, oil and gas, and growth in the recreational and tourist industries all contribute to the stresses brought to bear on the coastal ecosystem. Each participates in the significant loss of natural habitats and related impacts on biodiversity.

However, there has also been a significant shift in the approach being taken to the management of coastal resources. This has arisen from the growing appreciation that these resources cannot be managed or regulated in isolation from one another. Similarly, it is widely accepted that it is neither desirable nor possible to consider the management of coastal resources and activities in isolation from the physical and biological environment.

The result has been an aspiration for a more rounded and holistic approach to the management and sustainable use of coastal resources. The concept of integrated coastal management (ICM) developed rapidly during the 1990s, and has been widely embraced around the world. Concern for the environment and the conservation of biodiversity has also had a significant influence on the development of coastal policy. Consequently, in conjunction with ICM, another important element being promoted by scientists and managers is the potential value of marine protected areas (MPAs) for the conservation and sustainable use of coastal resources.

It is within this policy environment that the management of aquaculture must now be viewed. The aim of the present chapter is to examine the influence of ICM policies on an aquaculture industry that continues to expand.

8.2 The need for an integrated approach to the management of aquaculture

A common theme in the regulation of the aquaculture industry, in many locations, has been the inadequacy of existing legislation to properly manage growth in the industry as well as to take account of other activities and users of the coastal zone. Frequently, rapid growth, coupled with poor regulatory measures, has become a constraint within the industry itself; for example, in Tasmania, where demand for new sites began to outstrip supply in the early 1990s (Mcloughlin, 1997). However, the consequences are more often felt well beyond the industry. Some of the wider issues arising from two examples of aquaculture's growth serve to demonstrate the formidable challenges facing the sustainable management of coastal resources: the loss of mangrove habitat to tropical shrimp and fish farming; and, the interactions between farmed salmon and wild fish stocks in Europe.

8.2.1 The loss of mangrove habitat to tropical shrimp and fish farming

There is a demand for shallow coastal pools for shrimp and fish farming in Asia and other tropical regions. Such pools are commonly established in former mangrove habitats that have been cleared specifically for aquaculture or following clearance for less profitable agriculture prior to the change in use to aquaculture. The resultant loss of mangrove habitat in Asia has been significant, with some regions experiencing a reduction in excess of 70% (Phillips *et al.*, 1993; Bardach, 1997). This loss of habitat has a direct negative effect on biodiversity, but the effects can be more widespread. Mangrove is a rich and productive habitat that plays a crucial role in the tropical inshore ecosystem. Amongst other things, mangrove provides essential nursery habitat for fish species that are the subject of fisheries. In addition, mangrove is closely linked to the productivity of other coastal habitats, such as coral reefs and sea-grass beds. Naylor *et al.* (2000) have estimated that the consequences of loss of mangrove habitat are such that every kilogram of farmed shrimp produced costs 434 g of lost biomass in the coastal fisheries. Clearly, the effects are more widespread than was initially apparent and may even compromise the sustainable utilisation of the coastal resource (for further discussion of the impacts and sustainability of tropical shrimp production, see Kaiser, Chapter 3, Choo, Chapter 4, and Black Chapter 9 of this volume).

8.2.2 Interactions between farmed salmon and wild fish stocks in Europe

In temperate European waters, the farming of Atlantic salmon (*Salmo salar*) raises similarly complex issues both for the capture fishery and the conservation of biodiversity. The economic importance of the salmon farming industry is extremely high, often providing jobs and revenue in remote areas where employment opportunities are scarce. This is reflected in the continuing growth in farmed salmon production from the principal growers in Europe: namely, Norway, Scotland and Ireland. In Norway, for example, the resulting emissions of nitrogen and phosphorus from salmon farms in 1996 were 16,000 and 3500 tonnes, respectively. In many counties, the salmon industry is now the greatest anthropogenic source of these nutrients in coastal waters (Directorate for Nature Management, 1999). Wild salmonids (salmon and sea trout, *Salmo trutta*) are also of immense economic, cultural and nature conservation value. It is the target of a long-standing sport fishery that also generates significant income directly and through related tourism. However, this is threatened by the decline in salmon populations that is being suffered throughout their geographical range. While the cause of this decline is most likely a combination of factors, there is growing evidence that salmon farming may contribute, through interactions between wild and escaped fish and the increased prevalence of salmonid-specific ectoparasites in farming areas (see Pearson and Black, Chapter 1 of this volume).

Escapes, as a result, among other things, of bad weather and predator attacks, are inevitable where large numbers of salmon are intensively grown in sea cages. Studies have revealed substantial numbers of salmon of cultivated origin in rivers and coastal waters. In Norway, catches of salmon on the coast have comprised up to 54% escaped cultivated fish. Lower numbers are seen in the rivers, where the proportion in catches made by anglers ranges from 4–7% (Directorate for Nature Management, 1999). Both genetic and ecological interactions are likely between escaped salmon and wild stocks (Youngson, 1996).

Copepod parasites (predominantly *Lepeophtheirus salmonis*), commonly referred to as sea lice, are one of the most serious problems for the salmon-farming industry. There is also the likelihood that sea lice infestations on salmon farms cause elevated populations of sea lice in adjacent waters. These sea lice are, in turn, an additional stress to wild salmonids leaving or returning to their natal rivers. Studies in Scotland and Ireland have compared salmon-farming and non-salmon-farming areas. In both cases, the heaviest infestations of lice on wild fish and the greater prevalence of sea lice larvae were seen in the salmon-farming areas (Costelloe *et al.*, 1998; MacKenzie *et al.*, 1998).

While the extent of negative interactions between wild and farmed salmonids is not yet fully understood, it is clear that decisions regarding the siting of salmon farms should not be made on the basis of salmon farming considerations alone.

Both salmon farming and the tropical aquaculture/mangrove example above, clearly demonstrate the pitfalls for sectoral regulatory mechanisms that cannot take the wider issues into consideration. The implications for coastal management are clear. There are dangers in compartmentalising the management of any given activity and not recognising the complex interactions that exist between the aquaculture industry and other coastal activities, such as inshore fisheries and the conservation of biodiversity.

8.3 Integrated coastal management (ICM)

Within much the same time-frame as the global expansion in aquaculture, there has been a fundamental shift in international policy on the environment and sustainable development (Costa-Pierce, 1996). There has been a widespread acceptance of the fact that short-term, narrow-sectoral development planning in the coastal zone should be replaced by a more strategic, long-term and holistic approach. Major international initiatives have been pivotal in firmly establishing the concept of ICM during the last decade.

8.3.1 United Nations Conference on Environment and Development 1992

The adoption of Agenda 21 (The Program of Action for Sustainable Development) at the United Nations Conference on Environment and Development (UNCED) in Rio de Janeiro, 1992—the so-called 'Earth Summit'—was a critical point in the development of the principles of ICM. Chapter 17 of Agenda 21 concerns the oceans and coasts, and stresses both their importance in the global life-support system and the positive opportunity for sustainable development they represent. Chapter 17 includes seven major programme areas, the following of which are of particular relevance to coastal aquaculture:

- Integrated management and sustainable development of coastal areas, including Exclusive Economic Zones
- Marine environmental protection
- Sustainable use and conservation of marine living resources under national jurisdiction
- Addressing critical uncertainties for the management of the marine environment and climate change
- Strengthening international (including regional) cooperation and coordination

The concept of integrated management underpins the other sections of Chapter 17, becoming central to the management of coastal zones and ocean areas under national jurisdiction. The content of Chapter 17 stresses the need for integrated policy and decision-making processes and institutions, and suggests that each coastal state should: '*consider establishing, or where necessary strengthening, appropriate coordinating mechanisms (such as a high-level policy planning body) for integrated management and sustainable development of coastal and marine areas, at both the local and national levels*'. Thereafter, actions are suggested which include:

- Preparation of coastal/marine use plans
- Environmental impact assessment and monitoring
- Conservation and restoration of critical habitats
- Inclusion of sectoral programmes (including aquaculture) within an integrated framework

8.3.2 FAO Code of Conduct for Sustainable Fisheries

Further impetus to develop integrated planning frameworks to facilitate sustainable aquaculture development has been provided by the Food and Agriculture Organization of the United Nations (FAO), which has incorporated the Chapter 17 concepts within the Code of Conduct for Sustainable Fisheries (FAO, 1995). The Code aims to take account of the complex interactions that exist between aquaculture, the environment and numerous other activities in the coastal zone, as illustrated by the following relevant statements:

> '*9.1.1 States should establish, maintain and develop an appropriate legal and administrative framework, which facilitates the development of responsible aquaculture*'.
> '*9.1.3 States should produce and regularly update aquaculture development strategies and plans, as required, to ensure that aquaculture development is ecologically sustainable and to allow the rational use of resources shared by aquaculture and other activities*'.

In the aftermath of UNCED, a variety of ICM projects have been initiated, allowing Cicin-Sain *et al.* (1995) to distil a consensus on the set of guidelines for ICM from five international sources: World Bank, World Coast Conference, United Nations Environmental Programme (UNEP), Organization for Economic Cooperation and Development (OECD) and International Union for the Conservation of Nature (IUCN) (see Table 8.1). These guidelines characterize the ICM concept as it has come to be broadly understood.

From amongst these guidelines, there are two fundamental aspects underpinning ICM that are of particular relevance to the sustainable management of aquaculture: the precautionary approach and the ecosystem approach. With respect to the former, the UNCED (1992) declaration stated that:

> '*In order to protect the environment, the precautionary approach shall be widely applied by States according to their capabilities. Where there are threats of serious or irreversible damage, lack of full scientific certainty shall not be used as a reason for postponing cost-effective measures to prevent environmental degradation*'.

Table 8.1 A Consensus set of ICM guidelines (Cicin-Sain et al., 1995)

Purpose of ICM	The aim of ICM is to guide coastal area development in an ecologically sustainable fashion.
Principles	ICM is guided by the Rio Principles, with special emphasis on the principle of intergenerational equity, the precautionary principle and the polluter pays principle. ICM is holistic and interdisciplinary in nature, especially with regard to science and policy.
Functions	ICM strengthens and harmonizes sectoral management in the coastal zone. It preserves and protects the productivity and biological diversity of coastal ecosystems and maintains amenity values. ICM promotes the rational economic development and sustainable utilization of coastal and ocean resources and facilitates conflict resolution in the coastal zone.
Spatial integration	An ICM program embraces all of the coastal and upland areas, the uses of which can affect the coastal waters and the resources therein, and extends seaward to include that part of the coastal ocean which can affect the land of the coastal zone. The ICM program may also include the entire ocean area under national jurisdiction (Exclusive Economic Zone), over which national governments have stewardship responsibilities both under the Law of the Sea Convention and UNCED.
Horizontal and vertical integration	Overcoming the sectoral and intergovernmental fragmentation that exists in today's coastal management efforts is a prime goal of ICM. Institutional mechanisms for effective coordination among various sectors active in the coastal zone and between the various levels of government operating in the coastal zone are fundamental to the strengthening and rationalisation of the coastal management process. From the variety of options available, the coordination and harmonization mechanism must be tailored to fit the unique aspects of each particular national government setting.
The use of science	Given the complexities and uncertainties that exist in the coastal zone, ICM must be built upon the best science (natural and social) available. Techniques such as risk assessments, resource accounting, benefit-cost analysis and outcome-based monitoring should all be built into the ICM process, as appropriate.

Abbreviations: ICM, integrated coastal management; UNCED, United Nations Conference on Environment and Development.

The precautionary approach is clearly applicable to the management of aquaculture, where it remains true to say that there are considerable uncertainties because of the relatively recent history of this industry and our lack of understanding of how it is interacting with many aspects of the environment. Accordingly, the countries bordering the North Sea have recognised that there is the need for further implementation of the precautionary approach as a basis for the management of aquaculture in the North Sea (IMM, 1997).

The ecosystem approach concerns the importance of maintaining ecosystem composition, structure and function and stresses the complexity of different components of the ecosystem and their interdependencies. This 'systems' focus encourages a more holistic viewpoint and can, therefore, be extended beyond the context of ecological systems to the way that we manage the different modes of exploitation in the coastal zone. While currently being developed with particular emphasis on the management of fisheries, the principles of ecosystem management are particularly important in the context of ICM (ICES, 1994).

8.4 Progress towards the management of aquaculture in an ICM environment

In a detailed appraisal of aquaculture and integrated coastal management, Chua (1997) concluded that the primary purpose of ICM is to achieve the following sustainable development goals:

- Ensuring the rational use of natural resources
- Minimising conflicts over use
- Protecting and preserving the functional integrity of coastal ecosystems
- Equitable distribution of the economic benefits derived

Broadly speaking, Chua (1997) defined the process of ICM as identifying management issues that threaten sustainability and addressing these issues collectively through policy, management and technological intervention.

The adoption of these aspects of the integrated and holistic management of coastal areas has profoundly influenced the policy environment in which aquaculture planning is developing around the world. While there has been a proliferation of coastal zone management (CZM) initiatives around the world (Sorensen, 1993), it is still relatively early days. Due to the long-term nature of ICM programmes, there are currently relatively few examples (where aquaculture is a factor) to draw upon. However, it is possible to illustrate some of the progress that has been made and examine how the management and planning of aquaculture is developing in this age of integration.

Not surprisingly, the adoption of integrated coastal management principles has proceeded to varying degrees in different countries. Some countries have moved down the road of vertical integration by developing broad national policies on the oceans; among the best examples are Canada and Australia.

Canada published its Oceans Act in 1997. This Act paves the way for the development of a comprehensive Oceans Strategy, based on the principles of integrated management, shared stewardship, the precautionary approach and sustainable development. It also provides the Minister of Fisheries and Oceans with the authority to coordinate the Can$ 4 billion of federal spending on oceans-related issues.

In Australia, the protection of marine biodiversity and the sustainable use of marine resources have become the subject of national strategies, including the National Strategy for Ecologically Sustainable Development (1992) and the National Strategy for the Conservation of Australia's Biological Diversity (1996). More recently, in 1998, Australia's Oceans Policy was launched to provide a framework for the integration of environmental, economic, social and cultural ocean uses (Commonwealth of Australia, 1998). The Oceans Policy has established the broad principles and management approaches necessary to achieve ecologically sustainable development. At the core of the Oceans Policy is a move towards integrated and ecosystem-based planning and management that will be binding on all relevant agencies. The Policy will be delivered through the development of Regional Marine Plans for areas that are defined on the basis of large marine ecosystems (ANZECC, 1999).

While New Zealand has yet to prepare a comparable policy on the oceans, the Parliamentary Commissioner for the Environment (PCE) has recently published a wide-ranging review and report on coastal issues (PCE, 2000). New Zealand is, in many respects, a leader in the management of its coastal resources, and the management of aquaculture is already carried out within a relatively integrated planning framework (discussed later in this section). However, the report recognises that, amongst other things, *'there is no overarching framework or strategy to guide the many stakeholders towards sustainable management of the marine environment'*.

In Europe, the European Community (EC) has encouraged member states to develop integrated coastal zone management and has recently reported on the lessons gained from a demonstration programme of 35 integrated coastal management projects distributed throughout Europe. With respect to aquaculture, the report encourages, amongst other things, an ecosystems approach to aquaculture management and recognises that the viability of aquaculture is dependent on the quality and availability of environmental resources. In addition, the report concludes that, in order to advance the objectives of ICM (with a sustainable development perspective), there is a need for fisheries/aquaculture policy to be more effectively articulated with environmental policy (European Commission, 1999).

While relatively few countries have developed a fully-integrated coastal or oceans policy, there is a general trend towards greater horizontal integration and movement away from wholly sectoral management. Here, existing legislation, management mechanisms and institutions relevant to aquaculture have been modified or expanded to allow for the integration of, for example, environmental objectives and/or interdependencies with other sectors. This enhancement of sectoral mechanisms has often been a reflection of the fact that aquaculture has tended to develop and expand rapidly where existing legislation to manage the development of the industry was lacking.

In many countries, the existing planning regimes have a distinctly terrestrial bias and are frequently not intended to deal with developments that occur beyond the shore. In Scotland, for example, the principal planning tool is the Town and Country Planning (Scotland) Act 1972. However, this effectively excludes aquaculture, as it only gives local planning authorities jurisdiction down to the low water mark. Consequently, another route has had to be followed. The seabed of the UK is owned by the Crown and managed by the Crown Estate, with any resulting revenue going to the Treasury. Development control for aquaculture has been achieved via the allocation of seabed leases issued by the Crown Estate, which have effectively resulted in a quasi-planning role for the Crown Estate with respect to aquaculture. The Crown Estate undertook to accommodate the views of other coastal interests through the introduction, in 1986, of an interagency consultation exercise for each new application. However, the potential conflict of interest that the Crown Estate had as both planning authority and landlord, and a perceived lack of public accountability, attracted criticism (Ross, 1997; SOAEFD, 1997).

In 1998, the UK government therefore decided to bring aquaculture under the jurisdiction of local planning authorities. Although the legislation required has not yet been completed, it is anticipated that this new mechanism will enable a more integrated approach to aquaculture development. To facilitate the process, national locational siting guidelines for aquaculture have been published for the first time in the UK (Scottish Executive, 1999). This guidance provides a broad siting policy in the context of UK and international legislative obligations. Towards the goal of greater integration, the guidelines encourage the planning authorities to develop the fine detail of aquaculture siting policy in consultation with other users of coastal resources, through the publication of detailed aquaculture framework plans.

The experience in the aquacultural regions of Australia has been somewhat similar. The aquaculture industry has expanded rapidly over the past couple of decades and, in areas such as Tasmania and New South Wales, the demand for growth was outstripping the supply of available coastal sites by the early 1990s. It became clear that the existing legislation was inadequate to properly manage growth in the industry as well as account for other coastal zone users and environmental concerns (Scarsbrick, 1996; Preston *et al.*, 1997; Carvalho and Clarke, 1998). This has led to the widespread introduction of aquaculture and marine farming industry development plans, which are subject to extensive public consultation. The Tasmania Marine Farming Planning Act 1995, for example, provides for the preparation of such plans that include:

- A sectoral environmental impact statement
- Identification of marine farming zones which are similar in legal status to those on land
- Management controls and operational constraints affecting the zones, including monitoring (Mcloughlin, 1997)

New Zealand, by contrast, already has a relatively integrated planning framework for aquaculture. The Resource Management Act (1991) was introduced to replace over 20 major statutes covering various aspects of marine resource exploitation and use. Under the Resource Management Act a national Coastal Policy Statement has set down the broad policy and principles. Thereafter, regional councils are responsible for the sustainable management of their coastal marine areas, that is, the foreshore and the seabed of the territorial sea, implementing the national policy via Regional Coastal Plans. Aquaculture is one of a range of activities that may be granted exclusive occupation of coastal space, and the Regional Coastal Plans are developed in consultation with other stakeholders to ensure that the effects of aquaculture (and other activities) are avoided, remedied or mitigated (PCE, 2000).

8.5 The implications of marine protected areas for aquaculture planning

The maintenance of a high quality environment is widely acknowledged as the essential prerequisite for sustainable aquaculture. As mentioned above, the consensus of opinion on ICM favours an ecosystem-based approach to coastal management, with the aim of conserving marine biodiversity and maintaining ecological processes. It is increasingly recognised that this is best achieved through strategic regional planning that provides for the establishment and effective management of a representative system of marine protected areas (MPAs) and the complementary sustainable management of adjoining waters. The term MPA, in fact, covers a wide variety of designations, but the International Union for the Conservation of Nature (IUCN) definition of a marine protected area has been broadly accepted, as:

> '*An area of land and/or sea especially dedicated to the protection and maintenance of biological diversity, and of natural and associated cultural resources, and managed through legal or other effective means*' (IUCN, 1994).

The establishment and management of MPAs is likely to become an increasingly common feature of ICM (Gubbay, 1995) and, as such, will have considerable implications for the development of aquaculture management. On the one hand, MPAs have tended to be viewed with some suspicion by the industry, for perhaps denying access to certain areas and thus acting as a source of competition for coastal space. On the other hand, within the context of ICM, MPAs have a crucial role to play in the maintenance of the environmental quality on which a sustainable aquaculture industry depends. Therefore, it is worth giving some consideration to the management of aquaculture in the context of MPAs.

Firstly, the IUCN have defined several categories of MPA (Table 8.2), which cover a whole range, from those established to give very strict protection to particular features to those where the emphasis is on ensuring sustainable use. In fact, the number of sites that are covered by the strictest protection tend to be few in number and limited in size. For example, New Zealand is one of the relatively few countries that have established such strict reserves.

Table 8.2 Summary of IUCN Guidelines for Protected Area Management Categories (IUCN, 1994)

Category Ia	Strict nature reserve	Protected area managed mainly for science
Category Ib	Wilderness area	Protected area managed mainly for wilderness protection
Category II	National park	Protected area managed mainly for ecosystem conservation and recreation
Category III	Natural monument	Protected area managed for conservation of specific natural features
Category IV	Habitat/species management area	Protected area managed mainly for conservation through management intervention
Category V	Protected landscape/seascape	Protected areas managed mainly for landscape/seascape conservation and recreation
Category VI	Managed resource protected areas	Protected areas managed mainly for the sustainable use of natural ecosystems

Abbreviation: IUCN, International Union for the Conservation of Nature.

However, it has only 16 Marine Reserves, which have been established for the purpose of scientific study only, under the Marine Reserves Act 1971. In taking a strategic perspective on coastal management, it makes sense to maintain at least some areas free from development, for example, to provide reference points for environmental monitoring. In addition, MPAs have been considered as a way of implementing the precautionary approach, where they may serve as buffers against some management miscalculations and unforeseen conditions (Gubbay, 1995).

The more common experience, however, appears to be with MPAs which fall within the IUCN Categories IV and VI, that is, where MPAs are not exclusion zones but are being used to provide a management focus on the conservation of biodiversity and/or implement the ecosystem approach to coastal management. Canada has committed itself to developing a system of MPAs under the National Marine Conservation Areas (NMCA) Policy (1994) and the National Marine Parks Policy (1986). Delivery of the Canadian MPA system has been set out in the National Marine Conservation Areas System Plan (Mercier and Mondor, 1995). The coast (including the Great Lakes) has been split into 29 distinct 'natural regions', within which a representative selection of areas will be designated as MPAs and which will be managed primarily for sustainable use. However, each MPA may, where appropriate, contain smaller areas of higher protection for features of particular note or sensitivity. In this way, aquaculture operations may be influenced by the existence of an MPA but will not necessarily be excluded.

Recently, Australia has also launched its Strategic Plan of Action for the National Representative System of Marine Protected Areas (NRSMPA) (ANZECC, 1999). In a similar manner to Canada, an important feature of this scheme is the fact that the network of MPAs is being set within an ecosystem-based regionalisation of Australian coastal waters. A total of 60 different regions on the continental shelf have been identified on the basis of distinct biological and physical characteristics. This ecosystem-based (rather than political) regionalisation provides the first layer in a broad ecological framework for planning and management. The primary goal of the NRSMPA is to establish and manage a comprehensive and representative system of MPAs to contribute to the long-term ecological viability of marine and estuarine systems, to maintain ecological processes and systems, and to protect biological diversity at all levels. Secondary goals include: promoting the development of MPAs within the framework of integrated ecosystem management; and providing a formal management framework for a broad spectrum of human activities (including fisheries and aquaculture). Therefore, it is envisaged that within this system the MPAs will range from highly-protected areas to sustainable, multiple-use areas accommodating human enterprises.

Similarly, the European Community is currently in the process of establishing the Natura 2000 network of protected areas for the conservation of biodiversity (SNH *et al.*, 1997). Two types of European Marine Site are involved:

- Special Areas of Conservation (SAC) under Council Directive 92/43/EEC (the Habitats Directive), which may be designated where the site supports certain rare, endangered or vulnerable species of plants or animals (other than birds) or if the area supports outstanding examples of habitats characteristic of the region
- Special Protection Areas (SPAs) under Council Directive 79//409/EEC (the Birds Directive), which may be designated if an area supports significant numbers of wild birds and/or the habitats which are necessary for some aspect of the birds' life-cycle

While the Directives provide the basic principles, each country of the European Community is implementing them via its own domestic legislation. It is implicit among the major requirements of the directives that the management of European Marine Sites should be fully integrated and should take account of the economic, cultural, social and recreational needs of the local people. In the UK, this means that the management schemes being developed involve the participation of the relevant agencies and stakeholders to enable the conservation of important features to be achieved, whilst facilitating the continued sustainable use of the area (Baxter and Davison, 1999). This inclusive and participatory approach to the management of MPAs is very much in keeping with the principles of ICM (Cicin-Sain *et al.*, 1995), and is also being seen in the approach taken in Canada and New Zealand, where widespread consultation and involvement of stakeholders is a feature both of MPAs and aquaculture area development plans (Mercier and Mondor, 1995; Mcloughlin, 1997; Scarsbrick, 1997; ANZECC, 1999).

One further common factor that is emerging in schemes being developed for the management of coastal resources and within MPAs is the use of zoning.

Chua (1997) considered the allocation and management of natural resources for specific users, that is, functional zones for different activities, to be a significant feature of ICM. Indeed, such zonation has been advocated and increasingly employed in the management both of MPAs (Laffoley, 1995; Baxter and Davison, 1999) and marine farming (Mcloughlin, 1997) in various locations. Within the context of a well-developed and integrated management scheme, zonation can provide a powerful tool to reconcile the exploitation of resources with the protection of biodiversity.

Just as aquaculture cannot be managed in isolation from other activities and the environment, the conservation of biodiversity cannot be achieved through the designation and management of MPAs alone. MPAs alone cannot provide sufficient protection for the environment and cannot exist in isolation from all critical impacts. Their constituent habitats and communities are strongly influenced by the highly variable conditions of the water masses that constantly flow through them. An MPA boundary offers no protection from external threats, such as contamination from pollutants. Moreover, to a much greater extent than is seen in terrestrial systems, the scales of processes such as population replenishment are often too large for scattered MPAs to encompass. Therefore, if consideration is not given to biodiversity conservation in the seas outwith MPAs, the effectiveness of MPAs themselves will be compromised. This demonstrates the requirement to build biodiversity conservation goals into a broader, comprehensive management framework, such as that provided by ICM. Then, even in areas where conservation of biodiversity is not the primary management focus, environmental and biodiversity objectives are not forsaken (Gubbay, 1995; Allison *et al.*, 1998; Ray, 1999).

8.6 Practical considerations for managing aquaculture within an ICM environment

8.6.1 Estimating environmental capacity

The concept of environmental (or carrying) capacity is of fundamental importance within ICM. In relation to aquaculture, this is essentially the level of production that a given area (or water body) can accommodate without an adverse impact on the environment or other resource users. GESAMP (1986) defined environmental capacity as: '*a property of the environment and its ability to accommodate a particular activity or rate of an activity... without unacceptable impact*'.

Recent progress has been made in the field of modelling the dispersion and deposition of waste products from aquaculture. This provides an important step towards determining carrying capacity and is the subject of discussion elsewhere in this book (Silver and Cromey, Chapter 7). However, many existing regulatory mechanisms take insufficient account of carrying capacity by considering developments on a case-by-case basis only. It is necessary, instead, to take the cumulative and combined nature of impacts into account and dispersion/deposition models will need to accommodate this requirement.

Furthermore, environmental capacity tends to be considered within the context of biological and chemical parameters, such as:

- *'The rate at which nutrients are added without triggering eutrophication*
- *The rate of organic flux to the benthos without major disruption to natural benthic processes, or*
- *The rate of dissolved oxygen depletion that can be accommodated without mortality of the indigenous biota'* (GESAMP, 1996)

In addition to this biological/chemical perspective on environmental capacity, it will be necessary to include other parameters, such as landscape, that is, where landscapes of different character vary in their ability to accommodate multiple developments without significant visual impact.

8.6.2 Mapping coastal resources

To implement ICM successfully, it is also necessary to have a knowledge of the natural resources of the managed area, including the distribution of seabed habitats and their associated plants and animal communities. In most countries, however, even this seemingly basic information is often lacking because of the relative expense of gathering it. Fortunately, considerable progress has recently been made in mapping the broad-scale distribution of coastal habitats by employing relatively cost-effective remote survey methodologies, including sonar and video (e.g. Sotheran *et al.*, 1997; Davies *et al.*, 1998; Pinn *et al.*, 1998; Downie *et al.*, 1999; Foster-Smith *et al.*, 2000). Because this information is analysed and disseminated within a geographic information system (GIS) environment, it offers coastal planners the opportunity to easily integrate the natural resource characteristics of a given area with the results of aquaculture dispersion/deposition modelling, or data on the distribution and extent of other coastal activities.

8.7 Conclusion

In the preceding pages an attempt is made to construct a broad overview of the background to ICM and its current status, with particular reference to the management and regulation of aquaculture. Although the broadly accepted concept of ICM has now been circulating for more than a decade, and despite the fact that a multitude of ICM initiatives have been started, their long-term nature means that few have matured sufficiently to provide results to learn from. While some countries have embraced ICM and indulged in fairly dramatic restructuring of coastal management, others have only tinkered with their existing mechanisms. It is tempting, therefore, to speculate about the most valuable facets of any future ICM scheme.

Consequently, the desirable management framework should:

- Carefully match levels of exploitation and usage to the available resource.
- Provide clear goals for both the management of the environment and the relevant resources.
- Be based on a sound knowledge of the resource base.
- Encourage a long-term perspective on resource use and discourage short-term exploitation.
- Promote rapid identification of adverse interactions between activities.
- Help to avoid or reduce conflict amongst stakeholders.
- Make relevant information readily available to managers and users.
- Provide a balance between 'top-down' agency regulation and 'bottom-up' participatory management schemes which gives meaningful involvement to all of the key stakeholders.
- Provide for a network of protected areas which work to conserve bio-diversity and maintain a high quality environment for the benefit of all stakeholders.
- Provide for the maintenance of ecosystem structure and function throughout the managed area.
- Encourage a precautionary approach to be taken where there are uncertainties.

It is almost certain that global aquaculture production will continue to grow in the immediate future. It is also likely that the similarly upward trend in other coastal activities will continue. The potential for conflict, over-exploitation of resources and continued pressure on the environment does not look set to diminish. With so much at stake, the value of an integrated approach to the management of coastal resources as a way out of a potentially downward spiral becomes vividly clear.

References

Allison, G.W., Lubchenco, J. and Carr, M.H. (1998) Marine reserves are necessary but not sufficient for marine conservation. *Ecological Applications*, **8** s79-s92.

ANZECC-Australian and New Zealand Environment and Conservation Council Task Force on Marine Protected Areas. (1999) Strategic Plan of Action for the National Representative System of Marine Protected Areas: A Guide for Action by Australian Governments. Environment Australia, Canberra.

Bardach, J.E. (1997) Aquaculture, pollution and biodiversity, in *Sustainable Aquaculture* (ed. J.E. Bardach), John Wiley and Sons, New York, pp. 87-99.

Baxter, J.B. and Davison, A. (1999) Management of Scotland's marine and coastal special areas of conservation, in *Scotland's Living Coastline* (eds. J.M. Baxter, K. Duncan, S. Atkins and G. Lees), Stationary Office, London, pp. 111-122.

Carvalho, P. and Clarke, B. (1998) Ecological sustainability of the South Australian coastal aquaculture management policies. *Coastal Management*, **26** 281-290.

Chua, T.-E. (1997) Sustainable aquaculture and integrated coastal zone management, in *Sustainable Aquaculture* (ed. J.E. Bardach), John Wiley and Sons, New York, pp. 177-199.

Chua, T.-E. and Graces, L.R. (1994) Marine living resources management in the Asian region: lessons learned and the integrated management approach. *Hydrobiologia*, **285** 257-270.

Cicin-Sain, B., Knecht, R.W. and Fisk, G.W. (1995) Growth in capacity for integrated coastal management since UNCED: an international perspective. *Ocean and Coastal Management*, **29** 93-123.

Commonwealth of Australia (1998) Australia's Oceans Policy, Environment Australia, Canberra.

Costa-Pierce, B.A. (1996) Environmental impacts of nutrients from aquaculture: towards the evolution of sustainable aquaculture systems, in *Aquaculture And Water Resource Management* (eds. D.J. Baird, M.C.M. Beveridge, L.A. Kelly and J.F. Muir), Blackwell Science, pp. 81-113.

Costelloe, M., Costelloe, J., Coghlan, N., O'Donohue, G. and O'Connor, B. (1998). Distribution of the larval stages of *Lepeophtheirus salmonis* in three bays on the west coast of Ireland. *ICES Journal of Marine Science*, **55** 181-187.

Davies, J., Forster-Smith, R.L., Sotheran, I.S., Walton, R. and Donnan, D. (1998) Post-processing acoustic ground discrimination data for detailed biological resource mapping, in *Proceedings of the Canadian Hydrographic Conference: Turning Data into Dollars*. Victoria, Canada, pp. 432-444.

Directorate for Nature Management (1999) Environmental objectives for Norwegian aquaculture: new environmental objectives for 1998-2000. DN-rapport 1999-1b.

Downie, A.J., Donnan, D.W. and Davison, A. (1999) A review of Scottish Natural Heritage's work in subtidal marine biotope mapping using remote sensing. *International Journal of Remote Sensing*, **20** 585-592.

European Commission (1999) Lessons from the European Commission's Demonstration Programme on Integrated Coastal Zone Management (ICZM), Office for Official Publications of the European Communities, Luxembourg.

FAO (1995) Code of Conduct for Responsible Fisheries. Food and Agriculture Organisation, Rome.

FAO (1999a) Aquaculture Production Statistics 1988-1997, Food and Agriculture Organisation, Rome.

FAO (1999b) The State of World Fisheries and Aquaculture 1998, Food and Agriculture Organisation, Rome.

Foster-Smith, R.L., Davies, J. and Sotheran, I. (2000) Broad scale remote survey and mapping of sublittoral habitats and biota: technical report of the Broad Scale Mapping Project. Scottish Natural Heritage Research, Survey and Monitoring Report No. 167, Edinburgh.

GESAMP (IMO/FAO/Unesco-IOC, WMO/WHO/IAEA/UN/UNEP Joint Group of Experts on the Scientific Aspects of Marine Environmental Protection) (1986) Environmental capacity: an approach to marine pollution prevention. *GESAMP Reports and Studies*, **30**, IMO, London, 49 pp.

GESAMP (IMO/FAO/Unesco-IOC, WMO/WHO/IAEA/UN/UNEP Joint Group of Experts on the Scientific Aspects of Marine Environmental Protection) (1996) Monitoring the ecological effects of coastal aquaculture wastes, *GESAMP Reports and Studies*, **57**, IMO, London, 38 pp.

Gubbay, S. (1995) Marine Protected Areas: principles and techniques for management. Chapman and Hall, London.

ICES (1994) Report of the Study Group on Ecosystems: Effects of Fishing Activities. ICES CM 1994/Assess/Env.1, Copenhagen.

IMM (1997) Statement of Conclusions. Intermediate Ministerial Meeting on the Integration of Fisheries and Environmental Issues, 13-14 March 1997, Bergen, Norway. Ministry of Environment, Norway.

IUCN (1994) Guidelines for Protected Area Management Categories. Commission on National Parks and Protected Areas with the assistance of the World Conservation Monitoring Centre, Gland, Switzerland.

Laffoley, D. (1995) Techniques for managing marine protected areas: zoning, in *Marine Protected Areas: Principles and Techniques for Management* (ed. S. Gubbay), Chapman and Hall, London.

MacKenzie, K., Longshaw, M., Begg, G.S. and McVicar, A.H. (1998) Sea lice (Copepoda: Caligidae) on wild sea trout (*Salmon trutta*) in Scotland. *ICES Journal of Marine Science*, **55** 151-162.

Mcloughlin, R. (1997) Coastal planning and aquaculture: marine farming development plans for Tasmania, in *Developing and Sustaining World Fisheries Resources: The State of Science and Management*, CSIRO, Collingwood (Australia), pp. 455-461.

Mercier, F.M. and Mondor, C.A. (1995) Sea to Sea to Sea, Canada's Marine Conservation Areas System Plan, Parks Canada, Quebec.

Naylor, R.L., Goldburg, R.J., Primavera, J.H., Kautsky, N., Beveridge, M.C.M., Clay, J., Folke, C., Lubchenco, J., Mooney, H. and Troell, M. (2000) Effect of aquaculture on world fish supplies. *Nature*, **405** 1017-1024.

PCE—Parliamentary Commissioner for the Environment (2000) Setting Course for a Sustainable Future: The Management of New Zealand's Marine Environment. Wellington, New Zealand.

Phillips, M.J., Kweilin, M.C. and Beveridge, M. (1993) Shrimp culture and the environment: lessons from the world's rapidly expanding warm water aquaculture sector, in *Conference Proceedings 31* (eds. R.S.V. Pullin, H. Rosenthal and J.L. Maclean), ICLARM, Manila, Philippines, pp. 171-197.

Pinn, E.H., Robertson, M.R., Shand, C.W. and Armstrong, F. (1998) Broadscale benthic community analysis in the Greater Minch (Scottish West Coast) using remote and non-destructive techniques. *International Journal of Remote Sensing*, **19** 3030-3054.

Preston, N., Macleod, I., Rothlisberg, P. and Long, B. (1997) Environmentally sustainable aquaculture production: an Australian perspective, in *Developing and Sustaining World Fisheries Resources: The State of Science and Management*. CSIRO, Collingwood (Australia) pp. 471-477.

Ray, G.C. (1999) Coastal-marine protected areas: agonies of choice. *Aquatic Conservation: Marine and Freshwater Ecosystems*, **9** 607-614.

Ross, A. (1997) Leaping in the dark: a review of the environmental impacts of marine salmon farming in Scotland and proposals for change. Scottish Wildlife and Countryside Link, Perth.

Scarsbrick, J. (1997) Towards sustainability of the aquaculture industry in Jervis Bay, in *Developing and Sustaining World Fisheries Resources: The State of Science and Management*, CSIRO, Collingwood (Australia) pp. 520-524.

Scottish Executive (1999) Locational guidelines for the authorisation of marine fish farms in Scottish waters. Scottish Executive, Edinburgh.

SNH, EN, EHS(DOE NI), CCW and JNCC (1997) Natura 2000: European marine sites, an introduction to management. SNH, Perth.

SOAEFD (1997) *Report of the Scottish Salmon Strategy Task Force* (eds. A. McLay and K. Gordon-Rogers), Scottish Office Agriculture, Environment and Fisheries Department, Edinburgh.

Sorensen, J. (1993) The international proliferation of integrated coastal zone management efforts. *Ocean and Coastal Management*, **21** 45-80.

Sotheran, I.S., Forster-Smith, R.L. and Davies, J. (1997) Mapping of marine benthic habitats using image-processing techniques within a Raster-based Geographic Information System. *Estuarine, Coastal and Shelf Science*, **44** (Suppl. A) 25-31.

Youngson, A. (1996) Escaped farmed salmon in western Scotland, in *Aquaculture and Sealochs* (ed. K.D. Black), SAMS, Oban, pp. 89-93.

9 Sustainability of aquaculture
K.D. Black

9.1 Introduction

Sustainability is where environmental effects meet socioeconomics and markets. There are various definitions of sustainability but, in essence, these condense around concepts relating to stewardship. It is perfectly acceptable to exploit the environment, provided that this is done in a way which:

(a) does not significantly interfere with the commercial or amenity use of that environment by others (although those others must also utilise the environment in a sustainable manner to preserve equity);
(b) does not reduce the scope for future users to benefit from the environmental resource; and
(c) does not significantly alter or diminish environmental quality and biodiversity *per se*.

The latter point acknowledges the fact that an environment or ecosystem may have an inherent value, without necessarily being currently utilised or exploited.

In the present volume, the sustainability of aquaculture is a recurring theme addressed by each of the chapter authors, which indicates the importance of this concept across all fields of aquaculture and the increasing necessity to justify all human activities in these terms. Many environmentally-minded bodies that express concerns on a whole range of issues have, especially in the last few years, turned their attention to aquaculture.

Intensive aquaculture, particularly those forms dependent on fishmeal for the feeding of carnivorous species, is the form of culture most questioned in terms of sustainability (Folke *et al.*, 1994, 1997; Naylor *et al.*, 1998, 2000) and is, therefore, the main area addressed by this chapter. The culture of seaweeds and filter-feeding shellfish leads to a net reduction of nutrients and energy from the ecosystem and, unless practised in some extreme sense, is unlikely to have sustainability consequences beyond the purely local scale. These forms of aquaculture are, however, commonly spread over quite large areas of relatively important habitat in the intertidal zone or on the coastal fringe, and may seriously interfere or conflict with other users (for a discussion of Coastal Zone Management, see Donnan, Chapter 8 in this volume). The culture of herbivorous fish species, with only limited additions of external energy to the system, is generally regarded as being inherently sustainable although, again, local effects on other interests may be significant, for example, when water is a shared resource. Therefore, even where the culture type tends towards sustainability, management of resource use will always be required in order to ensure the protection of all legitimate users.

The concept of sustainability is steeped in the future. Clearly aquaculture practices that are unsustainable will not be sustained. The question is, will they do serious and/or irreversible environmental or socioeconomic damage prior to their decline? In the following sections, the broad impact types are discussed in the context of sustainability. The chapter concludes with the author's personal prediction of how aquaculture will develop in the future.

9.2 Assessing sustainability

Given the range of factors that might influence the sustainability of an activity, assessing sustainability objectively is not easy. The ecological footprint approach (for a review, see Folke *et al.*, 1998) is introduced by Kelly in Chapter 2 of this volume. This method attempts to determine the area that must be used to provide ecological services for a particular activity. For example, if you know the volume of feed required for a fish farm over a defined period of time, the area of ocean that might be used to provide that quantity of fish can be estimated using mean production rates. Another example is the required oxygen supply: if you can estimate the oxygen requirements of a culture activity then, from knowledge of mean photosynthesis rates for an environment (and mean gas exchange rates with the atmosphere), the area of water required to provide this oxygen can be estimated. The sum of all the areas required to supply environmental goods and services constitutes the ecological footprint. There are, however, a number of problems with this approach, which its proponents accept:

1) The areas required for ecological services may overlap, or even interact, within the system or externally. For example, the area used for assimilation of waste nutrients may also provide the oxygen supply. In this case, the areas could not be added together. In addition, there may be areas where exclusive use of the resource may be required by the farmer and other areas where a shared use would be the norm. Therefore, the ecological footprints of several activities will normally overlap in a complex way. In some cases, the farms use of ecological services could be of benefit to other users, for example, to anglers catching wild fish, which may be plentiful near fish cages in marine and freshwater.
2) The areas may be very distinct spatially. For example, the area required for fishmeal production could be anywhere where such a fishery exists, but the area required for local services, for example, for oxygen production, must be adjacent to the farm. When the sum is done, we are

therefore adding areas that may be large distances apart and may have inherently different values. The impact that an activity has can only be truly determined in context. For example, a farm positioned near a diverse reef is likely to attract more opposition than one positioned over a ubiquitous, muddy sediment. That is because the area used to assimilate waste for the reef site is inherently much more valuable in terms of biodiversity than that of the muddy site. A simple comparison of footprints might lead to the conclusion that each site used equal resources, as both would be dominated by the large area required to produce fishmeal.

It would be useful if the ecological footprint approach could be modified to work in an absolute sense, by giving each of the summed areas some sort of multiplier in order to reflect value. This has not, to the author's knowledge, been attempted.

These criticisms accepted, the method is still a good way to get an idea of the relative differences between culture types, as shown in Table 9.1. Although, in the author's view, the absolute figures should be treated with caution, the relative order appears correct and largely reflects the dependence on capture fisheries. The semi-intensive farm in Table 9.1 has a particularly low support area requirement because it is proposed that all external feeding comes from locally available fishery by-products.

The proponents of this method used it to assess the support capacity of the Columbian shrimp farming industry and concluded that that environment is close to utilising its full capacity. This is a very interesting result as: a) objective methods for determining regional carrying capacity are lacking; and b) the authors asserted that this conclusion was not immediately apparent from a visual assessment. This may explain why other shrimp farming regions have experienced boom and bust scenarios, as it may not have been obvious that

Table 9.1 Areas of ecological footprint for various culture types

Culture type	Area of ecosystem support required per unit farm area	Source
Baltic salmon farms	40000–50000	Folke, 1988 (cited in Folke and Kautsky, 1992)
Intensive tilapia cage culture in Lake Kariba, Zimbabwe	10900	Berg et al., 1996 (revised figures from Kautsky et al., 1997)
Shrimp pond culture, Columbia	35–190	Kautsky et al., 1997
Baltic mussel longline culture	20	Folk and Kautsky, 1989 (cited in Kautsky et al., 1997)
Semi-intensive pond farming in Lake Kariba, Zimbabwe	1	Berg et al., 1996

regional carrying capacity was being exceeded. This method could not, as it stands, be used to estimate regional carrying capacity for intensive cage culture, unless the fishmeal source was within the region (as was proposed for the intensive tilapia farm in Table 9.1). This is because the oceanic support area required for fishmeal production swamps the results, although it could be used for particular components, for example, to estimate capacity in terms of oxygen supply.

9.3 Impact types and sustainability

9.3.1 Wild seed collection

The collection of wild seed is widespread in aquaculture and operates on a range of scales. Where spat are collected passively for bivalve culture, on a small scale in comparison to the total spat abundance, there is likely to be comparatively little impact on sustainability. On the other hand, broodstock for some species are either difficult or expensive to maintain and seed supplies are harvested from the wild either as larvae, fry or berried females. Capture methods are often indiscriminate and large numbers of by-catch are killed in the process (Bagarinao and Taki, 1986; Islam *et al.*, 1996, both cited from Naylor *et al.*, 2000), which may have significant implications for fisheries and ecosystems more generally. This process may also lead to scarcity of local seed, necessitating the importation of foreign stock with the attendant risks of disease and parasite transfer.

9.3.2 Feedstocks

Fishmeal and fish oil are key constituents of pelleted diets for intensive production of carnivorous species. The effects of aquaculture on world fish supplies have been reviewed (Naylor *et al.*, 2000, 91 references cited) but the essential facts are presented here. World capture fishery production has plateaued (against a background of increasing fishing effort) at around 95 Mt, of which around 30 Mt is currently used for the production of fish meal. As fisheries for large, high-value carnivorous species become increasingly fully- or over-exploited, the proportion of smaller pelagic species in the catch increases. Some of these species could be used directly for human consumption. Around a third of the fishmeal product is currently used in aquaculture diets, with the remainder used for livestock, including pigs and poultry, but this proportion increases annually as aquaculture production continues to expand.

Feed conversion ratios (FCRs) are continuously improving as feeds are increasingly tailored to the needs of the cultured species and as feed wastage is reduced due to economic and, to some extent, environmental pressures. This is particularly the case in the highly-regulated northern salmon industries, where FCRs now approach 1:1, that is, one kilogram fish product (whole fish, wet weight) per kilogram feed (compound feed, typically around 10% water). Despite these improvements, it still requires 2–5 kg of wild fish to produce 1 kg of fishmeal-fed cultured fish.

This apparently wasteful use of the fish resource is, to some extent, mitigated by the fact that only some of the fish used for fishmeal production are fit, or appropriate, for human consumption. In addition, the conversion of low commercial value small pelagic fish into high value carnivorous species is probably more efficient in culture than in the wild, where there is likely to be a much lower transmission of energy between trophic levels. It is also possible that competition with cheap, farmed fish may reduce fishing effort, thus protecting endangered stocks, although this is offset by the fact that it may be possible to sell the wild product at an increased price to a sophisticated market.

There can be no argument that the availability of fishmeal limits the sustainable growth of those forms of aquaculture that depend on this resource. Furthermore, as pressure on fish stocks for fishmeal production grows, the vulnerability of an aquaculture industry dependent on these stocks is high. It is recognised that climatic oscillations, such as El Ninõ, can have a major effect on fishery production. A drastic reduction in fishmeal supply would increase prices of feed, so that industries where profits are often low, such as salmon culture in Europe, could collapse.

There is good evidence that cultured fish do best on diets closely approximating those normally experienced in the wild at each life stage (Bell, 1998). There are, however, many examples where marine-derived fish meal is used unnecessarily to feed freshwater fish, for example, salmon juveniles (Bell *et al.*, 1997). In addition, fishmeal is increasingly being added to the feed of herbivorous and omnivorous fish being cultured in high density.

Substitutes for fishmeal protein (e.g. Wm Kissel *et al.*, 2000) and marine fish oils (for a review of fish lipid nutrition, see Bell, 1998) are continuously being sought and some progress is being made. The basic problem in using vegetable substitutes is their lack of essential amino acids (e.g. lycine and methionine) and essential fatty acids, particularly eicosapentaenoic acid and docosahexaenoic acid. The present author believes that the production of fishmeal/fish oil substitutes from agricultural crops is a vital target for research, including transgenic research, provided that the environmental safety of such crop production can be ensured and that human health concerns are met. At the very least, it is likely that diets could be reformulated to replace part of the fish-derived ingredients with vegetable products.

9.3.3 Soluble wastes and water resources

In many areas of the world, freshwater resources are under increasing pressure. Certain types of aquaculture are major water users and it is likely that, in some areas, water resources may become a crucial limiting factor. Thus, the continuing expansion of aquaculture depends on improvements in water conservation and recycling (this aspect is discussed by Waller in Chapter 5 of the present volume).

Waste waters from aquaculture can contain high concentrations of organic and inorganic nutrients. In intensive culture with fishmeal-based feeds, it is clear that aquaculture transfers nutrients from the ocean to the coastal zone, where slow exchange with oceans may lead to local increases in nutrient concentration and thence to eutrophication. Eutrophication has several possible definitions, but that of Nixon (1995) seems most satisfactory: '*Eutrophication (noun)—an increase in the rate of supply of organic matter to an ecosystem*'.

Thus, no difference is implied between imported (allochthonus) carbon (e.g. from the fish farm or from production outside the system) and production of (autochthonus) carbon within the system (e.g. utilising nutrients from run-off or from aquaculture effluents). It is the effect produced by the addition of nutrients, that is, increase of the rate of carbon supply, which is of concern rather than the input of these nutrients *per se*. Whether eutrophication will occur as a consequence of nutrient addition will depend entirely on the state of the receiving system, which may vary spatially, over short time-scales or seasonally, depending on which factors limit primary production. Secondly, whether eutrophication is of significant consequence to the ecosystem will depend on a number of factors, including its intensity, its duration and, importantly, the trophic status of the receiving environment. The trophic status categories given by Nixon (1995) for marine systems are presented in Table 9.2.

Thus, were there to be an increase in supply rate of carbon of say $20 \text{ g m}^{-2} \text{ yr}^{-1}$ to an area as a consequence of nutrient inputs from fish farming, this would have a much higher significance in an oligotrophic environment, such as the eastern Mediterranean, than in a mesotrophic area, such as the European Atlantic coast. It is possible, however, that small changes to input might result in large changes to the ecosystem—if an ecosystem is already under pressure, a small addition may result in system collapse.

It has been argued that the input of nutrients from marine cage culture in the coastal zone has reached unsustainable levels in many areas, as severe ecosystem perturbation is now occurring or is likely. Others have argued that aquaculture contributes only a fraction of the total nutrients added to the coastal area, and that the system is well below carrying capacity. Part of the problem comes

Table 9.2 Trophic classification of marine waters (Nixon, 1995)

Trophic status	Organic carbon supply $\text{g m}^{-2} \text{ yr}^{-1}$
Oligotrophic	<100
Mesotrophic	100–300
Eutrophic	301–500
Hypertrophic	>500

from the difficulty in making an accurate assessment of nutrient inputs from a wide range of sources. While it is relatively easy to calculate budgets from fish farming, nutrients which enter the environment from natural water-courses are hard to measure accurately as they will be dependent on freshwater input, which may vary with season and interannually. It is also particularly difficult to make accurate estimations of atmospheric inputs, which may be significant (Enell, 1995). Further research to determine total regional nutrient budgets is urgently required, so that the proportion attributable to intensive aquaculture can be accurately assessed.

What is crucial, however, is not simply the input of nutrients but the degree to which these nutrients are assimilated by biota, thus increasing biomass and so carbon supply to the environment. If this assimilation takes place in such a way as to contribute to the nutrients in the oceanic nutrient pool (through biogeochemical recycling) rather than to biomass held in the coastal zone, it is unlikely that it will have any measurable effect, as it would only be a minor contribution to the global nutrient budget.

Nutrient levels from marine cage culture have been compared to sewage effluents in terms of human population equivalents (PEs). In these terms, the nutrients generated by a 100 tonne fish farm corresponded to the nutrients produced by 2800–3200 PE using 1974 conversion efficiency data, falling to 850–1950 PE by 1994 (Folke et al., 1994). Folke and co-workers then went on to assign a monetary value to the discharge in terms of the costs of implementing a reduction in nutrient discharges to the Baltic from point sources. These comparisons with waste from domestic or industrial sources have been challenged, on the basis that the chemical nature of the nutrient is in a different form compared with fish farm effluents (Black et al., 1997); this is also disputed (see Folke et al., 1997).

The main difficulty with this approach is, however, not that the forms of nutrient are different but that society approaches these different forms of pollution rather differently. In the case of human effluent, the main reasons for increasing the level of treatment are not only to reduce nutrients but also to reduce the level of particulate carbon and associated human pathogens entering the environment. There are several health- and aesthetic-related reasons for wishing to reduce the direct disposal of sewage into watercourses and the coastal zone. In the case of industrial wastes, there is clear public concern that toxic materials should not be dumped in the coastal zone, again for a combination of environmental, health and aesthetic reasons.

Folke et al. (1997) argue that dilution is not the answer for pollution. This, however, remains an aspiration; although there are increased pressures to reduce the total amount of waste, the normal method of pollution management is dilution. In the case of nutrients originally sourced from fishmeal, it could be argued that to dilute these in the ocean merely completes a natural cycle, all be it through a convoluted path from restricted coastal areas to the ocean.

This is more than can be said for most agricultural fertilisers (sourced from mining or from industrial fixation), the usage of which is likely to continue to increase, especially as more intensive agricultural production escalates in developing countries (Nixon, 1995). This will be exacerbated by increased urbanisation and construction of sewage systems in these countries, allowing human effluents easy access to watercourses and thence to the coastal zone. In developed countries, there is increased pressure to incinerate sewage sludge, in other words to dilute wastes into the environment. In spite of these arguments, there will be increasing pressure to reduce the total amount of nutrient released to the environment by more efficient collection and recycling systems.

In addition to the effects of eutrophication, alterations of nutrient ratios are often held responsible for changes to pelagic ecosystems that may lead to increased occurrence of harmful blooms (Folke *et al.*, 1997). The scientific community remains divided on this issue and further experimental study and fieldwork is urgently required. Folke *et al.* (1997) assert that, as a link between nutrient ratios and harmful blooms has not been disproved, the precautionary principle should be invoked. Presumably, this would involve a moratorium on future expansion in areas where there was any suspicion that nutrients from fish farms contributed to algal toxicity.

However, the basic argument that carnivorous culture redistributes nutrients from the oceans to the coast, where they have the potential to cause eutrophication while at the same time making a marginal contribution to world fish supplies, cannot be challenged. Thus, the case for the sustainability of intensive culture must involve a balance between negative environmental effects and positive socioeconomic benefits, but with a clear drive to employ mitigating technologies to reduce environmental effects.

9.3.4 Solid wastes

The effects of solid wastes from cage culture are well described in marine culture, at least for temperate areas (see Pearson and Black, Chapter 1 of this volume), and less well understood for freshwater systems (see Kelly and Elberizon, Chapter 2 of this volume). In both cases, and for obvious reasons, these wastes have their initial major effect on the local environment but will also make a partial contribution to the nutrient budget through sediment biogeochemical cycles. Provided that the areas of impact are confined to the immediate locality, there should be no significant sustainability issue as the total area of seabed so affected will usually be insignificant in regional terms.

9.3.5 Waste chemicals

The use of large amounts of chemicals in aquaculture is usually a sign of crisis and poor husbandry. Where such operations discharge significant amounts of hazardous chemicals to the aquatic environment, it is likely that the operation is unstable and may be unsustainable. Stressed animals are more likely to succumb to disease (for a review of stress in farmed fish, see Pickering, 1998). However, an improved understanding of the environmental requirements of the cultured species, together with immunological developments, such as vaccines, can significantly reduce chemical usage.

For example, environmentally-unsound shrimp farming practices requiring high chemical inputs and led to massive premature abandonment of intensive culture facilities in Indonesia, the Philippines and Taiwan. This was not experienced in Thailand, due to the adoption of more environmentally-friendly, locally- adapted methods involving reduced water exchange and reduced chemical usage (Kongkeo, 1997; see also Choo, Chapter 4 of this volume for a review).

In northern Europe, frequent outbreaks of furunculosis in Atlantic salmon, only partially controlled by the use of massive amounts of antibiotics, came close to destroying the industry in the early 1990s. This disease was overcome by the development of vaccines, resulting in a reduced use of antibiotics in that industry at the present time.

Currently, the use of a variety of chemical agents to control sea lice is a major environmental concern in European salmon culture, although attitudes vary regionally, with much less emphasis being placed on the potential environmental effects of these chemicals in Norway compared to the UK. Whether the use of such chemicals constitutes a major ecological threat (i.e. is sustainable) at either regional or local levels, is currently being actively researched, as is the search for an effective immunological solution (Woo, 1997).

9.3.6 Containment of cultured species, disease and parasites

The benefits and purposes of containment are discussed by Waller, Chapter 5 and Kapuscinski and Brister, Chapter 6 of this volume. The transmission of diseases and parasites between cultured stocks, and between cultured and wild stocks (in both directions), is clearly of paramount importance to the aquaculture industry. In northern temperate cage culture of Atlantic salmon, as a consequence of increased regulation and reporting obligations, the industry is well aware of problems associated with escapes of farmed fish. Briefly, escapes may be highly significant in terms of reducing diversity at the genetic level and may present serious damage to wild stocks by reducing effective population sizes and diluting genes with local adaptations (reviewed by Beardmore et al., 1997). Escapes continue in relatively open cage systems and it could certainly be argued that these constitute a significant threat to biodiversity and, therefore, are unsustainable without further efforts towards mitigation.

Mitigation measures are, in principle, available. These include compulsory use of sterile or sterilised stocks for on-growing, improved technology for containment in cages or a move towards tank culture. Each of these has a cost that would reduce the profit and therefore the socioeconomic benefits of this type of culture.

As has been stated previously, transmission of parasites and disease constitutes a major threat both to fish farming and to wild populations. If major disease episodes become more frequent as a consequence of transmission by wild or cultured vectors, the local or regional industry may not have sufficient economic stability to be sustained. In addition, if parasites or pathogens from culture make a large impact on local wild populations, significant impacts on biodiversity and on commercial or leisure fisheries may result in an industry being deemed unsustainable. However, in most cases it is difficult to establish causal relationships but, where the circumstantial evidence convinces a significant number of informed experts, perhaps it is necessary to invoke the precautionary principle.

9.3.7 Fisheries

The use of mangroves for shrimp farming may have serious consequences for coastal fisheries through loss of nursery areas and fishing grounds after habitat destruction. This subject is discussed by Choo in Chapter 4 of this volume. For each kilogram of shrimp produced from coastal ponds developed in mangroves, 447 g are lost from capture fisheries (Naylor *et al.*, 2000). This is particularly significant, as it increases the net drain on fisheries from using fishmeal and at the same time changes a common asset into private ownership.

9.4 Benefits of aquaculture

9.4.1 Environmental benefits

The reduction in nutrients and energy from aquatic systems as a consequence of herbivore/omnivore culture may, in some cases, be regarded as beneficial in terms of ecosystem function (e.g. in integrated polyculture systems), but there are few convincing cases of environmental benefits from intensive fish culture. It has often been suggested that fish in cages act as a monitor of environmental hazard by analogy with the canary used by miners to give early warning of poisonous gases. Although there have certainly been cases when fish have been killed by external factors that could not be associated with the farm, it is likely that, unlike the miner's canary, farmed fish can contribute to the reduction in environmental quality manifested by their poor performance or mortality.

It could be argued that cage arrays in the marine environment act as artificial reefs, which may enhance the local diversity of fish species, particularly in oligotrophic areas, but it is likely that any positive effects are compensated by a localised loss of species adapted to oligotrophy.

The intensive cultivation of broodstock as suppliers of eggs or juveniles for restocking programmes may also be regarded as a positive benefit, but this is only likely to be the case where native stocks are extinct or unviable. This widely-practised technique is, in principle, only likely to be effective over the long term (i.e. sustainable) if the reasons for the decline of the target population are understood and remedies are in place. In any case, this represents only a tiny fraction of aquaculture production.

9.4.2 Socioeconomic benefits

Intensive aquaculture increases the supply of aquatic food types that have strong and expanding markets in developed countries. This is particularly true of salmon and shrimps. These products compete directly in such markets with more traditional protein sources, such as Gadoid fish (cod, haddock, etc.), beef and lamb. Gadoid fisheries are generally over-fished and their continued exploitation cannot be sustained. In the UK, livestock are kept at historically high levels although demand is slipping, farm gate prices are low and farmers require massive public subsidy to stay in business. It is clearly unsustainable to pay farmers large amounts to keep vast herds of often intensively-farmed animals for which markets are shrinking. It is, therefore, important to consider sustainability in context, so that fish farmers at least have a level playing field.

Continued over-fishing of ground fish is clearly unsustainable: as well as causing huge pressure on populations with potential ecosystem-wide effects, fishing gears often cause major benthic impact. A comparison of the areas of seabed damaged by trawling gear with that impacted by solids output from cage farming would reveal a massive imbalance. Pelagic fisheries also have their environmental impacts, with well-recognised risks to cetaceans and seabirds (Moore and Jennings, 2000).

From the farmers' point of view, therefore, it would seem fair that a proper assessment of the relative sustainabilities of competing activities, such as livestock farming and capture fisheries, be made prior to fish culture operations being condemned on sustainability grounds.

Intensive culture brings money to often remote or regionally undeveloped areas, with a range of social benefits. This is particularly true when local companies retain profits. Unfortunately, because of low profit margins, there is a tendency for companies to merge and for control to be removed from the local area along with any profits but with only some of the risk. Nevertheless, it may be possible that larger companies are able to offer more long-term employment opportunities, thus increasing the stability of local communities. The true economic benefits have been much argued and are beyond the abilities of the present author to discuss further, but governments seem to be persuaded by them, as do regional development organisations, although such organisations may often be influenced by short-term economic benefits.

9.5 Growth and global carrying capacity

Some economists downplay the environmental consequences of growth on the basis that increased prosperity inevitably leads to improvements in the environment. This is justified on the basis of an apparent U-shaped relationship between income and environmental concern: as economies develop, the environment is sacrificed in favour of income generation but, on attaining higher income levels, society is more concerned with environmental quality and is prepared to invest in improved processes and mitigating technologies (Arrow *et al.*, 1995).

This may be true of some local pollutants where the effects are clearly visible, such as toxic discharges to rivers, but is less obviously accurate when the pollutant is long term or diffuse (Arrow *et al.*, 1995). For example, greenhouse gas emissions rise with income. Increased concern about pollution may simply lead to export of pollution-causing processes to less-developed countries, and the argument does not hold well for exploitation of biological resources (e.g. forestry and fisheries). Arrow *et al.* (1995) argue that exploited ecosystems might show resilience up to a point before irreversibly flipping to a new degraded stable state, thus preventing the completion of the U-shape. Economic growth and liberalisation is not, therefore, a substitute for environmental policy. Such policies must ensure incentives to protect the resilience of the environment, particularly when related to use of a resource stock.

Concerns relating to fishmeal-fed aquaculture are, in this context, identical to those of fisheries in general. Given projected levels of demand for fishmeal by aquaculture, it is clear that aquaculture is an important player in the pressures put on fisheries and, therefore, aquaculturists must be positively engaged by policies which incentivise reductions in fishmeal usage.

9.6 Conclusion and future predictions

Where aquaculture is identified as having sustainability weaknesses, in many cases these can be reduced by progressive legislation which incentivises reduction in impacts; where markets are global, it may be difficult to ensure that environmental standards are uniformly high. Global regulation is difficult and protracted but there is growing evidence that consumers are capable of making environmentally-sound choices, even against a cost differential.

The major reason for improvements in sustainability will, however, be increases in resource price. If the oceanic fishery is over-exploiting stocks, fishmeal is a finite resource that will grow in price in the long term, being particularly susceptible to climatic fluctuations. Freshwater will become a better-managed resource and culture operations will increasingly have to reduce their water requirements by applying improved technologies. Vegetable oils and meals will substitute for fishmeal, as increasing fishmeal costs incentivise academic and commercial research and innovation.

Increased regulation in the developed world will drive pollution to developing countries in lower latitudes. This could have a major effect on global biodiversity, which increases towards the equator (Beardmore et al., 1997). The culture of high-value, luxury species is the least sustainable form of aquaculture. However, the fact that these products are being sold into a sophisticated market allows the opportunity for market forces to drive continuing improvements in environmental performance. This is not always the case in developing countries, where provision of cheap nutrition or ability to supply an export market may be seen as of paramount importance. However, in the developed world, public awareness of global environmental issues is increasing and this may result in international market pressures to improve the sustainability of aquaculture practices in countries that export luxury products.

Acknowledgement

The author is grateful to Max Troell, who provided useful comment on the manuscript.

References

Arrow, K., Bolin, B., Costanza, R., Dasgupta, P., Folke, C., Holling, C.S., Jansson, B.O., Levin, S., Maler, K.G., Perrings, C. and Pimentel, D. (1995) Economic growth, carrying-capacity and the environment. *Science*, **268** 520-521.

Beardmore, J.A., Mair, G.C. and Lewis, R.I. (1997) Biodiversity in aquatic systems in relation to aquaculture. *Aquaculture Research*, **28** 829-839.

Bell, J.G. (1998) Current aspects of lipid nutrition in fish farming, in *Biology of Farmed Fish* (eds. K.D. Black and A.D. Pickering), Sheffield Academic Press, Sheffield, pp. 114-145.

Bell, J.G., Tocher, D.R., Farndale, B.M., Cox, D.I., McKinney, R.W. and Sargent, J.R. (1997) The effect of dietary lipid on polyunsaturated fatty acid metabolism in Atlantic salmon (*Salmo salar*) undergoing parr-smolt transformation. *Lipids*, **32** 515-525.

Berg, H., Michelsen, P., Troell, M., Folke, C. and Kautsky, N. (1996) Managing aquaculture for sustainability in tropical Lake Kariba, Zimbabwe. *Ecological Economics*, **18** 141-159.

Black, E., Gowen, R., Rosenthal, H., Roth, E., Stechy, D. and Taylor, F.J.R. (1997) The costs of eutrophication from salmon farming: implications for policy—a comment. *Journal of Environmental Management*, **50** 105-109.

Enell, M. (1995) Environmental impact of nutrients from Nordic fish farming. *Water Science and Technology*, **31** 61-71.

Folke, C. and Kautsky, N. (1992) Aquaculture with its environment—prospects for sustainability. *Ocean and Coastal Management*, **17** 5-24.

Folke, C., Kautsky, N. and Troell, M. (1994) The costs of eutrophication from salmon farming: implications for policy. *Journal of Environmental Management*, **40** 173-182.

Folke, C., Kautsky, N. and Troell, M. (1997) Salmon farming in context: response to Black et al., *Journal of Environmental Management*, **50** 95-103.

Folke, C., Kautsky, N., Berg, H., Jansson, A. and Troell, M. (1998)The ecological footprint concept for sustainable seafood production: a review. *Ecological Applications*, **8** S63-S71.

Kautsky, N., Berg, H., Folke, C., Larsson, J. and Troell, M. (1997) Ecological footprint for assessment of resource use and development limitations in shrimp and tilapia aquaculture. *Aquaculture Research*, **28** 753-766.

Kongkeo, H. (1997) Comparison of intensive fish farming systems in Indonesia, Philippines, Taiwan and Thailand. *Aquaculture Research*, **28** 789-796.

Moore and Jennings (eds.) (2000) *Commercial Fishing: the Wider Ecological Aspects.* Blackwell Science, London, 66 pp.

Naylor, R.L., Goldburg, R.J., Mooney, H., Beveridge, M., Clay, J., Folke, C., Kautsky, N., Lubchenco, J., Primavera, J. and Williams, M. (1998) Ecology—Nature's subsidies to shrimp and salmon farming. *Science*, **282** 883-884.

Naylor, R.L., Goldburg, R.J., Primavera, J.H., Kautsky, N., Beveridge, M.C.M., Clay, J., Folke, C., Lubcheno, J., Mooney, H. and Troell, M. (2000) Effects of aquaculture on world fish supplies. *Nature*, **405** 1017-1024.

Nixon, S.W. (1995) Coastal marine eutrophication: a definition, social causes and future concerns. *Ophelia*, **41** 199-219.

Pickering, A.D. (1998) Stress responses of farmed fish, in *Biology of Farmed Fish* (eds. K.D. Black and A.D. Pickering), Sheffield Academic Press, Sheffield, pp. 222-255.

Wm Kissil, G., Lupatsch, I., Higgs, D.A. and Hardy, R.W. (2000) Dietary substitution of soy and rapeseed protein concentrates for fish meal and their effects on growth and nutrient utilization in gilthead seabream, *Sparus aurata* L. *Aquaculture Research*, **31** 595-601.

Woo, P.T.K. (1997) Immunization against parasitic diseases of fish. *Fish Vaccinology*, **90** 233-241.

Index

aggressive behaviour differences 142
alien introductions 57-58
    see also exotics
ammonia 3, 6, 18 62, 80, 118
ammonium oxidation 10
antibiotic 56-57, 85-87, 110, 174-175

bacterial
    biomass 4
    mats 18, 62
Beggiatoa 12, 62, 164
benthic fauna 9-16, 58-62
    succession of 13
    shellfish harvesting 64
benthic recovery 18-20, 65
biosafety 129
bioturbation 15, 164
birds
    mangroves 68
    marine fish culture 23
    intertidal shellfish seed collection 55
    harvesting shellfish 63
Bonamia 57

Capitella 15-16, 25, 171
carbaryl 60
carbon 16-17, 113, 163, 204
carrying capacity 9, 53, 194-195, 210
catchment management 33
chemicals 8, 46, 81, 85-87, 110, 206-207
containment 99, 207
    biological barriers 133-136
    mechanical barriers 133
    physical barriers 43-44, 133

dams for aquaculture production 41
Decision Support Systems (DSS) 176-177
degassing 119
denitrification 10, 118
dichlorvos 21
disease 89, 173
    see also pathogen
dissolved organic
    carbon 3

    nitrogen 3
    phosphorus 3
diversity 3, 12
dredging of shellfish 54-55, 64

ecological footprint 39-40, 200-202
entry potential 143
epibenthos 17
escapes 22, 89-90, 101, 128-147, 184
eutrophication 4-5, 35-38, 68, 83, 165, 204, 206
exotics 22, 88-89

faecal contamination 53
fallowing 21
feed conversion ratio 36, 167, 202-203
fertilisation of ponds 80
fish mills 109
fishmeal 16, 17, 39-40, 199, 202-203, 205
    substitutes 203
flow through 112
flushing time 3-4, 163
furunculosis 22

gas exchange in freshwater 38
gastropods 52
genetic impacts 89-90
genetically engineered organism (GEO) 129-133
Geographical Information Systems (GIS) 175, 195
Gyrodactylus salaris 20

harmful algal blooms 84, 165-166, 206
human health 80
hydrography 13, 23, 42-43, 55, 58, 61
hypernutrification 3, 83

infectious salmonid anaemia 22
Integrated Coastal Management (ICM) guidelines 187
interbreeding 136-145
interspecific hybridisation 136
introgression potential 143

iron reduction 10
isotopic ratio 17
lipids 3, 5, 17

malachite green 35
mangroves 68, 68-70, 81-83, 183
marine protected areas (MPA) 182, 191-194
Mediterranean 5, 18, 23-24
methanogenesis 10
minimisation *or* mitigation of impact 25-26, 42-43, 67, 90-92, 133-136, 207-208
models 194
    definition 155
    deposition 168-170
    Dillon and Rigler 44, 46
    epidemiological 173-174
    fish farm impacts 159
    fish growth 164
    individual fish 160-162
    management 9
    water quality 166-167

nanoflagellates 4
Nile perch 22
nitrate reduction 10
nitrogen (N) 3, 4, 58, 68, 113, 184
    budget 5-7, 162

outbreeding depression 137, 139, 141, 144
    definition of 147
oxygen demand from sediments 9, 164
    consumption by fish 114

parasites 20
particle separation 115-117
pathogen 20, 68
pelagic ecosystem 3
phosphorus (P) 3, 4, 45, 58, 68, 80, 113, 118, 184
    budget 7
    iron phosphates 7
photoperiod 106
physical transport 162
population structure 138

random walk 170
recirculating systems 58, 111, 114-119
red tides 66
    see also harmful algal blooms
redox potential 8-10
resuspension 18, 162, 164, 171
Rio Earth Summit 77, 185

sea lice 20-22, 184, 207
seagrass 24
sediment traps 17, 42
seed collection 54-58, 83, 202
silicon (Si) 4
    budget 7-8
socioeconomics 32, 40-41, 51-52, 76, 99-100, 209
spat collection 55-56
sterility induction 135
sulphate reduction 10, 17-18
sustainability 33, 38-41, 78, 192-211
swimming behaviour 107-109

temperature preferendum 107
toxic algae 4, 5
    see also harmful algal blooms
transgenic
    definition of 147
    medaka 132
    salmon 131, 135
trash fish 6, 16
triploidy 134-135
Trojan gene effect 132

Contemporary Clinical Neuroscience

Contemporary Clinical Neurosciences bridges the gap between bench research in the neurosciences and clinical neurology work by offering translational research on all aspects of the human brain and behavior with a special emphasis on the understanding, treatment, and eradication of diseases of the human nervous system. These novel, state-of-the-art research volumes present a wide array of preclinical and clinical research programs to a wide spectrum of readers representing the diversity of neuroscience as a discipline. Volumes in the series have focused on Attention Deficit Hyperactivity Disorder, Neurodegenerative diseases, G Protein Receptors, Sleep disorders, and Addiction issues.

More information about this series at http://www.springer.com/series/7678

Pasquale Striano
Editor

Epilepsy Towards the Next Decade

New Trends and Hopes in Epileptology

Editor
Pasquale Striano
DINOGMI
University of Genoa
Genova
Genova
Italy

ISBN 978-3-319-12282-3   ISBN 978-3-319-12283-0 (eBook)
Contemporary Clinical Neuroscience
DOI 10.1007/978-3-319-12283-0

Library of Congress Control Number: 2014956378

Springer Cham Heidelberg New York Dordrecht London
© Springer International Publishing Switzerland 2015
This work is subject to copyright. All rights are reserved by the Publisher, whether the whole or part of the material is concerned, specifically the rights of translation, reprinting, reuse of illustrations, recitation, broadcasting, reproduction on microfilms or in any other physical way, and transmission or information storage and retrieval, electronic adaptation, computer software, or by similar or dissimilar methodology now known or hereafter developed.
The use of general descriptive names, registered names, trademarks, service marks, etc. in this publication does not imply, even in the absence of a specific statement, that such names are exempt from the relevant protective laws and regulations and therefore free for general use.
The publisher, the authors and the editors are safe to assume that the advice and information in this book are believed to be true and accurate at the date of publication. Neither the publisher nor the authors or the editors give a warranty, express or implied, with respect to the material contained herein or for any errors or omissions that may have been made.

Printed on acid-free paper

Springer is part of Springer Science+Business Media (www.springer.com)

Preface

Epilepsy is a common chronic neurological disorder affecting approximately 0.5–1% of the population worldwide (~50,000,000 people) and the main goal of the treatment is to eliminate seizures without producing significant side effects. The drug therapy of epilepsy has evolved tremendously in the last twenty years and several antiepileptic drugs have been approved and marketed, offering a good number of options for treatment a large variety of seizure types and epilepsy syndromes. Nevertheless, despite optimal medical treatment, up to 30% of patients continue to experience recurrent seizures, which may lead to a severe medically, physically, and socially disabling condition. The intent of *Epilepsy Towards the Next Decade: New Trends and Hopes in Epileptology* is to provide a comprehensive overview of recent advances in the field of epileptology as well as of the recent advances and current knowledge regarding epilepsy research from leading experts in the field. This book aims to provide a handy and updated reference for most recent knowledge regarding the biological basis and the modern clinical approach to epilepsy, bridging the gap between fundamental aspects and clinical implications.

Epilepsy accounts for a variety of neurological disorders characterized by recurrent seizures. More than half of all epilepsies have some genetic basis and single gene defects in ion channels or neurotransmitter receptors are associated with inherited forms of epilepsy. In the last ten years, advances in the genetic techniques including oligonucleotide array and the following large scale studies have yielded to the identification of recurrent copy number variants associated with epilepsy. The book starts by reviewing the increasingly reported copy number variants in association with distinct epileptic phenotypes (Chap. 1), delineating emerging epileptic syndromes. Once that the features and prognosis of these conditions have been completely delineated a proposal for inclusion within the International Classification of the Epileptic Syndromes should be considered. Moreover, in the last few years, genetic research in the field of epilepsy disorders is increasing in term of testing platform for the investigation of sequence and structural variation. In particular, epileptogenic mutations have been identified in several ion channel genes, leading to the concept that several epilepsies can be considered channelopathics. Functional studies have in some cases provided significant advances in the understanding of the molecular and cellular dysfunctions caused by mutations. However, the relationships between molecular deficits and clinical phenotypes are still unclear. Moreover, mutant channels that cause a distinct epilepsy syndrome show functional heterogeneity, which is in part produced by the different experimental conditions used in the studies: cell background, cDNA from other species or isoforms, and splicing variants. This aspects are fully reviewed in Chap. 2.

Although ion channels play an important role in genetic epilepsies, we should not overlook the fact that other pathways can lead to neuronal hyperexcitability. Mutations in the Leucine rich glioma inactivated 1 (LGI-1 or epitempin), a nonion channel gene that is implicated in autosomal dominant lateral temporal lobe epilepsy, a rare syndrome whose symptoms usually begin in adolescence. The molecular mechanisms for LGI1-mediated epilepsy are very complex and largely unknown. However, it seems that the defects in this gene can arrest the developmental maturation of excitatory circuits results in heightened seizure susceptibility. These data could also have clinical implications as pathways linked to LGI1 might become targets for epilepsy therapy (Chap. 3).

The aetiology of epilepsy is extremely complex and heterogeneous and both genetic and acquired factors can be responsible of this condition. Symptomatic epilepsies have mainly acquired causes, including malformations of cortical development, tumours, and metabolic diseases. The cellular mechanisms underlying the epileptogenicity of glioneuronal tumors depend on tumor histology, integrity of the blood-brain barrier, characteristics of the peritumoral environment, circuit abnormalities, or cellular and molecular defects. An evolving understanding of the mechanisms of tumor-related epileptogenicity may lead to improve surgical treatment and to identify more effective therapeutic strategies (Chap. 4). The main inborn errors of metabolism associated with epilepsy are reviewed in the Chap. 5. The diagnosis of a genetic defect or an inborn error of metabolism often results in requests for a vast array of biochemical and molecular tests leading to an expensive workup. However, a specific diagnosis of metabolic disorders in epileptic patients may provide the possibility of specific treatments that can improve seizures.

Although the diagnosis of epilepsy remains mainly clinical, Magnetic Resonance Imaging plays a crucial role in the detection of lesions that can cause epilepsy, with high impact on the diagnostic work-up as well as on therapeutic planning (reviewed in Chap. 6). Morphologic MR imaging is still the main technique for identifying lesions responsible for the epilepsy, providing images with high spatial resolution, excellent soft-tissue contrast, and multiplanar view. Functional MR imaging is used for lateralizing language functions, and also for surgical planning predicting functional deficits following epilepsy surgery. Functional imaging and other methods have contributed to understanding how these seizures arise, as observed in patients with reflex seizures, which are provoked by specific external stimuli and that are important clues for investigating complex mechanisms of epileptogenesis (Chap. 7). Future technical progress will hopefully offer the opportunity for further investigating cortical areas and brain networks involved in cerebral functions and in epileptic discharges, thus contributing to the comprehension of mechanisms of epileptogenesis.

A large section of the book is then dedicated to pathophysiological aspects of epilepsy and related conditions as well as the implications for the quest of new therapies. Insights into commonalities in the pathophysiology of epilepsy and other paroxysmal disorders may suggest new treatment approaches. Of special interest is the association between epilepsy and migraine (Chap. 8). The link between these conditions has been matter of debate for over many decades. However, new data have been now emerging in favour of a non-random relationship between these two entities and it has been also suggested that a headache may be sometimes the isolated ictal manifestation of an epileptic seizure, namely, "ictal epileptic headache", a new entity that has recently been quoted in the last International classification of headache disorders (ICHD-III). Another intriguing link is that between epilepsy and immune system. It is widely acknowledged that immune system influences several aspects of the central nervous system. Indeed, very recent evidences of specific antibodies found in epileptic encephalitis, the good response to immune therapy in refractory epileptic syndromes and the strong relationship between systemic autoimmune disease and epilepsy suggest a plausible role for the immune system also in paroxysmal neurological disorders (Chap. 9). This 'autoimmune hypothesis' represents a new potential approach to target antiepileptic therapy and deserves special attention the next years. In fact, still nowadays up to 30 % of patients continue to present recurrent seizures and the challenge for new more efficacious and better tolerated drugs is continuing. Advances in understanding of pathophysiology of epilepsy and in the physiology of ion channels and other molecular targets provide opportunities to create new and improved therapies. Potentially interesting molecular targets include KCNQ-type K+ channels, SV2A synaptic vesicle protein, ionotropic and metabotropic glutamate receptors. The pipeline for the development of new AEDs with novel mechanisms of action is narrowing with only a few interesting compounds on the immediate horizon. Chapter 10 reviews the available information on various classes of molecules that are in the pipeline for the treatment of epilepsy. There is also increasing interested about the use of possible alternative treatments. Reproductive hormones have for years suggested for this purpose (Chap. 11). Progesterone has been reported as effective, but only in studies coming from a single research group. More recently, synthetic neurosteroids have been proposed as a possible treatment devoid of unwanted side effects associated with natural steroids. In spite of the long standing interest in this therapeutic approach, clinical experience from controlled studies is at the present time still very limited.

The final section of the book (Chaps. 12 and 13) discusses the role of surgical neuromodulation for epilepsy treatment, i.e., procedures involving the electrical stimulation of cortical, diencephalic, cerebellar and peripheral targets, e.g., vagus nerve. Stereotactic radiosurgery also provides a neuromodulatory approach, affecting the discharging behavior of epileptic neurons in absence of evident target necrosis. Electrical stimulation and stereotactic radiosurgery are emerging procedures for the treatment of medically refractory epilepsy in patients not amenable to resective surgery due to inability to map the focus, presence of multiple epileptogenic foci and/or involvement of eloquent cortex.

I warmly thank all the authors and friends for their timely and insightful contributions, the series editor Dr. Mario Manto for suggesting this book, Ms. Simina Calin and Jacob Rosati at Springer for keeping things moving along. I also thank my coworkers Maria Stella Vari, Giovanna Giudizioso, and Francesca Pinto for their input into the chapters and my beloved family for being so patient with me.

I hope this book will serve as a helpful guide for adult and pediatric neurologists, including those beginning their careers and hone their skills, as well as for medical students and residents, and sophisticated patients and other lay persons who want to learn more about the pathophysiology, epidemiology and burden, comorbidities, treatment, research, and recourses for the management of persons with epilepsy.

Pasquale Striano MD, PhD

Contents

**Copy Number Variants and Epilepsy:
New Emerging Syndromes** 1
Antonietta Coppola and Maurizio Elia

Mutations of Ion Channels in Genetic Epilepsies 15
Massimo Mantegazza, Raffaella Rusconi and Sandrine Cestèle

LGI1 Dysfunction in Inherited and Acquired Epileptic Disorders 35
Carlo Nobile

**Glioneuronal Tumors and Epilepsy: Clinico-Diagnostic Features
and Surgical Strategies** 47
Alessandro Consales, Paolo Nozza, Maria Luisa Zoli,
Giovanni Morana and Armando Cama

Metabolic Causes of Epilepsy 71
Laura Papetti, Francesco Nicita, Stella Maiolo, Vincenzo Leuzzi
and Alberto Spalice

**New Insights into Mechanisms Underlying Generalized
Reflex Seizures** 101
Edoardo Ferlazzo, Domenico Italiano, Sara Gasparini,
Giovanbattista Gaspare Tripodi, Tiziana D'Agostino
and Umberto Aguglia

**Current Status and Future Prospective
of Neuroimaging for Epilepsy** 109
F. Caranci, F. D'Arco, A. D'Amico, C. Russo, F. Briganti,
M. Quarantelli and E. Tedeschi

The Complex Relationship Between Epilepsy and Headache and the Concept of Ictal Epileptic Headache 139
Pasquale Parisi

Epilepsy and Immune System: A Tour Around the Current Literature 163
Laura Mumoli, Angelo Labate, Antonietta Coppola,
Giovambattista De Sarro, Emilio Russo and Antonio Gambardella

Novel Molecular Targets for Drug-Treatment of Epilepsy 183
Vincenzo Belcastro and Alberto Verrotti

Reproductive Hormones in Epilepsy Therapy: From Old Promises to New Hopes 201
Alberto Verrotti, Giovanni Prezioso, Claudia D'Egidio and Vincenzo Belcastro

Neuromodulation for the Treatment of Drug-Resistant Epilepsy 213
Pantaleo Romanelli and Alfredo Conti

New Radiosurgical Paradigms to Treat Epilepsy Using Synchrotron Radiation 231
Pantaleo Romanelli, Alberto Bravin, Erminia Fardone and Giuseppe Battaglia

Index 237

Copy Number Variants and Epilepsy: New Emerging Syndromes

Antonietta Coppola and Maurizio Elia

Abstract In the last 10 years, advances in the genetic techniques including oligonucleotide array and the following large scale studies have yielded to the identification of recurrent copy number variants (CNVs) associated with epilepsy. Among these a small number has been increasingly reported in association with a distinct epileptic phenotype, delineating emerging epileptic syndromes. To date, none of these CNVs underlying a specific epileptic condition has been included in the ILAE Classification of the Epileptic Syndromes as a distinct form. However once the features and prognosis of these conditions have been completely delineated a proposal for new epileptic syndromes should be considered.

Introduction

The classification of epilepsies of the International League against epilepsy includes the categories of genetic, structural-metabolic and unknown causes (Berg et al. 2010). Many studies have proven that genetic plays a major role in epilepsy, mainly by identifying the involvement of ion channels subunits and neurotransmitters mutations. More recently, thanks to the whole genome higher definition technique, some CNVs have been increasingly reported in association with a peculiar epilepsy phenotype or syndrome. Again genes codifying for ion channels subunits or neurotransmitter protein are often the candidate genes within these segments (Mulley and Mefford 2011; Poduri and Lowenstein 2011). Even if rare, the clinical features and prognosis of these conditions are now better known defining new emerging syndromes. Here we describe an overview of the most reported CNVs associated syndromes presenting with a distinctive epilepsy phenotype.

A. Coppola (✉)
Epilepsy Centre, Neurology Department, Federico II University, Naples, Italy
e-mail: antonietta.coppola1@gmail.com

M. Elia
Oasi Institute for Research on Mental Retardation and Brain Aging, Troina, EN, Italy

© Springer International Publishing Switzerland 2015
P. Striano (ed.), *Epilepsy Towards the Next Decade*,
Contemporary Clinical Neuroscience, DOI 10.1007/978-3-319-12283-0_1

Genetic Method for Detecting Copy Number Variation

Routine cytogenetic analysis, namely G-banding karyotype and fluorescence in situ hybridization, have been implemented by "molecular karyotype" techniques. These refer to the evaluation of chromosome content using DNA hybridization, such as array-comparative genomic hybridization (CGH), multiplex ligation-dependent probe amplification (MLPA), single nucleotide polymorphisms (SNP) array, rather than direct observation of chromosome under the microscope, allowing clinicians and researchers to investigate the entire genome for CNVs in one experiment. The defects are then referred as genomic coordinates from the human DNA reference sequence. These have allowed the identification of micro-rearrangements including apparent "balanced" reciprocal translocation as diagnosed by light microscopy. However classical cytogenetic analysis cannot be completely replaced by "molecular karyotype", because the latter cannot detect truly balanced translocations or inversion.

Emerging Epilepsy Syndrome Associated to CNVs

Among the genome, some regions are more prone to micro-rearrangements because of the presence of breakpoint (BP) hotspots (blocks of segmental duplications that flank the deleted or duplicated sequence). Micro-rearrangements usually recur within these regions through a mechanism called nonallelic homologous recombination (NAHR). These CNVs have thus the same size and are known as "recurrent". In the last ten years many recurrent micro-rearrangements have been reported in association with epileptic phenotype (Striano et al. 2011a; Scheffer and Berkovic 2010; Heinzen et al. 2010; Dibbens et al. 2009). Among these, some present with distinct epileptic features or a specific epilepsy syndrome as the hallmark of the clinical condition. We will discuss here the emerging phenotype of epileptic conditions underlined by a micro-rearrangement including 2q24.4 deletion, 5q14.3 deletion, 6q terminal deletion, 14q12 deletion and duplication, 15q13.3 deletion and Xp11.2 duplication. Table 1 summarizes the clinical features of these emerging phenotypes. Other CNVs associated with epilepsy of minor interest are also reported (2q23.1 deletion and 7q11.23 deletion) (Mizugishi et al. 1998; Marshall et al. 2008; van Bon et al. 2010).

2q24.4 Deletion

Patients with a clinical and EEG picture of Dravet syndrome (DS) who result negative for SCN1A mutations may present SCN1A exonic or larger deletions involving SCN1A and contiguous genes (Madia et al. 2006; Mulley et al. 2006;

Table 1 Summary of the clinical features of the emerging CNVs syndromes associated to epilepsy

CNV (candidate gene)	Epileptic phenotype/ syndrome	Seizure's onset/type	EEG features	Dysmorphisms	Other neurological abnormalities	Brain abnormalities	Other associated condition	Ref
2q23.1 del (EPC2 and MBD5)	Rett-like	0–10 yo/ absences	Not reported	Broad forehead, microcephaly, brachycephaly, synophrys, arched eyebrows, long palpebral fissures, full everted lower lip, broad chin, downturned corners of the mouth, teeth anomalies	Severe MR, absent speech, ataxia, disturbed sleep, behavioral problems	Slight atrophy or normal	Hirsutism, hyperphagia	(van Bon et al 2010)
2q24.4 del (SCN1A and contiguous genes)	Dravet-like	2–10 mo/ absences, GTCS, myoclonic, febrile seizures, febrile SE, CPS	Those typically reported in DS	Microcephaly, bitemporal narrowing or frontal bossing, down-slanting or short palpebral fissures, bulbous nose or broad nasal bridge, low-implanted ears, thick helix, bow-shaped mouth, anterior open bite, single palmar creases bilaterally, and partial syndactyly of 2nd–3rd toes	Mild to severe MR, ataxia, muscle hypotonia, autistic behavior	One patient with diffuse lesions in the periventricular white matter and basal ganglia (postmortem abnormalities as seen in Leigh syndrome: spongiosis and increased gliosis of the internal and external pallidum, and less pronounced lesions in the brainstem	Not reported	(Madia et al. 2006; Suls et al. 2006; Davidsson et al. 2008; Marini et al. 2009)
5q14.3 (MEF2C)	Rett-like	1–10 mo/ febrile, CPS, spasms, myoclonic	Not reported	Broad, high forehead, relatively large, backward rotated ear lobes, mildly upwards-slanting palpebral fissures and cupid bowed or tented upper lip	Severe MR, hypotonia, strabism	Periventricular heterotopia, mild undermyelinisation	Not reported	(Le Meur et al. 2010; Engels et al. 2009; Zweier et al. 2010)

Table 1 (continued)

CNV (candidate gene)	Epileptic phenotype/ syndrome	Seizure's onset/type	EEG features	Dysmorphisms	Other neurological abnormalities	Brain abnormalities	Other associated condition	Ref
6qter del	Occipital epilepsy	4 mo–4 yo/CPS	Posterior SW complexes, more pronounced during sleep	Low frontal hairline, abnormal hair pattern, asymmetric face, bilateral epicanthus, horizontal or upslanting or short palpebral fissures, broad nasal bridge, micrognathia, high palate, a large gap between upper central incisors, posteriorly rotated or low-set ears, short neck, camptodactyly, phimosis, hypospadia, flat feet with valgus position of the calcaneus	Mild-moderate MR, hypotonia, diffuse joint laxity, strabism	Colpocephaly, dysgenesis of the corpus callosum and brainstem, hypertrophic massa intermedia	Not reported	(Elia et al. 2006; Striano et al. 2006; Bertini et al. 2006; Lee et al. 2011)
7q11.23-q21.11 del (MAGI2)	IS	Infancy/ spasms focal seizures	Hypsarrhytmia	Same as in Williams-Beuren syndrome; microcephaly	Hypotonia, severe MR	Not reported		(Mizugishi et al. 1998; Marshall et al. 2008)
14q12 del (FOXg1)	Rett-like	Infancy/ mainly generalized	Multifocal pattern with spikes and sharp waves	Microcephaly, downslanting palpebral fissures, bilateral epicanthic folds, depressed nasal bridge, bulbous nasal tip, tented upper lip, everted lower lip and large ears	Stereotypies; severe MR	Thin CC	Feeding problems	(Mencarelli et al. 2009)
14q12 dup (FOXg1)	IS	3–8 m/IS	Hypsarrhytmia, modified hypsarrhytmia	Frontal hairline, deep set eyes, hypotelorism	Delay/absent speech, MR	Thin CC	Non reported	(Brunetti-Pierri et al. 2011; Striano et al. 2011b)

Table 1 (continued)

CNV (candidate gene)	Epileptic phenotype/ syndrome	Seizure's onset/type	EEG features	Dysmorphisms	Other neurological abnormalities	Brain abnormalities	Other associated condition	Ref
15q13.3 (CHRNA7)	"Idiopathic generalized epilepsy"	Infancy/ absences, myoclonic absences, GTCS	Generalized S, PS, SW	Hypertelorism, upslanting palpebral fissures, prominent philtrum with full everted lips, clinodactyly	Mild MR, psychiatric problems	Absent	Not reported	(Dibbens et al. 2009; Sharp et al. 2008; Masurel-Paulet et al. 2010; Helbig et al. 2009; Muhle et al. 2011)
Xp11.22-11.23 dup	ESES	6 mo–12 yo/ absences, myoclonic, GTCS	Focal SW, generalized SW, PSW	Non specific facial dysmorphic features, lower-extremity anomalies, (flat or arched feet, fifth-toe hypoplasia, and syndactyly)	Borderline functioning to severe MR, speech delay, poor speech articulation, hoarse and/or nasal voice	Lateral ventricles and subarachnoid spaces dilation, slight peritrigonal hyperintensity	early puberty, overweight	(Giorda et al. 2009; Broli et al. 2011)

CC corpus callosum, *CPS* complex partial seizures, *DS* Dravet syndrome, *ESES* electrical status epilepticus during slow sleeps, *GTCS* generalized tonic clonic seizures, *IS* infantile spasms, *MR* mental retardation, *PS* polispike, *S* spikes, *PSW* polyspike-and-wave, *SE* status epilepticus, *SW* spike-and-wave, *yo* years old

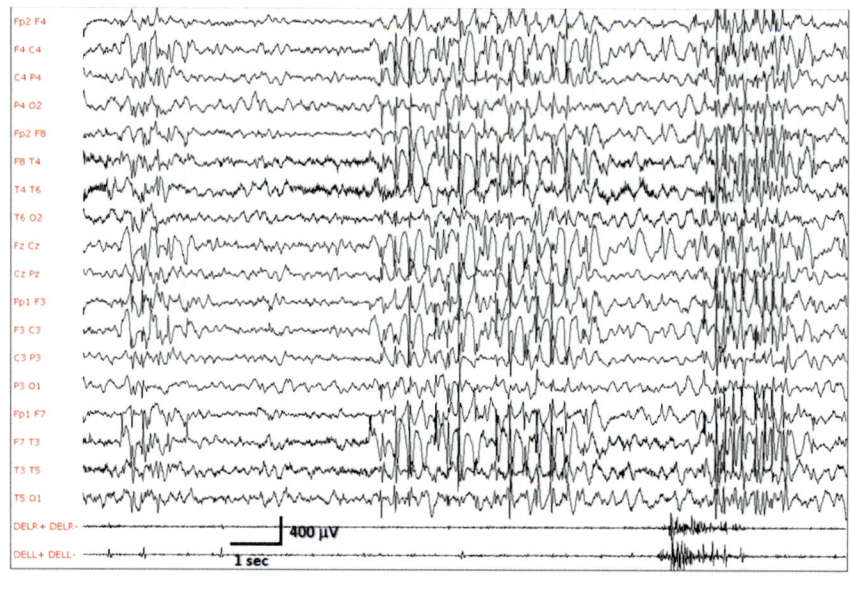

Fig. 1 Wakefulness EEG of a 4-year-old male with 2q24.4 deletion showing some generalized spike-and wave discharges

Suls et al. 2006; Davidsson et al. 2008) . These deletions account for 2–3 % of all DS cases and for about 12.5 % of patients with DS who are negative for mutations on sequencing (Marini et al. 2009).

Deletions extending beyond SCN1A and including variable numbers of contiguous genes can be associated with additional dysmorphic features, depending on the genes involved (Madia et al. 2006), or with a more severe epilepsy phenotype when other voltage-gated sodium channels (VGSC) α subunit genes clustered on chromosome 2q such as SCN2A, SCN3A, SCN7A, and SCN9A are involved (Davidsson et al. 2008; Pereira et al. 2004).

In particular, facial features were reported, such as microcephaly, bitemporal narrowing or frontal bossing, down-slanting or short palpebral fissures, bulbous nose or broad nasal bridge, low-implanted ears, thick helix, bow-shaped mouth, anterior open bite, single palmar creases bilaterally, and partial syndactyly between the second and third toes (Madia et al. 2006; Pereira et al. 2004). Davidsson et al. 2008 reviewed 43 previously published cases with a del(2)(q24.3q31.1). For the 22 seizure-positive cases, 2q24.3 region constituted the smallest commonly deleted region among the majority of the cases, where 2q22.1 and 2q33.3 regions represented the most proximal and distal breakpoint, respectively. The most common dysmorphic features were ear abnormalities, microcephaly, micrognatia and brachysyndactyly.

Seizures start always in the first year of life. The clinical and EEG picture (Fig. 1) is that typical of the classical DS with severe drug resistance, mild to severe mental retardation (MR), autistic behavior, ataxia, muscle hypotonia.

However, other studies did not find significant clinical differences between DS patients with deletions involving only SCN1A and those with deletions of contiguous genes (Mulley et al. 2006; Suls et al. 2006; Marini et al. 2009). It has been suggested that the usually severe DS phenotype might mask subtle clinical differences which may be determined by contiguous genes (Marini et al. 2009).

MRI is usually normal, although Suls et al. (2006) reported one patient who at 14 months of age showed diffuse lesions in the periventricular white matter and basal ganglia. Postmortem brain examination showed abnormalities as seen in Leigh syndrome, with spongiosis and increased gliosis of the internal and external pallidum, and less pronounced lesions in the pons and mesencephalon (central tegmental tract).

Such deletions can be easily identified by means of multiplex ligation-dependent probe amplification (MPLA) and array-comparative genomic hybridization (CGH) is able to determine the size of the abnormality and the additional genes involved. Haplotype analysis with microsatellite markers and single nucleotide polymorphisms (SNPs) can also be used to identify small chromosomal abnormalities affecting SCN1A (Madia et al. 2006; Suls et al. 2006; Marini et al. 2009).

5q14.3 Deletion

Microdeletions within chromosomal bands 5q14.3q15 were recently identified as a current cause of a Rett-like phenotype. Most of these patients do not show classical Rett syndrome with acquired microcephaly and developmental regression after an initial normal interval, but show primary hypotonia, severe mental and motor retardation, early onset seizures, and occasionally autistic behavior, stereotypic hand movements and episodic hyperventilation (Le Meur et al. 2010; Cardoso et al. 2009; Engels et al. 2009). Seizures onset between 1 and 10 months of age, usually with infantile spasms. Febrile seizures, myoclonic and complex partial seizures may also occur. These patients can also present with distinctive dysmorphic features including broad, high forehead, relatively large, backward rotated ear lobes, mildly upward-slanting palpebral fissures and cupid bowed or tented upper lip (Zweier et al. 2010) . The finding of periventricular heterotopia on a brain MRI scan is also possible (Cardoso 2009). MEF2C is the candidate gene for this phenotype. It encodes for a transcriptor factor and its activity relies on the recruitment of, and cooperation with, many other transcription factors, as well as on translational and posttranscriptional modifications (Potthoff and Olson 2007). Its deleterious activity resulting in a Rett-like phenotype may be explained by a common pathway among MEF2C, MECP2 and CDKL5 which are the main mutations responsible for classical Rett syndrome and atypical Rett syndrome respectively. Indeed patients with MEF2C defects showed diminished MECP2 and CDKL5 expression, and MEF2C mutations in vitro resulted in diminished transactivation of both the MECP2 and CDKL5 promoters (Zweier et al. 2010). A mutational screening for MEF2C microdeletion can be considered in patients with early onset Rett-like phenotype and negative for MECP2, CDKL5 and FOXG1 deletion.

6q Terminal Deletion

In 2006 a peculiar clinical, EEG, and neuroradiologic pattern was reported in five unrelated patients with a 6q terminal deletion ranging between 9 and 16 Mb (Elia et al. 2006).

The phenotype was characterized by low frontal hairline, abnormal hair pattern, asymmetric face, bilateral epicanthus, horizontal or upslanting or short palpebral fissures, broad nasal bridge, micrognathia, high palate, a large gap between upper central incisors, posteriorly rotated or low-set ears, short neck, camptodactyly, phimosis, hypospadia, flat feet with valgus position of the calcaneus.

Neurological picture was characterized by mild-moderate mental retardation, hypotonia, diffuse joint laxity, strabismus.

Interestingly, epilepsy was a feature of 6q terminal deletions in all patients. In particular, they shared a distinct clinical and EEG pattern not previously reported in this condition. Epilepsy started in the first or second decade of life. In all cases, seizures had a focal onset, characterized by the ictal signs of vomiting, cyanosis, and head and eye version with or without loss of consciousness. No status epilepticus or prolonged seizures occurred. Prognosis of epilepsy was generally good in our patients, in terms of both seizure control and evolution. In all subjects but one, interictal EEG was characterized by posterior spike-and-wave complexes which became more pronounced during sleep. The ictal signs and the EEG patterns in our patients suggested that the seizures originated in the occipital lobes. Given the early onset of seizures (between ages 4 months and 4 years), it was conceivable that an age-related low threshold of emetic centers caused the ictal vomiting, as occurs in Panayiotopoulos-type occipital epilepsy.

However, the occipital epilepsy in these patients should be considered symptomatic because of the mild/moderate mental retardation and brain anomalies.

In four of five cases, MRI revealed colpocephaly and dysgenesis of the corpus callosum and of the brainstem; three patients also had a hypertrophic massa intermedia (or interthalamic adhesion).

Subsequently, other seven patients with 6q subtelomere deletions and a similar clinical and EEG pattern were ascertained with a size of the deletion ranging from 3 to 13 Mb (Striano et al. 2006; Bertini et al. 2006).

In a recent review, 28 cases with pure 6q terminal deletion were counted (Lee et al. 2011). The most common breakpoint found in 14 cases was in the 8.0–9.0 Mb interval from the 6q terminus. There are approximately 34 known genes and six OMIM morbid genes in this ~9.0 Mb region.

A comparison of the case with the smallest deletion (~ 0.4 Mb; 3 known genes) reported to date and the case that has the largest deletion (<11 Mb;>34 known genes) showed no specific phenotype differences, with respect to developmental delay, intellectual disability, dysmorphic features, hypotonia, microcephaly, seizure and brain anomalies. The region of greatest interest resulted the smallest overlapped portion of the most distal part of chromosome 6q. The genes located in the region within 0.4 Mb of the 6q terminus were PSMB1, TBP and PDCD2. The TBP gene has been implicated as a candidate gene for phenotypes in patients with 6q terminal deletion; mutations of TBP gene are associated with spinocerebellar ataxia 17. However, the possibility of other genes playing a role in the phenotypes resulting from this deleted region cannot be ruled out.

The mechanism of chromosome 6q terminal deletion, in some cases, may be related to the fragile site, FRA6E.

It has been estimated that 6q terminal deletion is present in ~0.05% of patients with intellectual disability and/or development delay (Lee et al. 2011). 6q terminal deletion should be suspected in children with occipital epilepsy, MR, colpocephaly and malformative features, and in these cases multiprobe FISH, MLPA or array-CGH should be included in the diagnostic work-up.

14q12 Deletion and Duplication

14q12 microdeletion has been claimed as responsible for the congenital variant of Rett syndrome (Ariani et al. 2008). This CNV encompasses FOXG1 gene, a brain specific forkhead-box transcriptional repressor which is now considered the third gene responsible for Rett syndrome (Mencarelli et al. 2009) . The phenotype is characterized by early deterioration, postnatal microcephaly, hypotonia, seizures, stereotypies (Ariani et al. 2008; Mencarelli et al. 2009). Patients with 14q12 deletion differ from MECP2 mutated patients because they tend to exhibit abnormal development in early infancy, while also lacking the characteristic autonomic disturbances of Rett syndrome (Brunetti-Pierri et al. 2011). Seizures onset is in infancy with mainly generalized tonic-clonic convulsions. EEG can show multifocal pattern with spikes and sharp waves (Ariani et al. 2008). The clinical picture also includes mild dysmorphic features such as downslanting palpebral fissures, bilateral epicanthic folds, depressed nasal bridge, bulbous nasal tip, tented upper lip, everted lower lip and large ears (Mencarelli et al. 2009). Corpus callosum thinning can be reported (Mencarelli et al. 2009). FOXG1 deletion should be considered in children with atypical presentation of Rett syndrome including early onset infantile spasms, severe developmental delay, postnatal microcephaly, stereotypies and absence of autonomic features.

14q12 duplications have been reported in patients with various degree of developmental delay, delayed/absent speech and epilepsy, especially infantile spasms (Brunetti-Pierri et al. 2011; Striano et al. 2011b). The phenotype is different from that showed by patients with 14q12 deletions and inactivating mutations of FOXG1. Patient with duplication show a major involvement of speech and an early onset (3–8 months) developmental epilepsy mainly characterized by infantile spasms (Brunetti-Pierri et al. 2011; Striano et al. 2011b). This can be accompanied or not by hypsarrhytmic pattern at the EEG (Brunetti-Pierri et al. 2011). Dysmorphic features, when present, are mild and include high frontal hairline, deep set eyes, hypotelorism (Brunetti-Pierri et al. 2011).

15q13.3 Deletion

Among the CNV associated epileptic phenotype the 15q13.2-13.3 microdeletion has raised the highest interest in the field. It has been first described by Sharp as a recurrent CNV associated with mental retardation and epilepsy (Sharp et al. 2008). Soon after many reports described this abnormality together with a broad variety of phenotype including schizophrenia, severe neurodevelopmental disorders, autism and epilepsy (Dibbens et al. 2009; Ben-Shachar et al. 2009; Masurel-Paulet et al. 2010; van Bon et al. 2009; Shinawi et al. 2009). The different clinical manifestations have been correlated to the different breakpoints where the interruption occurs, and the breakpoints 4 and 5 (BP4 and BP5) have been associated to the epilepsy phenotype (Sharp et al. 2008). Furthermore a major severe phenotype has been mainly reported in the homozygous loss situation (Lepichon et al. 2010; Endris et al. 2010).

In the last few years the 15q13.3 CNV has been widely reported as a common risk factor for epilepsy, being detected in about 1% of patients with idiopathic generalized epilepsy with or without other neurological manifestations (Dibbens et al. 2009; Helbig et al. 2009). Interestingly, the 15q13.3 deletion resulted to be negative in a wide group of 3000 patients with partial epilepsies, suggesting that the deletion may confer a specific risk for generalized epilepsy (Heinzen et al. 2010; Kasperaviciute et al. 2010). Dibbens et al indicated a complex inheritance and incomplete penetrance for the IGE component of the phenotype in multiplex families (Dibbens et al. 2009). Among the idiopathic generalized epileptic syndromes, absences have been recently described in three out of 570 children with epilepsy and a various degree of intellectual disability (Muhle et al. 2011).

No brain abnormalities have been reported in association to this CNV. Dysmorphisms are not always present and the most common are hypertelorism, upslanting palpebral fissures, prominent philtrum with full everted lips, clinodactyly (Sharp et al. 2008).

Haploinsufficiency of CHRNA7, which encodes for the α7 subunit of the acetylcholine receptor is postulated as the most likely responsible factor for this phenotype (Dibbens et al. 2009; Mulley and Dibbens 2009). We recently observed two families with multiple affected individuals presenting with absences or myoclonic absences associated to mild intellectual disability, confirming the data reported by Muhle et al (Coppola et al. 2013). The ictal EEG showed generalized polispike and wave or spike and waves discharges (Fig. 2). Apparently these patients presented with a common idiopathic generalized epilepsy; however their seizures tended to persist in the elderly and were difficult to control requiring an association of at least two AEDs (Coppola et al. 2013). These features together with the occurrence of intellectual disability make the diagnosis of idiopathic generalized epilepsy unlikely. Thus in the presence of apparently generalized idiopathic epilepsy with absences and mild intellectual disability clinicians should consider a genetic test for 15q13.3 deletion.

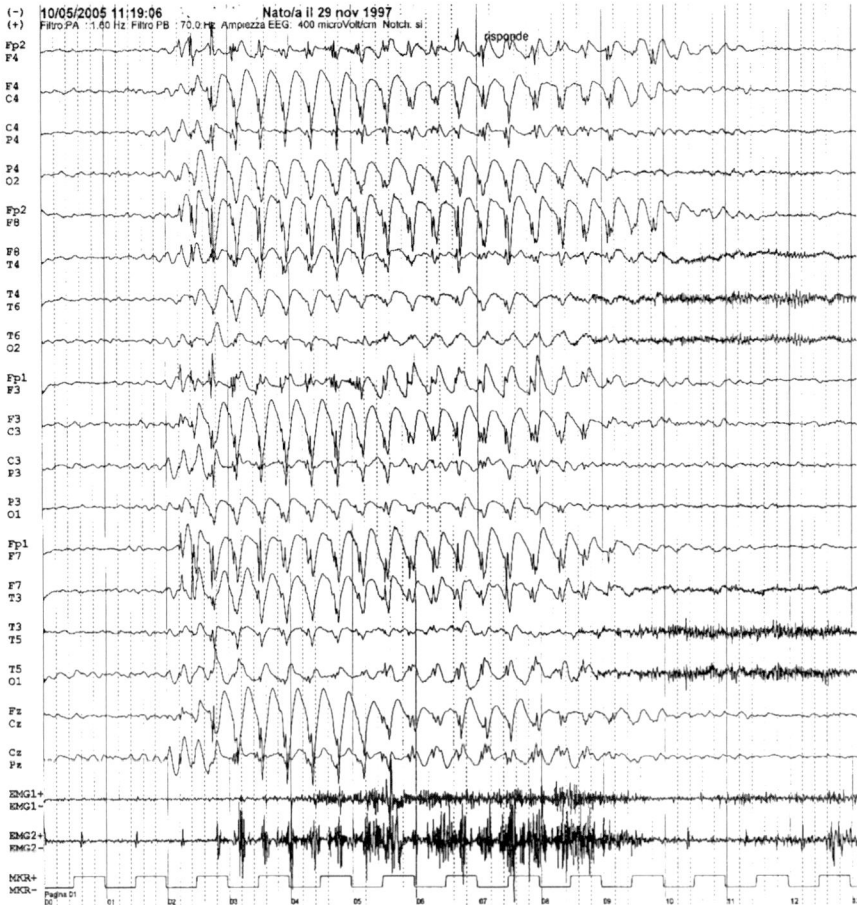

Fig. 2 Critical EEG of a 15q13.3 deletion patient showing generalized spike and wave activity

Xp11.22–11.23 Duplication

Recently a group of 9 subjects (2 males and 7 females) with a microduplication at Xp11.22-11.23 were identified at a diagnostic genome array screening of 2400 subjects with MR, either isolated or associated with a more complex phenotype. The duplication was either familial or sporadic (Giorda et al. 2009).

The clinical phenotype is characterized by a cognitive disturbance (from borderline functioning to severe MR), speech delay, poor speech articulation, hoarse and/or nasal voice, early puberty, overweight, non specific facial dysmorphic features, lower-extremity anomalies, including flat or arched feet, fifth-toe hypoplasia, and syndactyly.

A subsequent study was aimed to better define the neurological phenotype of this new syndrome (Broli et al. 2011). Electrical status epilepticus during sleep (ESES)

was observed in five of nine patients, particularly in younger ones (from 5 to 13 years), and was associated with speech delay in all cases. ESES was controlled by antiepileptic drugs in three out of five patients; the other two patients remained untreated. Neuroimaging did not disclose specific abnormalities. Therefore, the speech delay may be correlated to the ESES, able to cause local EEG slow-wave activity damage, impairing the plastic changes associated with language development.

Epilepsy was also reported in three of nine patients, with different types of seizures starting in infancy or in childhood, such as clonic jerks of the limbs and stare, generalized tonic-clonic seizures during sleep, absences. Outcome was benign.

Xp11.2 is a gene-rich, rearrangement-prone region within the critical linkage interval for several neurogenetic disorders harboring X-linked mental retardation (MRX) genes which could be responsible for the syndrome phenotype (Giorda et al 2009).

In the presence of MR, speech delay and ESES, array-CGH array should be performed in order to disclose Xp11.22-11.23 duplication.

Conclusions

The availability of whole genome high resolution techniques has shed light on new emerging syndromes underlined by chromosomal micro-rearrangements. Epileptic seizures are often a symptom of these complex conditions and sometimes epilepsy is the distinctive feature of the clinical phenotype. In few such cases the natural history, drug responsiveness and prognosis can resemble well known syndromes. However the association with other neurological symptoms and/or somatic defects and/or dysmorphic features render these conditions unique arising the question whether they can be considered new distinctive syndromes. At least for the conditions here reported it appears that a sufficient knowledge has been reached to consider them unique entities.

A genotype-phenotype correlation of further population studies can help delineating new epileptic phenotypes that recur in such genetic conditions. To this purpose it can be also helpful to report sporadic but well characterized cases. This should allow the clinicians to address the genetic test only in candidate patients avoiding expensive and time consuming analysis where not necessary.

Lastly, the study of the genes involved in these CNVs can be also crucial to better understand the physiopathology underlying some epileptic syndromes (i.e. ESES) or simply epileptic seizures.

In the future, the up coming newer techniques, namely high resolution customized array-CGH and next generation sequencing, will hopefully allow the identification of a greater number of CNV-related syndromes and the role of the clinicians will be in parallel important to validate the genetic data.

References

Ariani F et al (2008) FOXG1 is responsible for the congenital variant of Rett syndrome. Am J Hum Genet 83(1):89–93

Ben-Shachar S et al (2009) Microdeletion 15q13.3: a locus with incomplete penetrance for autism, mental retardation, and psychiatric disorders. J Med Genet 46(6):382–388

Berg AT et al (2010) Revised terminology and concepts for organization of seizures and epilepsies: report of the ILAE Commission on Classification and Terminology, 2005–2009. Epilepsia 51(4):676–685

Bertini V et al (2006) Isolated 6q terminal deletions: an emerging new syndrome. Am J Med Genet A 140(1):74–81

Broli M et al (2011) Definition of the neurological phenotype associated with dup (X)(p11.22-p11.23). Epileptic Disord 13(3):240–251

Brunetti-Pierri N et al (2011) Duplications of FOXG1 in 14q12 are associated with developmental epilepsy, mental retardation, and severe speech impairment. Eur J Hum Genet 19(1):102–107

Cardoso C et al (2009) Periventricular heterotopia, mental retardation, and epilepsy associated with 5q14.3-q15 deletion. Neurology 72(9):784–792

Coppola A et al (2013) Different electroclinical picture of generalized epilepsy in two families with 15q13.3 microdeletion. Epilepsia 54(5):e69–e73

Davidsson J et al (2008) Deletion of the SCN gene cluster on 2q24.4 is associated with severe epilepsy: an array-based genotype-phenotype correlation and a comprehensive review of previously published cases. Epilepsy Res 81(1):69–79

Dibbens LM et al (2009) Familial and sporadic 15q13.3 microdeletions in idiopathic generalized epilepsy: precedent for disorders with complex inheritance. Hum Mol Genet 18(19):3626–3631

Elia M et al (2006) 6q terminal deletion syndrome associated with a distinctive EEG and clinical pattern: a report of five cases. Epilepsia 47(5):830–838

Endris V et al (2010) Homozygous loss of CHRNA7 on chromosome 15q13.3 causes severe encephalopathy with seizures and hypotonia. Am J Med Genet A 152A(11):2908–2911

Engels H et al (2009) A novel microdeletion syndrome involving 5q14.3-q15: clinical and molecular cytogenetic characterization of three patients. Eur J Hum Genet 17(12):1592–1599

Giorda R et al (2009) Complex segmental duplications mediate a recurrent dup(X)(p11.22-p11.23) associated with mental retardation, speech delay, and EEG anomalies in males and females. Am J Hum Genet 85(3):394–400

Heinzen EL et al (2010) Rare deletions at 16p13.11 predispose to a diverse spectrum of sporadic epilepsy syndromes. Am J Hum Genet 86(5):707–718

Helbig I et al (2009) 15q13.3 microdeletions increase risk of idiopathic generalized epilepsy. Nat Genet 41(2):160–162

Kasperaviciute D et al (2010) Common genetic variation and susceptibility to partial epilepsies: a genome-wide association study. Brain 133(Pt 7):2136–2147

Le Meur N et al (2010) MEF2C haploinsufficiency caused by either microdeletion of the 5q14.3 region or mutation is responsible for severe mental retardation with stereotypic movements, epilepsy and/or cerebral malformations. J Med Genet 47(1):22–29

Lee JY, Cho YH, Hallford G (2011) Delineation of subtelomeric deletion of the long arm of chromosome 6. Ann Hum Genet 75(6):755–764

Lepichon JB et al (2010) A 15q13.3 homozygous microdeletion associated with a severe neurodevelopmental disorder suggests putative functions of the TRPM1, CHRNA7, and other homozygously deleted genes. Am J Med Genet A 152A(5):1300–1304

Madia F et al (2006) Cryptic chromosome deletions involving SCN1A in severe myoclonic epilepsy of infancy. Neurology 67(7):1230–1235

Marini C et al (2009) SCN1A duplications and deletions detected in Dravet syndrome: implications for molecular diagnosis. Epilepsia 50(7):1670–1678

Marshall CR et al (2008) Infantile spasms is associated with deletion of the MAGI2 gene on chromosome 7q11.23-q21.11. Am J Hum Genet 83(1):106–111

Masurel-Paulet A et al (2010) Delineation of 15q13.3 microdeletions. Clin Genet 78(2):149–161

Mencarelli MA et al (2009) 14q12 Microdeletion syndrome and congenital variant of Rett syndrome. Eur J Med Genet 52(2–3):148–152

Mizugishi K et al (1998) Interstitial deletion of chromosome 7q in a patient with Williams syndrome and infantile spasms. J Hum Genet 43(3):178–181

Muhle H et al (2011) Absence seizures with intellectual disability as a phenotype of the 15q13.3 microdeletion syndrome. Epilepsia 52(12):e194–e198

Mulley JC, Dibbens LM (2009) Chipping away at the common epilepsies with complex genetics: the 15q13.3 microdeletion shows the way. Genome Med 1(3):33

Mulley JC, Mefford HC (2011) Epilepsy and the new cytogenetics. Epilepsia 52(3):423–432

Mulley JC et al (2006) A new molecular mechanism for severe myoclonic epilepsy of infancy: exonic deletions in SCN1A. Neurology 67(6):1094–1095

Pereira S et al (2004) Severe epilepsy, retardation, and dysmorphic features with a 2q deletion including SCN1A and SCN2A. Neurology 63(1):191–192

Poduri A, Lowenstein D (2011) Epilepsy genetics—past, present, and future. Curr Opin Genet Dev 21(3):325–332

Potthoff MJ, Olson EN (2007) MEF2: a central regulator of diverse developmental programs. Development 134(23):4131–4140

Scheffer IE, Berkovic SF (2010) Copy number variants–an unexpected risk factor for the idiopathic generalized epilepsies. Brain 133(Pt 1):7–8

Sharp AJ et al (2008) A recurrent 15q13.3 microdeletion syndrome associated with mental retardation and seizures. Nat Genet 40(3):322–328

Shinawi M et al (2009) A small recurrent deletion within 15q13.3 is associated with a range of neurodevelopmental phenotypes. Nat Genet 41(12):1269–1271

Striano P et al (2006) Clinical phenotype and molecular characterization of 6q terminal deletion syndrome: Five new cases. Am J Med Genet A 140(18):1944–1949

Striano P et al (2011a) Clinical significance of rare copy number variations in epilepsy: A case-control survey using microarray-based comparative genomic hybridization. Arch Neurol 69(3):322–30

Striano P et al (2011b) West syndrome associated with 14q12 duplications harboring FOXG1. Neurology 76(18):1600–1602

Suls A et al (2006) Microdeletions involving the SCN1A gene may be common in SCN1A-mutation-negative SMEI patients. Hum Mutat 27(9):914–920

van Bon BW et al (2009) Further delineation of the 15q13 microdeletion and duplication syndromes: a clinical spectrum varying from non-pathogenic to a severe outcome. J Med Genet 46(8):511–523

van Bon BW et al (2010) The 2q23.1 microdeletion syndrome: clinical and behavioural phenotype. Eur J Hum Genet 18(2):163–170

Zweier M et al (2010) Mutations in MEF2C from the 5q14.3q15 microdeletion syndrome region are a frequent cause of severe mental retardation and diminish MECP2 and CDKL5 expression. Hum Mutat 31(6):722–733

Mutations of Ion Channels in Genetic Epilepsies

Massimo Mantegazza, Raffaella Rusconi and Sandrine Cestèle

Abstract Epileptogenic mutations have been identified in several ion channel genes, leading to the concept that several epilepsies can be considered channelopathies. However, increasing number of genes involved in a diversity of functional and developmental processes are being recognized through whole exome or genome sequencing, confirming that there is remarkable complexity underlying epileptogenesis. Additionally, recent studies of large cohorts of patients suggest that many patient-specific mutations in several genes are important for generating a particular phenotype, rather than mutations in a few genes common to most of the patients.

We will review the epilepsy syndromes linked to ion channel gene mutations and the main results of genetic and functional studies, highlighting that also other genes can be important but stressing the central role of ion channels in the pathophysiology of genetic epilepsies. Although the picture is becoming more complex than previously thought, the identification of epileptogenic mutations in patients before epilepsy onset and the possibility to develop therapeutic strategies tested in experimental models may facilitate experimental approaches that prevent epilepsy or decrease its severity.

Because neuronal excitability depends on the activity of voltage-dependent or receptor-activated membrane ion channels, their dysfunctions have been hypothesized to have a central role in epilepsy. In fact, early observations have shown that epileptiform neuronal activity can be induced by spontaneous or experimental modifications of the properties of ion channels leading to alterations of neuronal excitability or synaptic transmission (McCormick and Contreras 2001; Avanzini and Franceschetti 2003).

However, the first demonstration that a human disorder of excitability is caused by a genetic mutation of an ion channel came from the identification of a $Na_V1.4$ Na^+ channel α subunit mutant (gene *SCN4A*), causing a skeletal muscle disease: hyperkalemic periodic paralysis (Ptacek et al. 1991). Since then, mutations of ion

---

M. Mantegazza (✉) · R. Rusconi · S. Cestèle
Institute of Molecular and Cellular Pharmacology (IPMC),
CNRS UMR7275 and University of Nice-Sophia Antipolis,
660 Route des Lucioles, 06560 Valbonne, France
e-mail: mantegazza@ipmc.cnrs.fr

© Springer International Publishing Switzerland 2015
P. Striano (ed.), *Epilepsy Towards the Next Decade*,
Contemporary Clinical Neuroscience, DOI 10.1007/978-3-319-12283-0_2

channels have been identified in several diseases whose pathological mechanism involves defects of cellular excitability, the "channelopathies" (Ptacek 1997), including hemiplegic migraine, episodic ataxias, myotonias, hyperekplexia and cardiac syndromes (Kass 2005; Ashcroft 2006; Kullmann 2010). These diseases, similarly to idiopathic epilepsies, show acute and transient presentation of symptoms in individuals that otherwise appear normal. Indeed, few years after the discovery of hyperkalemic periodic paralysis mutations, the first epileptogenic mutation of an ion channel was identified in the α4 subunit of the neuronal acetylcholine receptor of patients affected by autosomal dominant nocturnal frontal lobe epilepsy (ADNFLE) (Steinlein et al. 1995). Since then numerous other mutations and genetic variants of ion channel genes have been identified in different forms of epilepsy (Fig. 1; Avanzini et al. 2007; Helbig et al. 2008; Reid et al. 2009; Kullmann 2010; Mantegazza et al. 2010b; Guerrini et al. 2014), but the picture is becoming less clear than it was foreseen.

Several genes that do not codify for ion channels and sometimes have still unidentified functions have been implicated in genetic epilepsy. Moreover, phenotypic variability is often large, making more difficult the correlation between mutations in specific genes and specific epileptic syndromes, and complicating early diagnosis and genetic counseling. Phenotypic variability has been ascribed to genetic modifiers: polymorphisms or other genetic variants that can modulate the effect of the mutation. Notably, mutations are defined as modifications in the sequence of a gene that are clearly identified as the cause of the disease, thus the term should be used for mendelian monogenic disorders. Genetic variants are instead modifications that contribute to disease susceptibility, and their implication in disease is often inferred from the fact that the variants are mainly found in patients and that they induce functional effects. In polygenic epilepsies, a specific epilepsy phenotype can be generated by the combination of less penetrant alleles with large effect and of polygenic alleles with small effect. Novel technologies have allowed the sequencing of whole genomes (whole genome sequencing, WGS) or of their coding part (whole exome sequencing, WES), identifying an increasing number of genetic variants. However, the identification of their importance for determining a specific phenotype is not straightforward. Moreover, in many cases both genetic and acquired factors can contribute to the determinism of epilepsy, and environmental factors can have an important role in determining phenotypes also in forms with Mendelian pattern of inheritance (Berkovic et al. 2006).

However, despite these complications, several studies have indisputably linked ion channels to specific epileptic syndromes and pathogenic mechanisms.

Mutations of Ion Channels in Genetic Epilepsies

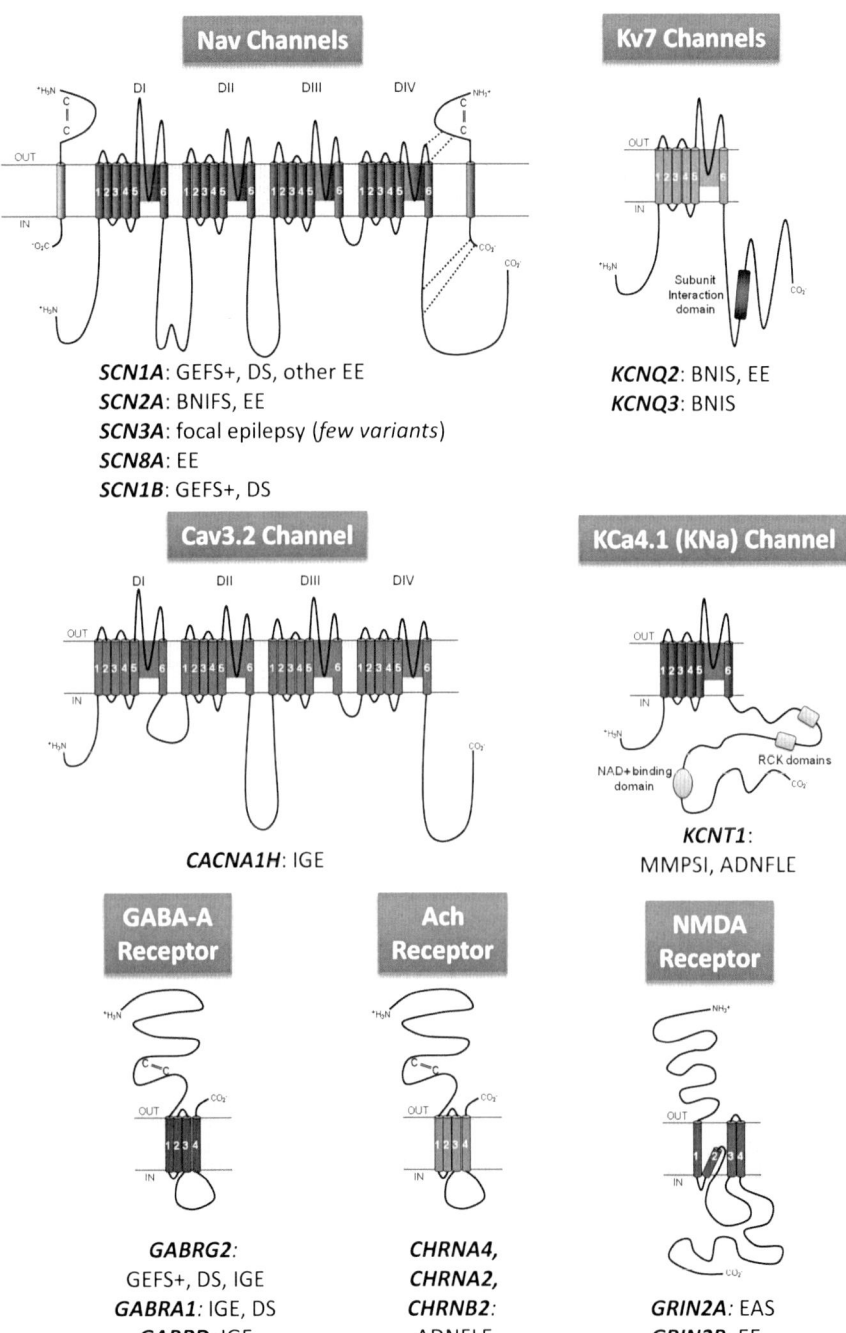

Fig. 1 Basic structure of main voltage- and ligand-gated ion channel proteins involved in genetic epilepsy. The structure of the subunits targeted by mutations/variants identified in genetic forms of epilepsy are shown. The names of the genes and the forms of epilepsy in which they are involved

Voltage-gated Na⁺ Channel SCN1A Gene-related Epilepsies and Epileptic Encephalopathies: A Prototypical Spectrum of Severity

Voltage gated Na⁺ channels (Na_V) are essential for the generation of cellular excitability, target of antiepileptic drugs and their mutations are important causes of genetic epilepsy (Mantegazza et al. 2010a; Marini and Mantegazza 2010; Catterall 2012). Na_V are composed by a principal pore-forming α subunit (nine isoforms: $Na_V1.1$-$Na_V1.9$ for the proteins, SCN1A-SCN11A for the genes), and by auxiliary β subunits (four isoforms: β1-β4 for the proteins, SCN1B-SCN4B for the genes) (Mantegazza and Catterall 2012). The primary sequence of α subunits contains four homologous domains (DI-DIV), each comprising six predicted transmembrane segments (S1-S6) that form voltage-sensing modules (S1-S4; S4 is the voltage sensor) and pore modules (S5-S6) in each domain. The β subunits contain a single transmembrane segment.

SCN1A/$Na_V1.1$ is one of the most clinically relevant epilepsy genes (Marini and Mantegazza 2010; Guerrini et al. 2014), with hundreds of mutations reported thus far in different epilepsy syndromes characterized by variable phenotypes, and is also the target of some familial hemiplegic migraine (FHM-type III) mutations; see www.molgen.ua.ac.be/SCN1AMutations and http://www.scn1a.info/ for *SCN1A* variant databases. The most severe epileptic phenotype associated with $Na_V1.1$ mutations is Dravet syndrome (DS), also known as Severe Myoclonic Epilepsy of Infancy (SMEI), an extremely severe epileptic encephalopathy (i.e. a disorder in which it is hypothesized that epileptic seizures and epileptiform activity impair brain function, although this causal link has not been clearly demonstrated yet). DS is characterized by onset in the first year of life as prolonged seizures triggered by fever and later appearance of severe afebrile seizures of various type, drug resistance, ataxia, delayed psychomotor development, cognitive impairment and behavioral dysfunctions (Dravet et al. 2005). In general, it is caused by de novo deletions or missense mutations of $Na_V1.1$ (Claes et al. 2001; Depienne et al. 2009), which are found in >80% of patients.

Genetic (Generalized) Epilepsy with Febrile Seizures Plus (GEFS+) patients shows large phenotypic heterogeneity in families, including febrile seizures (FS) and febrile seizures plus (FS+: FS after 6 years of age). The course and response

---

are indicated below the diagram of the protein. Nav voltage-gated Na+ channels, Cav3.2 voltage-gated Ca2+ channels, T-type-1H, Kv7 voltage-gated K+ channels, M-type, KCa4.1 Na+-activated K+ (KNa) channel, SLACK-SLO2.2 type; GABA-A gamma-aminobutyric acid receptor, type A, Ach nicotinic acetylcholine receptor, NMDA glutamate receptor, N-methyl-D-aspartate (NMDA) type, DS Dravet syndrome, GEFS+ generalized (genetic) epilepsy with febrile seizures plus, EE epileptic encephalopathies, BFNIS benign familial neonatal-infantile seizures, BNIS benign neonatal familial seizures, IGE idiopathic generalized epilepsies, MMPSI malignant migrating partial seizures of infancy, ADNFLE autosomal dominant nocturnal frontal lobe epilepsy, EAS epilepsy-aphasia syndromes

to antiepileptic drugs may be considerably variable within the same family: some patients experience rare febrile or non-febrile seizures that remit after a few years, while others, even within the same family, have drug resistant epilepsy, with Dravet syndrome as the extreme of the spectrum (Scheffer and Berkovic 1997). GEFS+ was originally recognized thanks to large autosomal dominant pedigrees with 60–70 % penetrance, but it is likely that most cases occur in small families or are sporadic. It is caused in general by missense $Na_V1.1$ mutations, which are found in about 10 % of families (Marini and Mantegazza 2010).

$Na_V1.1$ mutations have also been identified in some patients presenting with different epileptic encephalopathies, ranging from Lennox-Gastaut syndrome to epilepsy aphasia syndrome (Depienne et al. 2009; Marini and Mantegazza 2010; Carvill et al. 2013b; Guerrini et al. 2014). One of the mildest epileptic phenotypes associated with missense $Na_V1.1$ mutations is benign simple febrile seizures (sFS) (Mantegazza et al. 2005), although some patients of this family developed also mesial temporal lobe epilepsy with hippocampal sclerosis (MTLE&HS). Interestingly, genome-wide association studies have linked *SCN1A* single nucleotide polymorphisms (rs7587026 and rs11692675) to development of MTLE&HS upon a history of febrile seizures (Kasperaviciute et al. 2013), and a further *SCN1A* polymorphism has been identified as risk factor for idiopathic/genetic generalized epilepsies (rs11890028) (Steffens et al. 2012). Familial hemiplegic migraine (FHM) is a rare severe autosomal dominant inherited subtype of migraine with aura characterized by hemiparesis during the attacks (Vecchia and Pietrobon 2012). Some FHM families carry missense $Na_V1.1$ mutations, in some cases without any signs of epileptic phenotypes (Cestele et al. 2013a).

In vitro functional studies of missense epileptogenic $Na_V1.1$ mutations in heterologous systems have initially shown controversial results (Mantegazza et al. 2010b; Mantegazza 2011), revealing both gain- and loss-of-function effects, but loss-of function seems to be the predominant mechanism of action of both truncations and missense mutations. In fact, consistently with the phenotype of DS patients, mouse models of $Na_V1.1$ truncating DS mutations exhibit spontaneous seizures, cognitive impairment and reduction of Na^+ current selectively in GABAergic inhibitory interneurons, causing reduction of their excitability and of GABAergic inhibition, leading to network hyperexcitability (Yu et al. 2006; Ogiwara et al. 2007; Han et al. 2012; Ito et al. 2012; Liautard et al. 2013). $Na_V1.1$ truncations lead to haploinsufficiency without negative dominance (Bechi et al. 2012). Studies of animal models have confirmed that also GEFS+ mutations cause loss of function of $Na_V1.1$ and induce reduced excitability of GABAergic neurons (Tang et al. 2009; Martin et al. 2010).

Although some light has been shed on the pathomechanism of $Na_V1.1$ mutations, the causes of the striking phenotypic variability observed in some patients are still not clear. For instance, some $Na_V1.1$ mutations can cause phenotypes extending from different types of epilepsy to familial hemiplegic migraine (Cestele et al. 2013b). Some of the phenotypic variability can be linked to the combined action of mutations/variants in different genes. It has been shown that a $Na_V1.1$ missense mutant, identified in families with extreme phenotypes comprising Dravet syndrome, causes loss of function because of folding defects that can be rescued by molecular interactions with associated proteins or pharmacological chaperones (Cestele et al. 2008). These results have been confirmed and extended to typical GEFS+ families and Dravet syndrome patients with de novo mutations (Rusconi et al. 2009; Sugiura et al. 2012; Thompson et al. 2012). This mechanism may generate phenotypic variability because of genetic background-dependent variability in rescue, and also be possibly used in the development of therapeutic approaches. As recently shown for a missense *SCN1A* mutant identified in a family with pure FHM (Cestele et al. 2013a), rescue of folding defects may also transform non-functional loss of function mutants (effect that is consistent with severe epilepsy) into gain of function ones, effect that is consistent with familial hemiplegic migraine (Cestele et al. 2008). Genetic background can modulate also the effect of $Na_V1.1$ truncating mutations, because there are reports of mild phenotypes or no phenotype in some individuals carrying these mutations (Orrico et al. 2009; Klassen et al. 2011).

Rare mutations in the *SCN1B* gene, coding for the Na^+ channel β1 subunit have been identified in both GEFS+ (Wallace et al. 1998; Meadows et al. 2002) and DS (Patino et al. 2009).

Similarly, few mutations of the *GABRG2* gene (coding the γ2 subunit of the gamma-aminobutyric acid receptor type A, GABA-A, ionotropic heteropentameric receptor of the main inhibitory neutrotransmitter of the brain) can cause GEFS+, sometimes with DS phenotypes (Baulac et al. 2001). Functional expression of some *GABRG2* mutations, identified in patients with GEFS+, revealed a pronounced loss-of-function by altered gating or defective trafficking and reduced surface expression as a common pathogenic mechanism (Macdonald et al. 2010). Knock-in mouse models of the R43Q *GABRG2* mutation shows generalized seizures and reduced GABAergic inhibition (Chiu et al. 2008). Recently, also mutations of the GABRA1 gene (coding for the α1 subunit of the GABA-A receptor) have been identified in few Dravet syndrome patients (Carvill et al. 2014). Hence, these mutations could reduce GABAergic neurotransmission, the main mechanism for neuronal inhibition in the brain, similarly to *SCN1A* mutations, which may explain the occurrence of seizures.

Mutations of the SCN2A Voltage Gated Na^+ Channel and of KCNQ2-KCNQ3 K^+ Channels: An Unexpected Spectrum of Severity

SCN2A/$Na_V1.2$, KCNQ2 and KCNQ3 have been initially involved in mild benign epilepsies of newborns and infants, but more recently their mutations have also been linked to severe epileptic encephalopathies, showing a spectrum of severity that is similar to that observed for $Na_V1.1$ mutations.

K^+ channels are composed by four subunits forming the ion-conducting pore and generate repolarizing currents that oppose depolarizing currents generated by, e.g. Na^+ channels. KCNQ channels, which consist of homomeric or heteromeric tetramers, are responsible for the so called M-current (muscarinic receptor regulated), which is a non-inactivating K$^+$ current that activates at subthreshold membrane potentials counteracting membrane depolarizations that would lead to action potential generation. Thus, it plays an important role in influencing neuronal firing activity, limiting spiking frequency and reducing the responsiveness to synaptic inputs. *KCNQ2* and *KCNQ3* form a heteromeric K$^+$ channel (Fig. 1), which is particularly important in the axon initial segment and nodes of Ranvier of glutamatergic neurons (Delmas and Brown 2005).

Mutations or deletions/duplications involving one or more exons of *KCNQ2* and, in a smaller number of patients, mutations of KCNQ3 have been identified in benign familial neonatal seizures (BFNS) (Biervert et al. 1998; Singh et al. 1998; Singh et al. 2003). BFNS is characterized by clusters of seizures that appear from the first days of life up to the third month to spontaneously disappear after weeks to months. Seizures have focal onset, often with hemi-tonic or hemiclonic symptoms or apneic spells, or can clinically appear as generalized. Interictal EEG is usually normal. The risk of seizures recurring later in life is about 15% (44). Functional studies in in-vitro systems performed co-expressing heteromeric wild-type and mutant KCNQ2/3 channels revealed a reduction of about 20–30% in the resulting K$^+$ current, which is apparently sufficient to cause BFNS (Maljevic et al. 2008). Although the reduction of the K$^+$ current can cause epileptic seizures by subthreshold membrane depolarization, which increases neuronal firing, it is not fully understood why seizures preferentially occur in neonates (Weber and Lerche 2008). It is possible that the neonatal brain is more vulnerable to changes, even small, of neuronal excitability. Alternatively, *KCNQ2* and *KCNQ3* channels, when mutated, might be replaced by other K$^+$ channels that become functional after the first months of life. Transgenic and knock-in BFNS mice have been generated. Transgenic mice expressing a dominant negative *KCNQ2* mutant (Peters et al. 2005) show spontaneous partial and generalized tonic-clonic seizures, but also pronounced hyperactivity and cell loss in the hippocampus, with impaired hippocampus-related memory, which are not consistent with the typical BNFS human phenotype. Knock-in mice of *KCNQ2* A306T or G311V mutations (Singh et al. 2008) show spontaneous tonic-clonic seizures and, consistently with BNFS, no hippocampal neurodegeneration; however, only homozygous mice manifest epilepsy and they also tend to have seizures in adulthood.

Mutations in *SCN2A*/Na$_V$1.2 have been identified in benign familial neonatal-infantile seizures (BFNIS) (Berkovic et al. 2004). BFNIS are characterized by seizures similar to those observed in children with BFNS, but age at seizure onset ranges from the neonatal period to infancy in different family members, with a mean onset age of 3 months. In general, remission occurs by 12 months with a very low risk of later seizures. Na$_V$1.2 (Fig. 1) is particularly important for the excitability of the axon initial segment and nodes of Ranvier in glutamatergic neurons early in development. Mutations causing BFNIS have been studied in both transfected neocortical neurons and cell lines identifying gain of function effects (Scalmani et al. 2006; Liao et al. 2010) consistent with hyperexcitability of excitatory neurons. The remission may depend on a developmental switch between Na$_V$1.2 and Na$_V$1.6

in myelinated axons that occur at early developmental stages (Scalmani et al. 2006; Liao et al. 2010). No animal models reproducing a BFNIS mutation have been developed yet.

Similarly to $Na_V1.1$ mutations, *KCNQ2* and $Na_V1.2$ mutations can cause a wide phenotypic spectrum that includes severe epileptic encephalopathies. In particular, *KCNQ2* mutations have been identified in severe neonatal epileptic encephalopathies associated with intellectual disability and motor impairment, with a burst-suppression EEG pattern or multifocal epileptiform activity, but also in milder forms (Weckhuysen et al. 2012; Weckhuysen et al. 2013). Functional studies of these mutations have been recently performed in Xenopus Oocytes, showing that they can have more severe loss of function than those causing BFNS (Miceli et al. 2013) or, in some cases, they can cause negative dominance inhibiting function of wild type *KCNQ2* (Orhan et al. 2014). A transgenic mouse model expressing a *KCNQ2* dominant negative mutant may model these forms (Peters et al. 2005).

Similarly, de novo $Na_V1.2$ truncating and missense mutations have been identified in early onset intractable childhood epilepsies, with features ranging from Ohtahara syndrome to Dravet syndrome (Lossin et al. 2012; Nakamura et al. 2013; Touma et al. 2013). Contrary to BFNIS mutants, functional analysis of some of these mutants using in-vitro expression systems has shown loss of function (Lossin et al. 2012), although it is not yet clear how loss-of-function in a Na^+ channel predominantly expressed in excitatory neurons would lead to network hyperexcitability. Notably, a loss of function *Scn2a* knockout mouse does not show an overt epileptic phenotype, although this has not been studied in detail (Planells-Cases et al. 2002).

Neuronal Nicotinic Acetylcholine Receptors and KCNT1 K^+ Channel Mutations in Autosomal Dominant Nocturnal Frontal Lobe Epilepsy

Neuronal nicotinic acetylcholine receptors (nAchR) have important neuromodulatory functions (including modulation of GABA and glutamate release, the main inhibitory and excitatory neurotransmitters of the brain, respectively) and consist of homo- or heteromeric pentamers of various combinations of at least 17 subunits: α1—10, β1—4, δ, ε and γ. The α4-β2 combination is the most common in the thalamus and cerebral cortex. A mutation in the gene *CHRNA4*, encoding the α4-subunit of a neuronal nicotinic acetylcholine receptor (nAchR), was the first ion channel mutation found in an inherited form of epilepsy (Steinlein et al. 1995): autosomal dominant nocturnal frontal lobe epilepsy (ADNFLE). About 15 mutations in *CHRNA4*, five in *CHRNB2*, which encodes the β2-subunit of nAchR, and one in *CHRNA2*, encoding the neuronal nAchR α2-subunit, have been reported so far and account for < 10% of the tested ADNFLE families (Ferini-Strambi et al. 2012). All the identified mutations are located in the pore-forming M2 transmembrane segments.

ADNFLE includes frequent, usually brief, seizures with onset on average at around 10 years of age, with hyperkinetic or tonic manifestations, typically in clusters at night during slow-wave sleep. Paroxysmal arousals, dystonia-like attacks and epileptic nocturnal wanderings are also part of the phenotype. Functional studies of nAchR have produced controversial results, which make the underlying pathogenic mechanisms still unclear. Expression of α4 mutants in heterologous systems resulted in various effects consistent with either gain or loss of function: increased sensitivity to acetylcholine (gain of function); decreased Ca^{2+} potentiation or accelerated desensitization (loss of function) (Bertrand et al. 2002). Mutations of β2 showed gain of function by increased sensitivity to acetylcholine or slower desensitization (De Fusco et al. 2000) and the α2 mutation showed gain of function by increased sensitivity to acetylcholine (Aridon et al. 2006). The α4 mutations S252F and +L264 engineered in knock-in mice (Klaassen et al. 2006) induced spontaneous seizures of various types, in some cases similar to those of the human phenotype, but no paroxysmal arousal and dystonia-like manifestations; whereas the α4 subunit S248F knock-in mice show no spontaneous seizures but nicotine-induced dystonic attacks (Teper et al. 2007). α4-subunit S284L transgenic rats (Zhu et al. 2008) show a more complete ADNFLE phenotype, with spontaneous attacks during slow-wave sleep, comprising of paroxysmal arousals (frightened behavior), dystonic activity and epileptic wandering. Notably, the pathogenic mechanisms are also different in these models: upon application of nicotine, GABAergic inhibition is increased in frontal cortex of S252F and +L264 knock-in mice (Klaassen et al. 2006), whereas it is reduced in somatosensory cortex of S284L transgenic rats (Zhu et al. 2008).

Missense mutations in the Na^+-activated K^+ channel gene *KCNT1* (KCa4.1-SLACK-SLO2.2 type; Fig. 1) have been recently reported in 4 unrelated families with a severe form of ADNFLE (Heron et al. 2012). *KCNT1* is activated by the inward flux of Na^+ ions during neuronal firing and its activity contributes to the slow hyperpolarization that follows repetitive firing. Thus, its action negatively modulates high frequency firing. There is no information about the effect of its ADNFLE mutations because functional analysis was not performed. Patients had an earlier mean age of onset (6 years) compared to other ADNFLE forms and frequently exhibited psychiatric features and intellectual disability. Thus, the phenotype has some features that are typical of epileptic encephalopathies and, interestingly, mutations of KCNT1 have also been identified in the epileptic encephalopathy malignant migrating partial seizures in infancy (see below).

Therefore, genetic variability is evident in this clinically relatively homogenous epileptic form.

Ion Channel Mutations Recently Identified with Whole Exome Sequencing Studies

Current efforts of whole exome sequencing (WES) are generating a great amount of information that is improving our understanding of the pathophysiology of genetic epilepsies, in particular for rare epileptic encephalopathies. Most of the mutations that have been thus far interpreted as causative arise as de novo mutations or are inherited in an autosomal recessive fashion, often as compound heterozygous mutations. Notably, mutations in these new epilepsy genes associated with epileptic encephalopathies with variable phenotypes, occasionally resembling known syndromes including Dravet syndrome, are found in a small number of patients (Allen et al. 2013; Carvill et al. 2013b; Suls et al. 2013). Although mutations in non-ion channel genes have been identified, genes coding for ion channels are still common ones. For instance, mutations of the *HCN1* gene (coding for the type 1 subunit of the hyperpolarization-activated, cyclic nucleotide-gated channel that contributes to the cationic Ih current in neurons and can regulate the excitability of neuronal networks) have been identified in patients with Dravet syndrome-like phenotypes (Nava et al. 2014). Mutations of $SCN8A$/Na$_V$1.6 are associated with other types of early onset epileptic encephalopathies (109, 112); this is the fifth Na$^+$-channel gene to be mutated in epilepsy when variants of $SCN3A$/Na$_V$1.3 are included (Vanoye et al. 2014; Fig. 1).

Mutations in the GRIN2B gene (encoding the NR2B subunit of the heterotetrameric N-methyl-D-aspartate, NMDA, receptor, a ionotropic receptor of the main excitatory neurotransmitter of the brain: glutamate) have also been recently identified in 3 patients with phenotypes that include West syndrome and focal epilepsy with intellectual disability (Lemke et al. 2014).

Mutations in the *GRIN2A* gene encoding the NR2A subunit of the NMDA glutamate receptor have been identified in epilepsy-aphasia syndromes (EAS), which are a group of severe epileptic encephalopathies with a characteristic EEG pattern and developmental regression particularly affecting language that include Landau-Kleffner Syndrome (LKS) and continuous spike-waves in slow sleep (CSWS); functional analysis of these mutations using in-vitro systems has shown gain of function (Carvill et al. 2013a; Lemke et al. 2013; Lesca et al. 2013). Phenotypes of GRIN2A patients appear to be very different from those of GRIN2B mutations.

WES and targeted sequencing studies have also showed that ion channel genes previously associated with milder phenotypes can also cause epileptic encephalopathies in some patients (Carvill et al. 2013b), including *GABRA1* (coding for the α1 subunit of the GABA-A receptor), which has been previously involved in mild idiopathic generalized epilepsy (see below) and, as highlighted above, *SCN2A* Na$^+$ channel and *KCNQ2* K$^+$ channel genes. Similarly, de-novo gain of function mutations in the *KCNT1* gene have been identified in patients with malignant migrating partial seizures of infancy (MMPSI), a rare syndrome with infantile onset intractable and migrating focal seizures with severe impairment of psychomotor development (Barcia et al. 2012), but KCNT1 mutations have also been associated to other and very different epilepsy phenotypes, including autosomal dominant nocturnal frontal lobe epilepsy (Heron et al. 2012) (see above).

Idiopathic Generalized Epilepsies with Complex Inheritance

The idiopathic generalized epilepsies (IGE) represent 20–30% of all epilepsies and are formed by a group of syndromes including childhood absence epilepsy (CAE), juvenile absence epilepsy (JAE), juvenile myoclonic epilepsy (JME), and epilepsy with generalized tonic-clonic seizures alone. Among epilepsies with complex inheritance, IGEs have long been considered to be particularly suitable for genetic studies because they are common, have a relatively well-defined phenotype and often occur in familial clusters. It has been initially suggested that they might result from the interaction of two or few genes (Berkovic and Scheffer 2001), but a high degree of genetic complexity appears from large scale exome sequencing of ion channels, which showed that rare missense variations in known Mendelian disease genes are equally prevalent in healthy individuals and in those with idiopathic generalized epilepsy, revealing that even probably deleterious ion channel mutations confer an uncertain risk to an individual, depending on the other variants with which they are combined (Klassen et al. 2011). Linkage studies on a large number of families with IGE have suggested several susceptibility loci (Sander et al. 2000; Durner et al. 2001) and some variants have been identified. In rare families pathogenic mutations in single ion channel genes have been reported. Mutations in the *GABRG2* gene (coding the γ2 subunit of the GABA-A receptor, which is mutated also in GEFS+, see above) were identified in families with febrile seizures and childhood absence epilepsy and cause loss of function by reduced membrane targeting (trafficking defects) or by accelerated desensitization and reduced sensitivity to benzodiazezine, consistently with reduced GABAergic inhibition (Wallace et al. 2001; Kang et al. 2006). Mutations in the *GABRA1* gene (coding the α1 subunit of the GABA-A receptor) were identified in families with juvenile myoclonic epilepsy or childhood absence epilepsy, and cause loss of function by protein truncation or folding/trafficking defects (Cossette et al. 2002; Gallagher et al. 2007). Mutations of the *GABRB3* gene, coding for the GABA-A receptor β3 subunit have been identified in small families showing childhood absence epilepsy, and cause loss of function but with an unclear mechanism (Tanaka et al. 2008). Mutations in *CLCN2* (coding the ClC2 Cl⁻ channel) were identified in families with heterogeneous IGE phenotypes, including childhood absences epilepsy (D'Agostino et al. 2004). Variants in *CACNA1H*, coding for the Cav3.2 T-type voltage-gated Ca^{2+} channel (Fig. 1) that is mainly targeted to neuronal cell bodies and dendrites (highly expressed in thalamic neurons), have been identified in childhood absence epilepsy and other generalized epilepsy phenotypes; they cause various modifications in gating properties often consistent with gain of function, but differences in functional effects have been observed between rat and human clones, and between splicing variants (Chen et al. 2003; Khosravani et al. 2004). Consistently with the human findings, it has been shown that the *Cacna1h* variant R1584P is a susceptibility factor in a spontaneous rat model of absences (GAERS) (Powell et al. 2009).

Other Ion Channels Involved in Genetic Epilepsy

Other mutations or variants of other ion channels have been identified in few families or sporadic patients. Some of them show mendelian inheritance with high penetrance, but many are probably limited to few families.

For instance, a gain of function mutation of the *KCNMA1* gene coding the KCa1.1 large conductance Ca^{2+} activated K^+ channel (BK), which is activated by depolarizations and intracellular Ca^{2+} and is implicated in action potential repolarization and after-hyperpolarizations, has been identified in a family with generalized epilepsy and paroxysmal dyskinesia. Gain of function may increase excitability by inducing more rapid repolarization and increasing firing frequency (Du et al. 2005).

A loss of function mutations of the *KCNA1* gene, coding for the Kv1.1 subunit of delayed rectifier voltage gated K^+ channels that forms homo- or hetero-tetramers with other Kv1 subunits, have been identified in a family with partial epilepsy associated with myokymia (Eunson et al. 2000).

Loss of function mutations of the *KCNJ10* gene, coding for the Kir4.1 glial ATP-dependent inwardly rectifier K^+ channel implicated in K^+ buffering in the brain have been identified in complex phenotypes comprising generalized epilepsy, ataxia, sensori-neural deafness, tubulopathy and mental retardation; they probably impair K^+ buffering (Bockenhauer et al. 2009; Scholl et al. 2009).

Mutations of the *CACNA1A* gene, coding for the $Ca_v2.1$ P/Q high voltage activated (HVA) Ca^{2+} channel that is mainly targeted to presynaptic terminals, have been found in few families showing childhood absence epilepsia and ataxia, and cause loss of function consistent with decreased neurotransmission (Imbrici et al. 2004). Interestingly, mutations of $Ca_v2.1$ have been also involved in familial hemiplegic migraine, and data from animal models show that they induce gain of function selectively in glutamatergic neurons (Vecchia and Pietrobon 2012). Mutations of *CACNB4* coding for the β4 auxiliary subunit of HVA Ca^{2+} channels (Cav1.x, Cav2.x) have been identified in few families showing IGE and episodic ataxia, but their functional effect was consistent with gain of function, differently than for epileptogenic $Ca_v2.1$ mutations (Escayg et al. 2000).

A recessive loss-of-function missense mutation in the HCN2 gene (coding for the type2 subunit of the hyperpolarization-activated, cyclic nucleotide-gated channel) has been found in a patient with sporadic idiopathic generalized epilepsy; the proband was the only affected member of the family and homozygous for the mutation (DiFrancesco et al. 2011). Functional analysis revealed that the homomeric mutants, but not heteromeric wild-type/mutant channels, show loss of function and induce hyperexcitability in transfected cultured neurons.

Among intracellular channels, mutations of the ATP synthase proton channel (respiratory chain complex V) have been definitively identified as the cause of NARP, in which seizures are part of a complex phenotype comprising neuropathy, ataxia and retinitis pigmentosa (DiMauro and Schon 2008).

Beyond Channelopathies

Although mutations of ion channels are a main pathomechanism of genetic epilepsies, as outlined above, several genes that do not encode for ion channels have been implicated as well. For instance, mutations of protocadherin delta-2 subclass of the cadherin super family (PCDH19), Aristaless-related homeobox (ARX), cyclin-dependent, kinase-like 5 (CDKL5) and syntaxin binding protein 1 (STXBP1) genes have been identified in monogenic epilepsies without brain malformations that often manifest as severe forms, with features of epileptic encephalopathy, often having early seizure onset and developmental delay, even if with diverse and distinctive phenotypes, sometimes resembling Dravet syndrome (Guerrini et al. 2014).

Other examples are mutations of the *LGI1* gene (leucine-rich, glioma-inactivated 1) that have been associated to autosomal dominant temporal lobe epilepsy (ADTLE), a form of autosomal dominant partial epilepsy associated to auditory symptoms and audiogenic seizures (Kalachikov et al. 2002; Morante-Redolat et al. 2002). The pathogenetic mechanism related to *LGI1* mutations remains to be clarified. Results obtained from animal models are more consistent with a direct presynaptic mechanism of hyperexcitability involving modulation of K^+ channels, whereas postsynaptic effects may be more involved in glutamatergic synapse maturation and dendritic pruning (Schulte et al. 2006; Zhou et al. 2009).

Recent studies have identified in some families presenting with familial focal epilepsy with variable foci mutations in Dishevelled, Egl-10 and Pleckstrin domain-containing (DEPDC5) protein, whose function is still unclear but might be involved in membrane trafficking, G protein signaling and/or modulation of the mTOR complex 1 (Dibbens et al. 2013; Picard et al. 2014). Most mutations resulted in a truncated protein and are consistent with loss of function.

De-novo mutations of CHD2 (encoding chromodomain helicase DNA binding protein 2) have been recently identified in three patients with a fever-sensitive myoclonic epileptic encephalopathy sharing features with Dravet syndrome (Suls et al. 2013).

Conclusions

Numerous epileptogenic mutations have been identified in plasma-membrane ion channel genes in part also because of the bias induced by the prior knowledge of their importance for neuronal excitability. Recent unbiased research efforts have identified epileptogenic mutations also in genes that do not code for ion channels and in the future the list of these genes will probably be expanded, also expanding the type and number of epileptogenic mechanisms. However, it is feasible that for several of them a role in ion channel regulation, targeting or expression will be discovered.

The hypothesis that mutations in a 'few and common genes' might cause phenotypes with overlapping clinical features needs to be revised, because genetic findings emerging from WES studies of large cohorts of patients, realized by consortia of laboratories, rather suggest several patient-specific mutations in several genes to be at play (Allen et al. 2013). The proteins codified by the genes whose mutations/variants contribute to a specific phenotype might be localized within a so-called "functional network system".

Thus, phenotypic and genetic heterogeneity are common in genetic epilepsies and may be explained by pleiotropic expression of a single-gene mutation, modifying genes, or by several genes producing a similar phenotype, at times because they affect the same developmental or functional pathway. Thus, it will be essential to compare clinical and experimental studies in order to better disclose pathogenic mechanisms that lead to a particular form of epilepsy.

References

Allen AS et al (2013) De novo mutations in epileptic encephalopathies. Nature 501:217–221

Aridon P, Marini C, Di Resta C, Brilli E, De Fusco M, Politi F, Parrini E, Manfredi I, Pisano T, Pruna D, Curia G, Cianchetti C, Pasqualetti M, Becchetti A, Guerrini R, Casari G (2006) Increased sensitivity of the neuronal nicotinic receptor alpha 2 subunit causes familial epilepsy with nocturnal wandering and ictal fear. Am J Hum Genet 79:342–350

Ashcroft FM (2006) From molecule to malady. Nature 440:440–447

Avanzini G, Franceschetti S (2003) Cellular biology of epileptogenesis. Lancet Neurol 2:33–42

Avanzini G, Franceschetti S, Mantegazza M (2007) Epileptogenic channelopathies: experimental models of human pathologies. Epilepsia 48(Suppl 2):51–64

Barcia G, Fleming MR, Deligniere A, Gazula VR, Brown MR, Langouet M, Chen H, Kronengold J, Abhyankar A, Cilio R, Nitschke P, Kaminska A, Boddaert N, Casanova JL, Desguerre I, Munnich A, Dulac O, Kaczmarek LK, Colleaux L, Nabbout R (2012) De novo gain-of-function KCNT1 channel mutations cause malignant migrating partial seizures of infancy. Nat Genet 44:1255–1259

Baulac S, Huberfeld G, Gourfinkel-An I, Mitropoulou G, Beranger A, Prud'homme JF, Baulac M, Brice A, Bruzzone R, Leguern E (2001) First genetic evidence of GABA(A) receptor dysfunction in epilepsy: a mutation in the gamma2-subunit gene. Nat Genet 28:46–48

Bechi G, Scalmani P, Schiavon E, Rusconi R, Franceschetti S, Mantegazza M (2012) Pure haploinsufficiency for Dravet syndrome Na(V)1.1 (SCN1A) sodium channel truncating mutations. Epilepsia 53:87–100

Berkovic SF, Scheffer IE (2001) Genetics of the epilepsies. Epilepsia 42(Suppl 5):16–23

Berkovic SF, Heron SE, Giordano L, Marini C, Guerrini R, Kaplan RE, Gambardella A, Steinlein OK, Grinton BE, Dean JT, Bordo L, Hodgson BL, Yamamoto T, Mulley JC, Zara F, Scheffer IE (2004) Benign familial neonatal-infantile seizures: characterization of a new sodium channelopathy. Ann Neurol 55:550–557

Berkovic SF, Mulley JC, Scheffer IE, Petrou S (2006) Human epilepsies: interaction of genetic and acquired factors. Trends Neurosci 29:391–397

Bertrand D, Picard F, Le Hellard S, Weiland S, Favre I, Phillips H, Bertrand S, Berkovic SF, Malafosse A, Mulley J (2002) How mutations in the nAChRs can cause ADNFLE epilepsy. Epilepsia 43(Suppl 5):112–122

Biervert C, Schroeder BC, Kubisch C, Berkovic SF, Propping P, Jentsch TJ, Steinlein OK (1998) A potassium channel mutation in neonatal human epilepsy. Science 279:403–406

Bockenhauer D et al (2009) Epilepsy, ataxia, sensorineural deafness, tubulopathy, and KCNJ10 mutations. N Engl J Med 360:1960–1970

Carvill GL et al (2013a) GRIN2A mutations cause epilepsy-aphasia spectrum disorders. Nat Genet 45:1073–1076

Carvill GL et al (2013b) Targeted resequencing in epileptic encephalopathies identifies de novo mutations in CHD2 and SYNGAP1. Nat Genet 45:825–830

Carvill GL et al (2014) GABRA1 and STXBP1: novel genetic causes of Dravet syndrome. Neurology 82:1245–1253

Catterall WA (2012) Voltage-gated sodium channels at 60: structure, function and pathophysiology. J Physiol 590:2577–2589

Cestele S, Scalmani P, Rusconi R, Terragni B, Franceschetti S, Mantegazza M (2008) Self-limited hyperexcitability: functional effect of a familial hemiplegic migraine mutation of the Nav1.1 (SCN1A) Na^+ channel. J Neurosci 28:7273–7283

Cestele S, Schiavon E, Rusconi R, Franceschetti S, Mantegazza M (2013a) Nonfunctional NaV1.1 familial hemiplegic migraine mutant transformed into gain of function by partial rescue of folding defects. Proc Natl Acad Sci USA 110:17546–17551

Cestele S, Labate A, Rusconi R, Tarantino P, Mumoli L, Franceschetti S, Annesi G, Mantegazza M, Gambardella A (2013b) Divergent effects of the T1174S SCN1A mutation associated with seizures and hemiplegic migraine. Epilepsia 54:927–935

Chen Y, Lu J, Pan H, Zhang Y, Wu H, Xu K, Liu X, Jiang Y, Bao X, Yao Z, Ding K, Lo WH, Qiang B, Chan P, Shen Y, Wu X (2003) Association between genetic variation of CACNA1H and childhood absence epilepsy. Ann Neurol 54:239–243

Chiu C, Reid CA, Tan HO, Davies PJ, Single FN, Koukoulas I, Berkovic SF, Tan SS, Sprengel R, Jones MV, Petrou S (2008) Developmental impact of a familial GABAA receptor epilepsy mutation. Ann Neurol 64:284–293

Claes L, Del Favero J, Ceulemans B, Lagae L, Van Broeckhoven C, De Jonghe P (2001) De novo mutations in the sodium-channel gene SCN1A cause severe myoclonic epilepsy of infancy. Am J Hum Genet 68:1327–1332

Cossette P, Liu L, Brisebois K, Dong H, Lortie A, Vanasse M, Saint-Hilaire JM, Carmant L, Verner A, Lu WY, Wang YT, Rouleau GA (2002) Mutation of GABRA1 in an autosomal dominant form of juvenile myoclonic epilepsy. Nat Genet 31:184–189

D'Agostino D, Bertelli M, Gallo S, Cecchin S, Albiero E, Garofalo PG, Gambardella A, St Hilaire JM, Kwiecinski H, Andermann E, Pandolfo M (2004) Mutations and polymorphisms of the CLCN2 gene in idiopathic epilepsy. Neurology 63:1500–1502

De Fusco M, Becchetti A, Patrignani A, Annesi G, Gambardella A, Quattrone A, Ballabio A, Wanke E, Casari G (2000) The nicotinic receptor beta 2 subunit is mutant in nocturnal frontal lobe epilepsy. Nat Genet 26:275–276

Delmas P, Brown DA (2005) Pathways modulating neural KCNQ/M (Kv7) potassium channels. Nat Rev Neurosci 6:850–862

Depienne C, Trouillard O, Saint-Martin C, Gourfinkel-An I, Bouteiller D, Carpentier W, Keren B, Abert B, Gautier A, Baulac S, Arzimanoglou A, Cazeneuve C, Nabbout R, LeGuern E (2009) Spectrum of SCN1A gene mutations associated with Dravet syndrome: analysis of 333 patients. J Med Genet 46:183–191

Dibbens LM et al (2013) Mutations in DEPDC5 cause familial focal epilepsy with variable foci. Nat Genet 45:546–551

DiFrancesco JC, Barbuti A, Milanesi R, Coco S, Bucchi A, Bottelli G, Ferrarese C, Franceschetti S, Terragni B, Baruscotti M, DiFrancesco D (2011) Recessive loss-of-function mutation in the pacemaker HCN2 channel causing increased neuronal excitability in a patient with idiopathic generalized epilepsy. J Neurosci 31:17327–17337

DiMauro S, Schon EA (2008) Mitochondrial disorders in the nervous system. Annu Rev Neurosci 31:91–123

Dravet C, Bureau M, Oguni H, Fukuyama Y, Cokar O (2005) Severe myoclonic epilepsy in infancy: Dravet syndrome. Adv Neurol 95:71–102

Du W, Bautista JF, Yang H, Diez-Sampedro A, You SA, Wang L, Kotagal P, Luders HO, Shi J, Cui J, Richerson GB, Wang QK (2005) Calcium-sensitive potassium channelopathy in human epilepsy and paroxysmal movement disorder. Nat Genet 37:733–738

Durner M, Keddache MA, Tomasini L, Shinnar S, Resor SR, Cohen J, Harden C, Moshe SL, Rosenbaum D, Kang H, Ballaban-Gil K, Hertz S, Labar DR, Luciano D, Wallace S, Yohai

D, Klotz I, Dicker E, Greenberg DA (2001) Genome scan of idiopathic generalized epilepsy: evidence for major susceptibility gene and modifying genes influencing the seizure type. Ann Neurol 49:328–335

Escayg A, De Waard M, Lee DD, Bichet D, Wolf P, Mayer T, Johnston J, Baloh R, Sander T, Meisler MH (2000) Coding and noncoding variation of the human calcium-channel beta4-subunit gene CACNB4 in patients with idiopathic generalized epilepsy and episodic ataxia. Am J Hum Genet 66:1531–1539

Eunson LH, Rea R, Zuberi SM, Youroukos S, Panayiotopoulos CP, Liguori R, Avoni P, McWilliam RC, Stephenson JB, Hanna MG, Kullmann DM, Spauschus A (2000) Clinical, genetic, and expression studies of mutations in the potassium channel gene KCNA1 reveal new phenotypic variability. Ann Neurol 48:647–656

Ferini-Strambi L, Sansoni V, Combi R (2012) Nocturnal frontal lobe epilepsy and the acetylcholine receptor. Neurologist 18:343–349

Gallagher MJ, Ding L, Maheshwari A, Macdonald RL (2007) The GABAA receptor alpha1 subunit epilepsy mutation A322D inhibits transmembrane helix formation and causes proteasomal degradation. Proc Natl Acad Sci USA 104:12999–13004

Guerrini R, Marini C, Mantegazza M (2014) Genetic Epilepsy Syndromes Without Structural Brain Abnormalities: Clinical Features and Experimental Models. Neurotherapeutics 11:269–285

Han S, Tai C, Westenbroek RE, Yu FH, Cheah CS, Potter GB, Rubenstein JL, Scheuer T, de la Iglesia HO, Catterall WA (2012) Autistic-like behaviour in Scn1a+/− mice and rescue by enhanced GABA-mediated neurotransmission. Nature 489:385–390

Helbig I, Scheffer IE, Mulley JC, Berkovic SF (2008) Navigating the channels and beyond: unravelling the genetics of the epilepsies. Lancet Neurol 7:231–245

Heron SE, Smith KR, Bahlo M, Nobili L, Kahana E, Licchetta L, Oliver KL, Mazarib A, Afawi Z, Korczyn A, Plazzi G, Petrou S, Berkovic SF, Scheffer IE, Dibbens LM (2012) Missense mutations in the sodium-gated potassium channel gene KCNT1 cause severe autosomal dominant nocturnal frontal lobe epilepsy. Nat Genet 44:1188–1190

Imbrici P, Jaffe SL, Eunson LH, Davies NP, Herd C, Robertson R, Kullmann DM, Hanna MG (2004) Dysfunction of the brain calcium channel CaV2.1 in absence epilepsy and episodic ataxia. Brain 127:2682–2692

Ito S, Ogiwara I, Yamada K, Miyamoto H, Hensch TK, Osawa M, Yamakawa K (2012) Mouse with Na(v)1.1 haploinsufficiency, a model for Dravet syndrome, exhibits lowered sociability and learning impairment. Neurobiol Dis 49C:29–40

Kalachikov S, Evgrafov O, Ross B, Winawer M, Barker-Cummings C, Martinelli BF, Choi C, Morozov P, Das K, Teplitskaya E, Yu A, Cayanis E, Penchaszadeh G, Kottmann AH, Pedley TA, Hauser WA, Ottman R, Gilliam TC (2002) Mutations in LGI1 cause autosomal-dominant partial epilepsy with auditory features. Nat Genet 30:335–341

Kang JQ, Shen W, Macdonald RL (2006) Why does fever trigger febrile seizures? GABAA receptor gamma2 subunit mutations associated with idiopathic generalized epilepsies have temperature-dependent trafficking deficiencies. J Neurosci 26:2590–2597

Kasperaviciute D et al (2013) Epilepsy, hippocampal sclerosis and febrile seizures linked by common genetic variation around SCN1A. Brain 136:3140–3150

Kass RS (2005) The channelopathies: novel insights into molecular and genetic mechanisms of human disease. J ClinInvest 115:1986–1989

Khosravani H, Altier C, Simms B, Hamming KS, Snutch TP, Mezeyova J, McRory JE, Zamponi GW (2004) Gating effects of mutations in the Cav3.2 T-type calcium channel associated with childhood absence epilepsy. J Biol Chem 279:9681–9684

Klaassen A, Glykys J, Maguire J, Labarca C, Mody I, Boulter J (2006) Seizures and enhanced cortical GABAergic inhibition in two mouse models of human autosomal dominant nocturnal frontal lobe epilepsy. Proc Natl Acad Sci USA 103:19152–19157

Klassen T, Davis C, Goldman A, Burgess D, Chen T, Wheeler D, McPherson J, Bourquin T, Lewis L, Villasana D, Morgan M, Muzny D, Gibbs R, Noebels J (2011) Exome sequencing of ion channel genes reveals complex profiles confounding personal risk assessment in epilepsy. Cell 145:1036–1048

Kullmann DM (2010) Neurological channelopathies. Annu Rev Neurosci 33:151–172
Lemke JR et al (2013) Mutations in GRIN2A cause idiopathic focal epilepsy with rolandic spikes. Nat Genet 45:1067–1072
Lemke JR, Hendrickx R, Geider K, Laube B, Schwake M, Harvey RJ, James VM, Pepler A, Steiner I, Hortnagel K, Neidhardt J, Ruf S, Wolff M, Bartholdi D, Caraballo R, Platzer K, Suls A, De Jonghe P, Biskup S, Weckhuysen S (2014) GRIN2B mutations in West syndrome and intellectual disability with focal epilepsy. Ann Neurol 75:147–154
Lesca G et al (2013) GRIN2A mutations in acquired epileptic aphasia and related childhood focal epilepsies and encephalopathies with speech and language dysfunction. Nat Genet 45:1061–1066
Liao Y, Deprez L, Maljevic S, Pitsch J, Claes L, Hristova D, Jordanova A, Ala-Mello S, Bellan-Koch A, Blazevic D, Schubert S, Thomas EA, Petrou S, Becker AJ, De Jonghe P, Lerche H (2010) Molecular correlates of age-dependent seizures in an inherited neonatal-infantile epilepsy. Brain 133:1403–1414
Liautard C, Scalmani P, Carriero G, de Curtis M, Franceschetti S, Mantegazza M (2013) Hippocampal hyperexcitability and specific epileptiform activity in a mouse model of Dravet syndrome. Epilepsia 54:1251–1261
Lossin C, Shi X, Rogawski MA, Hirose S (2012) Compromised function in the Na(v)1.2 Dravet syndrome mutation R1312T. Neurobiol Dis 47:378–384
Macdonald RL, Kang JQ, Gallagher MJ (2010) Mutations in GABAA receptor subunits associated with genetic epilepsies. J Physiol 588:1861–1869
Maljevic S, Wuttke TV, Lerche H (2008) Nervous system KV7 disorders: breakdown of a subthreshold brake. J Physiol 586:1791–1801
Mantegazza M (2011) Dravet syndrome: insights from in vitro experimental models. Epilepsia 52(Suppl 2):62–69
Mantegazza M, Catterall WA (2012) Voltage-gated Na$^+$ channels: structure, function, and pathophysiology
Mantegazza M, Gambardella A, Rusconi R, Schiavon E, Annesi F, Cassulini RR, Labate A, Carrideo S, Chifari R, Canevini MP, Canger R, Franceschetti S, Annesi G, Wanke E, Quattrone A (2005) Identification of an Nav1.1 sodium channel (SCN1A) loss-of-function mutation associated with familial simple febrile seizures. Proc Natl Acad Sci USA 102:18177–18182
Mantegazza M, Curia G, Biagini G, Ragsdale DS, Avoli M (2010a) Voltage-gated sodium channels as therapeutic targets in epilepsy and other neurological disorders. Lancet Neurol 9:413–424
Mantegazza M, Rusconi R, Scalmani P, Avanzini G, Franceschetti S (2010b) Epileptogenic ion channel mutations: from bedside to bench and, hopefully, back again. Epilepsy Res 92:1–29
Marini C, Mantegazza M (2010) Sodium channelopathies and epilepsy: recent advances and new perspectives. Expert Rev Clinical Pharmacology 3:371–384
Martin MS, Dutt K, Papale LA, Dube CM, Dutton SB, de Haan G, Shankar A, Tufik S, Meisler MH, Baram TZ, Goldin AL, Escayg A (2010) Altered function of the SCN1A voltage-gated sodium channel leads to gamma-aminobutyric acid-ergic (GABAergic) interneuron abnormalities. J Biol Chem 285:9823–9834
McCormick DA, Contreras D (2001) On the cellular and network bases of epileptic seizures. Annu Rev Physiol 63:815–846
Meadows LS, Malhotra J, Loukas A, Thyagarajan V, Kazen-Gillespie KA, Koopman MC, Kriegler S, Isom LL, Ragsdale DS (2002) Functional and biochemical analysis of a sodium channel beta1 subunit mutation responsible for generalized epilepsy with febrile seizures plus type 1. J Neurosci 22:10699–10709
Miceli F, Soldovieri MV, Ambrosino P, Barrese V, Migliore M, Cilio MR, Taglialatela M (2013) Genotype-phenotype correlations in neonatal epilepsies caused by mutations in the voltage sensor of K(v)7.2 potassium channel subunits. Proc Natl Acad Sci USA 110:4386–4391
Morante-Redolat JM et al (2002) Mutations in the LGI1/Epitempin gene on 10q24 cause autosomal dominant lateral temporal epilepsy. Hum Mol Genet 11:1119–1128
Nakamura K et al (2013) Clinical spectrum of SCN2A mutations expanding to Ohtahara syndrome. Neurology 81:992–998

Nava C et al (2014) De novo mutations in HCN1 cause early infantile epileptic encephalopathy. Nat Genet 46:640–645

Ogiwara I, Miyamoto H, Morita N, Atapour N, Mazaki E, Inoue I, Takeuchi T, Itohara S, Yanagawa Y, Obata K, Furuichi T, Hensch TK, Yamakawa K (2007) Na(v)1.1 localizes to axons of parvalbumin-positive inhibitory interneurons: a circuit basis for epileptic seizures in mice carrying an Scn1a gene mutation. J Neurosci 27:5903–5914

Orhan G, Bock M, Schepers D, Ilina EI, Reichel SN, Loffler H, Jezutkovic N, Weckhuysen S, Mandelstam S, Suls A, Danker T, Guenther E, Scheffer IE, De Jonghe P, Lerche H, Maljevic S (2014) Dominant-negative effects of KCNQ2 mutations are associated with epileptic encephalopathy. Ann Neurol 75:382–394

Orrico A, Galli L, Grosso S, Buoni S, Pianigiani R, Balestri P, Sorrentino V (2009) Mutational analysis of the SCN1A, SCN1B and GABRG2 genes in 150 Italian patients with idiopathic childhood epilepsies. Clin Genet 75:579–581

Patino GA, Claes LR, Lopez-Santiago LF, Slat EA, Dondeti RS, Chen C, O'Malley HA, Gray CB, Miyazaki H, Nukina N, Oyama F, De Jonghe P, Isom LL (2009) A functional null mutation of SCN1B in a patient with Dravet syndrome. J Neurosci 29:10764–10778

Peters HC, Hu H, Pongs O, Storm JF, Isbrandt D (2005) Conditional transgenic suppression of M channels in mouse brain reveals functions in neuronal excitability, resonance and behavior. Nat Neurosci 8:51–60

Picard F et al (2014) DEPDC5 mutations in families presenting as autosomal dominant nocturnal frontal lobe epilepsy. Neurology 82:2101–2106

Planells-Cases R, Caprini M, Zhang J, Rockenstein EM, Rivera RR, Murre C, Masliah E, Montal M (2000) Neuronal death and perinatal lethality in voltage-gated sodium channel alpha(II)-deficient mice. Biophys J 78:2878–2891

Powell KL, Cain SM, Ng C, Sirdesai S, David LS, Kyi M, Garcia E, Tyson JR, Reid CA, Bahlo M, Foote SJ, Snutch TP, O'Brien TJ (2009) A Cav3.2 T-type calcium channel point mutation has splice-variant-specific effects on function and segregates with seizure expression in a polygenic rat model of absence epilepsy. J Neurosci 29:371–380

Ptacek LJ (1997) Channelopathies: ion channel disorders of muscle as a paradigm for paroxysmal disorders of the nervous system. Neuromuscul Disord 7:250–255

Ptacek LJ, George AL Jr, Griggs RC, Tawil R, Kallen RG, Barchi RL, Robertson M, Leppert MF (1991) Identification of a mutation in the gene causing hyperkalemic periodic paralysis. Cell 67:1021–1027

Reid CA, Berkovic SF, Petrou S (2009) Mechanisms of human inherited epilepsies. Prog Neurobiol 87:41–57

Rusconi R, Combi R, Cestele S, Grioni D, Franceschetti S, Dalpra L, Mantegazza M (2009) A rescuable folding defective Nav1.1 (SCN1A) sodium channel mutant causes GEFS+: common mechanism in Nav1.1 related epilepsies? Hum Mutat 30:E747–E760

Sander T et al (2000) Genome search for susceptibility loci of common idiopathic generalised epilepsies. Hum Mol Genet 9:1465–1472

Scalmani P, Rusconi R, Armatura E, Zara F, Avanzini G, Franceschetti S, Mantegazza M (2006) Effects in neocortical neurons of mutations of the Na(v)1.2 Na$^+$ channel causing benign familial neonatal-infantile seizures. J Neurosci 26:10100–10109

Scheffer IE, Berkovic SF (1997) Generalized epilepsy with febrile seizures plus. A genetic disorder with heterogeneous clinical phenotypes. Brain 120(Pt 3):479–490

Scholl UI, Choi M, Liu T, Ramaekers VT, Hausler MG, Grimmer J, Tobe SW, Farhi A, Nelson-Williams C, Lifton RP (2009) Seizures, sensorineural deafness, ataxia, mental retardation, and electrolyte imbalance (SeSAME syndrome) caused by mutations in KCNJ10. Proc Natl Acad Sci USA 106:5842–5847

Schulte U, Thumfart JO, Klocker N, Sailer CA, Bildl W, Biniossek M, Dehn D, Deller T, Eble S, Abbass K, Wangler T, Knaus HG, Fakler B (2006) The epilepsy-linked Lgi1 protein assembles into presynaptic Kv1 channels and inhibits inactivation by Kvbeta1. Neuron 49:697–706

Singh NA, Charlier C, Stauffer D, DuPont BR, Leach RJ, Melis R, Ronen GM, Bjerre I, Quattlebaum T, Murphy JV, McHarg ML, Gagnon D, Rosales TO, Peiffer A, Anderson VE, Leppert

M (1998) A novel potassium channel gene, KCNQ2, is mutated in an inherited epilepsy of newborns. Nat Genet 18:25–29

Singh NA, Westenskow P, Charlier C, Pappas C, Leslie J, Dillon J, Anderson VE, Sanguinetti MC, Leppert MF (2003) KCNQ2 and KCNQ3 potassium channel genes in benign familial neonatal convulsions: expansion of the functional and mutation spectrum. Brain 126:2726–2737

Singh NA, Otto JF, Dahle EJ, Pappas C, Leslie JD, Vilaythong A, Noebels JL, White HS, Wilcox KS, Leppert MF (2008) Mouse models of human KCNQ2 and KCNQ3 mutations for benign familial neonatal convulsions show seizures and neuronal plasticity without synaptic reorganization. J Physiol 586:3405–3423

Steffens M et al (2012) Genome-wide association analysis of genetic generalized epilepsies implicates susceptibility loci at 1q43, 2p16. 1, 2q22.3 and 17q21.32. Hum Mol Genet 21:5359–5372

Steinlein OK, Mulley JC, Propping P, Wallace RH, Phillips HA, Sutherland GR, Scheffer IE, Berkovic SF (1995) A missense mutation in the neuronal nicotinic acetylcholine receptor alpha 4 subunit is associated with autosomal dominant nocturnal frontal lobe epilepsy. Nat Genet 11:201–203

Sugiura Y, Ogiwara I, Hoshi A, Yamakawa K, Ugawa Y (2012) Different degrees of loss of function between GEFS+ and SMEI Nav 1.1 missense mutants at the same residue induced by rescuable folding defects. Epilepsia 53:e111–114

Suls A et al (2013) De novo loss-of-function mutations in CHD2 cause a fever-sensitive myoclonic epileptic encephalopathy sharing features with Dravet syndrome. Am J Hum Genet 93:967–975

Tanaka M, Olsen RW, Medina MT, Schwartz E, Alonso ME, Duron RM, Castro-Ortega R, Martinez-Juarez IE, Pascual-Castroviejo I, Machado-Salas J, Silva R, Bailey JN, Bai D, Ochoa A, Jara-Prado A, Pineda G, Macdonald RL, Delgado-Escueta AV (2008) Hyperglycosylation and reduced GABA currents of mutated GABRB3 polypeptide in remitting childhood absence epilepsy. Am J Hum Genet 82:1249–1261

Tang B, Dutt K, Papale L, Rusconi R, Shankar A, Hunter J, Tufik S, Yu FH, Catterall WA, Mantegazza M, Goldin AL, Escayg A (2009) A BAC transgenic mouse model reveals neuron subtype-specific effects of a Generalized Epilepsy with Febrile Seizures Plus (GEFS+) mutation. Neurobiol Dis 35:91–102

Teper Y et al (2007) Nicotine-induced dystonic arousal complex in a mouse line harboring a human autosomal-dominant nocturnal frontal lobe epilepsy mutation. J Neurosci 27:10128–10142

Thompson CH, Porter JC, Kahlig KM, Daniels MA, George AL Jr (2012) Nontruncating SCN1A mutations associated with severe myoclonic epilepsy of infancy impair cell surface expression. J Biol Chem 287:42001–42008

Touma M, Joshi M, Connolly MC, Grant PE, Hansen AR, Khwaja O, Berry GT, Kinney HC, Poduri A, Agrawal PB (2013) Whole genome sequencing identifies SCN2A mutation in monozygotic twins with Ohtahara syndrome and unique neuropathologic findings. Epilepsia 54:e81–85

Vanoye CG, Gurnett CA, Holland KD, George AL Jr, Kearney JA (2014) Novel SCN3A variants associated with focal epilepsy in children. Neurobiol Dis 62:313–322

Vecchia D, Pietrobon D (2012) Migraine: a disorder of brain excitatory-inhibitory balance? Trends Neurosci 35:507–520

Wallace RH, Wang DW, Singh R, Scheffer IE, George AL Jr, Phillips HA, Saar K, Reis A, Johnson EW, Sutherland GR, Berkovic SF, Mulley JC (1998) Febrile seizures and generalized epilepsy associated with a mutation in the Na$^+$-channel beta1 subunit gene SCN1B. Nat Genet 19:366–370

Wallace RH, Marini C, Petrou S, Harkin LA, Bowser DN, Panchal RG, Williams DA, Sutherland GR, Mulley JC, Scheffer IE, Berkovic SF (2001) Mutant GABA(A) receptor gamma2-subunit in childhood absence epilepsy and febrile seizures. Nat Genet 28:49–52

Weber YG, Lerche H (2008) Genetic mechanisms in idiopathic epilepsies. Dev Med Child Neurol 50:648–654

Weckhuysen S et al (2012) KCNQ2 encephalopathy: emerging phenotype of a neonatal epileptic encephalopathy. Ann Neurol 71:15–25

Weckhuysen S et al (2013) Extending the KCNQ2 encephalopathy spectrum: clinical and neuroimaging findings in 17 patients. Neurology 81:1697–1703

Yu FH, Mantegazza M, Westenbroek RE, Robbins CA, Kalume F, Burton KA, Spain WJ, McKnight GS, Scheuer T, Catterall WA (2006) Reduced sodium current in GABAergic interneurons in a mouse model of severe myoclonic epilepsy in infancy. Nat Neurosci 9:1142–1149

Zhou YD, Lee S, Jin Z, Wright M, Smith SE, Anderson MP (2009) Arrested maturation of excitatory synapses in autosomal dominant lateral temporal lobe epilepsy. Nat Med 15:1208–1214

Zhu G et al (2008) Rats harboring S284L Chrna4 mutation show attenuation of synaptic and extrasynaptic GABAergic transmission and exhibit the nocturnal frontal lobe epilepsy phenotype. J Neurosci 28:12465–12476

LGI1 Dysfunction in Inherited and Acquired Epileptic Disorders

Carlo Nobile

Abstract LGI1 is a multifunctional brain protein whose dysfunction is related to several neurologic disorders as diverse as autosomal dominant lateral temporal epilepsy (ADLTE), autoimmune limbic encephalitis (LE), and glioma tumor progression. ADLTE is a genetic focal epilepsy characterized by auditory or aphasic aura and onset in infancy/adolescence, whereas autoimmune LE occurs in adult life and is characterized by amnesia, confusion, and seizures. The complex molecular mechanisms underlying these epileptic conditions are largely unknown. In this chapter, I outline the clinical features, the genetic or autoimmune causes, and a molecular mechanism possibly underlying both ADLTE and autoimmune LE.

Introduction

The leucine-rich, glioma inactivated 1 (LGI1) gene has been associated with clinical phenotypes as different as malignant glioma, autosomal dominant lateral temporal epilepsy (ADLTE), a rare genetic focal epilepsy syndrome, and autoimmune limbic encephalitis (LE), an acquired immunological disorder of the brain. Although very different in nature, these brain disorders result from a reduction of the physiological level of the Lgi1 protein.

In 1998, LGI1 was cloned due to its rearrangements in the T98G glioblastoma multiforme (GBM) cell line and was found to be downregulated in many malignant gliomas, suggesting a possible tumor suppressor function (Chernova et al. 1998). Subsequent studies failed to reveal point mutations in the LGI1 coding sequence and differential methylation of its core promoter region in GBM tumors, arguing against a role of LGI1 as a tumor suppressor gene (Somerville et al. 2000; Piepoli et al. 2006). However, LGI1 has been shown to control proliferation and invasiveness of glioma cell lines by regulating expression of the matrix

---

C. Nobile (✉)
CNR-Neuroscience Institute, Padova, Italy
e-mail: nobile@bio.unipd.it

© Springer International Publishing Switzerland 2015
P. Striano (ed.), *Epilepsy Towards the Next Decade,*
Contemporary Clinical Neuroscience, DOI 10.1007/978-3-319-12283-0_3

metalloproteinases MMP1 and MMP3, suggesting that LGI1 may serve as a tumor metastasis suppressor gene (Kunapuli et al. 2003; 2004). In 2002, LGI1 heterozygous mutations were shown to cause ADLTE in several American and European families (Kalachikov et al. 2002; Morante-Redolat et al. 2002). Subsequent studies showed that LGI1 mutations account for about 50% of ADLTE families (Michelucci et al 2003; Ottman et al. 2004). ADLTE is clinically characterized by auras with auditory features and is also named autosomal dominant partial epilepsy with auditory features (ADPEAF). Finally, in 2010, LGI1 was implicated in acquired autoimmune limbic encephalitis, a neurological disorder of adulthood (Irani et al. 2010; Lai et al. 2010). Patients with autoantibodies directed against the Lgi1 protein suffer from psychiatric symptoms, including memory loss and confusion, as well as epilepsy.

The LGI1 gene consists of eight exons with a coding region of 1671 bp. It is expressed mainly in neurons, particularly in the neocortex and limbic regions (Kalachikov et al. 2002; Senechal et al. 2005) and encodes a protein of 557 amino acids with no similarities to ion channels. The Lgi1 protein is secreted (Senechal et al. 2005), and its structure consists of an N-terminal signal peptide and two distinct structural domains: the N-terminal region contains four leucine-rich repeats (LRR) flanked by conserved cysteine clusters (Kobe and Kajava 2001), whereas the C-terminal region consists of seven copies of a repeat named EPTP (Staub et al. 2002), which form a beta-propeller structural domain (Paoli 2001). Both LRR and beta-propeller motifs mediate protein-protein interactions (Kobe and Kajava 2001; Paoli 2001).

In this chapter I describe the clinical features of ADLTE and LE, the role of LGI1 as a cause of these syndromes, and outline the current views about the possible functions of LGI1.

ADLTE

Clinical Features

ADLTE is a rare familial condition characterized by focal seizures with prominent ictal auditory symptoms, negative MRI findings, and relatively benign evolution (Ottman et al. 1995; Michelucci et al. 2009). Its prevalence is unknown but it may account for about 19% of genetic focal epilepsies (Ottman et al. 2004). Since the first description of the syndrome (Ottman et al. 1995), 39 families, most of which with unique LGI1 mutations have been reported. The syndrome segregates with an autosomal dominant inheritance pattern with reduced penetrance. Familial diagnosis is based on the existence of at least two cases with unprovoked focal or secondarily generalized seizures whose symptoms suggest a lateral temporal lobe onset.

The age of onset ranges between 1 and 60 years with a mean of 18 years. The ictal semiology includes focal seizures and secondarily generalized tonic-clonic seizures. Focal seizures are characterized by auditory auras in about 2/3 of the cases. In most cases, auditory symptoms are described as simple sounds (such as humming, buzzing, ringing); complex hallucinations (i.e. music, voices) are less frequent, and, sometimes, the aura is characterized by a distortion of sounds (becoming louder and louder or suddenly low). Aphasic seizures associated with auditory phenomena are reported, but may also be the only clinical symptom in some pedigrees (see Michelucci et al. 2003). Other less frequent auras include complex visual, psychic, autonomic, vertiginous, and other sensory symptoms, usually accompanying auditory phenomena.

Secondary tonic-clonic seizures are common, occurring in 90% of cases, both during wakefulness and sleep, and may unmask an otherwise undiagnosed history of elementary focal seizures with auditory symptoms. Interictal EEGs show temporal abnormalities (described as mild slow/sharp waves) in about half of the patients, with a clear left predominance of the abnormalities in some families (Brodtkorb et al. 2005; Pisano et al. 2005). Standard MRI shows no abnormalities, but a study of 8 ADLTE patients with LGI1 mutations performed by voxel-based MRI and diffusion tensor imaging demonstrated a cluster of fractional anisotropy in the left lateral temporal cortex, suggesting a malformative origin of the abnormality (Tessa et al. 2007).

Genetic studies have revealed mutations of LGI1 in about 30–50% of the families, providing evidence for genetic heterogeneity (Michelucci et al. 2003; 2013). Detailed analysis of families with and without LGI1 mutations showed no phenotypic differences between the pedigrees.

Sporadic, non-familial cases with auditory seizures have been reported (Bisulli et al. 2004a). Despite their negative family history, they had a clinical picture indistinguishable from that of ADLTE patients, characterized by focal seizures with auditory auras, a mean age of onset of 19 years, a high rate of secondary generalized tonic-clonic seizures, low seizure frequency, good response to antiepileptic treatment, unrevealing EEGs and normal MRIs. *De novo* LGI1 mutations have been found in two sporadic LTE cases (see below).

LGI1 Pathogenic Mutations

To date, a total of 37 LGI1 mutations have been described, either segregating in ADLTE families or occurring *de novo* in sporadic LTE patients (Fig. 1 and Table 1; nucleotide numbering uses the A of the ATG translation initiation start site as nucleotide +1). Thirty-six mutations segregate in 39 affected families, whereas two, c.406C>T and c.1420C>T, are *de novo* mutations identified in non-familial cases, the latter occurring in both a family and a sporadic case (Morante-Redolat et al. 2002; Bisulli et al. 2004b). Of these mutations, 23 allow single amino acid substitutions, 13 result in protein truncation due to frameshift deletions or insertion, and to non-sense or splice site mutations, whereas one in frame deletion mutation (c.377–379delACA) results in the deletion of an asparagine in the encoded protein. In addition, an internal deletion of exons 3–4, resulting from altered splicing

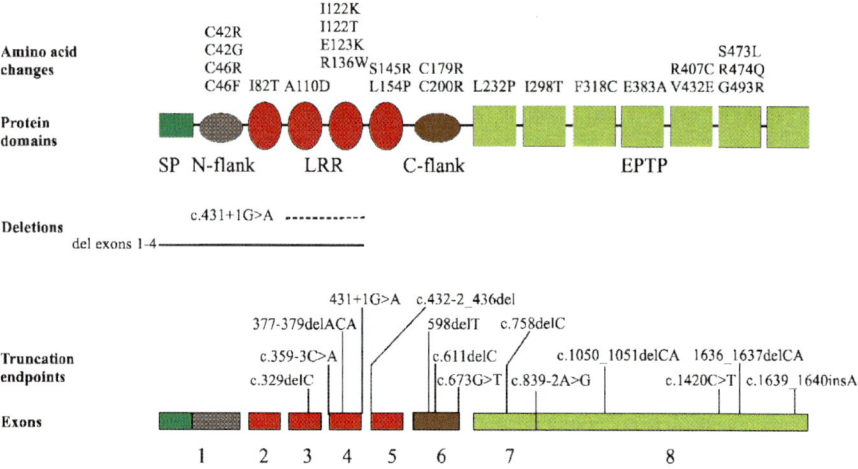

Fig. 1 Distribution of the pathogenic LGI1 mutations described so far. Most of them are single amino acid substitutions. Mutations are uniformly distrubuted along the gene, though a higher density of mutations appears to occur in exons 3–5 , which correspond to the N-terminal LRR repeat 2–4

(c.431+1G>A), and a genomic microdeletion spanning the first 4 LGI1 exons have been identified (Table 1). Thus, the majority of the mutations as yet identified are missense nucleotide changes. The mutation distribution along the gene is somewhat uniform, though a higher density of mutations appears to occur in exons 3–5, which correspond to the N-terminal LRR repeat 2–4 (Ho et al. 2012; see Fig. 1). The overall penetrance estimate of LGI1 mutations was 66% (Rosanoff and Ottman 2008) and the proportion of families with penetrance>=75% was similar among those with missense vs truncation mutations (Ho et al. 2012).

The Lgi1 protein is expressed mainly in neurons and is secreted into the extracellular compartment (Senechal et al. 2005; Fukata et al. 2010). All but one LGI1 mutations reported so far inhibit protein secretion (Nobile et al. 2009), suggesting a loss-of-function effect of mutations. Only one missense mutation (R407C) has been shown to allow protein secretion (Striano et al. 2011), and its effect in the extracellular comparment of the brain is unclear. A three-dimensional protein model predicts that this mutation may perturb extracellular interactions of Lgi1 with other proteins, and that this effect may not be limited to this mutation (Leonardi et al. 2011).

Lgi1 Protein Function

The function of LGI1 and the mechanisms of LGI1-related epilepsy remain unclear. Three main functions have been proposed for LGI1 in the CNS: (1) inhibition of inactivation of the presynaptic voltage-gated potassium channel Kv1.1

Table 1 LGI1 mutations reported in literature

Nucleotide change	Gene region	Predicted effect	Reference
c.124T>C	Exon 1	p.C42R	Ottman et al. (2004)
c.124T>G	Exon 1	p.C42G	Berkovic et al. (2004)
c.136T>C	Exon 1	p.C46R	Gu et al. (2002); Pizzuti et al. (2005)
c.137G>T	Exon 1	p.C46F	Lee et al. (2014)
c.245T>C	Exon 2	p.I82T	Sadleir et al. (2013)
c.329C>A	Exon 3	p.A110D	Ottman et al. (2004)
c.329delC	Exon 3	Truncation	Hedera et al. (2004)
c.359-3C>A	Intron 3	Truncation	Kalachikov et al. (2002)
c.365T>A	Exon 4	p.I122K	Striano et al. (2008)
c.365T>C	Exon 4	p.I122T	Di Bonaventura et al. (2011)
c.367G>A	Exon 4	p.E123K	Di Bonaventura et al. (2009)
c.377-379delACA	Exon 4	p.Asn126del	de Bellescize et al. (2009)
c.406C>T	Exon 4	p.R136W	Di Bonaventura et al. (2011); Michelucci et al. (2007)
Genomic deletion	Exon 1–4	Deletion	Fanciulli et al. (2012)
c.431+1G>A	Intron 4	Deletion	Chabrol et al. (2007); Berghuis et al. (2013)
c.432-2_436del	Exon 5	Truncation	Sadleir et al. (2013)
c.435C>G	Exon 5	p.S145R	Hedera et al. (2004)
c.461T>C	Exon 5	p.L154P	Pisano et al. (2005)
c.535T>C	Exon 6	p.C179R	Di Bonaventura et al. (2011)
c.598T>C	Exon 6	p.C200R	Michelucci et al. (2003)
c.598delT	Exon 6	Truncation	Heiman et al. (2010)
c.611delC	Exon 6	Truncation	Kalachikov et al. (2002)
c.673G>T	Exon 6	Truncation	Sadleir et al. (2013)
c.695T>C	Exon 7	p.L232P	Chabrol et al. (2007)
c.758delC	Exon 7	Truncation	Morante-Redolat et al. (2002)
c.839-2A>G	Intron 7	Truncation	Kobayashi et al. (2003)
c.893T>C	Exon 8	p.I298T	Ottman et al. (2004)
c.953T>G	Exon 8	p.F318C	Fertig et al. (2003)
c.1050_1051delCA	Exon 8	Truncation	Kalachikov et al. (2002)
c.1148A>C	Exon 8	p.E383A	Kalachikov et al. (2002)
c.1219C>T	Exon 8	p.R407C	Striano et al. (2011)
c.1295T>A	Exon 8	p.V432E	Michelucci et al. (2003)
c.1418C>T	Exon 8	p.S473L	Berkovic et al. (2004); Kawamata et al. (2010)

Table 1 (continued)

Nucleotide change	Gene region	Predicted effect	Reference
c.1420C>T	Exon 8	Truncation	Morante-Redolat et al. (2002); Bisulli et al. (2004)
c.1421G>A	Exon 8	p.R474Q	Kawamata et al. (2010)
c.1477G>A	Exon 8	p.G493R	Heiman et al. (2010)
c.1636_1637delCA	Exon 8	Truncation	Heiman et al. (2010)
c.1639_1640insA	Exon 8	Truncation	Kalachikov et al. (2002)

LGI1 GenBank reference sequence: NM_005097.2. Nucleotide numbering uses the A of the ATG translation initiation start site as nucleotide +1

(Schulte et al. 2006); (2) potentiation of AMPA receptor-mediated synaptic transmission in the hippocampus through interaction with the transmembrane receptors ADAM22 and ADAM23 (Fukata et al. 2006; 2010; Ohkawa et al. 2013); (3) postnatal maturation of glutamatergic synapses, regulation of spine density, and dendritic pruning (Zhou et al. 2009). Remarkably, Lgi1- knockout mice (Fukata et al. 2010; Chabrol et al. 2010; Yu et al. 2010) display spontaneous seizures. Homozygous Lgi1 knockout (KO) mice have spontaneous seizures with onset at postnatal day 10 and all pups die before the end of the third postnatal week, while heterozygous Lgi1+/− mice exhibit increased susceptibility to sound-induced or pentylenetetrazole-induced seizures (Chabrol et al. 2010; Fukata et al. 2010). The "intermediate" phenotype of heterozygous +/− mice supports the loss-of-function model, whereas the phenotypic features of mice overexpressing a truncated protein in the presence of normal endogenous Lgi1 suggest that some mutations may have a dominant negative effect (Zhou et al. 2009).

Acquired Immune-Mediated Disorders

LE

The pivotal role of LGI1 in epileptic disorders has been expanded with the recent finding of autoantibodies against Lgi1 in patients with autoimmune LE, a neurological disorder of adulthood characterized by amnesia, confusion, seizures, which mostly involve the temporal lobes, and personality change or psychosis (Irani et al. 2010; Lai et al. 2010). Autoimmune LE belongs to the group of autoimmune synaptic encephalopathies, in which patients develop antibodies against synaptic proteins, including the excitatory N-Methyl-D-aspartate (NMDA) and alpha-amino-3-hydroxy-5-methyl-4-isoxazolepropionic acid (AMPA) receptors and voltage-gated potassium channel (VGKC)-complexes (Kv1). Because of the crucial role of these receptors in synaptic transmission and plasticity, the autoimmunity usually causes seizures and neuropsychiatric symptoms, ranging from alterations in memory, behaviour, and cognition, to psychosis. The resulting disorders are severe but treatable, and some of them can be individually classified on the basis of a particular immune response (for example, anti-NMDA receptor encephalitis (Dalmau et al. 2008; Gable et al. 2009); or anti-VGKC LE (Vincent et al. 2004; Tan et al. 2008). The VGKC complex antibodies have traditionally been measured by immunoprecipitation of VGKC Kv1 subunits solubilized in detergent from rabbit brain tissue and radioactively labeled by 125I-dendrotoxin, a snake toxin that binds very strongly to Kv1. Recently, a series of experiments has shown that the antibodies do not bind directly to Kv1 subunits but rather to other molecules that are complexed with the Kv1 in brain extracts. The most frequently associated molecules identified so far are Lgi1 and Contactin-associated protein 2 (CASPR2), with the great majority of the LE patients having antibodies to Lgi1 (Lai et al 2010). Commonly, these antibodies are detected using an immunofluorescence cell-based method which detects the binding of patients' sera to the surface of cells transfected with cDNA encoding the protein.

LE patients with autoantibodies directed against Lgi1 protein suffer from psychiatric symptoms and epilepsy. Clinical seizures, generalized or involving the temporal lobes, occur in about 80% of LE patients, whereas 40% have myoclonus; 60% of patients have hyponatraemia On MRI, increased T2 signal involving one or both medial temporal lobes are frequently observed (Lai et al. 2010). Lgi1-positive LE usually is not associated with tumors. Immunotherapy, consisting of intravenous immunoglobulin, glucocorticoids, plasma exchange, or a combination thereof, results in full recovery or mild residual memory impairment in most cases. Relapses only occur in a minority of cases.

In a proportion of patients, autoimmune LE is associated with faciobrachial dystonic seizures (FBDS), which is characterized by typical episodes of facial grimacing and ipsilateral arm dystonia (Irani et al. 2008; 2013). Seizures occur at high frequency (up to 360 episodes per day). The onset is variable (22–83 years) and is sometimes triggered by auditory stimuli or high emotion. EEG abnormalities, i.e. rhythmic frontotemporal spikes, are detectable in about one-fourth of FBDS cases. This syndrome has been found consistently associated with antibodies against Lgi1 (Irani et al. 2008; 2011; Vincent et al. 2011). Response to anti-epileptic drugs is poor whereas immunotherapy yields excellent results. Most FBDS patients subsequently develop the full LE phenotype, whereas in a minority of cases FBDS occur in isolation or follows LE. Early immunotherapy for faciobrachial dystonic seizures may postpone or even prevent progression to the cognitive impairment that characterizes LE (Irani et al. 2011).

Neuromyotonia and Morvan Syndrome

Lgi1 autoantibodies are also found in acquired immune-mediated peripheral nerve disorders neuromyotonia and Morvan syndrome, though often associated with other antibodies. Acquired neuromyotonia is characterized by the presence of spontaneous activity including positive sharp-waves and fibrillation potentials, fasciculations, myokymia, multiple discharges, muscle cramps and repetitive after-discharges in response to a voluntary contraction (Warmolts and Mendell 1980; Hart et al. 1997). Acquired neuromyotonia may be clinically misdiagnosed as amyotrophic lateral sclerosis (ALS), particularly in the early stages of ALS where widespread fasciculations may be evident in the absence of other clinical features of ALS.

Morvan's syndrome has been recognized as a rare constellation of peripheral nerve neuromyotonia combined with sensory abnormalities and central nervous system (CNS) features, such as dysautonomia and encephalopathy with marked insomnia (Hart et al. 2002). Some of the patients had a thymoma, but many do not have a tumor. In patients with neuromyotonia or Morvan syndrome CASPR2 antibodies are mainly detected, whereas Lgi1 antibodies are much less frequent and, when occur, are frequently associated with CASPR2 antibodies (Irani et al. 2012).

Whether Lgi1 autoantibodies are pathogenic or represent an epiphenomenon is still uncertain. However, immunotherapies can lead to a marked improvement of these syndromes paralleling with antibody titer decrease, suggesting a possible direct role of antibodies on neuronal function.

LGI1 Functional Models

Despite definitive genetic evidence, the pathophysiological function of LGI1 in the brain remains controversial. Several studies have shown that the secreted Lgi1 protein binds to ADAM22 and ADAM23 receptors on the surface of neuronal cells and that these protein complexes exert various functions during neuronal maturation and synaptic transmission (Fukata et al. 2006; 2010; Owuor et al. 2009). The involvement of both ADAM22 and ADAM23 in epilepsy is suggested by studies of knock-out mice, showing that lack of expression of either of these genes results in spontaneous seizures (Owuor et al. 2009; Sagane et al. 2005). It has been shown that the ligand-receptor complexes between Lgi1 and ADAM22/ADAM23 regulates AMPA receptor-mediated synaptic transmission in the hippocampus, suggesting a possible molecular mechanism underlying ADLTE (Fukata 2006; 2010). This mechanism is further supported by recent findings showing that Lgi1 antibodies associated with LE neutralize the specific protein-protein interaction between LGI1 and ADAM22/ADAM23, and that disruption of this complex is sufficient to reduce AMPA receptors in rat hippocampal neurons (Ohkawa 2013). Thus, a unifying view is emerging, suggesting that reduced binding of Lgi1 to ADAM22/23 may be a pathogenic mechanism for both genetically inherited ADLTE and acquired LE, providing further support to the central role of LGI1 in mechanisms for regulating brain function and excitability.

References

Berghuis B, Brilstra EH, Lindhout D et al (2013) Hyperactive behavior in a family with autosomal dominant lateral temporal lobe epilepsy caused by a mutation in the LGI1/epitempin gene. Epilepsy Behav 28:41–46

Berkovic SF, Izzillo P, McMahon JM et al (2004) LGI1 mutations in temporal lobe epilepsies. Neurology 62:1115–1119

Bisulli F, Tinuper P, Avoni P et al (2004a) Idiopathic partial epilepsy with auditory features (IPE-AF): a clinical and genetic study of 53 sporadic cases. Brain 127:1343–1352

Bisulli F, Tinuper P, Scudellaro E et al (2004b) A de novo LGI1 mutation in sporadic partial epilepsy with auditory features. Ann Neurol 56:455–456

Brodtkorb E, Michler RP, Gu W et al (2005) Speech-induced aphasic seizures in epilepsy caused by LGI1 mutation. Epilepsia 46:963–966

Chabrol E, Popescu C, Gourfinkel-An I et al (2007) Two novel epilepsy-linked mutations leading to a loss of function of LGI1. Arch Neurol 64:217–222

Chabrol E, Navarro V, Provenzano G et al (2010) Electroclinical characterization of epileptic seizures in leucine-rich, glioma-inactivated 1- deficient mice. Brain 133:2749–2762

Chernova OB, Somerville RP, Cowell JK (1998) A novel gene, LGI1, from 10q24 is rearranged and downregulated in malignant brain tumors. Oncogene 17:2873–2881

Dalmau J, Gleichman AJ, Hughes EG et al (2008) Anti-NMDA-receptor encephalitis: case series and analysis of the effects of antibodies. Lancet Neurol 7:1091–1098

de Bellescize J, Boutry N, Chabrol E et al (2009) A novel three base-pair LGI1 deletion leading to loss of function in a family with autosomal dominant lateral temporal epilepsy and migraine-like episodes. Epilepsy Res 85:118–122

Di Bonaventura C, Carni M, Diani E et al (2009) Drug resistant ADLTE and recurrent partial status epilepticus with dysphasic features in a family with a novel LGI1 mutation: electroclinical, genetic, and EEG/fMRI findings. Epilepsia 50:2481–2486

Di Bonaventura C, Operto FF, Busolin G et al (2011) Low penetrance and effect on protein secretion of LGI1 mutations causing autosomal dominant lateral temporal epilepsy. Epilepsia 52:1258–1264

Fertig E, Lincoln A, Martinuzzi A et al (2003) Novel LGI1 mutation in a family with autosomal dominant partial epilepsy with auditory features. Neurology 60:1687–1690

Fanciulli M, Santulli L, Errichiello L et al (2012) LGI1 microdeletion in autosomal dominant lateral temporal epilepsy. Neurology 78:1299–1303

Fukata Y, Adesnik H, Iwanaga T et al (2006) Epilepsy-related ligand/receptor complex LGI1 and ADAM22 regulate synaptic transmission. Science 313:1792–1795

Fukata Y, Lovero KL, Iwanaga T et al (2010) Disruption of LGI1-linked synaptic complex causes abnormal synaptic transmission and epilepsy. Proc Natl Acad Sci U S A 107:3799–3804

Gable MS, Gavali S, Radner A et al (2009) Anti-NMDA receptor encephalitis: report of ten cases and comparison with viral encephalitis. Eur J Clin Microbiol Infect Dis 28:1421–1429

Gu W, Brodtkorb E, Steinlein OK (2002) LGI1 is mutated in familial temporal lobe epilepsy characterized by aphasic seizures. Ann Neurol 52:364–367

Hart IK, Waters C, Vincent A et al (1997) Autoantibodies detected to expressed Kþ channels are implicated in neuromyotonia. Ann Neurol 41:238–246

Hart IK, Maddison P, Newsom-Davis J et al (2002) Phenotypic variants of autoimmune peripheral nerve hyperexcitability. Brain 125:1887–1895

Hedera P, Abou-Khalil B, Crunk AE et al (2004) Autosomal dominant lateral temporal epilepsy: two families with novel mutations in the LGI1 gene. Epilepsia 45:218–222

Heiman GA, Kamberakis K, Gill R et al (2010) Evaluation of depression risk in LGI1 mutation carriers. Epilepsia 51:1685–1690

Ho YY, Ionita-Laza I, Ottman R (2012) Domain-dependent clustering and genotype-phenotype analysis of LGI1 mutations in ADPEAF. Neurology 78:563–568

Irani SR, Buckley C, Vincent A et al (2008) Immunotherapy-responsive seizure-like episodes with potassium channel antibodies. Neurology 71:1647–1648

Irani SR, Alexander S, Waters P et al (2010) Antibodies to Kv1 potassium channel-complex proteins leucine-rich, glioma inactivated 1 protein and contactin-associated protein-2 in limbic encephalitis, Morvan's syndrome and acquired neuromyotonia. Brain 133:2734–2748

Irani SR, Michell AW, Lang B et al (2011) Faciobrachial dystonic seizures precede Lgi1 antibody limbic encephalitis. Ann Neurol 69:892–900

Irani SR, Pettingill P, Kleopa KA et al (2012) Morvan syndrome: Clinical and serological observations in 29 cases. Ann Neurol 72:241–255

Irani SR, Charlotte JS, Schott JM et al (2013) Faciobrachial dystonic seizures: the influence of immunotherapy on seizure control and prevention of cognitive impairment in a broadening phenotype. Brain 136:3151–3162

Kalachikov S, Evgrafov O, Ross B et al (2002) Mutations in LGI1 cause autosomal-dominant partial epilepsy with auditory features. Nat Genet 30:335–341

Kawamata J, Ikeda A, Fujita Y et al (2010) Mutations in LGI1 gene in Japanese families with autosomal dominant lateral temporal lobe epilepsy: the first report from Asian families. Epilepsia 51:690–693

Kobe B, Kajava AV (2001) The leucine-rich repeat as a protein recognition motif. Curr Opin Struct Biol 11:725–732

Kunapuli P, Chitta KS, Cowell JK (2003) Suppression of the cell proliferation and invasion phenotypes in glioma cells by the LGI1 gene. Oncogene 22:3985–3991

Kunapuli P, Kasyapa CS, Hawthorn L et al (2004) LGI1, a putative tumor metastasis suppressor gene, controls in vitro invasiveness and expression of matrix metalloproteinases in glioma cells through the ERK1/2 pathway. J Biol Chem 279:23151–23157

Leonardi E, Andreazza S, Vanin S et al (2011) A computational model of the LGI1 protein suggests a common binding site for ADAM proteins. Plos One 6:e18142

Lai M, Huijbers MGM, Lancaster E et al (2010) Investigation of LGI1 as the antigen in limbic encephalitis previously attributed to potassium channels: a case series. Lancet Neurol 9:776–785

Lee MK, Kim SW, Lee JH et al (2014) A Newly discovery of LGI1 mutation in a Korean family with autosomal dominant lateral temporal lobe epilepsy. Seizure 23:69–73

Michelucci R, Poza JJ, Sofia V et al (2003) Autosomal dominant lateral temporal epilepsy: clinical spectrum, new epitempin mutations, and genetic heterogeneity in seven European families. Epilepsia 44:1289–1297

Michelucci R, Mecarelli O, Bovo G et al (2007) A de novo LGI1 mutation causing idiopathic partial epilepsy with telephone-induced seizures. Neurology 68:2150–2151

Michelucci R, Pasini E, Nobile C (2009) Lateral temporal lobe epilepsies: clinical and genetic features. Epilepsia 50(Suppl 5):52–54

Michelucci R, Pasini E, Malacrida S et al (2013) Low penetrance of autosomal dominant lateral temporal epilepsy in Italian families without LGI1 mutations. Epilepsia 54:1288–1297

Morante-Redolat JM, Gorostidi-Pagola A, Piquer-Sirerol S et al (2002) Mutations in the LGI1/Epitempin gene on 10q24 cause autosomal dominant lateral temporal epilepsy. Hum Mol Genet 11:1119–1128

Nobile C, Michelucci R, Andreazza S et al (2009) LGI1 mutations in autosomal dominant and sporadic lateral temporal epilepsy. Hum Mutat 30:530–536

Ohkawa T, Fukata Y, Yamasaki M et al (2013) Autoantibodies to Epilepsy-Related LGI1 in Limbic Encephalitis Neutralize LGI1-ADAM22 Interaction and Reduce Synaptic AMPA Receptors. J Neurosci 33:18161–18174

Owuor K, Harel NY, Englot DJ et al (2009) LGI1-associated epilepsy through altered ADAM23-dependent neuronal morphology. Mol Cell Neurosci 42:448–457

Ottman R, Risch N, Hauser WA et al (1995) Localization of a gene for partial epilepsy to chromosome 10q. Nat Genet 10:56–60

Ottman R, Winawer MR, Kalachikov S et al (2004) LGI1 mutations in autosomal dominant partial epilepsy with auditory features. Neurology 62:1120–1126

Paoli M (2001) Protein folds propelled by diversity. Prog Biophys Molec Biol 76:103–130

Piepoli T, Jakupoglu C, Gu W et al (2006) Expression studies in gliomas and glial cells do not support a tumor suppressor role for LGI1. Neuro Oncol 8:96–108

Pisano T, Marini C, Brovedani P et al (2005) Abnormal phonologic processing in familial lateral temporal lobe epilepsy due to a new LGI1 mutation. Epilepsia 46:118–123

Pizzuti A, Flex E, Di Bonaventura C et al (2003) Epilepsy with auditory features: a LGI1 gene mutation suggests a loss-of-function mechanism. Ann Neurol 53:396–399. Erratum in. Ann Neurol 2003;54:137

Rosanoff MJ, Ottman R (2008) Penetrance of LGI1 mutations in autosomal dominant partial epilepsy with auditory features. Neurology 71:567–571

Sadleir LG, Agher D, Chabrol E et al (2013) Seizure semiology in autosomal dominant epilepsy with auditory features, due to novel LGI1 mutations. Epilepsy Res 107:311–317

Sagane K, Hayakawa K, Kaj J et al (2005) Ataxia and peripheral nerve hypomyelination in ADAM22-deficient mice. BMC Neurosci 6:33

Schulte U, Thumfart JO, Klocker N et al (2006) The epilepsy-linked Lgi1 protein assembles into presynaptic kv1 channels and inhibits inactivation by Kvbeta1. Neuron 49:697–706

Senechal KR, Thaller C, Noebels JL (2005) ADPEAF mutations reduce levels of secreted LGI1, a putative tumor suppressor protein linked to epilepsy. Hum Mol Genet 14:1613–1620

Somerville RP, Chernova O, Liu S et al (2000) Identification of the promoter, genomic structure, and mouse ortholog of LGI1. Mamm Genome 11:622–627

Staub E, Perez-Tur J, Siebert R et al (2002) The novel EPTP repeat defines a superfamily of proteins implicated in epileptic disorders. Trends Biochem Sci 27:441–444

Striano P, de Falco A, Diani E et al (2008) A novel loss-of-function LGI1 mutation linked to autosomal dominant lateral temporal epilepsy. Arch Neurol 65:939–942

Striano P, Busolin G, Santulli L et al (2011) Familial temporal lobe epilepsy with psychic auras associated with a novel LGI1 mutation. Neurology 76:1173–1176

Tan KM, Lennon VA, Klein CJ et al (2008) Clinical spectrum of voltage-gated potassium channel autoimmunity. Neurology 70:1883–1890

Tessa C, Michelucci R, Nobile C et al (2007) Structural anomaly of left lateral temporal lobe in epilepsy due to mutated LGI1. Neurology 69:1298–1300

Vincent A, Buckley C, Schott JM et al (2004) Potassium channel antibodyassociated encephalopathy: a potentially immunotherapy-responsive form of limbic encephalitis. Brain 127:701–712

Vincent A, Bicn CG, Irani SR et al (2011) Autoantibodies associated with diseases of the CNS: new developments and future challenges. Lancet Neurol 10:759–772

Warmolts JR, Mendell JR (1980) Neurotonia: impulse-induced repetitive discharges in motor nerves in peripheral neuropathy. Ann Neurol 7:245–250

Yu YE, Wen L, Silva J et al (2010) Lgi1 null mutant mice exhibit myoclonic seizures and CA1 neuronal hyperexcitability. Hum Mol Genet 19:1702–1711

Zhou YD, Lee S, Jin Z et al (2009) Arrested maturation of excitatory synapses in autosomal dominant lateral temporal lobe epilepsy. Nat Med 15:1208–1214

Glioneuronal Tumors and Epilepsy: Clinico-Diagnostic Features and Surgical Strategies

Alessandro Consales, Paolo Nozza, Maria Luisa Zoli, Giovanni Morana and Armando Cama

Abstract Glioneuronal tumors (GNTs) are a set of tumors of the Central Nervous System (CNS) composed entirely or partially of cells with neuronal differentiation. The description applies to several tumors referred to in the Classification of the World Health Organization (WHO) as neuronal and mixed neuronal-glial tumors. Some of them arise typically in the cerebral cortex and represent a common cause of drug-resistant focal epilepsies in children and young adults. Three groups of tumors are of considerable relevance: ganglion cell tumors (GCTs), ganglioglioma (GG) being the typical example, dysembryoplastic neuroepithelial tumors (DNTs) and neurocytic tumors, namely extraventricular neurocytoma. GNTs commonly arise from a cortex housing developmental malformations, that are able to provoke seizures, such as focal cortical dysplasia (FCD) and, less frequently, hyppocampal sclerosis (HS). Management of patients with GNTs includes dealing with both tumor and epilepsy. From the neuro-oncological point of view, consistently with the long clinical history, these tumors are generally considered as low-grade gliomas (LGGs). They correspond to grades I or II of the WHO classification and their therapy relies mainly on surgery. However, tumor progression or transformation into higher grade tumors may occur.

From the neurological point of view, it is noteworthy that seizures are likely to respond very well to surgical treatment. Despite the favorable seizure outcome, the best surgical strategy has not been fully established yet. Indeed, while some authors regard tumor resection (the so-called lesionectomy) alone as sufficient for complete

---

A. Consales (✉) · A. Cama
Department of Neurosurgery, Istituto Giannina Gaslini, Via Gerolamo Gaslini, 5, 16148 Genova, Italy
e-mail: alessandroconsales@ospedale-gaslini.ge.it

P. Nozza
Department of Pathology, Istituto Giannina Gaslini, Via Gerolamo Gaslini, 5, 16148 Genova, Italy

M. L. Zoli
Neurologist, ASL3 Genovese, Via Giovanni Maggio, 6, 16147 Genova, Italy

G. Morana
Department of Neuroradiology, Istituto Giannina Gaslini, Via Gerolamo Gaslini, 5, 16148 Genova, Italy

seizure control, other investigators also recommend the additional resection of peritumoral epileptogenic zones to maximize the seizure outcome.

Anyway, surgical treatment should be planned on the basis of anatomo-electro-clinical correlation defining the epileptogenic zone to be resected.

To conclude, in patients with GNTs-related epilepsies, surgery should be proposed in order to obtain complete seizure control with freedom from antiepileptic drugs (AEDs), and to prevent tumor growth and risk of malignant transformation.

List of abbreviations

ADC	Apparent Diffusion Coefficient
AED(s)	Antiepileptic Drug(s)
BBB	Blood Brain Barrier
CNS	Central Nervous System
Cho	Choline
CT	Computed Tomography
DWI	Diffusion Weighted Imaging
DNT(s)	Dysembryoplastic Neuroepithelial Tumor
ECoG	Electrocorticography
EEG	Electroencephalography
FCD	Focal Cortical Dysplasia
GCT(s)	Ganglion Cell Tumor(s)
GG(s)	Ganglioglioma(s)
GNT(s)	Glioneuronal Tumor(s)
H&E	Hematoxylin & Eosin
HS	Hyppocampal sclerosis
LGG(s)	Low-grade glioma(s)
mI	myo-inositol
MRI	Magnetic Resonance Imaging
MRS	Magnetic Resonance Spectroscopy
NAA	N-acetyl aspartate
PWI	Perfusion Weighted Imaging
rCBV	relative Cerebral Blood Volume
VFD(s)	Visual Field Defect(s)
WHO	World Health Organization

Glioneuronal tumors (GNTs) are a set of tumors of the Central Nervous System (CNS) composed entirely or partially of cells with neuronal differentiation, diagnosed mainly by means of histochemical and immunohistochemical stainings and, more rarely, by electron microscopy. The description applies to several tumors referred to in the Classification of the World Health Organization (WHO) as neuronal and mixed neuronal-glial tumors (Daumas-Duport et al. 1988; Louis et al. 2007a; Allende and Prayson 2009). Some of them arise tipically in the cerebral cortex and represent a common cause of drug-resistant focal epilepsies in children and young adults (Aronica et al. 2001; Benifla et al. 2006; Cataltepe et al. 2005; Consales et al. 2013; Clusmann et al. 2004; Daumas-Duport et al. 1988; Giulioni et al. 2005; Morris et al. 1998; Nolan et al. 2004). Three groups of tumors are of considerable relevance: ganglion cell tumors (GCTs), ganglioglioma (GG) being the typical example, dysembryoplastic neuroepithelial tumors (DNTs) and neurocytic tumors, namely extraventricular neurocytoma. However, this classification scheme is rather arbitrary and provisional. Tumors with composite features, such as GG and DNT may occur and new entities will probably be described. On the other hand, it could sometimes be difficult to distinguish a tumor from a malformation of cortical development (Prayson and Napekoski 2012). From a more general point of view, it should be kept in mind that also tumors not included in the chapter of neuronal and neuronal glial tumors, should be included in the more general classification of 'long-term epilepsy-associated tumors'. Some of them, such as the pleomorphic xanthoastrocytoma, may show neuronal differentiation, while some others, such as the pilocytic astrocytoma or 'isomorphic astrocytoma', are astrocytic tumors. All of them make clear however that, to induce long-term epilepsy, a tumor need not be composed of neuronal cells (Blümcke et al. 2004).

Furthermore, GNTs commonly arise from a cortex housing developmental malformations, that are able to provoke seizures, such as focal cortical dysplasia (FCD) (40–80% of cases) and, less frequently (up to 25% of cases), hyppocampal sclerosis (HS) (Aronica et al. 2001; Bilginer et al. 2009; Blümcke and Wiestler 2002; Cataltepe et al. 2005; Chang et al. 2010; Minkin et al. 2008; Nolan et al. 2004; Prayson et al. 2010; Sharma et al. 2009).

Therefore, management of these cases includes dealing with both tumor and epilepsy.

From the neuro-oncological point of view, consistently with the long clinical history, these tumors are generally considered as low-grade gliomas (LGGs). They correspond to grades I or II of the WHO classification and their therapy relies mainly on surgery (Becker et al. 2007; Daumas-Duport et al. 2007). However, tumor progression or transformation into higher grade tumors may occur (Aronica et al. 2001; Nolan et al. 2004; Hall et al. 1986; Hammond et al. 2000; Kim et al. 2003; Luyken et al. 2004).

From the neurological point of view, it is noteworthy that seizures are likely to respond very well to surgical treatment (Aronica et al. 2001; Cataltepe et al. 2005; Luyken et al. 2003; Clusmann et al. 2002; Clusmann et al. 2004).

Despite the favorable seizure outcome, the best surgical strategy has not been fully established yet. Indeed, while some authors regard tumor resection (the so-called *lesionectomy*) alone as sufficient for complete seizure control (Giulioni et al. 2005; Bourgeois et al. 2006; Iannelli et al. 2000; Montes et al. 1995), others recommend the resection of peritumoral epileptogenic zones (Aronica et al. 2001; Clusmann et al. 2004; Morris et al. 1998; Cossu et al. 2008; Kim et al. 2008; Luyken et al. 2003).

The aim of this chapter is to elaborate on the current concepts concerning the diagnosis and surgical management of patients with medically intractable focal epilepsy related to GNTs.

Epidemiology and Type of Tumors

GNTs account for 0.4–1.3% of all brain tumors (Becker et al. 2007; Daumas-Duport et al. 2007) and are an increasingly recognized cause of epilepsy in children and young adults (Aronica et al. 2001; Benifla et al. 2006; Cataltepe et al. 2005; Consales et al. 2013; Clusmann et al. 2004; Daumas-Duport et al. 1988; Giulioni et al. 2005; Morris et al. 1998; Nolan et al. 2004).

Their incidence in the pediatric population is about 8%, accounting for up to 30% of long-standing medically intractable epilepsies (Aronica et al. 2001; Daumas-Duport et al. 1988; Johnson et al. 1997; Obeid et al. 2009; Zaghloul and Schramm 2011). Among telencephalic GNTs (box 1), seizures are reported in up to 100% of DNTs and in 80–90% of GGs (Chang et al. 2008; Englot et al. 2011; van Breemen et al. 2007; Zaghloul and Schramm 2011).

Box 1

Neuronal and mixed neuronal-glial tumors (WHO Classification) (Louis et al. 2007b)

(in bold are tumors relevant for epilepsy surgeons)

Dysplastic gangliocytoma of cerebellum (Lhermitte–Duclos)	WHO Grade I
Desmoplastic infantile astrocytoma/ganglioglioma	WHO Grade I
Dysembryoplastic neuroepithelial tumor	**WHO Grade I**
Gangliocytoma	**WHO Grade I**
Ganglioglioma	**WHO Grade I**
Anaplastic ganglioglioma	WHO Grade III
Central neurocytoma	WHO Grade II
Extraventricular neurocytoma	**WHO Grade II**
Cerebellar liponeurocytoma	WHO Grade I
Papillary glioneuronal tumor	WHO Grade I
Rosette-forming glioneuronal tumor of the fourth ventricle	WHO Grade I
Paraganglioma	WHO Grade I

Mechanisms of Epileptogenesis Associated with GNTs

The pathophysiological mechanisms of epileptogenesis associated with brain tumors are not well understood. Several pathogenetic factors have been advocated to explain the tumor-related epileptogenesis.

Schematically, three types of mechanisms have been hypothesized, namely: (a) intrinsic to the lesion itself, including the expression of various ion channels and the relative proportion of different cell types within the tumor (Aronica et al. 2001;

Ventureyra et al. 1986; Rossi et al. 1999); (b) intrinsic to the tumor location, including peritumoral amino acid disturbances, local metabolic imbalance, cerebral edema, pH abnormalities, neuropil changes, neuronal/glial enzyme changes, altered protein expression and immunological activity (Beaumont and Whittle 2000); (c) unique to the patient harboring the disease (Gaggero et al. 2009; Wetjen et al. 2004).

There is significant evidence in the literature supporting the view that GNTs have intrinsic epileptogenic activities due to their neuronal and glial components (Aronica et al. 2001; Blümcke and Wiestler 2002; Daumas-Duport et al. 1988; Prayson et al. 1995; Prayson et al. 1996; Prayson 1999; Zentner et al. 1994). The neurochemical profile of GNTs shows some similarities in expression of various enzymes and receptors to that of neocortical neurons (Aronica et al. 2001; Aronica et al. 2007; de Groot et al. 2010; Lee et al. 2006). Moreover, the relatively low incidence of HS associated with temporal GNTs suggests that these tumors may be the primary source of epilepsy (Beaumont and Whittle 2000; Blümcke and Wiestler 2002).

The association between GNTs and other epileptogenic lesions (as FCD) has been frequently reported, however the specific role of each of them in determining epilepsy is not well understood (Beaumont and Whittle 2000; Louis et al. 2007b; Prayson et al. 2010). This implicates that simple tumor resection in patients affected also by FCD may determine unsatisfactory seizure outcome.

Finally, it is important to consider that the epileptogenic focus is not located within the tumor in about 30–40% of patients. The presence of a secondary epileptogenic focus must be postulated, particularly in patients affected by temporal tumors (van Breemen et al. 2007).

Electroclinical Features

Focal epilepsy represents the most common clinical feature. Seizures are often the only symptom. Neurological deficits are quite rare, probably because of the slow growth of GNTs. Epilepsy can present at any age, even if most cases are diagnosed during the adolescence. Seizure type depends on tumor location. Status epilepticus is rare. There does not seem to be differences ascribable to tumor type (DNT, GG, or rarer tumors). The response of GNTs-related epilepsy to antiepileptic drugs (AEDs) is often discouraging, with frequent development of drug-resistance. Short duration of epilepsy, partial seizures and lack of secondary generalization are the most important clinical prognostic factors for a successful seizure outcome. There are two differences between children and young adults concerning the characteristics of GNTs-related epilepsy: a) aura is diagnosed more frequently in young adults; b) mean age at seizure onset is lower in children. This can be explained, respectively, by the limited capability of children to describe their symptoms and their lower seizure threshold (Ozlen et al. 2010).

In the non-invasive diagnostic work-up of patients, long-term neurophysiological monitoring, such as Video-EEG, is very useful in recording seizures and identifying the epileptogenic zone.

Concerning EEG features, interictal data usually consist of spikes and/or sharp waves, sometimes intermixed with slow activities. There is also the possibility of a normal EEG. The abnormal features are usually lateralized to the tumor side. This does not guarantee good seizure outcome after tumor resection. In addition, it should be stressed that satisfactory seizure outcome can be achieved after tumor surgery even in patients whose EEG epileptiform activities are not localized to the tumor site (Morris et al. 1998).

When it is not possible to determine the lateralization of seizure focus by means on non-invasive diagnostic investigations, intracranial EEG (Stereo-EEG, depth electrodes, ECoG) may provide useful information. However, intracranial recordings are mainly aimed at mapping eloquent cortex when tumor resection in critical areas is planned. Thus, little is presently known, for example, about ECoG spike discharge patterns in patients with GNTs (Ferrier et al. 2006).

On the other hand, it is commonly acknowledged that patients with GNTs and FCDs may have similar invasive EEG patterns (Chassoux et al. 2000; Chassoux and Daumas-Duport 2013).

Ganglion Cell Tumors (GCTs)

GG and gangliocytoma are neuroepithelial tumors composed of ganglion cells (gangliocytoma) or both ganglion and glial cells (GG). The distinction seems slightly artificial and seldom acquires a deep, practical significance. Thus, these tumors can be lumped together under the designation 'GCTs'.

They are the most common tumors determining drug-resistant epilepsy and account for over 40 % of all epileptogenic tumors in most surgical series (Lawson and Duchowny 2004).

GCTs of the cerebral cortex mainly occur in the temporal lobe and, secondarily, in the frontal lobe (Casazza et al. 1989; Otsubo et al. 1990–1991). They can occur throughout the neuraxis (Becker et al. 2007).

Neuroimaging characteristics are variable (Fig. 1a, b, c, d and Fig. 2a, b) as these tumors may present as a mixed lesion (solid and cystic), as a purely solid mass or, less commonly, as a cyst (Zentner et al. 1994; Castillo et al. 1990; Raybaud et al. 2006). Classically, GG has been described as a cystic mass with a mural nodule in approximately 40 % of cases. On non contrast CT, most GGs are hypodense to gray matter, but some can have mixed or high density. Calcifications are relatively common and are reported in about 35–50 % of cases (Castillo 1998). Scalloping of the calvarium may be seen. Sometimes this neoplasm may be undetectable on CT. On MRI, the signal behavior is variable as well, with a spectrum of signal intensity depending on the neuropathological features of the different components. Solid portions are iso- to hypointense on T1-weighted images, with a variable degree of hyperintensity on T2-weighted images.

Cystic components, hypointense on T1- and hyperintense on T2-weighted images, may or may not show wall enhancement following gadolinium injection (Zhang et al. 2008). Enhancement of the solid portion is highly variable, ranging from no enhancement to intense homogeneity; grossly half of tumors display at least some degree of enhancement. There is usually little associated mass effect or surrounding

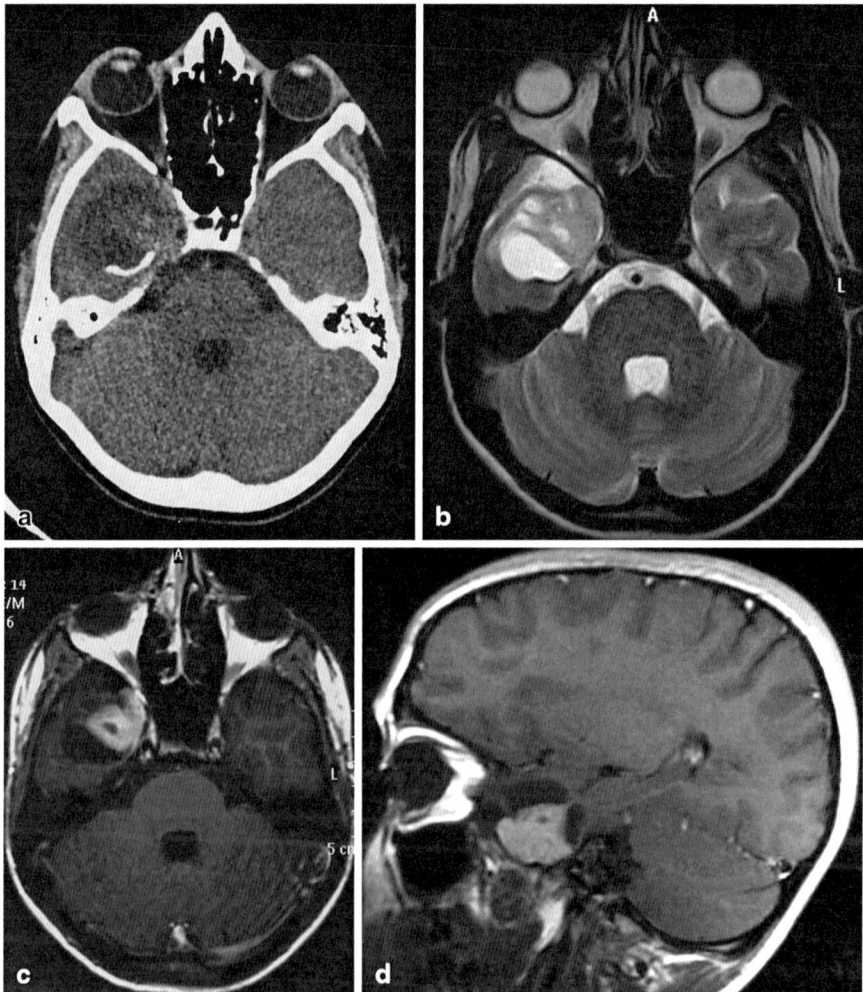

Fig. 1 GG in a 10-year-old girl. **a** Axial CT scan. **b** Axial T2-weighted image. **c** Axial Gd-enhanced T1-weighted image. **d** Sagittal Gd-enhanced T1-weighted image. CT shows an heterogeneous, prevailingly hypodense, mass lesion in the right temporal pole with gross, shell-like calcifications (**a**). The tumor has a cortical location and cystic and solid appearance (**b**). The solid component is mildly hyperintense on T2-weighted image (**b**) with marked enhancement following gadolinium injection (**c, d**)

vasogenic edema. Intratumoral hemorrhage is uncommon. Leptomeningeal or ependymal seeding may rarely occur (Adachi and Yagishita 2008).

Diffusion weighted imaging (DWI) shows increased diffusivity (high signal on ADC maps), whereas perfusion weighted imaging (PWI) demonstrates lower rCBV values compared to normal parenchyma. On magnetic resonance spectroscopy (MRS), most GGs display the features of a LGG, with decreased N-acetyl aspartate (NAA) and mildly elevated Choline (Cho).

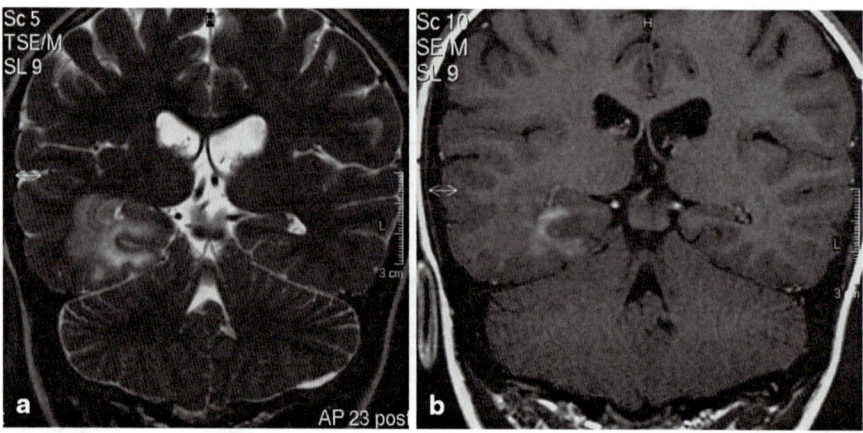

Fig. 2 GG in a 10-year-old boy. a Coronal T2-weighted image. b Coronal Gd-enhanced T1-weighted image. Lesion involving the right mesial temporal structures, with mild swelling of the parahippocampal region, hyperintense on T2-weighted image (**a**). Following gadolinium injection the lesion enhances. No cystic components are present

From the neuropathological point of view, the key feature is the presence of neoplastic ganglion cells. They differ from normal ganglion cells—large cells with vesicular nuclei, central and prominent nucleoli, and abundant cytoplasm—because of abnormalities of size, shape, cell processes, irregular accumulation of Nissl substance and cytoplasmic vacuolization. Cellular gigantism and bi- or multinucleation are a common finding (Fig. 3a). These cells do not achieve any degree of architectural uniformity.

Neurofibrillary tangles, granulo-vacuolar degeneration and basic cellular neuronal lesions typical of neurodegenerative diseases may be observed.

Neoplastic glial cells, whose presence is a requisite for the diagnosis of GG, have a variable morphology, which can make the tumor resemble a pilocytic astrocytoma,

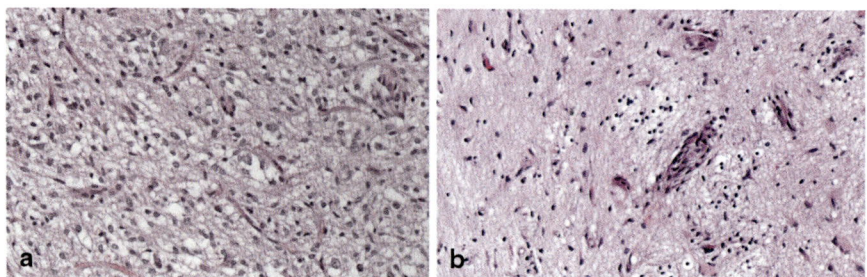

Fig. 3 GGa H&E, 200x. Gangliogliomas are typically non-infiltrating tumors; several multinucleated cells, which immunohistochemistry can prove to be neurons, are seen. **b** H&E. Lymphocytes are common

a diffusely infiltrating astrocytoma or even an oligodendroglial tumor. Melanotic cells may be found (Soffer et al. 1992).

The stroma is rich in reticular fibers, which encircle not only vessels, as in normal brain, but also neoplastic cells.

Vascular proliferation may be seen. The presence of hemosiderin deposits is an evidence of prior hemorrhage and, like contrast enhancement, is related to an impairment of the blood brain barrier (BBB). Intratumoral BBB dysfunction in concert with subsequent accumulation of albumin by neoplastic glial cells may represent an epileptogenic mechanism underlying tumor-associated long-term epilepsy (Schmitz et al. 2013).

Rosenthal fibers, basic astrocytic lesions, typical of fibrillary gliosis, and eosinophilic granular bodies are often observed. Lymphocytic infiltration is typical (Fig. 3b)

Necrosis is rare except in anaplastic cases.

It is interesting to note that temporal lobe GGs express CD34 glycoprotein, which apparently is not expressed in the frontal or parietal lobe tumors and it could be a marker of epileptogenic dysplasia (Kerkhof and Vecht 2013; Blümcke and Wiestler 2002).

GGs Grading

The third edition of the WHO Classification of CNS Tumors (Kleihues and Cavenee 2000) admitted the existence of atypical GGs, corresponding to grade II. Atypical tumors should have been distinguished from typical ones on the basis of cellularity, nuclear pleomorphism, microvascular proliferation, and proliferation index ($>5\%$). Furthermore, necrosis and a proliferation index ($>10\%$) should have been the hallmarks of anaplastic tumors. Unluckily, these criteria are unable to reliably predict the clinical behavior (Wolf et al. 1994; Luyken et al. 2004; Selvanathan et al. 2011). The fourth edition of the Classification includes no longer grade II (Louis et al. 2007b).

Although microscopic signs of atypia or anaplasia mostly occur in the glial component, they may occur in neuronal cells, too (Jay et al. 1994). Although anaplastic tumors typically appear *de novo*, one tenth of them is the result of the transformation of a GG. Anaplastic transformation is more common in pediatric cases and is associated with previous subtotal tumor resection and radiotherapy (Im et al. 2002).

Dysembryoplastic Neuroepithelial Tumor (DNT)

Although they may occur throughout the CNS, DNTs are cortical-based tumors, that typically affect children and young adults with long-standing pharmacoresistant epilepsy (Daumas-Duport et al. 1988, 2007).

DNTs of the cerebral cortex typically appear in the *isocortex* or in the *allocortex* of the temporal lobe; these tumors can arise in the frontal lobe and, sporadically, in the parietal and occipital lobes (Daumas-Duport et al. 1988; Thom et al. 2011).

Although there seems to be a relationship between DNTs and the development of the cerebral cortex, and cortical malformations may coexist alongside, from the neuro-oncologic point of view DNTs are considered as LGGs corresponding to grade I of the WHO Classification (Thom et al. 2011; Louis et al. 2007b).

Initially, the cells of the subpial granular layer were considered the source of this tumor(Daumas-Duport et al. 1988).

DNT coexistence with FCD (Daumas-Duport et al. 2007), neuronal heterotopia (Honavar et al. 1999), microdysgenesis (Rojiani et al. 1996) and neurofibromatosis type 1 (Lellouch-Tubiana et al. 1995) supports the hypothesis of a developmental origin.

DNTs show minimal or no mass effect and absence of peritumoral edema. On CT scan, they appear as a hypoattenuating lesion that may occasionally present areas of calcification and remodeling of the adjacent inner table of the skull. On MRI, the typical appearance consists of a well demarcated pseudocystic lesion, strongly hyperintense on T2- and hypointense on T1-weighted images, with variable FLAIR signal (hypo-isointense or hyperintense). DNTs may have a gyriform or a triangular-shaped pattern with the base pointing to the cortical surface. Hyperintense stripes on FLAIR images are visible both along the surface (bright rim) and on thin septa, resulting in a multicystic, bubbly appearance. Additional small cysts are often located in the vicinity, separated from the main mass.

Some lesions may show a more heterogeneous signal consistent with solid, cystic or semiliquid structures. Solid tissue is usually interspersed between the pseudocysts and it is often found in the adjacent subcortical white matter. Contrast enhancement is rare, variable, and often ring-like (Fig. 4a, b, c, d and Fig. 5a, b, c). Hemorrhage is also uncommon (Daumas-Duport et al. 1988; Daumas-Duport 1993; Ostertun et al. 1996; Campos et al. 2009). Diffusion and perfusion weighted images show respectively increased diffusivity and low rCBV values. The MRS pattern is nonspecific with increase in myo-inositol (mI) and slight reduction of NAA. Lactate and lipids are usually absent (Bulakbasi et al. 2007).

The neuropathological key feature is a microscopic structure, the *specific glioneuronal element*, which assumes the leading role in the *simple* forms, sharing the stage with the *glial nodules* in the *complex* forms. A highly controversial set of neuroepithelial tumors lacking the specific glioneuronal element has been classified as nonspecific or diffuse forms (Thom et al. 2011; Honavar et al. 1999; Bodi et al. 2012).

Simple and complex DNTs are generally circumscribed cystic lesions corresponding to type 1 of some neuroimaging classifications. The diffuse forms of the *neocortex* are quite nodular and sometimes calcific (type 2), the diffuse forms of the *allocortex* (mesial temporal lobe) tend to have dim outlines (type 3) (Chassoux and Daumas-Duport 2013; Chassoux et al. 2012).

The glioneuronal element is characterized by a mucoid matrix in which cells resembling oligodendrocytes align in a columnar fashion along bundles of axons and capillaries arrayed perpendicular to the pial surface. Mature neurons seem to float among the columns (Fig. 6a, b, c). Binucleate cells can be present; perineuronal satellitosis is not generally seen.

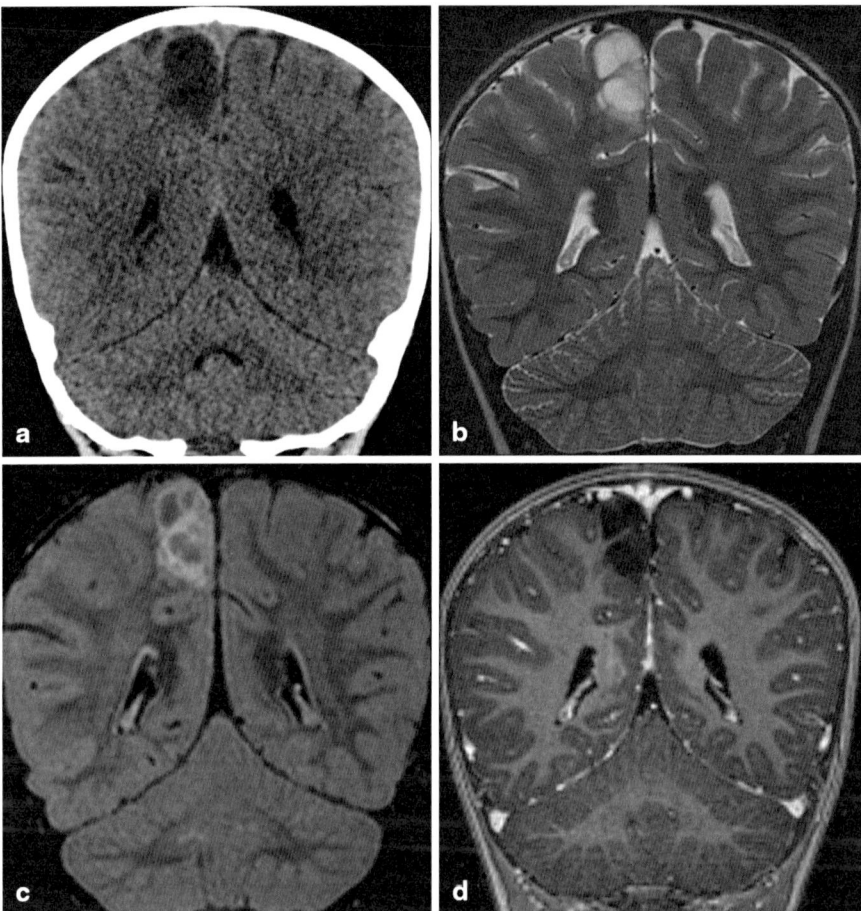

Fig. 4 DNT in a 6-year-old boy. **a** Coronal CT scan. **b** Coronal T2-weighted image. **c** Coronal FLAIR image. **d** Coronal Gd-enhanced T1-weighted image. Right paracentral lobule, cortical-subcortical, lesion whose density (**a**) and signal intensity on T2 (**b**) is cystlike. Corresponding FLAIR image (**c**) shows bright internal septa and lesion periphery producing a multicystic "bubbly" appearance. Following gadolinium administration enhancement is absent (**d**)

The glial nodules, which may contain neurons, are made up of astrocytes and oligodendrocytes, showing a variable degree of differentiation or pleomorphism. The nodules may resemble pilocytic astrocytoma or 'diffuse gliomas' (Fig. 6d).

Calcifications may be observed (Daumas-Duport et al. 1988; Thom et al. 2011); ossification is an exceptional finding (Thom et al. 2011).

These tumors have the capability both to cross the cortical sulci and to extend into the white matter and the leptomeninges (Thom et al. 2011; Zhang et al. 2013). White matter may show loss of myelin, gliosis and mycrocysts (Thom et al. 2011). HS, more commonly atypical (end-folium sclerosis) than classic, is common (Thom et al. 2011). Neurofibrillary tangles may be observed. The presence of intracellular pigments, such as iron or melanin, is quite common (Thom et al. 2011).

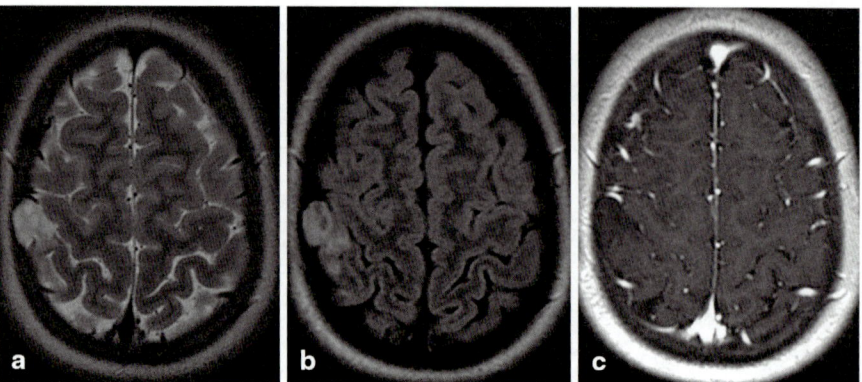

Fig. 5 DNT in a 9-year-old girl. **a** Axial T2-weighted image. **b** Axial FLAIR image. **c** Axial Gd-enhanced T1-weighted image. Superficially located tumor involving the right postcentral gyrus with remodeling of the adjacent calvarium (**a**). Two small internal pseudocysts are visibile on FLAIR image (**b**). There is no enhancement on post contrast T1-weighted image (**c**)

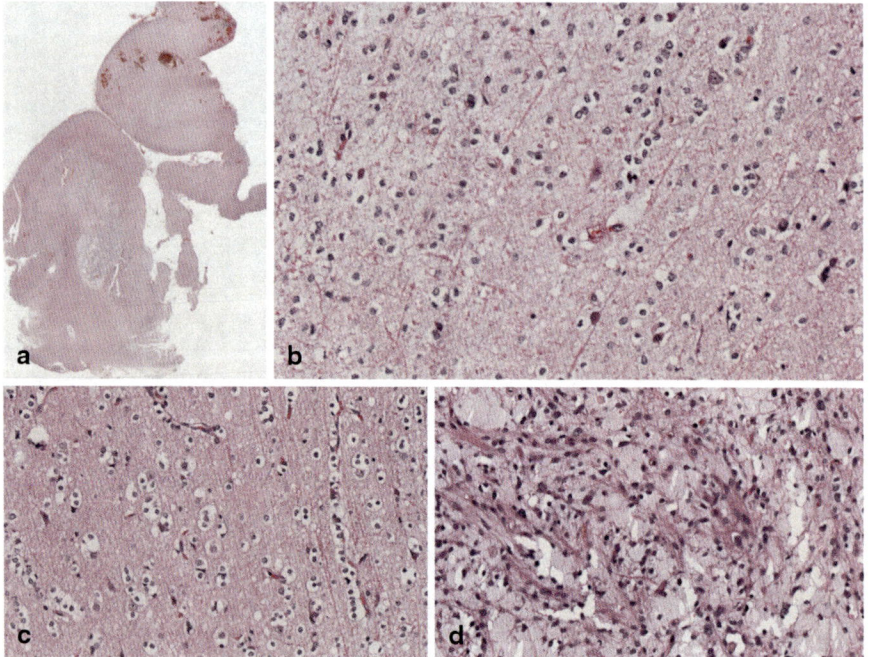

Fig. 6 DNT **a** Low-power of a DNT. **b** H&E, 200x. Bundles of axons and capillaries cross a mucoid matrix, in which some pyramidal neurons float. **c** H&E, 200x. Columns of cells resembling oligodendrocytes line up around axons. **d** H&E, 200x. Some areas may resemble to an pilocytic astrocytoma or ganglioglioma

Mitotic figures are very rare in the typical cases and the proliferative index is low. Although prominent vessels can be seen, endothelial proliferation is generally absent.

Necrosis is unexpected.

These tumors are able to regrow even after apparently complete resections showing a typical or atypical morphology (Daghistani et al. 2013; Schittenhelm et al. 2007). In some cases tumors seem to turn into something similar to anaplastic astrocytomas (Ray et al. 2009), glioblastoma (Duggal et al. 2008; Chuang et al. 2013) or oligoastrocytoma (Gonzales et al. 2007).

Neurocytic Tumors (Extraventricular Neurocytoma)

Neurocytic tumors of the cerebral cortex may cause focal drug-resistant epilepsy (Giulioni et al. 2011). These rare tumors, corresponding to grade II of the WHO Classification, typically appear in the *frontal lobe*, although they can arise in the parietal lobe and, sporadically, in the temporal and occipital lobes (Brat et al. 2001).

Temporal lobe tumors may coexist with FCD (Giulioni et al. 2011).

Neurocytic tumors are well demarcated lesions with variable and nonspecific appearance on neuroimaging studies depending on cellularity and anatomic location. On plain CT scan, they can appear as hypodense or isodense to gray matter; areas of patchy calcifications are common. On MRI, they have been described as hypointense or isointense on T1- and iso- to hyperintense on T2-weighted images with heterogeneous enhancement following the administration of a contrast agent. They commonly display cystic areas and mild to moderate peritumoral edema; intratumoral hemorrhage can occur. DWI studies showed nonspecific features with variable degree of diffusivity. On MRS, strongly decreased or no NAA peak and prominent Cho peak have been reported (Liu et al. 2013; Tortori-Donati et al. 1999; Han et al. 2013; Patil et al. 2014).

The key neuropathological feature is the presence of neurocytes, which are small round cells with modest quantity of cytoplasm, which can look clear, and nuclei with finely granular chromatin and one or more small nucleoli (Fig. 7a, b). Larger ganglioid cells or even larger ganglion cells may be found (Giangaspero et al.

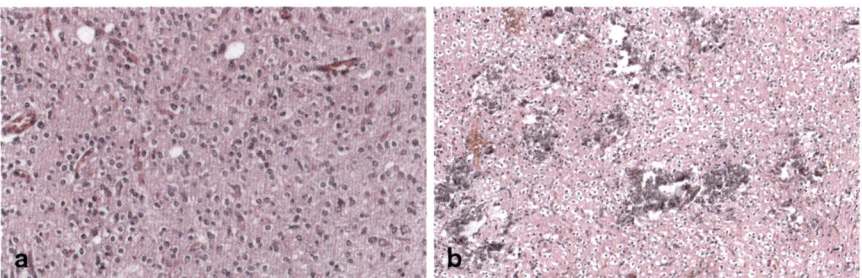

Fig. 7 Neurocytoma. **a** H&E, 200x. Neurocytomas are typically composed of uniform cells, whose clear halo may be prominent. **b** H&E, 100x. Calcifications are common

1997; Tortori-Donati et al. 1999). Tumor cells arrange in sheets, clusters, ribbons, or rosettes, with fine neuropil separating cell aggregates (Scheithauer et al. 2001). Immunohistochemical stainings consistently demonstrate synaptophysin, a major synaptic vesicle protein, in the neurocytes.

Some tumors (ganglioglioneurocytomas) harbor an astrocytic component.

The mitotic index is generally inconspicuous. Atypical features (high mitotic index, vascular proliferation, necrosis) and high cell proliferation may herald an aggressive clinical behavior.

Multinodular and Vacuolating Neuronal Tumor (MVNT)

This new entity is a purely neuronal tumor of the cerebral hemispheres, which seems to arise mainly from the temporal lobe, displaying a benign clinical behavior. This tumor has been associated with intractable epilepsy (Bodi et al. 2014; Huse et al. 2013).

Surgical Strategies and Complications

Surgery can be considered the treatment of choice for GNTs-related epilepsy. Focal epilepsies caused by GNTs are scarcely responsive to antiepileptic drugs and the surgical treatment of these epileptogenic tumors can offer seizure-freedom in up to 90% of cases (Daumas-Duport et al. 1988; Giulioni et al. 2005; Kirkpatrick et al. 1993).

Furthermore, surgical resection prevents tumor growth and the risk of anaplastic transformation. Last, the surgical approach leads to neuropathological diagnosis.

Although surgery in patients with GNTs-related epilepsy yields a good seizure outcome, the optimal surgical strategy for these tumors has not been fully established. In fact, some authors consider that tumor resection alone (the so-called "lesionectomy") is sufficient to achieve complete seizure control (Bourgeois ct al. 2006; Giulioni et al. 2005; Iannelli et al. 2000; Montes et al. 1995). On the other hand, other investigators also recommend the additional resection of peritumoral epileptogenic zones to maximize the seizure outcome (Aronica et al. 2001; Clusmann et al. 2004; Cossu et al. 2008; Morris et al. 1998; Kim et al. 2008; Luyken et al. 2003).

In the surgical management of GNTs, it is of paramount importance to consider some differences between these tumors and other groups of low-grade intra-axial neoplasms. First of all, GNTs are benign tumors in which epilepsy is generally the only clinical feature. Conversely, other types of LGG may present many other clinical findings (intracranial hypertension, focal neurological deficits, etc.). Second, unlike other LGG, GNTs are often located on temporal or temporo-mesial regions. Third, the pathophysiological mechanism of epilepsy in GNTs differs greatly compared to other LGG. GNTs have a so-called intrinsic epileptogenicity due to their neuronal and glial components. In addition, other epileptogenic diseases (above all, FCD) may be associated with GNTs, may contribute to determine the epileptogenic zone in some patients and are not always well distinguishable from the tumoral tissue on MRI. Finally, in temporal GNTs, the hippocampus may contribute to the epileptogenesis of these lesions even without apparent involvement at MRI or pathological examination (Chang et al. 2008; Englot et al. 2011; Smits and Duffau 2011; van Breemen et al. 2007; Blümcke et al. 2011; Tassi et al. 2002; Giulioni et al. 2009; Schramm and Aliashkevich 2008; Morioka et al. 2007).

Considering these features, the target of surgery for GNTs to obtain the best seizure outcome is not only the anatomical (tumoral) lesion, but the entire epileptogenic zone, which sometimes is larger than the simple structural lesion. Thus, the surgical strategy must be based on the anatomo-electro-clinical correlations that are necessary to identify the epileptogenic zone (Lüders et al. 2006). This principle is particularly important especially for lesions located in the temporal lobe, where the epileptogenic network is usually much more complex and extended than the tumoral area.

Thus, from the surgical point of view, distinctive "philosophies" of treatment may be adopted, usually depending on the location of the tumors.

For lesions located in extratemporal regions the target of surgical resection is the tumoral lesion only. In other words, only a so-called "lesionectomy" is performed (Bourgeois et al. 2006; Montes et al. 1995).

For lesions located in the temporal lobe (the most frequent site of GNTs), several approaches have been described, including lesionectomy, extended lesionectomy, tailored resection and anterior temporal lobectomy (Casazza et al. 1997; Clusmann et al. 2004; Cossu et al. 2008; Jooma et al. 1995; Quarato et al. 2005).

To our knowledge, presently few studies correlate the best surgical strategy to obtain the best seizure outcome to tumor location (Cataltepe et al. 2005; Giulioni et al. 2009; Luyken et al. 2003). There is some consensus that the best seizure outcome for epileptogenic GNTs located in extratemporal and temporo-lateral regions is provided by simple lesionectomy. For temporo-mesial GNTs, the results of a simple lesionectomy are not particularly encouraging, while more extended surgeries (tailored resection) offer better seizure outcomes (Giulioni et al. 2009).

Finally, another important issue concerning the surgical strategy for GNTs is the additional resection of the hippocampal-parahippocampal complex when it is not invaded by GNTs and does not show other signal abnormalities on MR imaging. In fact, if a more extended surgery than simple lesionectomy for temporo-mesial GNTs is a generally accepted concept (Schramm 2008), the correct balance between the extent of resection necessary to provide the best seizure outcome and the avoidance of neuropsychological deficits is still an open problem deserving further studies (Helmstaedter et al. 2011).

Neurosurgery has become safer in the last decades, thanks to the improvement of diagnostic and operative techniques and anaesthesiological procedures.

However, craniotomy and brain surgery imply the risk of complications, whose rate is quite low; mortality is between 0.5 and 1 % (Pilcher and Rusyniak 1993).

The neurosurgical complications of GNTs surgery can be subdivided into craniotomy-related complications (general neurosurgical complications) and complications more specifically related to resective brain surgery.

The first group is represented by hematomas (extradural, subdural, and intracerebral), infections, cerebrospinal fluid leak and, rarely, air embolism and pulmonary embolism. These complications are responsive to appropriate treatment and do not usually influence surgical outcome.

Concerning the second group of complications, we can distinguish between major complications (when severe neurological deficit or daily activity impairment occur) and minor complications (disappearing within 3 months).

Neurological complications depend on the site of resection. Considering that lesionectomy and tailored resection are the surgical strategies commonly adopted in the management of GNTs and these neoplasms are preferentially located in the temporal lobe, typical major neurological complications of temporal lobe surgery (TLS) include contralateral hemiparesis or hemiplegia, homolateral third nerve palsy or paresis, contralateral visual field defects (VFDs) and speech disturbance (Pilcher and Rusyniak 1993; Polkey 2004).

The hemiplegia or hemiparesis can be related mainly to vascular causes (manipulation of middle cerebral artery branches and/or other arteries) (Polkey 2004).

VFD consists classically in superior quadrantanopia contralateral to the resection (Egan et al. 2000). This finding is caused by injury of Meyer's loop (the most anterior part of the optic radiation). Other VFDs in temporal lobe surgery can also occur because of injuries of the optic tract or lateral geniculate body. However VFDs should be considered as an expected event in TLS. Today recent advances in neuroimaging techniques as diffusion tensor imaging (DTI) allowing a better anatomic definition of the optic radiation could contribute to reduce post-resection VFDs.

Other rare complications of TLS include hemorrhage distant from the site of surgery (Toczek et al. 1996; Giulioni et al. 2006; Yacubian et al. 1999), probably due to transmural venous pressure variations (Giulioni and Martinoni 2011) and middle fossa cyst causing raised pressure (Weaver et al. 1996). The global incidence of major neurological complications of TLS is 0.37–4 % (Pilcher and Rusyniak 1993; Polkey 2004).

Surgery-related complications in frontal lobe GNTs include hemiparesis if the resection encroaches upon the gyrus anterior to the precentral gyrus. Broca's area must be preserved in the dominant hemisphere. Some authors recommended such resections should be performed under local anaesthesia (Polkey 2004) and/or using a certain number of localizing techniques such as functional MRI and transcranial magnetic stimulation for motor cortex localization (Macdonell et al. 1999).

In spite of these precautions, any resection centered on the central area (primary motor and sensory area) carries the risk of loss of its function.

Resections for lesions located in parietal and posterior temporal regions of dominant hemisphere can determine receptive aphasia, dyslexia, disgraphia and dyscalculia.

Finally, for occipital resections, some degree of VFD(s) can be expected, even if some authors noted that the patients adapted to their VFD within 1 year after surgery (Williamson et al. 1992).

Seizure Outcome

The post-operative seizure outcome is currently assessed according to Engel's Classification (Engel 1996).

The rate of seizure-free outcomes (Engel Class I) in GNTs –related focal epilepsies is about 90%.

Several factors have been correlated with seizure outcome: extension of surgical resection, histotype, duration of epilepsy, age at surgery, association of the tumor with perilesional cortical dysplasia (Aronica et al. 2001; Daumas-Duport et al. 1988; Im et al. 2002; Morris et al. 1998; Nolan et al. 2004).

The main predictor of excellent seizure outcome is the complete removal of the tumor (Morris et al. 1998; Nolan et al. 2004; Khajavi et al. 1995; Kim et al. 2001).

Concerning the completeness of resection, it must be emphasized that only few studies attempted to establish a correlation between tumor site (i.e., temporo-mesial versus temporo-lateral and extra-temporal) and surgical strategy (lesionectomy versus resection of the epileptogenic focus, including the tumor).

Giulioni et al. (2005, 2006) observed that patients affected by extra-temporal and temporo-lateral GNTs had better seizure outcome than those with temporo-mesial GNTs after lesionectomy alone. In addition, in a study comparing retrospectively seizure outcome between two homogeneous series of temporo-mesial epileptogenic GNTs treated respectively by lesionectomy alone and tailored surgery, the same authors reported a much better seizure outcome (93% vs 42.8%) in patients who underwent tailored resection (Giulioni et al. 2009).

These observations and other evidences concerning the role of the temporo-polar cortex in temporal lobe seizures (Chabardès et al. 2005) and the frequent association of GNTs with other epileptogenic abnormalities (Daumas-Duport et al. 2007) suggest that tailored resection for temporo-mesial GNTs is to be preferred to obtain a better seizure outcome.

Future Trends

In the future, optimal therapeutic management of GNTs will need a better understanding of the epileptogenic mechanisms related to these neoplasms.

In addition, further studies are required to elucidate GNTs preference for the temporal lobe.

Neuronal precursors in the subgranular zone of the dentate gyrus and the occurrence of postnatal life neurogenesis have been recently described (González-Martínez et al. 2007; Paradisi et al. 2010; Siebzehnrubl and Blümcke 2008). These data could contribute to explain more complex epileptogenic mechanisms and/or the occurrence of GNTs.

Moreover, the well known association between GNTs and FCD needs further investigations. If common precursor cells for these two diseases can be hypothesized, either at an embryological stage or during postnatal life, it still remains to be clarified whether FCD coexists with GNTs or has some potential to change into neoplastic cells (Daumas-Duport et al. 2007).

Conclusions

GNTs are one of the most common causes of drug-resistant epilepsy in children and young adults. GNTs-related epilepsies usually respond very well to surgical treatment with an extremely favorable seizure outcome and important effects on cognitive development and patients' quality of life.

Surgical treatment should be planned on the basis of anatomo-electro-clinical correlation defining the epileptogenic zone (which may involve a larger area than the tumor) to be resected.

Tumor location is another important factor (temporo-mesial, temporo-lateral, extratemporal site) to be considered in planning surgical resection.

Finally, even in the event of sporadic seizures and/or seizures responsive to AEDs, surgery should be proposed in order to obtain complete seizure control with freedom from AEDs, and to prevent tumor growth and risk of malignant transformation.

References

Adachi Y, Yagishita A (2008) Gangliogliomas: characteristic imaging findings and role in the temporal lobe epilepsy. Neuroradiology 50:829–834

Allende DS, Prayson RA (2009) The expanding family of glioneuronal tumors. Adv Anat Pathol 16:33–39

Aronica E, Leenstra S, van Veelen CW, van Rijen PC, Hulsebos TJ, Tersmette AC, Yankaya B, Troost D (2001) Glioneuronal tumors and medically intractable epilepsy: a clinical study with long-term follow-up of seizure outcome after surgery. Epilepsy Res 43:179–191

Aronica E, Redeker S, Boer K, Spliet WG, van Rijen PC, Gorter JA, Troost D (2007) Inhibitory networks in epilepsy-associated gangliogliomas and in the perilesional epileptic cortex. Epilepsy Res 74:33–44 (Epub 30 Jan 2007)

Beaumont A, Whittle IR (2000) The pathogenesis of tumour associated epilepsy. Acta Neurochir (Wien) 142:1–15

Becker AJ, Wiestler OD, Figarella-Branger D, Blümcke I (2007) Ganglioglioma and gangliocytoma. In: Louis DN, Ohgaki H, Wiestler OD, Cavenee WK (eds) WHO classification of tumours of the central nervous system. International Agency for Research on Cancer, Lyon, pp. 103–105

Benifla M, Otsubo H, Ochi A, Weiss SK, Donner EJ, Shroff M, Chuang S, Hawkins C, Drake JM, Elliott I, Smith ML, Snead OC 3rd, Rutka JT (2006) Temporal lobe surgery for intractable epilepsy in children: an analysis of outcomes in 126 children. Neurosurgery 59:1203–1214

Bilginer B, Yalnizoglu D, Soylemezoglu F, Turanli G, Cila A, Topçu M, Akalan N (2009) Surgery for epilepsy in children with dysembryoplastic neuroepithelial tumor: clinical spectrum, seizure outcome, neuroradiology, and pathology. Childs Nerv Syst 25:485–491. doi:10.1007/s00381-008-0762-x (Epub 5 Dec 2008)

Blümcke I, Wiestler OD (2002) Gangliogliomas: an intriguing tumor entity associated with focal epilepsies. J Neuropathol Exp Neurol 61:575–584

Blümcke I, Luyken C, Urbach H, Schramm J, Wiestler OD (2004) An isomorphic subtype of long-term epilepsy-associated astrocytomas associated with benign prognosis. Acta Neuropathol 107:381–388

Blümcke I, Thom M, Aronica E, Armstrong DD, Vinters HV, Palmini A, Jacques TS, Avanzini G, Barkovich AJ, Battaglia G, Becker A, Cepeda C, Cendes F, Colombo N, Crino P, Cross JH, Delalande O, Dubeau F, Duncan J, Guerrini R, Kahane P, Mathern G, Najm I, Ozkara C, Raybaud C, Represa A, Roper SN, Salamon N, Schulze-Bonhage A, Tassi L, Vezzani A, Spreafico R (2011) The clinicopathologic spectrum of focal cortical dysplasias: a consensus classification proposed by an ad hoc Task Force of the ILAE Diagnostic Methods Commission. Epilepsia 52:158–174. doi:10.1111/j.1528-1167.2010.02777.x (Epub 10 Nov 2010)

Bodi I, Selway R, Bannister P, Doey L, Mullatti N, Elwes R, Honavar M (2012) Diffuse form of dysembryoplastic neuroepithelial tumour: the histological and immunohistochemical features of a distinct entity showing transition to dysembryoplastic neuroepithelial tumour and ganglioglioma. Neuropathol Appl Neurobiol 38:411–425

Bodi I, Curran O, Selway R, Elwes R, Burrone J, Laxton R, Al-Sarraj S, Honavar M (2014) Two cases of multinodular and vacuolating neuronal tumour. Acta Neuropathol Commun 2:7

Bourgeois M, Di Rocco F, Sainte-Rose C (2006) Lesionectomy in the pediatric age. Childs Nerv Syst 22:931–935 (Epub 5 July 2006)

Brat DJ, Scheithauer BW, Eberhart CG, Burger PC (2001) Extraventricular neurocytomas: pathologic features and clinical outcome. Am J Surg Pathol 25:1252–1260

Bulakbasi N, Kocaoglu M, Sanal TH, Tayfun C (2007) Dysembryoplastic neuroepithelial tumors: proton MR spectroscopy, diffusion and perfusion characteristics. Neuroradiology 49:805–812

Campos AR, Clusmann H, von Lehe M, Niehusmann P, Becker AJ, Schramm J, Urbach H (2009) Simple and complex dysembryoplastic neuroepithelial tumors (DNT) variants: clinical profile, MRI, and histopathology. Neuroradiology 51:433–443

Casazza M, Avanzini G, Broggi G, Fornari M, Franzini A (1989) Epilepsy course in cerebral gangliogliomas: a study of 16 cases. Acta Neurochir Suppl (Wien) 46:17–20

Casazza M, Avanzini G, Ciceri E, Spreafico R, Broggi G (1997) Lesionectomy in epileptogenic temporal lobe lesions: preoperative seizure course and postoperative outcome. Acta Neurochir Suppl 68:64–69

Castillo M (1998) Gangliogliomas: ubiquitous or not? AJNR Am J Neuroradiol 19:807–809

Castillo M, Davis PC, Takei Y, Hoffman JC (1990) Intracranial ganglioglioma: MR, CT, and clinical findings in 18 patients. AJNR Am J Neuroradiol 11:109–114

Cataltepe O, Turanli G, Yalnizoglu D, Topçu M, Akalan N (2005) Surgical management of temporal lobe tumor-related epilepsy in children. J Neurosurg 102:280–287

Chabardès S, Kahane P, Minotti L, Tassi L, Grand S, Hoffmann D, Benabid AL (2005) The temporopolar cortex plays a pivotal role in temporal lobe seizures. Brain 128:1818–1831 (Epub 27 April 2005)

Chang EF, Potts MB, Keles GE, Lamborn KR, Chang SM, Barbaro NM, Berger MS (2008) Seizure characteristics and control following resection in 332 patients with low-grade gliomas. J Neurosurg 108:227–235. doi:10.3171/JNS/2008/108/2/0227

Chang EF, Christie C, Sullivan JE, Garcia PA, Tihan T, Gupta N, Berger MS, Barbaro NM (2010) Seizure control outcomes after resection of dysembryoplastic neuroepithelial tumor in 50 patients. J Neurosurg Pediatr 5:123–130. doi:10.3171/2009.8.PEDS09368

Chassoux F, Daumas-Duport C (2013) Dysembryoplastic neuroepithelial tumors: where are we now? Epilepsia 54:129–134. doi:10.1111/epi.12457

Chassoux F, Devaux B, Landré E, Turak B, Nataf F, Varlet P, Chodkiewicz JP, Daumas-Duport C (2000) Stereoelectroencephalography in focal cortical dysplasia: a 3D approach to delineating the dysplastic cortex. Brain 123:1733–1751

Chassoux F, Rodrigo S, Mellerio C, Landré E, Miquel C, Turak B, Laschet J, Meder JF, Roux FX, Daumas-Duport C, Devaux B (2012) Dysembryoplastic neuroepithelial tumors: an MRI-based scheme for epilepsy surgery. Neurology 79:1699–1707

Chuang NA, Yoon JM, Newbury RO, Crawford JR (2013) Glioblastoma multiforme arising from dysembryoplastic neuroepithelial tumor in a child in the absence of therapy. J Pediatr Hematol Oncol (4 Dec) [Epub ahead of print]

Clusmann H, Schramm J, Kral T, Helmstaedter C, Ostertun B, Fimmers R, Haun D, Elger CE (2002) Prognostic factors and outcome after different types of resection for temporal lobe epilepsy. J Neurosurg 97:1131–1141

Clusmann H, Kral T, Gleissner U, Sassen R, Urbach H, Blümcke I, Bogucki J, Schramm J (2004) Analysis of different types of resection for pediatric patients with temporal lobe epilepsy. Neurosurgery 54:847–860

Consales A, Striano P, Nozza P, Morana G, Ravegnani M, Piatelli G, Pavanello M, Zoli ML, Baglietto MG, Cama A (2013) Glioneuronal tumors and epilepsy in children: seizure outcome related to lesionectomy. Minerva Pediatr 65:609–616

Cossu M, Lo Russo G, Francione S, Mai R, Nobili L, Sartori I, Tassi L, Citterio A, Colombo N, Bramerio M, Galli C, Castana L, Cardinale F (2008) Epilepsy surgery in children: results and predictors of outcome on seizures. Epilepsia 49:65–72 (Epub 21 July 2007)

Daghistani R, Miller E, Kulkarni AV, Widjaja E (2013) Atypical characteristics and behavior of dysembryoplastic neuroepithelial tumors. Neuroradiology 55:217–224

Daumas-Duport C (1993) Dysembryoplastic neuroepithelial tumors. Brain Pathol 3:255–268

Daumas-Duport C, Scheithauer BW, Chodkiewicz JP, Laws ER Jr, Vedrenne C (1988) Dysembryoplastic neuroepithelial tumor: a surgically curable tumor of young patients with intractable partial seizures. Neurosurgery 23:545–556

Daumas-Duport C, Pietsch T, Hawkins C, Shankar SK (2007) Dysembryoplastic neuroepithelial tumour. In: Louis DN, Ohgaki H, Wiestler OD, Cavenee WK (eds) WHO classification of tumours of the central nervous system. International Agency for Research on Cancer, Lyon, pp 99–102

de Groot M, Toering ST, Boer K, Spliet WG, Heimans JJ, Aronica E, Reijneveld JC (2010) Expression of synaptic vesicle protein 2A in epilepsy-associated brain tumors and in the peritumoral cortex. Neuro Oncol 12:265–273. doi:10.1093/neuonc/nop028 (Epub 6 Jan 2010)

Duggal N, Taylor R, Zou GY, Hammond RR (2008) Dysembryoplastic neuroepithelial tumours: clinical, proliferative and apoptotic features. J Clin Pathol 61:127–131

Egan RA, Shults WT, So N, Burchiel K, Kellogg JX, Salinsky M (2000) Visual field deficits in conventional anterior temporal lobectomy versus amygdalohippocampectomy. Neurology 55:1818–1822

Engel J Jr (1996) Surgery for seizures. N Engl J Med 334:647–652

Englot DJ, Berger MS, Barbaro NM, Chang EF (2011) Predictors of seizure freedom after resection of supratentorial low-grade gliomas. A review. J Neurosurg 115:240–244. doi:10.3171/2011.3.JNS1153 (Epub 29 April 2011)

Ferrier CH, Aronica E, Leijten FS, Spliet WG, van Huffelen AC, van Rijen PC, Binnie CD (2006) Electrocorticographic discharge patterns in glioneuronal tumors and focal cortical dysplasia. Epilepsia 47:1477–1486

Gaggero R, Consales A, Fazzini F, Mancardi MM, Baglietto MG, Nozza P, Rossi A, Pistorio A, Tumolo M, Cama A, Garrè ML, Striano P (2009) Epilepsy associated with supratentorial brain tumors under 3 years of life. Epilepsy Res 87:184–189. doi:10.1016/j.eplepsyres.2009.08.012 (Epub 23 Sept 2009)

Giangaspero F, Cenacchi G, Losi L, Cerasoli S, Bisceglia M, Burger PC (1997) Extraventricular neoplasms with neurocytoma features. A clinicopathological study of 11 cases. Am J Surg Pathol 21:206–212

Giulioni M, Martinoni M (2011) Postoperative intracranial haemorrhage and remote cerebellar haemorrhage. Neurosurg Rev 34:523–525. doi:10.1007/s10143-011-0335-4 (Epub 15 June 2011)

Giulioni M, Galassi E, Zucchelli M, Volpi L (2005) Seizure outcome of lesionectomy in glioneuronal tumors associated with epilepsy in children. J Neurosurg 102:288–293

Giulioni M, Gardella E, Rubboli G, Roncaroli F, Zucchelli M, Bernardi B, Tassinari CA, Calbucci F (2006) Lesionectomy in epileptogenic gangliogliomas: seizure outcome and surgical results. J Clin Neurosci 13:529–535

Giulioni M, Rubboli G, Marucci G, Martinoni M, Volpi L, Michelucci R, Marliani AF, Bisulli F, Tinuper P, Castana L, Sartori I, Calbucci F (2009) Seizure outcome of epilepsy surgery in focal epilepsies associated with temporomesial glioneuronal tumors: lesionectomy compared with tailored resection. J Neurosurg 111:1275–1282. doi:10.3171/2009.3.JNS081350

Giulioni M, Martinoni M, Rubboli G, Marucci G, Marliani AF, Battaglia S, Badaloni F, Pozzati E, Calbucci F (2011) Temporo-mesial extraventricular neurocytoma and cortical dysplasia in focal temporal lobe epilepsy. J Clin Neurosci 18:147–148

Gonzales M, Dale S, Susman M, Nolan P, Ng WH, Maixner W, Laidlaw J (2007) Dysembryoplastic neuroepithelial tumor (DNT)-like oligodendrogliomas or Dnts evolving into oligodendrogliomas: two illustrative cases. Neuropathology 27:324–330

González-Martínez JA, Bingaman WE, Toms SA, Najm IM (2007) Neurogenesis in the postnatal human epileptic brain. J Neurosurg 107:628–635

Hall WA, Yunis EJ, Albright AL (1986) Anaplastic ganglioglioma in an infant: case report and review of the literature. Neurosurgery 19:1016–1020

Hammond RR, Duggal N, Woulfe JM, Girvin JP (2000) Malignant transformation of a dysembryoplastic neuroepithelial tumor. Case report. J Neurosurg 92:722–725

Han L, Niu H, Wang J, Wan F, Shu K, Ke C, Lei T (2013) Extraventricular neurocytoma in pediatric populations: a case report and review of the literature. Oncol Lett 6:1397–1405

Helmstaedter C, Roeske S, Kaaden S, Elger CE, Schramm J (2011) Hippocampal resection length and memory outcome in selective epilepsy surgery. J Neurol Neurosurg Psychiatry 82:1375–1381 doi:10.1136/jnnp.2010.240176 (Epub 7 June 2011)

Honavar M, Janota I, Polkey CE (1999) Histological heterogeneity of dysembryoplastic neuroepithelial tumour: identification and differential diagnosis in a series of 74 cases. Histopathology 34:342–356

Huse JT, Edgar M, Halliday J, Mikolaenko I, Lavi E, Rosenblum MK (2013) Multinodular and vacuolating neuronal tumors of the cerebrum: 10 cases of a distinctive seizure-associated lesion. Brain Pathol 23:515–524

Iannelli A, Guzzetta F, Battaglia D, Iuvone L, Di Rocco C (2000) Surgical treatment of temporal tumors associated with epilepsy in children. Pediatr Neurosurg 32:248–254

Im SH, Chung CK, Cho BK, Wang KC, Yu IK, Song IC, Cheon GJ, Lee DS, Kim NR, Chi JG (2002) Intracranial ganglioglioma: preoperative characteristics and oncologic outcome after surgery. J Neurooncol 59:173–183

Jay V, Squire J, Becker LE, Humphreys R (1994) Malignant transformation in a ganglioglioma with anaplastic neuronal and astrocytic components. Report of a case with flow cytometric and cytogenetic analysis. Cancer 73:2862–2868

Johnson JH Jr, Hariharan S, Berman J, Sutton LN, Rorke LB, Molloy P, Phillips PC (1997) Clinical outcome of pediatric gangliogliomas: ninety-nine cases over 20 years. Pediatr Neurosurg 27:203–207

Jooma R, Yeh HS, Privitera MD, Gartner M (1995) Lesionectomy versus electrophysiologically guided resection for temporal lobe tumors manifesting with complex partial seizures. J Neurosurg 83:231–236

Kerkhof M, Vecht CJ (2013) Seizure characteristics and prognostic factors of gliomas. Epilepsia 54:12–17. doi:10.1111/epi.12437

Khajavi K, Comair YG, Prayson RA, Wyllie E, Palmer J, Estes ML, Hahn JF (1995) Childhood ganglioglioma and medically intractable epilepsy. A clinicopathological study of 15 patients and a review of the literature. Pediatr Neurosurg 22:181–188

Kim SK, Wang KC, Hwang YS, Kim KJ, Cho BK (2001) Intractable epilepsy associated with brain tumors in children: surgical modality and outcome. Childs Nerv Syst 17:445–452

Kim NR, Wang KC, Bang JS, Choe G, Park Y, Kim SK, Cho BK, Chi JG (2003) Glioblastomatous transformation of ganglioglioma: case report with reference to molecular genetic and flow cytometric analysis. Pathol Int 53:874–882

Kim SK, Wang KC, Hwang YS, Kim KJ, Chae JH, Kim IO, Cho BK (2008) Epilepsy surgery in children: outcomes and complications. J Neurosurg Pediatr 1:277–283. doi:10.3171/PED/2008/1/4/277

Kirkpatrick PJ, Honavar M, Janota I, Polkey CE (1993) Control of temporal lobe epilepsy following en bloc resection of low-grade tumors. J Neurosurg 78:19–25

Kleihues P, Cavenee WK (eds) (2000) Pathology and genetics of tumours of the nervous system. International Agency for Research on Cancer, Lyon

Lawson JA, Duchowny MS (2004) Paediatric epilepsy surgery. In: Shorvon SD, Perucca E, Fish DR, Dodson WE (eds) The treatment of epilepsy, 2nd edn. Blackwell Publishing, Oxford, pp 779–789

Lee MC, Kang JY, Seol MB, Kim HS, Woo JY, Lee JS, Jung S, Kim JH, Woo YJ, Kim MK, Kim HI, Kim SU (2006) Clinical features and epileptogenesis of dysembryoplastic neuroepithelial tumor. Childs Nerv Syst 22:1611–1618 (Epub 30 Aug 2006)

Lellouch-Tubiana A, Bourgeois M, Vekemans M, Robain O (1995) Dysembryoplastic neuroepithelial tumors in two children with neurofibromatosis type 1. Acta Neuropathol 90:319–322

Liu K, Wen G, Lv XF, Deng YJ, Deng YJ, Hou GQ, Zhang XL, Han LJ, Ding JL (2013) MR imaging of cerebral extraventricular neurocytoma: a report of 9 cases. AJNR Am J Neuroradiol 34:541–546

Louis DN, Ohgaki H, Wiestler OD, Cavenee WK, Burger PC, Jouvet A, Scheithauer BW, Kleihues P (2007a) The 2007 WHO classification of tumours of the central nervous system. Acta Neuropathol 114:97–109 (Epub 6 July 2007)

Louis DN, Ohgaki H, Wiestler OD, Cavenee WK (eds) (2007b) WHO classification of tumours of the central nervous system. International Agency for Research on Cancer, Lyon

Lüders HO, Najm I, Nair D, Widdess-Walsh P, Bingman W (2006) The epileptogenic zone: general principles. Epileptic Disord 8:S1–S9 (Erratum in: Epileptic Disord 2008, 10, 191)

Luyken C, Blümcke I, Fimmers R, Urbach H, Elger CE, Wiestler OD, Schramm J (2003) The spectrum of long-term epilepsy-associated tumors: long-term seizure and tumor outcome and neurosurgical aspects. Epilepsia 44:822–830

Luyken C, Blümcke I, Fimmers R, Urbach H, Wiestler OD, Schramm J (2004) Supratentorial gangliogliomas: histopathologic grading and tumor recurrence in 184 patients with a median follow-up of 8 years. Cancer 101:146–155

Macdonell RA, Jackson GD, Curatolo JM, Abbott DF, Berkovic SF, Carey LM, Syngeniotin A, Fabinyi GC, Scheffer IE (1999) Motor cortex localization using functional MRI and transcranial magnetic stimulation. Neurology 53:1462–1467

Minkin K, Klein O, Mancini J, Lena G (2008) Surgical strategies and seizure control in pediatric patients with dysembryoplastic neuroepithelial tumors: a single-institution experience. J Neurosurg Pediatr 1:206–210. doi:10.3171/PED/2008/1/3/206

Montes JL, Rosenblatt B, Farmer JP, O'Gorman AM, Andermann F, Watters GV, Meagher-Villemure K (1995) Lesionectomy of MRI detected lesions in children with epilepsy. Pediatr Neurosurg 22:167–173

Morioka T, Hashiguchi K, Nagata S, Miyagi Y, Yoshida F, Shono T, Mihara F, Koga H, Sasaki T (2007) Additional hippocampectomy in the surgical management of intractable temporal lobe epilepsy associated with glioneuronal tumor. Neurol Res 29:807–815

Morris HH, Matkovic Z, Estes ML, Prayson RA, Comair YG, Turnbull J, Najm I, Kotagal P, Wyllie E (1998) Ganglioglioma and intractable epilepsy: clinical and neurophysiologic features and predictors of outcome after surgery. Epilepsia 39:307–313

Nolan MA, Sakuta R, Chuang N, Otsubo H, Rutka JT, Snead OC 3rd, Hawkins CE, Weiss SK (2004) Dysembryoplastic neuroepithelial tumors in childhood: long-term outcome and prognostic features. Neurology 62:2270–2276

Obeid M, Wyllie E, Rahi AC, Mikati MA (2009) Approach to pediatric epilepsy surgery: State of the art, Part II: Approach to specific epilepsy syndromes and etiologies. Eur J Paediatr Neurol 13:115–127. doi:10.1016/j.ejpn.2008.05.003

Ostertun B, Wolf HK, Campos MG, Matus C, Solymosi L, Elger CE, Schramm J, Schild HH (1996) Dysembryoplastic neuroepithelial tumors: MR and CT evaluation. AJNR Am J Neuroradiol 17:419–430

Otsubo H, Hoffman HJ, Humphreys RP, Hendrick EB, Drake JM, Hwang PA, Becker LE, Chuang SH (1990–1991) Evaluation, surgical approach and outcome of seizure patients with gangliogliomas. Pediatr Neurosurg 16:208–212

Ozlen F, Gunduz A, Asan Z, Tanriverdi T, Ozkara C, Yeni N, Yalcinkaya C, Ozyurt E, Uzan M (2010) Dysembryoplastic neuroepithelial tumors and gangliogliomas: clinical results of 52 patients. Acta Neurochir (Wien) 152:1661–1671. doi:10.1007/s00701-010-0696-4 (Epub 5 June 2010)

Paradisi M, Fernández M, Del Vecchio G, Lizzo G, Marucci G, Giulioni M, Pozzati E, Antonelli T, Lanzoni G, Bagnara GP, Giardino L, CalzÃ L (2010) Ex vivo study of dentate gyrus neurogenesis in human pharmacoresistant temporal lobe epilepsy. Neuropathol Appl Neurobiol 36(6):535–550. doi:10.1111/j.1365-2990.2010.01102.x (Oct 2010)

Patil AS, Menon G, Easwer HV, Nair S (2014) Extraventricular neurocytoma, a comprehensive review. Acta Neurochir (Wien) 156: 349–354. doi:10.1007/s00701-013-1971-y (Epub 20 Dec 2013)

Pilcher WH, Rusyniak WG (1993) Complications of epilepsy surgery. Neurosurg Clin N Am 4:311–325

Polkey CE (2004) Complications of epilepsy surgery. In: Shorvon SD, Perucca E, Fish DR, Dodson WE (eds) The treatment of epilepsy, 2nd edn. Blackwell Publishing, Oxford, 849–860

Prayson RA (1999) Composite ganglioglioma and dysembryoplastic neuroepithelial tumor. Arch Pathol Lab Med 123:247–250

Prayson RA, Napekoski KM (2012) Composite ganglioglioma/dysembryoplastic neuroepithelial tumor: a clinicopathologic study of 8 cases. Hum Pathol 43:1113–1118

Prayson RA, Khajavi K, Comair YG (1995) Cortical architectural abnormalities and MIB1 immunoreactivity in gangliogliomas: a study of 60 patients with intracranial tumors. J Neuropathol Exp Neurol 54:513–520

Prayson RA, Morris HH, Estes ML, Comair YG (1996) Dysembryoplastic neuroepithelial tumor: a clinicopathologic and immunohistochemical study of 11 tumors including MIB1 immunoreactivity. Clin Neuropathol 15:47–53

Prayson RA, Fong J, Najm I (2010) Coexistent pathology in chronic epilepsy patients with neoplasms. Mod Pathol 23:1097–1103. doi:10.1038/modpathol.2010.94 (Epub 21 May 2010)

Quarato PP, Di Gennaro G, Mascia A, Grammaldo LG, Meldolesi GN, Picardı A, GiampÃ T, Falco C, Sebastiano F, Onorati P, Manfredi M, Cantore G, Esposito V (2005) Temporal lobe epilepsy surgery: different surgical strategies after a non-invasive diagnostic protocol. J Neurol Neurosurg Psychiatry 76:815–824

Ray WZ, Blackburn SL, Casavilca-Zambrano S, Barrionuevo C, Orrego JE, Heinicke H, Dowling JL, Perry A (2009) Clinicopathologic features of recurrent dysembryoplastic neuroepithelial tumor and rare malignant transformation: a report of 5 cases and review of the literature. J Neurooncol 94:283–292

Raybaud C, Shroff M, Rutka JT, Chuang SH (2006) Imaging surgical epilepsy in children. Childs Nerv Syst 22:786–809

Rojiani AM, Emery JA, Anderson KJ, Massey JK (1996) Distribution of heterotopic neurons in normal hemispheric white matter: a morphometric analysis. J Neuropathol Exp Neurol 55:178–183

Rossi GF, Pompucci A, Colicchio G, Scerrati M (1999) Factors of surgical outcome in tumoural epilepsy. Acta Neurochir (Wien) 141:819–824

Scheithauer BW, Eberhart CG, Burger PC (2001) Extraventricular neurocytomas: pathological features and clinical outcome. Am J Surg Pathol 25:1252–1260

Schittenhelm J, Mittelbronn M, Wolff M, Truebenbach J, Will BE, Meyermann R, Beschorner R (2007) Multifocal dysembryoplastic neuroepithelial tumor with signs of atypia after regrowth. Neuropathology 27:383–389

Schmitz AK, Grote A, Raabe A, Urbach H, Friedman A, von Lehe M, Becker AJ, Niehusmann P (2013) Albumin storage in neoplastic astroglial elements of gangliogliomas. Seizure 22:144–150

Schramm J (2008) Temporal lobe epilepsy surgery and the quest for optimal extent of resection: a review. Epilepsia 49:1296–1307. doi:10.1111/j.1528-1167.2008.01604.x (Epub 11 April 2008)

Schramm J, Aliashkevich AF (2008) Temporal mediobasal tumors: a proposal for classification according to surgical anatomy. Acta Neurochir (Wien) 150:857–864. doi:10.1007/s00701-008-0013-7 (Epub 23 Aug 2008)

Selvanathan SK, Hammouche S, Salminen HJ, Jenkinson MD (2011) Outcome and prognostic features in anaplastic ganglioglioma: analysis of cases from the SEER database. J Neurooncol 105:539–545

Sharma MC, Jain D, Gupta A, Sarkar C, Suri V, Garg A, Gaikwad SB, Chandra PS (2009) Dysembryoplastic neuroepithelial tumor: a clinicopathological study of 32 cases. Neurosurg Rev 32:161–170. doi:10.1007/s10143-008-0181-1 (Epub 23 Jan 2009)

Siebzehnrubl FA, Blümcke I (2008) Neurogenesis in the human hippocampus and its relevance to temporal lobe epilepsies. Epilepsia 49:55–65. doi:10.1111/j.1528-1167.2008.01638.x

Smits A, Duffau H (2011) Seizures and the natural history of World Health Organization Grade II gliomas: a review. Neurosurgery 68:1326–1333. doi:10.1227/NEU.0b013e31820c3419

Soffer D, Lach B, Constantini S (1992) Melanotic cerebral ganglioglioma: evidence for melanogenesis in neoplastic astrocytes. Acta Neuropathol 83:315–323

Tassi L, Colombo N, Garbelli R, Francione S, Lo Russo G, Mai R, Cardinale F, Cossu M, Ferrario A, Galli C, Bramerio M, Citterio A, Spreafico R (2002) Focal cortical dysplasia: neuropathological subtypes, EEG, neuroimaging and surgical outcome. Brain 125:1719–1732

Thom M, Toma A, An S, Martinian L, Hadjivassiliou G, Ratilal B, Dean A, McEvoy A, Sisodiya SM, Brandner S (2011) One hundred and one dysembryoplastic neuroepithelial tumors: an adult epilepsy series with immunohistochemical, molecular genetic, and clinical correlations and a review of the literature. J Neuropathol Exp Neurol 70:859–878

Toczek MT, Morrell MJ, Silverberg GA, Lowe GM (1996) Cerebellar hemorrhage complicating temporal lobectomy. Report of four cases. J Neurosurg 85:718–722

Tortori-Donati P, Fondelli MP, Rossi A, Cama A, Brisigotti M, Pellicanò G (1999) Extraventricular neurocytoma with ganglionic differentiation associated with complex partial seizures. AJNR Am J Neuroradiol 20:724–727

van Breemen MS, Wilms EB, Vecht CJ (2007) Epilepsy in patients with brain tumours: epidemiology, mechanisms, and management. Lancet Neurol 6:421–430

Ventureyra E, Herder S, Mallya BK, Keene D (1986) Temporal lobe gangliogliomas in children. Childs Nerv Syst 2:63–66

Weaver JP, Phillips C, Horowitz SL, Benjamin S (1996) Middle fossa cyst presenting as a delayed complication of temporal lobectomy: case report. Neurosurgery 38:1047–1051

Wetjen NM, Radhakrishnan K, Cohen-Gadol AA, Cascino G (2004) Resective surgery of neoplastic lesions for epilepsy. In: Shorvon SD, Perucca E, Fish DR, Dodson WE (eds) The treatment of epilepsy, 2nd edn. Blackwell Publishing, Oxford, pp 728–741

Williamson PD, Thadani VM, Darcey TM, Spencer DD, Spencer SS, Mattson RH (1992) Occipital lobe epilepsy: clinical characteristics, seizure spread patterns, and results of surgery. Ann Neurol 31:3–13

Wolf HK, Müller MB, Spänle M, Zentner J, Schramm J, Wiestler OD (1994) Ganglioglioma: a detailed histopathological and immunohistochemical analysis of 61 cases. Acta Neuropathol 88:166–173

Yacubian EM, de Andrade MM, Jorge CL, Valério RM (1999) Cerebellar hemorrhage after supratentorial surgery for treatment of epilepsy: report of three cases. Neurosurgery 45:159–162

Zaghloul KA, Schramm J (2011) Surgical management of glioneuronal tumors with drug-resistant epilepsy. Acta Neurochir (Wien) 153:1551–1559. doi:10.1007/s00701-011-1050-1 (Epub 21 May 2011)

Zentner J, Wolf HK, Ostertun B, Hufnagel A, Campos MG, Solymosi L, Schramm J (1994) Gangliogliomas: clinical, radiological, and histopathological findings in 51 patients. J Neurol Neurosurg Psychiatry 57:1497–1502

Zhang D, Henning TD, Zou LG, Hu LB, Wen L, Feng XY, Dai SH, Wang WX, Sun QR, Zhang ZG (2008) Intracranial ganglioglioma: clinicopathological and MRI findings in 16 patients. Clin Radiol 63:80–91

Zhang JG, Hu WZ, Zhao RJ, Kong LF (2013) Dysembryoplastic neuroepithelial tumor: a clinical, neuroradiological, and pathological study of 15 Cases. J Child Neurol (10 June) [Epub ahead of print]

Metabolic Causes of Epilepsy

Laura Papetti, Francesco Nicita, Stella Maiolo, Vincenzo Leuzzi and Alberto Spalice

Abstract Inborn errors of metabolism comprise a large class of genetic diseases involving disorders of metabolism. Presentation is usually in the neonatal period or infancy but can occur at any time, even in adulthood. Seizures are frequent symptom in inborn errors of metabolism, with no specific seizure types or EEG signatures. The diagnosis of a genetic defect or an inborn error of metabolism often results in requests for a vast array of biochemical and molecular tests leading to an expensive workup. However a specific diagnosis of metabolic disorders in epileptic patients may provide the possibility of specific treatments that can improve seizures. In a few metabolic diseases, epilepsy responds to specific treatments based on diet or supplementation of cofactors (vitamin-responsive epilepsies), but for most of them specific treatment is unfortunately not available, and conventional antiepileptic drugs must be used, often with no satisfactory success. In this review we present an overview of metabolic epilepsies based on various criteria such as treatability, age of onset, seizure type, and pathogenetic background.

Introduction

A very large number of inherited errors of metabolism (IEM) may occur with neurologic symptoms such as seizures, developmental delay, mental deterioration, cranial nerve deficits and movement disorders (Wolf et al. 2005). Epilepsy may dominate the clinical picture, especially in newborns and infants, or may be part of a larger clinical spectrum with other extraneurologic findings (osseous, cutaneous, visceral, endocrine, sensorial, and metabolic). Indeed, the presence of extraneurologic signs raises a strong probability of finding systemic metabolic disturbances (Wolf et al. 2009a, b). Epilepsy associated with IEM has usually the features of a "catastrophic encephalopathy" because seizures begin usually at early age, they are often refractory to conventional antiepileptic drugs (AEDs) and epileptic activity is associated

A. Spalice (✉) · L. Papetti · F. Nicita · S. Maiolo · V. Leuzzi
Department of Pediatrics and Child Neurology and Psychiatry, Division of Child Neurology, Sapienza University of Rome, Viale Regina Elena 324, 00161 Rome, Italy
e-mail: childneurology.sapienzaroma@live.it

© Springer International Publishing Switzerland 2015
P. Striano (ed.), *Epilepsy Towards the Next Decade*,
Contemporary Clinical Neuroscience, DOI 10.1007/978-3-319-12283-0_5

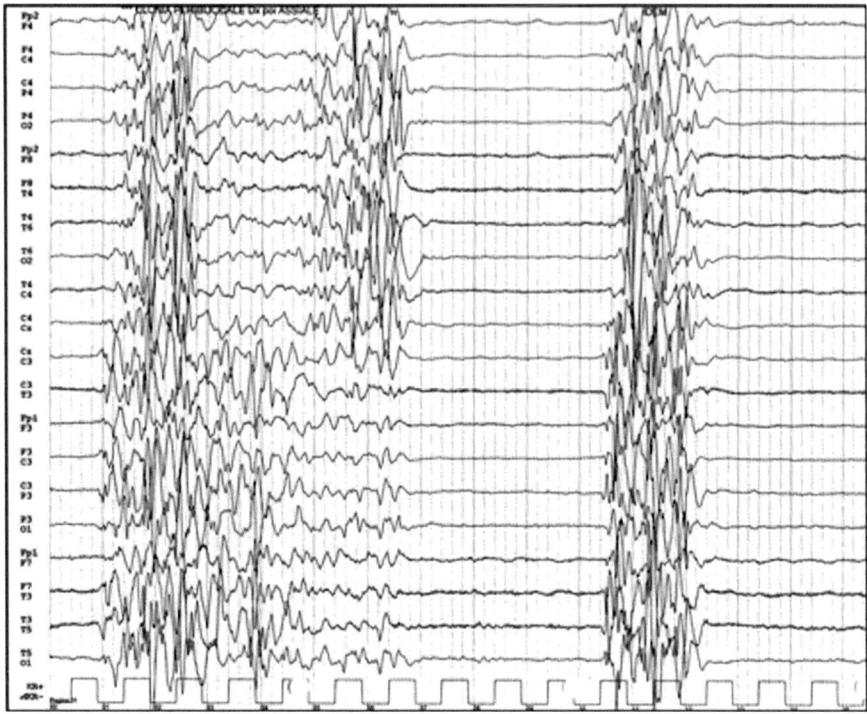

Fig. 1 EEG pattern of burst suppression

with severe cognitive, sensorial, and/or motor functions deterioration (Wolf et al. 2005, 2009a, b). Metabolic epileptic encephalopathies display an age dependent susceptibility and expression in the clinical phenotype. The seizure phenotype thus can be seen to evolve over time to fit descriptions of different epilepsy syndromes (Wolf et al. 2005). The EEG findings can be strikingly abnormal but they lack specificity and overlapping findings are frequent in different IEMs. EEG changes range from disorganized and slow background rhythms, focal and multifocal epileptiform patterns, generalized abnormalities as well as suppression-burst patterns (Wolf et al. 2005; Papetti et al. 2013; Fig. 1). The MRI findings may be normal or reveal associated structural abnormalities. MR spectroscopy is able to none invasively identify several metabolites peaks related to metabolic encephalopathies (Wolf et al. 2005). The diagnosis of metabolic disorders in epileptic patients may provide the possibility of specific treatments that can improve seizures. In a few metabolic diseases, epilepsy responds to specific treatments based on diet or supplementation of cofactors (vitamin responsive epilepsies), but for most of them specific treatment is unfortunately not available, and conventional antiepileptic drugs must be used, often with no satisfactory success (Wolf et al. 2009a, b). Neurometabolic epilepsies can be classified according to different criteria, i.e., type of biochemical defects and clinical presentation. More recently the age of onset of metabolic epilepsy has been considered for classification (Table 1; Papetti et al. 2013).

Table 1 Metabolic epilepsies according to age at onset

Neonatal onset	Infantile onset	Childhood onset
Pyridoxine dependent seizures; PNPO deficiency; Folinic acid response; Non-ketotic hyperglycinaemia; Organic acidemias; Urea cycle defects; Holocarboxylase synthase deficiency; GABA transaminase deficiency	GLUT1 deficiency; Non-ketotic hyperglycinaemia; Organic aciduria; Creatine deficiency; Biotinidase deficiency; Aminoacid disorders; Infantile neuronal lipofuscinosis; Late pyridoxine dependency; Ethylmalonic encephalopathy; Congenital disorders of glycosylation; Purine metabolism defects	Late infantile neuronal lipofuscinoses; Mitochondrial disorders; Storage disorders; Purine metabolism defects; Lafora disease; GLUT1 deficiency; Creatine transporter deficiency

In this chapter we will focus on diseases and conditions where epilepsy is the predominant clinical manifestation and especially where the disease course can be positively influenced by specific metabolic therapies (Table 2; Papetti et al. 2013; Pascual et al. 2008).

Neonatal Onset Seizures

Vitamin B Response Epileptic Seizures

Pyridoxal phosphate (PSP), the active form of vitamin B6, is the cofactor for over 100 enzyme-catalyzed reactions in the body, including many involved in the synthesis or catabolism of neurotransmitters (e.g., dopamine, serotonin, and inhibitory transmitter c-aminobutyric acid) (Gospe 2010). Inadequate levels of pyridoxal phosphate in the brain cause neurological dysfunction, particularly epilepsy (Plecko and Stöckler 2009). Three genetic epilepsies are recognized to be cause PSP deficiency: Pyridoxine-dependent seizures, Pyridoxal phosphate (PLP) dependent epilepsy and hypophosphatasia. They do so by different mechanisms: the first by inactivation, the second by blocking conversion of other forms of vitamin B6 to PLP, and the third by reducing transport into the brain and into cells (Baxter 2003).

Pyridoxine-dependent seizures (PDS) are due to an autosomal recessive inborn error of metabolism and they are characterized by neonatal seizures that are not controlled with anticonvulsants but that respond both clinically and electrographically to large daily supplements of pyridoxine (vitamin B6). The disorder may present within few hours of birth as an epileptic encephalopathy; sometimes intrauterine fetal seizures occur. Other cases may present with seizures at a later time during the first several weeks of life. In rare instances, children with this condition do not have seizures until before 2 years of age, and these are considered to be late-onset cases

Table 2 Diagnosis and treatment of metabolic epilepsies. a-AASA, a-amino adipic semialdehyde; iv, intravenous; PNPO, Pyridox(am)ine 5¢-phosphate oxidase; PLP pyridoxal 5¢-phosphate; HCS, holocarboxylase synthetase; EMA, ethylmalonic acid

Disease	Diagnosis	Specific therapy
Pyridoxine dependent seizures	α-AASA in urine and/or plasma and CSF	Single 100 mg dose of iv pyridoxine followed by oral maintenance dose of 5–15 mg/kg/day in two divided doses
PNPO deficiency	Reduced CSF PLP and monoamine metabolites; mutation analysis of the PNPO gene	Oral PLP 10 and 30 mg/kg/day
Folinic acid response seizures	CSF high-performance liquid chromatography	Folinic acid 3–5 mg/kg/day
HCS deficiency	Plasma ammonia; plasma and CSF lactate; plasma, urine and CSF organic acid; lymphocyte or fibroblast carboxylase activity; molecular analysis	Oral biotin 5–10 mg/day
Biotinidase deficiency	Biotin in plasma and urine; serum biotinidase enzyme activity; molecular analysis	Oral biotin 5–10 mg/day
Non-ketotic hyperglycinaemia	Plasma and CSF amino acids	Sodium benzoate and dextromethorphan
Serine biosynthesis defects	Plasma and CSF amino acids	Oral L-serine 200–600 mg/kg/dye
Organic acidemias	Urine organic acids, blood spot acylcamitine profile, and plasma and urine amino acids; enzymatic activity in fibroblasts; molecular analysis	Dietary restrictions; adjunctive compounds to (a) dispose of toxic metabolites or (b) increase activity of deficient enzymes
Urea cycle defects	Ammonemia, aminoacids, urinary orotic acid	Dietary restrictions; adjunctive compounds
Menkes disease	Serum copper and caeruloplasmin; ratio of urinary dopamine to noradrenaline; analysis of *ATP7A* gene	Subcutaneous injections of copper histidine
Peroxisomal disorders	Plasma VLCFA; biochemical studies in cultured fibroblasts; molecular analyses	Dietary restrictions; adjunctive compounds
GABA transaminase deficiency	Levels of GABA in the CSF or on brain MR spectroscopy; enzyme assay in cultured lymphocytes; analysis of *ABAT* gene	No specific therapy
GLUT1 deficiency	CSF glucose (CSF to plasma glucose ratio); analysis of *SLC2A1* gene	Ketogenic diet
Creatine deficiency	Guanidinoacetate, creatine, nd creatinine levels in plasma and urine; creatine levels on brain MR spectroscopy; analysis of *GAMT, GATM,* and *SLC6A8* genes	Oral creatine monohydrate 350–500 mg/kg/dye, dietary arginine restriction, ornithine supplementation

Table 2 (continued)

Disease	Diagnosis	Specific therapy
Neuronal ceroid lipofuscinosis	Enzyme activity in dried blood spots; molecular analysis of CLN genes	No specific therapy
Ethylmalonic encephalopathy	Plasma acylcarnitines; plasma and urine EMA; urine organic acids; skeletal muscle biopsy; analysis of ETHE1 gene	Carnitine and riboflavin
Congenital disorders of glycosylation	Isoelectric focusing of serum transferrin.	No specific therapy
Mitochondrial disorders	Plasma and CSF lactate, brain MR spectroscopy; skeletal muscle biopsy; molecular analysis	Adjunctive compounds
Purine metabolism defects	Urinary purine profile; molecular analysis	No specific therapy
Lysosomal storage disorders	Abnormal storage cells (lymphocytes, fibroblasts): molecular analysis (CLN3, NPC1 and NPC 2 genes)	Glycosphingolipid synthesis inhibitor in Niemann-Kck type C
Lafora disease	EPM1 gene	No specific therapy

(Gospe 2010). Affected newborn typically experience prolonged seizures which either recur serially or evolve into status epilepticus. Seizures generally include partial seizures, generalized tonic clonic seizures (GTCS), spasms and myoclonus. Additional features of pyridoxinedependent epilepsy include hypothermia, poor muscle tone, and neurodevelopment disabilities (Plecko and Stöckler 2009). These patients are not pyridoxine-deficient, but they are metabolically dependent on the vitamin, so that the institution of either parenteral or oral pyridoxine rapidly results in seizure control and improvement in the encephalopathy (Gospe 2010). The EEG is usually severely abnormal and the possible patterns include burst suppression, hypsarrhytmia and multiple spike-wave discharges (Fig. 2, Nabbout and Dulac 2008; Papetti et al. 2013). Imaging may be normal or may demonstrate cerebellar dysplasia, hemispheric hypoplasia or atrophy, neuronal dysplasia, periventricular hyperintensity or intracerebral hemorrhage (Baxter 2003).

PDS is caused by mutations in the *ALDH7A1 gene* that encodes the protein antiquitin (a-aminoadipic semialdehyde dehydrogenase), that functions within the cerebral lysine catabolism pathway (Scharer et al. 2010). The deficient activity of antiquitin results in the accumulation of a-aminoadipic semialdehyde (AASA) and piperideine-6-carboxylic acid (P6C). The P6C was shown to inactivate pyridoxal-phosphate (PLP), the active vitamer of pyridoxine, by a Knoevenagel condensation reaction, leading to severe secondary PLP deficiency. The PLP is a cofactor of various enzymes in the central nervous system, so that seizures in PDS are due to a decrease in GABA levels in the brain with an imbalance between the excitatory and inhibitory neurotransmitters (Mill et al. 2010; Scharer et al. 2010).

Diagnosis may be made by concurrently administering pyridoxine (100 mg) intravenously while monitoring the EEG, oxygen saturation, and vital signs. In individuals with PDS, clinical seizures generally cease over several minutes. If a

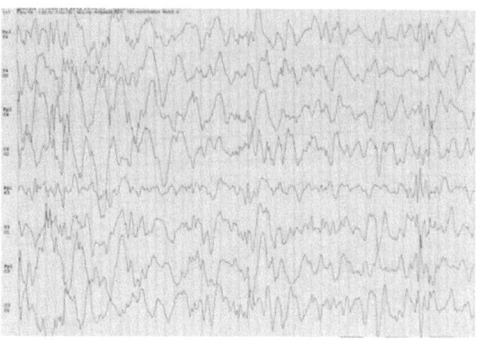

Fig. 2 EEG pattern of hypsarythmia

clinical response is not demonstrated, the dose should be repeated up to a maximum of 500 mg. An alternate diagnostic approach is suggested for patients who are experiencing frequent short anticonvulsant-resistant seizures. In those cases, oral pyridoxine (up to 30 mg/kg/day) should be prescribed, and patients with PDS should have a resolution of clinical seizures within 3–7 days. Biochemical tests include measurement of the specific biomarker a-AASA and pipecolid acid (PA) in the urine, plasma and CSF. Molecular genetic testing of *ALDH7A1* is also recommended as confirmatory testing (Gospe 2010). When the diagnosis of PDS is established, the lifelong therapy with pyridoxine should be instituted. The daily administration of 50–200 mg (given once daily or in two divided doses) is generally effective in preventing seizures in most patients (Gospe 2010).

Pyridoxal phosphate (PLP) dependent epilepsy is characterized by neonatal seizures refractory both to conventional AEDs and pyridoxine administration (Kuo and Wang 2002). Instead, individuals with this type of epilepsy are responsive to large daily doses of pyridoxal 5′-phosphate (30 mg/kg/day in three or four divided doses enterally) (Hoffmann et al. 2007). PLP dependent epilepsy is inherited with an autosomal recessive pattern. The gene involved is the PNPO gene that encodes an enzyme called pyridoxine 50-phosphate oxidase involved in the conversion of vitamin B6 derived from food (in the form of pyridoxine and pyridoxamine) to the active form of vitamin B6 that is PLP (Mills et al. 2005).

Affected babies are usually born prematurely and may have immediate signs of encephalopathy, lactic acidosis and hypoglycemia. Clinical seizures may consist of myoclonus, clonic movements and ocular, facial and other automatisms (Baxter 2010). Untreated, the disorder results either in death or in profound neurodevelopment impairment. In treated patients, particularly those in whom the disorder was recognized early, near normal development may be possible (Bagci et al. 2008).

CSF and urine analyses in affected children show evidence of secondary deficiencies of several PLP dependent enzymes including aromatic L-amino acid decarboxylase (decreased CSF concentrations of homovanillic acid and 5-hydroxyindoleacetic acid and increased L-DOPA and 3-methoxytyrosine as well as increased urinary concentrations of vanillactic acid) (Plecko and Stöckler 2009).

Chronic therapy for confirmed PLP dependent epilepsy consists of administration of PLP 30–50 mg/kg/day divided in four to six doses (Plecko 2005).

Hypophosphatasia is an inherited disorder that affects the development of bones and teeth. However it can also led to PDS in neonates as pyridoxal phosphate is not dephoshorilated and therefore cannot cross membranes (Plecko 2005). Biochemically, this disorder consists of deficient activity of the tissue non-specific isoenzyme of alkaline phosphatase. The enzymatic deficiency results from mutations in the liver/bone/kidney alkaline phosphatase gene *ALPL*. Laboratory diagnosis is confirmed by reduced levels of serum alkaline phosphatase, and raised levels of urinary phosphoethanolamine (PEA) (Balasubramaniam et al. 2010).

Folinic acid-responsive seizures are characterized by cessation of seizures after administration of folinic acid (3–5 mg/kg/day enterally, for 3 to 5 days) (Plecko 2005). Only a few affected infants were published. Patients present with seizures, either myoclonic or clonic, apnea and irritability within 5 days after birth (Gallagher et al. 2009). A characteristic pattern of peaks, reflecting two unidentified compounds, was recognized in the cerebrospinal fluid (CSF) when analyzed by high-performance liquid chromatography (HPCC) with electrochemical detection to quantify monoamine metabolites. Recently, it has been demonstrated that folinic acid responsive seizures are also caused by mutations in *ALDH7A1* and therefore should be treated with adequate doses of pyridoxine. Whether additional treatment of these children with folinic acid is of added benefit, remains to be shown (Gallagher et al. 2009).

Disorders of Amino Acid Metabolism

Some disorders of amino acid metabolism such as nonketotic (NKH), methylene tetrahydrofolate reductase (MTHFR) deficiency, GABA transaminase deficiency, serine deficiency, and congenital glutamine deficiency, can also give rise to this epileptic syndrome, each with its specific biochemical traits (Wolf et al. 2009a, b).

Nonketotic hyperglycinemia (NKH) is a metabolic disorder with autosomal recessive inheritance, causing severe, frequently lethal, neurological symptoms in the neonatal period. NKH derives from a defect of a larger enzyme complex, known as glycine cleavage enzyme (GCS) that is responsible for the glycine degradation. When glycine cleavage enzyme is defective, excess glycine can build up to toxic levels in the body's organs and tissues (Applegarth and Toone 2004). The three genes known to be associated with glycine encephalopathy are: *GLDC* (encoding the P-protein component of the GCS complex and accounting for 70–75% of disease), *AMT* (encoding the T-protein component of the GCS complex and accounting for ~20% of disease), and *GCSH* (encoding the H-protein component of the GCS complex and accounting for <1% of disease) (Hamosh et al. 2009).

The majority of glycine encephalopathy presents in the neonatal period (85% as the neonatal severe form and 15% as the neonatal mild form). Of those presenting in infancy, 50% have the infantile mild form and 50% have the infantile severe form (Rossi et al. 2009). Patients with classical neonatal NKH present in the first days of life with seizures or with encephalopathy, abnormal jerking movements, lethargy and severe hypotonia. Affected newborns will have repeated episodes of severe and prolonged apnea that require ventilatory support. Hiccuping is frequent and brain ultrasound scans may show defects of the corpus callosum (Hamosh et al. 2009). Epilepsy associated with NKH may reflects the early myoclonic encephalopathy (EME) with erratic or fragmentary myoclonus, simple focal seizures, focal tonic seizures, and tonic spasms, generally after 1 month of age and with an EEG patter of burst-suppression (SB) and progression towards hypsarrhytmia (Rossi et al. 2009). Untreated, the neonatal form of non-ketotic hyperglycinaemia is associated with death in the first months of life. Therapy with sodium benzoate and dextromethorphan may be helpful in some milder forms of the disease, alongside AEDs and general supportive care. The epilepsy remains drug resistant, infantile spasms may emerge, and the EEG evolves to hypsarrhythmia or multifocal discharges on a back- ground without normal activity. Atypical neonatal form of NKH can be similar to classical NKH, with hypotonia and apnea episodes that may require assisted ventilation, though seizures are less severe. Subsequently psychomotor development is significantly better than in the majority of patients with the classical NKH (Dinopoulos et al. 2005).

Atypical variants of NKH include also the infantile and late onset forms. Children with this condition develop normally until they are about 6 months old, when they experience delayed development and may begin having seizures. As they get older, many develop intellectual disability, abnormal movements, and behavioral problems. The late onset form is less common and more heterogeneous. The clinical presentation is after the second birthday and even during adulthood, mainly with mild cognitive decline and behavioral problems (Hamosh et al. 2009).

Transient neonatal hyperglycinemia (TNH) is characterized by elevated plasma and CSF glycine levels at births that are normalized within 2–8 weeks. TNH is clinically and biochemically indistinguishable from typical nonketotic hyperglycinemia at onset (Dinopoulos et al. 2005). The biochemical hallmark of NKH is increased glycine concentration in the plasma and, to an even greater extent, in the cerebrospinal fluid (CSF), with an abnormally high ratio between CSF and plasma levels. Confirmatory tests include enzymatic analysis in liver tissue and/or mutation analysis (Pascual 2003). Some babies have structural brain abnormalities evident on MRI, apparently as a result of the toxic effect of glycine on the developing brain. Treatment consists of reducing the intake of glycine and serine as well as improving its elimination by administering benzoate and by exchange transfusion (Pascual 2003).

Defects in the synthesis of *L-serine* lead to a syndrome of congenital microcephaly, neurodevelopmental disability, and epilepsy which may have neonatal onset (Pearl 2009). Two serine-deficiency syndromes have been described, namely 3-phosphoglycerate dehydrogenase (3-PGDH) deficiency and 3-phosphoserine phosphatase (3-PSP) deficiency (de Koning and Klomp 2004). The 3-PGDH deficiency is an autosomal recessive disorder characterized by neurological symptoms which dominate the clinical phenotype (i.e., microcephaly, seizures, and neurodevelopmental delay). Seizures either started as generalized tonic clonic seizures or as flexor spasms with West syndrome. In older patients, diagnosed at ages 5–9, tonic, atonic and myoclonic seizures as well as absences were described (Tabatabaie et al. 2010). The EEG of patients with 3-PGDH showed hypsarrhytmia or severe multifocal epileptic abnormalities with poor background activity (de Koning and Klomp 2004). The 3-PGDH gene is located on chromosome 1q12. Two different homozygous missense mutations were described. Both mutations lead to a significant reduction of enzyme activity after expression of the mutant enzymes in vitro (de Koning and Klomp 2004). Low concentrations of serine and, to a variable degree, of glycine in plasma and CSF, are the biochemical hallmark of the disease. Oral supplementation of l-serine is proved to be very effective in the treatment of seizures (Tabatabaie et al. 2010).

Phenylketonuria (PKU) is a disorder of phenylalanine metabolism that frequently results in epilepsy if a dietary restriction was not implemented at birth (Blau et al. 2010). Classical PKU is an autosomal recessive disorder, caused by mutations in both alleles of the gene for phenylalanine hydroxylase (PAH), found on chromosome 12. In the body, phenylalanine hydroxylase converts the amino acid phenylalanine to tyrosine. As consequence of mutations in both copies of the gene for PAH, the enzyme is inactive or is less efficient, and the concentration of phenylalanine in the body can build up to toxic levels. In some cases, mutations in PAH will result in a phenotypically mild form of PKU called hyperphenylalanemia. Both diseases are the result of a variety of mutations in the PAH locus; in those cases where a patient is heterozygous for two mutations of PAH (i.e., each copy of the gene has a different mutation), the milder mutation will predominate (Blau et al. 2010). A small minority of PKU cases results from defects in the metabolism of tetrahydrobiopterin, the obligate cofactor of PAH. The symptoms of untreated PKU, which manifest primarily in the brain, are diverse, and can range from mild cognitive impairment to severe mental retardation, with motor impairment and pyramidal signs (Martynyuk et al. 2007). Refractory epilepsy is common, with infantile spasms or GTCs; PKU has been also reported in patients with West syndrome. At present, for almost all patients with phenylketonuria, diagnosis and the start of treatment result from neonatal screening rather than clinical symptoms. Treatment consists of dietary restriction and L-dopa, 5-hydroxytryptophan and folinic acid supplements (Martynyuk et al. 2007).

Urea Cycle Disorders

Urea cycle disorders (UCD) represent a group of inborn errors of metabolism result from single gene defects involved in the detoxification pathway of ammonia to urea (Zhongshu et al. 2001). The components of the pathway are: carbamyl phosphate synthase I (CPSI); ornithine transcarbamylase (OTC); argininosuccinic acid synthetase (ASS); argininosuccinic acid lyase (ASL); arginase (ARG) and the cofactor, N-acetyl glutamate synthetase (NAGS). Deficiencies of CPSI, ASS, ASL, NAGS, and ARG are inherited in an autosomal recessive manner. OTC deficiency is inherited in an X-linked manner (Braissant 2010; Summar 2005). Infants with a UCD often appear normal initially but rapidly develop cerebral edema and the related signs of lethargy, anorexia, hyperventilation or hypoventilation, hypothermia, seizures, neurologic posturing, and coma. In milder (or partial) UCD, ammonia accumulation may be triggered by illness or stress at almost any time of life, resulting in multiple mild elevations of plasma ammonia concentration; the hyperammonemia is less severe and the symptoms more subtle. In individuals with partial enzyme deficiencies, the first recognized clinical episode might be delayed for months or years (Braissant 2010). Seizures are frequent during the early stages of hyperammonaemia, especially in newborns. The EEG may show variable pattern of epileptic discharge, i.e., multifocal independent spike- and sharp-wave discharges, repetitive paroxysmal activity, unusually low-voltage fast activity, and findings consistent with complex partial seizures (Summar 2005). In late-onset UCD cases, EEG may show continuous semirhythmic activity with sharp components, leading to diagnosis of complex partial status epilepticus (Gropman et al. 2007).

The therapy of UCD include dialysis to reduce plasma ammonia concentration, intravenous administration of arginine chloride and nitrogen scavenger drugs to allow alternative pathway excretion of excess nitrogen, restriction of protein for 24–48 h to reduce the amount of nitrogen in the diet, providing calories as carbohydrates and fat reduce catabolism, and physiologic stabilization with intravenous fluids and cardiac pressors to reduce the risk of neurologic damage (Clague 2002).

Organic Acidemias

Organic acidemias (OA) consist of a group of disorders characterized by the excretion of non-amino organic acids in urine. Most organic acidemias result from dysfunction of a specific step in amino acid catabolism, usually the result of deficient enzyme activity. The majority of the classic organic acid disorders are caused by abnormal amino acid catabolism of branched-chain amino acids or lysine. The main types of OA include maple syrup urine disease (MSUD), propionic acidemia, methylmalonic acidemia, isovaleric acidemia and glutaric acidemia type I (GA-1) (Clague and Thomas 2002).

A neonate affected with an OA is usually well at birth and for the first few days of life. The usual clinical presentation is that of toxic encephalopathy and includes vomiting, poor feeding, neurologic symptoms such as seizures and abnormal tone, and lethargy progressing to coma. Outcome is enhanced by diagnosis and treatment in the first 10 days of life. In the older child or adolescent, variant forms of the OAs can present as loss of intellectual function, ataxia or other focal neurologic signs, Reye syndrome, recurrent ketoacidosis, or psychiatric symptoms (Van Gosen 2008; Seashore 2009).

MSUD (OMIM 248600) is caused by a deficiency of the branched-chain alpha-keto acid dehydrogenase complex (BCKDC). The mammalian BCKD complex consists of three catalytic components: E1, E2, and E3, and two regulatory enzymes. Mutations in these regions lead to the accumulation of three branched-chain amino acids (BCAA) (leucine, isoleucine, and valine) and their toxic by-products in the blood and urine. The major clinical features of maple syrup urine disease are mental and physical retardation, feeding problems, and a maple syrup odor to the urine (Rahman et al. 2013)

There are presently five known clinical phenotypes for MSUD: classic, intermediate, intermittent, thiamin responsive, and dihydrolipoamide dehydrogenase (E3)-deficient, based on severity of the disease, response to thiamin therapy, and the gene locus affected (Wang et al. 2003).

Classic MSUD is the most frequent form and the affected newborns appear normal at birth, with symptoms developing between 4 and 7 days of age. The infants show lethargy, weight loss, metabolic derangement, and progressive neurologic signs of altering hypotonia and hypertonia, reflecting a severe encephalopathy. Seizures and coma usually occur, followed by death if untreated (Seashore 2009). The seizures can be of different types, with occasionally presenting with status epilepticus and early treatment may improve the prognosis (Wang et al. 2003). The EEG pattern is variable and it includes spikes, polyspikes, spike-wave complexes, triphasic waves, severe slowing and bursts of periodic suppression ad are not related to blood BCAA levels (Korein et al. 1994).

Propionic acidemia (PA) (OMIM 606054) is characterized by the accumulation of propionic acid due to a deficiency in Propionyl CoA Carboxylase, a biotin dependent enzyme involved in amino acid catabolism. Patients may present with vomiting, dehydration, lethargy, and encephalopathy. Among the neurological complication often observed, developmental delay, seizures, cerebral atrophy and EEG abnormalities have been the most prominent. Seizures generally have onset in the neonatal period and they may include focal seizures, spasms, and generalized tonic and myoclonic seizures (Haberlandt et al. 2009). In 40% of affected children, generalized convulsions and myoclonic seizures develop in later infancy, and older children may have atypical absence seizures (Aicardi 2007). Photosensitivity and fever induced seizures have also been described at the beginning in patients with PA. Intractable seizures may develop. The EEG pattern is variable and it may show hypsarrhythmia, burst suppression and diffuses delta wave activity with generalized or focal temporal spikes during the encephalopatic phase. MRI usually reveals alterated signal in the caudate, putamen, and globus pallidus. MRS shows decreased NNA and myo-inositol and increased glutamate/glutamine in the basal ganglia (Chemelli et al. 2000).

Glutaric aciduria type 1 (GA-1, OMIM 608801) is an autosomal recessive disease due to an inborn error of the metabolism of the amino acids lysine, hydroxylysine, and tryptophan due to mutations in the glutarylcoenzyme. A dehydrogenase gene (GCDH), on chromosome 19p13.2 (McClelland et al. 2009). Clinical expression usually involves an acute encephalopathic episode in infancy, followed by the development of severe dystonia-dyskinesia. Seizures may occur at presentation in the context of acute encephalopathy, but ongoing seizures are not common in GA-1 unless accompanied by severe brain damage (McClelland et al. 2009). Many children may have sudden dystonic spasms that could be mistaken for seizures. Although chronic epilepsy is rare, glutaric aciduria type I patients may present with epileptic seizures that are difficult to control with first- or second-line anticonvulsants as the sole clinical feature (Cerisola et al. 2009). This disorder can be identified by increased glutaryl (C5DC) carnitine on newborn screening. Urine organic acid analysis indicates the presence of excess 3-OH-glutaric acid, and urine acylcarnitine profile shows glutaryl carnitine as the major peak (McClelland et al. 2009).

The brain MRI is helpful for the diagnosis. Atrophy or hypoplasia of the frontotemporal regions of the cerebral hemispheres, enlarged pretemporal middle cranial fossa subarachnoid spaces, and cyst-like dilatation of the Sylvian fissures are often early findings in glutaric acidemia type I with "batwing" or "box-like" fissures (Neumaier-Probst et al. 2004).

The aim of therapy in OA is to restore biochemical and physiologic homeostasis. Neonates require emergency diagnosis and treatment depending on the specific biochemical lesion, the position of the metabolic block, and the effects of the toxic compounds. Treatment strategies include: (1) dietary restriction of the precursor amino acids and (2) use of adjunctive compounds to (a) dispose of toxic metabolites or (b) increase activity of deficient enzymes. Decompensation caused by catabolic stress (e.g., from vomiting, diarrhea, febrile illness, and decreased oral intake) requires prompt and aggressive intervention (Seashore 2009).

Peroxisomal Disorders

Zellweger syndrome (ZS, OMIM 214100) may be responsible for epilepsy in the neonatal period. Individuals with ZS develop signs and symptoms of the condition during the newborn period. These infants experience hypotonia, feeding problems, hearing loss, vision loss, and seizures. Children with ZS also develop life-threatening problems in other organs and tissues, such as the liver, heart, and kidneys. They may have skeletal abnormalities, including a large space between the bones of the skull (fontanels) and characteristic bone spots known as chondrodysplasia punctata that can be seen with an X-ray (Rahman et al. 2013). Affected individuals have distinctive facial features, including a flattened face, broad nasal bridge, and high forehead. Children with ZS typically do not survive beyond the first year of life (Steinberg et al. 2003). Areas of polymicrogyria are often frontal or opercular, resulting in a focal EEG and seizure semiology, and there are often focal motor seizures (Takahashi et al. 1997). Mutations in the *PEX1* gene are the most common cause of the Zellweger spectrum and are found in nearly 70% of affected individuals. The other genes associated with the Zellweger spectrum each account for a smaller percentage of cases of this condition (Steinberg et al. 2003).

Infantile Onset Seizures

Biotin Response Epileptic Seizures

Characteristic organic aciduria, cutaneous and neurologic symptoms with frequent seizures are present in holocarboxylase synthetase (HCS) and biotinidase deficiencies (BTD). Both disorders in biotin metabolism lead to multiple carboxylase deficiency and respond dramatically to biotin therapy (Joshi et al. 2010).

Biotinidase deficiency (BTD) (OMIM 253260) is an autosomal recessively inherited disorder in which the vitamin, biotin, cannot be appropriately recycled (Wolf 2011a, b). BTD is the only gene known to be associated with biotinidase deficiency. The BTD gene provides instructions for making an enzyme called biotinidase that removes biotin that is bound to proteins in food, leaving the vitamin in its free state. Mutations in the BTD gene reduce or eliminate the activity of biotinidase. Deficiency of BTD leads to decrease biotin available resulting in reduced conversion of apocarboxylases to holocarboxylases or multiple carboxylase deficiency (Wolf 2011a, b). This subsequently causes the accumulation of abnormally high concentrations of toxic metabolites (Pindolia et al. 2010). Profound biotinidase deficiency results when the activity of biotinidase is reduced to less than 10% of normal. Partial biotinidase deficiency occurs when biotinidase activity is reduced to between 10 and 30% of normal (Thoene and Wolf 1983). If untreated, young children with profound BTD deficiency usually exhibit neurologic abnormalities including seizures, hypotonia, ataxia, developmental delay, vision problems, hearing loss, and cutaneous abnormalities (Bhardwaj et al. 2010). Seizures are the presenting symptom in 38% of patients with biotinidase deficiency and are found in up to 55% of patients at some time before treatment. Seizures often start after the first 3 or 4 months of life, and often as infantile spasms or GCTS (tonic-clonic, clonic and myoclonic). The refractory seizures respond promptly to small doses of biotin (5–20 mg/day) (Zempleni et al. 2008).

Holocarboxylase synthetase (HCS) deficiency is an inherited disorder in which the body is unable to use the vitamin biotin effectively. Mutations in the HLCS gene cause holocarboxylase synthetase (Suzuki et al. 2005). The signs and symptoms of holocarboxylase synthetase deficiency typically appear within the first few months of life, but the age of onset varies. Symptoms are very similar to biotinidase deficiency and treatment (large doses of biotin) is also the same. Seizures are less frequent, occurring in 25–50% of all children (Pascual et al. 2008).

Disorders of Creatine Metabolism

Creatine deficiency syndromes represent a group of inborn errors of creatine metabolism that is responsible for mental retardation, language delay and early-onset epilepsy (Nasrallah et al. 2010). Three inherited defects in the biosynthesis and transport of creatine have been described. The biosynthetic defects include deficiencies of the enzymes L-arginine–glycine amidinotransferase (AGAT) and guanidinoacetate methyltransferase (GAMT) (Nasrallah et al. 2010). The third is the deficiency of the creatine transporter 1 (CT1). Epilepsy is one of the main symptoms in two of these conditions, GAMT and CT1 deficiency, whereas the occurrence of febrile convulsions in infancy is a relatively common presenting symptom in all (Leuzzi 2002).

Clinical presentation of GAMT deficiency is usually characterized by normal developmental milestones in the first months of life, which can be abruptly discontinued by arrest/regression of psychomotor development with or without seizures.

arrest/regression of psychomotor development with or without seizures. Epilepsy is the second most frequent symptom in GAMT deficiency after intellectual disability. Febrile seizures have often been reported in the early phase of the disease occurring during the first 24 months of life (mainly 3–6 months). The pattern of seizures is not consistent, and more than one type of seizures can occur in the same patient at different ages. Life-threatening tonic seizures with apnea or myoclonic seizures can be observed in the first months of life, whereas myoclonic astatic seizures, generalized tonic–clonic seizures, partial seizures with secondary generalization, drop attacks, absences, and staring episodes appear in early infancy or in adolescence. Febrile seizures, generalized tonic–clonic seizures, and myoclonic astatic seizures are the most commonly reported seizure types. No typical electroencephalography (EEG) pattern can be defined in GAMT deficiency. An early derangement of background organization and interictal multifocal spikes and slow wave discharges are frequently recorded. Focal EEG abnormalities, with a prominent involvement of frontal regions, have also been reported. Severe epilepsy has been reported in almost all cases. Movement disorders, such as athetosis, chorea, choreoathetosis, ballismus, and dystonia may be present (Leuzzi 2002).

The most typical neuroimaging alteration in GAMT deficiency is represented by bilateral pallidal lesions (hypointensity in T1-weighted and hyperintensity in T2-MRI images). In a few cases the lesion extended in the brainstem and selectively involved the white matter of the floor of fourth ventricle or the posterior pontine region (Leuzzi et al. 2013).

Biochemical findings associated with GAMT deficiency include the following: (1) reduced concentration of creatine in plasma, urine, cerebrospinal fluid (CSF), muscle, and brain; and (2) marked increase of guanidinoacetic acid (GAA) in all the biologic fluids, mainly in the CSF. High values of GAA can be detected also in dry blood spot since the first days of life. A mild increase of GAA over the normal range has been detected in blood and/or urine of some carriers of GAMT gene mutations. GAMT enzyme activity may be tested in liver tissue, skin fibroblasts, and lymphoblasts. The aim of treatment is to correct both the depletion of creatine/ creatine-phosphate pools and the accumulation of GAA. In GAMT deficiency, a lifelong oral supplementation with high doses of creatine monohydrate (350 mg/kg/ day–2 g/kg/day) has been shown, by plasma creatine assessment (muscle creatine pool) and brain H-MRS (brain creatine pool), to replenish body creatine pools. A further abating effect on AGAT activity can be obtained through a dietary restriction of arginine (15 mg/kg/day) coupled with ornithine supplementation (ornithine aspartate 350–800 mg/kg/day). Medicaments such as sodium benzoate and phenylbutyrate, which remove arginine and glycine, respectively, have also been proposed according to a similar substrate inhibition approach. Among the different clinical manifestations of GAMT deficiency, epilepsy is by far the most responsive to treatment (Nasrallah et al. 2010; Verhoeven et al. 2005).

CT1 deficiency is one of the main causes of X-linked mental retardation in males, and it is caused by *SLC6A8* gene mutations. Mental retardation and specific language derangement (oral-verbal dyspraxia of speech) are, in fact, the core symptoms of the disease. CT1 deficiency. It is rarely severe and it is usually responsive to conventional antiepileptic drugs. Its onset ranges between 16 months and 12 years. Febrile convulsions represent the first seizure-type in a number of subjects, and in a single case they led to subcontinuous generalized tonic–clonic seizures. Seizure pattern and EEG alterations can be extremely variable. Seizure-types include myoclonic seizures, generalized tonic–clonic seizures, convulsive status epilepticus, and partial seizures with secondary generalization. EEG recordings include normal tracing, diffuse slowing, aspecific sharp abnormalities, and focal/generalized paroxysmal or slow abnormalities, with or without sleep activation (Leuzzi et al. 2013). However, paroxysmal abnormalities are generally less severe as the child grows older. *SLC6A8* genotype is not associated with epilepsy, as exemplified by personal observations and cases from the literature. Neuroimaging and clinical features suggest in some patients a possible perinatal ischemic insult. This aspect may be confounding from the diagnostic point of view because clinical history rarely justifies this suspect. However, these lesions are congruent with the concept of creatine as a protective factor against potentially ischemic damage. Their possible role in the determinism of epilepsy needs to be elucidated. There are a few clinical reports on females carrying *SLC8A6* gene mutations. When symptomatic, they express a milder phenotype, including mild intellectual disability, behavioral disorders, problems of language development, learning difficulties, impairment of visual-constructional and fine motor skills, mild cerebellar symptoms, and constipation. Late occasional epileptic seizures have been described but not systematically studied (Leuzzi 2002).

The main biochemical alteration of patients with CT1 defect is the lack of brain creatine on H-MRS. Creatine is one of the major peaks in proton MR spectroscopy and is almost absent in all disorders of creatine synthesis and transport (Leuzzi 2002).

The urinary ratio creatine/creatinine (Cr/Crn) was proposed and confirmed as diagnostic marker of the disease. Diagnostic urinary Cr/Crn ratio ranged from 1.4 to 5.5 (reference values 0.006–1.2 in children under 4 years, 0.017–0.72 in patients between 4 and 12 years, 0.011–0.24 after 12 years of age). However, urinary Cr/Crn may be influenced by various nutritional and individual factors (Verhoeven et al. 2005).

Fibroblast and lymphoblast express SLC6A8 gene, and creatine uptake can be tested in these cells in patient with suspect CT1 defect. In contrast, muscle creatine is normal on both biochemical and H-MRS examination. No key clinical and/or neuropsychological cues have been identified to suggest the diagnosis of CT1 deficiency in girls with epilepsy and intellectual disability or learning difficulties. For these reasons gene sequencing seems to be the best diagnostic tool for females with a clinical suspect of CT1 (Verhoeven et al. 2005).

No effective treatment is available for males with CT1 defect. The supplementation of creatine, also at high dosages, does not improve H-MRS detectable brain creatine pool and/or clinical status. In contrast, creatine, as well as creatine precursor, supplementation is potentially effective in symptomatic females where the defect of CT1 is partial (Leuzzi et al. 2013).

Diagnosis is based on concentration of creatine and its precursors, measurement of enzyme activity for AGAT and GAMT, creatine uptake test for the diagnosis of CT1 defect, and mutation analysis.

The AGAT and GAMT deficiencies are inherited as an autosomal recessive trait; the *SLC6A8* deficiency is X-linked inherited. In humans, the AGAT protein is encoded by the gene *GATM* (15q21.1). The human *GAMT* gene is located on chromosome 19p13.3, while the gene CT1 (alias CRTR, SLC6A8) has been mapped to chromosome band Xq28 (Nasrallah et al. 2010).

Disorders of GABA Metabolism

Amino butyric acid (GABA) metabolism is associated with several disorders, including GABA-transaminase deficiency, and succinic semialdehyde dehydrogenase deficiency (SSADH) (Wolf et al. 2005).

GABA-transaminase deficiency is a rare disease with only few reported cases and it is characterized by abnormal development, seizures, and high levels of GABA in serum and cerebrospinal fluid (Jaeken 2002).

SSADH deficiency is an uncommon autosomal recessively inherited neurotransmitter disease involving GABA degradation. *ALDH5A1* is the only gene currently known to be associated with SSADH deficiency. SSADH is an enzyme that catalyzes the oxidation of succinate semialdehyde to succinate, the second and final step of the degradation of the inhibitory neurotransmitter GABA. Clinical manifestations in patients with SSADH deficiency are varied, and may range from mild mental retardation, speech delay, or behavioral problems, to severe psychomotor retardation with intractable seizures (Pearl and Gibson 2004).

Approximately half of patients with SSADH deficiency have epilepsy, usually with GTCS and also atypical absence and myoclonic seizures. The EEG may reveal background slowing and disorganization as well as diffused and multifocal epileptiform abnormalities. MRI shows increased T2-weighted signal involving the globus pallidi bilaterally and symmetrically, in addition to the cerebellar dentate nuclei and subthalamic nuclei (Pearl et al. 2007).

Vigabatrin, an irreversible inhibitor of GABAtransaminase, inhibits the formation of succinic semialdehyde and thus is one of the most widely prescribed AEDs in this disorder (Matern et al. 1996). However, vigabatrin has shown inconsistent results and MRI signal changes have been observed in patients treated with high doses (Pearl et al. 2009).

Glucose Transporter Deficiency

Glucose Transporter 1 (GLUT1) is a membrane transporter that facilitates glucose transport across the blood–brain barrier. GLUT1 deficiency syndrome (OMIM #606777) is disorder that primarily affects the brain. Glucose transporter-1 (GLUT1) is encoded by *SLC2A1* gene and usually mutations occur de novo, although the disease can also be inherited as autosomal dominant trait (Brockmann 2009). GLUT1 is highly expressed in the endothelial cells of erythrocytes and the blood–brain barrier and is exclusively responsible for glucose transport into the brain (Vannucci et al. 1997). Its deficiency leads to low glucose concentration in the cerebrospinal fluid (hypoglycorrhachia) (not associated with hypoglycaemia), in combination with a low to normal lactate in the cerebrospinal fluid. The classic patient with GLUT1 deficiency syndrome generally has drug resistant seizures beginning in the first year of life (Seidner et al. 1998).

Babies with GLUT1 deficiency syndrome have a normal head size at birth, but growth of the brain and skull is often slow, in severe cases resulting in microcephaly (Rotstein et al. 2010). Patients present with early-onset epilepsy, developmental delay, and a complex movement disorder as hypotonia, spasticity, ataxia and dystonia (Klepper and Leiendecker 2007; Schneider et al. 2009). The phenotype is highly variable and several atypical variants have been described (Klepper 2008). Seizures begin between age 1 and 4 months in 90% of cases. Apneic episodes and abnormal episodic eye movements simulating opsoclonus may precede the onset of seizures by several months. Five seizure types occur: generalized tonic or clonic, myoclonic, atypical absence, atonic, and unclassified (Wang et al. 2009). The interictal EEG may be normal. The ictal EEG may show focal slowing or discharges, including 2.5–4 Hz spike and wave. A striking difference between pre- and postprandial EEG may be seen, with a decrease in epileptic discharges following carbohydrate intake. GLUT1 deficiency is now known to be a cause of drug-resistant childhood absence epilepsy and of adult-onset absence epilepsy with a normal CSF glucose. Patients with a non-classical phenotype have been described, characterized by developmental delay and movement disorders without epilepsy or familial and sporadic paroxysmal exercise induced dyskinesia with or without epilepsy (Overweg-Plandsoen et al. 2003; Friedman et al. 2006; Klepper and Leiendecker 2007; Suls et al. 2008).

When GLUT1 deficiency syndrome is suspected, a lumbar puncture in the fasting state should be performed. Diagnosis is made by documenting CSF glucose levels below 40 mg/dl (2.5 mmol/l) and low CSF/blood glucose ratio (<0.50). CSF lactate is normal or low. The degree of hypoglycorrhachia and absolute 'cut-off' for a diagnosis of GLUT1 deficiency remain a source of debate, and mild clinical phenotypes have been reported with normoglycorrhachia and a normal CSF to blood glucose ratio; thus molecular genetic analysis of the SLC2A1 gene is considered the standard criterion for diagnosis. Approximately 80% of patients harbour pathogenic mutations (Wang et al. 2009).

Epilepsy in GLUT1 deficiency is drug resistant and may be aggravated by fasting and by AEDs that inhibit GLUT1 (phenobarbitone, valproate, diazepam). The ketogenic diet is highly effective in controlling the seizures and is generally well tolerated. However, neurobehavioral and motor deficits persist in most cases. This high-fat, low-carbohydrate diet provides an alternative source of energy for the brain as ketone bodies, which are produced in the liver and which can easily penetrate the blood–brain barrier (Klepper 2008; Rahman et al. 2013).

Defects of Purine and Pyrimidine Metabolism

Adenylosuccinate lyase (ADSL) deficiency is an autosomal recessive defect of purine metabolism causing serious neurological and physiological symptoms. ASL catalyzes two distinct reactions in the synthesis of purine nucleotides, both of which involve the b-elimination of fumarate. The deficiency of ADSL results in the accumulation of succinylpurines in CFS, plasma and urine (Spiegel et al. 2006).

The human *ADSL* gene has been mapped to chromosome 22q13.1–13.2. Most ADSL deficiency patients are compound heterozygotes and in cases in which the parents have been genotyped, each parent carries one mutant and one normal allele and is asymptomatic. No individuals with ADSL deficiency are completely lacking in enzyme activity; complete lack of ADSL activity in humans is probably incompatible with life (Spiegel et al. 2006). The potential mechanisms whereby ADSL may provoke neurological manifestations include deficiency of purine nucleotides, impairment of energy metabolism, and toxic effects by accumulated intermediates (Ciardo et al. 2001).

The clinical presentation is characterized by severe neurologic involvement including seizures, developmental delay, hypotonia, and autistic features. Neonatal seizures and a severe infantile epileptic encephalopathy are often the first manifestations of this disorder. The epileptic phenotype consists of myoclonias, partial epilepsy, GTCS, spasms and status epilepticus (Ciardo et al. 2001).

Epilepsy in ADSL deficiency is usually associated with psychomotor delay, autism and signs of cerebellar and pyramidal dysfunction (Wolf et al. 2005).

Mitochondrial Diseases

Mitochondrial diseases (MCDs) are a clinically heterogeneous group of disorders that arise as a result of dysfunction of the mitochondrial respiratory chain. They can be caused by mutations of nuclear or mitochondrial DNA (mtDNA) (Cree et al. 2009). Nuclear gene defects may be inherited in an autosomal recessive or autosomal dominant manner. Mitochondrial DNA defects are transmitted by maternal inheritance. Epilepsy is also a frequent CNS manifestation of MCDs. Seizure may start at infancy as infantile spasms, West syndrome, myoclonic jerks, astatic seizures, or juvenile myoclonic epilepsy. Epilepsy is particularly prevalent in patients with MELAS, MERRF, LS, or NARP (Finsterer 2006).

Several mitochondrial diseases have been linked to ineffective mtDNA replication by mitochondrial DNA polymerase gamma (POLG). Mutations in POLG, are associated with Alpers syndrome (and Alpers-like encephalopathy), childhood Myocerebrohepatopathy spectrum disorders, ataxia-neuropathy syndromes, myoclonus epilepsy myopathy sensory ataxia, and dominant and recessive forms of progressive external ophthalmalplegia (PEO) (Milone and Massie 2010).

Mitochondrial DNA depletion syndrome, also known as *Alpers syndrome* (OMIM #203700), is an autosomal recessive disorder characterized by a clinical triad of psychomotor retardation, intractable epilepsy, and liver failure. Seizures

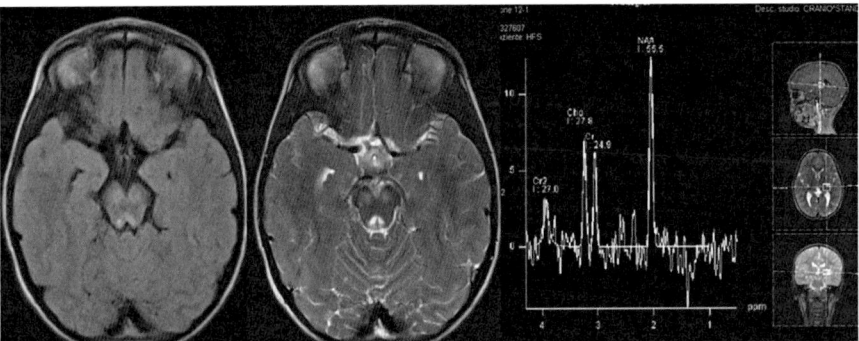

Fig. 3 MRI of a boy with Leigh phenotype with complex I deficiency. Axial T2-weighted images show focal bilateral lesions in the brainstem. The MR-spectroscopy show a lactate peak in deep gray matter

may initially be focal and subsequently generalize. Epilepsia partialis continua and convulsive status epilepticus are common. The disorder, diagnosed in infants and young children, is progressive and often leads to death from hepatic failure or status epilepticus before age 3 years. MtDNA in muscle and liver samples of Alpers syndrome patients is depleted (Milone and Massie 2010). The EEG features of posterior rhythmic high amplitude delta with superimposed polyspikes (RHADS) are very helpful although not mandatory in all cases. The course is usually rapidly progressive; most affected infants die before the age of 3 years (Wolf et al. 2009a, b).

Magnetic resonance imaging changes may be nonspecific, such as atrophy (both general and involving specific structures, such as cerebellum), more suggestive of particular disorders such as focal and often bilateral lesions confined to deep brain nuclei, or clearly characteristic of a given disorder such as stroke-like lesions that do not respect vascular boundaries in mitochondrial myopathy, encephalopathy, lactic acidosis, and stroke-like episode (MELAS). White matter hyperintensities with or without associated gray matter involvement may also be observed (Fig. 3; Friedman et al. 2010; Papetti et al. 2013).

Secondary mitochondrial dysfunction is also seen in a number of different genetic disorders, including *ethylmalonic aciduria* (EE, OMIM # 602473) (Tiranti et al. 2009). EE is an autosomal recessive metabolic disorder of infancy affecting the brain, the gastrointestinal tract and peripheral vessels. It is caused by a defect in the ETHE1 gene product, which a mitochondrial dioxygenase involved in hydrogen sulfide (H (2) S) detoxification. Patients present in infancy with psychomotor retardation, chronic diarrhea, orthostatic acrocyanosis and relapsing petechiae. High levels of lactic acid, ethylmalonic acid (EMA) and methylsuccinic acid (MSA) are detected in body fluids. The signs and symptoms of EE are apparent at birth or begin in the first few months of life. Seizures start early as tonic seizure, spasms and West syndrome (Fig. 2; Papetti et al. 2013). MRI generally shows symmetrical increased signals on T2-weighted images in the basal ganglia which correspond to symmetrical necrotic lesions (Fig. 4). They occasionally have signal anomalies in subcortical areas, white matter, and brainstem (Pigeon et al. 2009).

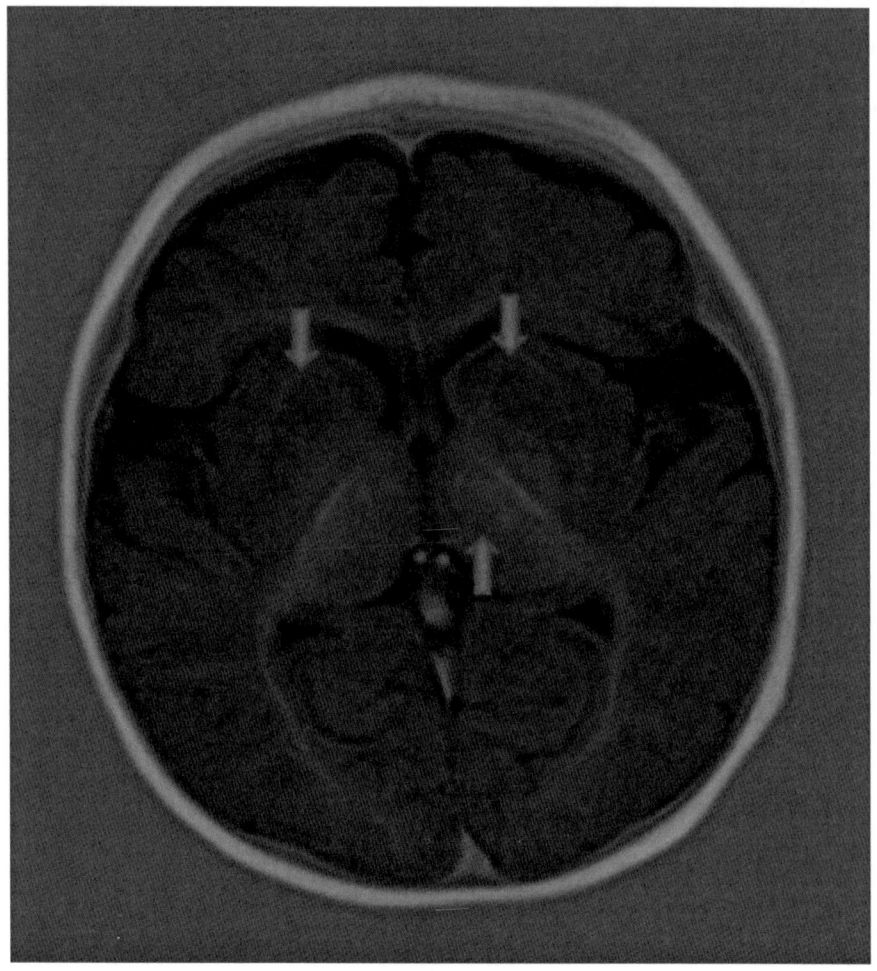

Fig. 4 MRI of 4 months-girl with Ethylmalonic aciduria. Axial flair- sequence image shows bilateral hypointense lesions of basal nuclei (n. caudati; n. lentiformi)

Diagnostic approach in MCDs should include patient and family history, laboratory examination, and neurological workup as initial workups, then specific biochemical studies, muscle biopsy, and molecular genetic studies as the further workups. The most useful basic test is to check serum lactate level (Kang et al. 2013). In patients with mitochondrial encephalopathy, when pyruvate oxidation in mitochondria is disturbed due to abnormalities in pyruvate dehydrogenase complex, Krebs cycle, or electron transport chain, excessive pyruvate can be either transformed into alanine or reduced to lactate, increasing blood lactate level. The ratio of lactate to pyruvate depends on the degree of oxidation–reduction in tissue. Since the increase in serum lactate level can be equivocal in patients with mitochondrial encephalopathy, it is sometimes helpful to evaluate lactate level of cerebrospinal fluid. More specifically, a particularly significant increase in the ratio of lactate to pyruvate and 3-hydroxybutyrate to acetoacetate may suggest respiratory chain defect. The further diagnostic approaches to mitochondrial diseases include morphological observation using microscopic tools, biochemical assays that measure enzyme activities of respiratory chain reaction in skeletal muscles (most commonly complex I defect), and molecular genetic studies to examine mtDNA or nuclear DNA mutation (Kang et al. 2013).

To treat epilepsy with mitochondrial diseases, general supportive care is first provided to treat multiorgan involvement. Then, the children are given antioxidants and respiratory chain cofactors, along with recommendation for diet change to the ketogenic diet or caloric restriction. Finally, antiepileptic drugs (AEDs) are used to control seizure (Kang et al. 2013).

Coenzyme Q10 has been reported to have two functions, as an electron carrier in the mitochondrial respiratory chain reaction and as a scavenger molecule. It is important to identify and treat disorders of CoQ10 biosynthesis, since these remain the only readily treatable forms of mitochondrial epilepsy (Steele et al. 2004; Rahman et al. 2010, 2013). Many present in infancy with a multisystem syndrome including epilepsy, frequently associated with sensorineural hearing loss and a prominent steroid- resistant nephropathy. Other neurological features in these patients include nystagmus, ataxia, spasticity, and dystonia. Mutations in five genes (COQ2, PDSS1, PDSS2, COQ9, and COQ6) have so far been reported to cause infantile onset of CoQ10 deficiency. Treatment is with oral CoQ10 supple mentation; 10–30 mg/kg/day in three divided doses is usually sufficient. The best outcome in this disorder was reported in a female who was diagnosed early because of an affected older sibling, and in whom treatment was initiated at the first manifestation of disease (Montini et al. 2008; Rahman et al. 2010). Riboflavin, tocopherol (vitamin E), succinate, ascorbate (vitamin C), menadione, and nicotinamide have also been used to treat mitochondrial diseases with deficiency in specific enzymes (Kang et al. 2013).

Males with the X-linked form of *pyruvate dehydrogenase complex (PDHc) deficiency* usually present with Leigh syndrome (OMIM 308930), but females who are heterozygous for a severe mutation in the PDHA1 gene can present in the first 6 months of life with infantile spasms, an EEG showing hypsarrhythmia, and developmental regression (West syndrome), or just with severe myoclonic seizures (OMIM 312170). MRI may show periventricular multicystic leukoencephalopathy and agenesis of the corpus callosum. CSF lactate is often elevated, usually with an elevation of blood lactate, and fibroblast studies show reduced pyruvate dehydrogenase complex activity. Some cases of pyruvate dehydrogenase complex deficiency respond well to treatment with thiamine and/or a ketogenic diet, and this reduction in seizure severity (Barnerias et al. 2010).

Childhood Onset Seizures

Storage Disorders

The neuronal ceroid-lipofuscinoses (NCLs) are a group of inherited, neurodegenerative, lysosomalstorage disorders characterized by progressive intellectual and motor deterioration, seizures, and early death. Visual loss is a feature of most forms. NCLs variants are classified by age of onset and order of appearance of the clinical features: infantile neuronal ceroid-lipofuscinosis (INCL), late-infantile (LINCL), juvenile (JNCL), adult (ANCL), and Northern epilepsy (NE, progressive epilepsy with intellectual disability) (Mole and Williams 2010). In INCL seizures start at the end of the first year of life, with myoclonus, atonic and GTCS followed by dementia and movement disorders. The EEG shows a depression of background activity (Pascual et al. 2008). Symptoms of LINCL generally appear after the second year of life. The seizures are GTCS, tonic clonic, atonic, myoclonic and myoclonic-astatic. The EEG can show epileptic discharges during intermittent photostimulation at 1 Hz (Wolf et al. 2009a, b). The JNCL form is characterized by seizures that typically appear between ages 5 and 18 years. Northern epilepsy is characterized by tonic-clonic or complexpartial seizures, intellectual disability, and motor dysfunction. Onset occurs between ages 2 and 10 years (Mole and Williams 2010).

The genes *PPT1 (CNL1), TPP1 (CNL2), CLN3, CLN5, CLN6, MFSD8 (CLN7), CLN8,* and *CTSD (CNL10)* are known to be associated with NCLs. In INCL, a lysosomal enzyme, palmitoyl protein thioesterase 1 (PPT1) is deficient. Patients with LINCL are deficient in a pepstatin-insensitive lysosomal peptidasetripeptidyl peptidase 1 (TTP1). JNCL is due to mutation of CLN3 gene that encodes a protein that is thought to be a part of the lysosomal membrane. The ANCL is associated with mutations of the CLN4 gene (not mapped yet). Mutations in another gene, CLN5 is associated with Finnish variant LINCL that occurs predominantly in the Finnish population (Mole and Williams 2010).

Myoclonic epilepsy of Lafora (EPM2, OMIM #254780) is a severe autosomal recessive disorder characterized by fragmentary, symmetric, or generalized myoclonus and/or GTCS, occipital seizures, and progressive neurologic degeneration including cognitive and/or behavioral deterioration, dysarthria, and ataxia beginning in previously healthy adolescents between ages 12 and 17 years. The frequency and intractability of seizures increase over time. Survival is short, less than 10 years after onset. The disease is characterized by intracellular polyglucosan inclusions (Lafora bodies) in the brain, liver, skin and muscles (Monaghan and Delanty 2010). Two genetic forms are known, one of which (EPM2A) is caused by mutations in the *laforin* gene and another (NHLRC1)–by mutations in the *malin* gene. The *EMP2A* encodes a protein phosphatase and *NHLRC1* encodes an ubiquitin ligase. These two proteins interact with each other and, as a complex, are thought to regulate critical neuronal functions (Singh and Ganesh 2009).

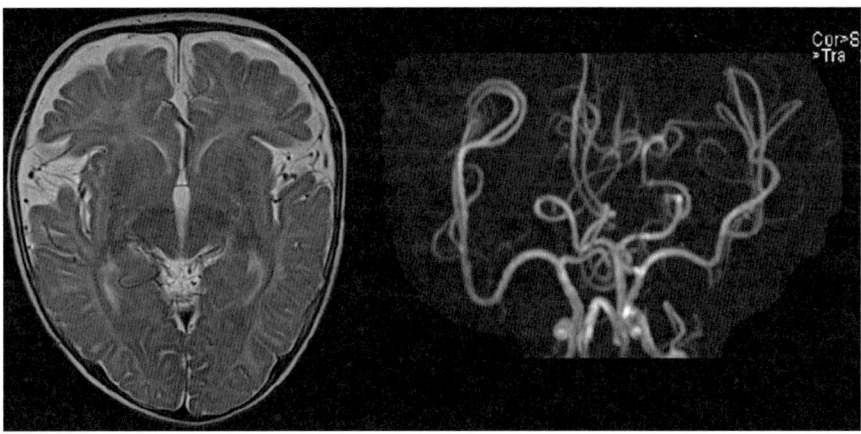

Fig. 5 MRI of 6 months-boy with Menkes disease. Axial T2-weighted image show dilatation of fronto-insular subarachnoid spaces and enlargement of anterior interhemispheric fissure. Axial MR angiography shows tortuous intracranial vessels

Copper Metabolism Errors

Menkes disease (MD) is a multisystemic disorder of copper metabolism. Progressive neurodegeneration and connective tissue disturbances, together with the peculiar 'kinky' hair are the main manifestations. MD is inherited as an X-linked recessive trait, and is due to mutations in the *ATP7A* gene. The vast majority of *ATP7A* mutations are intragenic mutations or partial gene deletions. ATP7A is an energy dependent transmembrane protein, which is involved in the delivery of copper to the secreted copper enzymes and in the export of surplus copper from cells (Tümer and Møller 2010). Seizures may occur during the first few months of life, although there are mild forms of the defect with later onset and include myoclonus, spasms and multifocal seizures. Three successive periods in the course of epilepsy have been observed: early focal status, then infantile spasms, and then myoclonic and multifocal epilepsy after age 2 years (Bahi-Buisson et al. 2006).

Radiological findings are various and combine cortical and cerebellar atrophy, delayed myelinisation, tortuosity and dilatation of the intra- and extracranial vessels, and subdural fluid collections (Fig. 5) (Bahi-Buisson et al. 2006; Papetti et al. 2013).

Congenital Disorders of Glycosylation (Cdg)

Congenital disorders of glycosylation (CDG) are a group of disorders of abnormal glycosylation of Nlinked oligosaccharides caused by deficiency in 21 different enzymes in the N-linked oligosaccharide synthetic pathway. Most commonly, the disorders begin in infancy; manifestations range from severe developmental delay and hypotonia with multiple organ system involvement to hypoglycemia and protein-losing enteropathy with normal development (Sparks and Krasnewich 2011). Mutations identified in the 15 genes (*PMM2, MPI, DPAGT1, ALG1, ALG2, ALG3, ALG9, ALG12, ALG6, ALG8, DOLK, DPM1, DPM3, MPDU1,* and *RFT1*) yield a deficiency of dolichol-linked oligosaccharide biosynthesis resulting in CDG (Haeuptle and Hennet 2009). Epilepsy associated with developmental delay, dysmorphisms and hypotonia has been described more frequently in CDG-1d. Seizures generally start in the infancy and the EEG revealed generalized epileptic changes, while brain MRI shows evidence of leukodystrophy (Grünewald et al. 2000).

Lysosomal Storage Disorder

Gaucher disease (GD) is a lipid storage disease characterized by the recessively inherited deficiency of lysosomal glucocerebrosidase, encoded by GBA (OMIM# 606463). GD manifests with diverse symptoms, and is commonly divided into three types, based on the absence (type 1) and rate of progression of neurological manifestations (types 2 and 3) (Mignot et al. 2006). GD3 comprises a heterogeneous group of patients suffering either mild or severe systemic disease combined with variable neurological involvement. Two further subtypes of GD3 patients have been reported differentiating those with mild systemic involvement associated with progressive myoclonic epilepsy (PME), called GD3a, from those with severe systemic involvement associated with oculomotor apraxia (OMA), called GD3b (Kraoua et al. 2011). In a recent perspective study conducted by the International Collaborative Gaucher Group, seizures were reported in 19 out of 122 patients (16%). Some patients had experienced more than one type of seizure. The types reported were tonic–clonic seizures (7 out of 122, 6%), clonic seizures (5 out of 122, 4%), tonic seizures (3 out of 120, 3%), myoclonic seizures (3 out of 121, 2%), typical absence seizures (2 out of 119, 2%), and atypical absence seizures (1 out of 119, 1%) (Tylki-Szymańska et al. 2010).

Niemann–Pick type C (NPC; OMIM 257220) is a progressive neurodegenerative disorder characterized by accumulation of free cholesterol, sphingomyelin, glycosphingolipids (GSLs) and sphingosine in lysosomes, mainly due to a mutation in the NPC1 gene. The clinical spectrum of the disease ranges from a neonatal rapidly fatal disorder to an adult-onset chronic neurodegenerative disease. Epilepsy is generally a late onset manifestation with partial, generalized tonic–clonic and atonic seizures (Sévin et al. 2007).

Peroxisomal Disorders

X-linked adrenoleukodystrophy (X-ALD, OMIM 300100) is an inherited, recessive, neurodegenerative disease affecting brain white matter, adrenal cortex and testis.

The disorder is caused by mutations in the *ABCD1* gene, which impair peroxisomal b-oxidation, resulting in the accumulation of very long-chain fatty acids (VLCFa) in plasma. There are several distinct clinical phenotypes ranging from cerebral forms, adrenomyeloneuropathy (AMN) to asymptomatic persons or isolated adrenal insufficiency without CNS involvement (Addison's disease only). Cerebral X-ALD is further divided into childhood (CCALD), adolescent (AdolCALD) and adult form (ADCALD). Affected boys with CCALD present before the age of 10 years a rapidly progressive disorder with ataxia, spasticity, deafness, visual deficits, personality changes and seizures. The less common adolescent form after the age of 10 years demonstrates similar course. Cerebral X-ALD is frequently associated with Addison's disease, but the primary adrenal insufficiency may precede, coexist or develop after neurological disturbances. In one large series, 20 out of 485 individuals presented with seizures: focal seizures in six males and generalized in the remainder, with four having status epilepticus (Stephensons et al. 2000). Typical MR findings in the brain of X-ALD patients have been well documented recently and consist of bilateral white matter abnormalities. Typically they occur initially in the posterior cerebral regions and progress to parietal, temporal and frontal lobes sequentially. Such a pattern is found in approximately 80 % of cases; therefore MR strongly suggests the diagnosis of X-ALD (Poll-The 2012).

Conclusions

Inborn errors of metabolism are a rare cause of epilepsy in pediatric age. The suspicious of metabolic epilepsy should rise if seizures are refractory to standard antiepileptic drugs, and if additional symptoms are present such as mental retardation, dysmorphism, movement disorders, and visceral abnormalities. The recognition of metabolic causes of epilepsy requires a multidisciplinary approach and several investigations such as EEG, evoked potentials, neuroimaging (MRI-spectroscopy, SPECT), blood exams, urinary exams and CSF analysis. The diagnosis of metabolic disorders in epileptic patients may provide the possibility of specific treatments that can improve seizures.

References

Aicardi J (2007) Epilepsy syndromes. In: Engel J, Pedley TA, Aicardi J (eds) Epilepsy a comprehensive textbook, vol 1. Lippincott Williams & Wilkins, Philadelphia, pp 2608–2609
Applegarth DA, Toone JR (2004) Glycine encephalopathy (nonketotic hyperglycinaemia): review and update. J Inherit Metab Dis 27:417–422
Bagci S, Zschocke J, Hoffmann GF, Bast T, Klepper J, Müller A et al (2008) Pyridoxal phosphate-dependent neonatal epileptic encephalopathy. Arch Dis Child Fetal Neonatal Ed 93:151–152
Bahi-Buisson N, Kaminska A, Nabbout R, Barnerias C, Desguerre I, De Lonlay P et al (2006) Epilepsy in Menkes disease: analysis of clinical stages. Epilepsia 47:380–386

Balasubramaniam S, Bowling F, Carpenter K, Earl J, Chaitow J, Pitt J et al (2010) Perinatal hypophosphatasia presenting as neonatal epileptic encephalopathy with abnormal neurotransmitter metabolism secondary to reduced co-factor pyridoxal-5'-phosphate availability. J Inherit Metab Dis. http://dx.doi.org/10.1007/s10545-009-9012-y

Barnerias C, Saudubray JM, Touati G et al (2010) Pyruvate dehydrogenase complex deficiency: four neurological phenotypes with differing pathogenesis. Dev Med Child Neurol 52:1–9

Baxter P (2003) Pyridoxine-dependent seizures: a clinical and biochemical conundrum. Biochim Biophys Acta 1647:36–41

Baxter P (2010) Recent insights into pre- and postnatal pyridoxal phosphate deficiency, a treatable metabolic encephalopathy. Dev Med Child Neurol 52:597–598

Bhardwaj P, Kaushal RK, Chandel A (2010) Biotinidase deficiency: a treatable cause of infantile seizures. J Pediatr Neurosci 5:82–83

Blau N, van Spronsen FJ, Levy HL (2010) Phenylketonuria. Lancet 376:1417–1427

Braissant O (2010) Current concepts in the pathogenesis of urea cycle disorders. Mol Genet Metab 100:3–12

Brockmann K (2009) The expanding phenotype of GLUT1-deficiency syndrome. Brain Dev 31:545–552

Cerisola A, Campistol J, Pérez-Dueñas B, Poo P, Pineda M, García-Cazorla A et al (2009) Seizures versus dystonia in encephalopathic crisis of glutaric aciduria type I. Pediatr Neurol 40:426–431

Chemelli AP, Schocke M, Sperl W, Trieb T, Aichner F, Felber S (2000) Magnetic resonance spectroscopy (MRS) in five patients with treated propionic acidemia. J Magn Reson Imaging 11:596–600

Ciardo F, Salerno C, Curatolo P (2001) Neurologic aspects of adenylosuccinate lyase deficiency. J Child Neurol 16:301–308

Clague A, Thomas A (2002) Neonatal biochemical screening for disease. Clin Chim Acta 315:99–110

Cree LM, Samuels DC, Chinnery PF (2009) The inheritance of pathogenic mitochondrial DNA mutations. Biochim Biophys Acta 1792:1097–1102

de Koning TJ, Klomp LW (2004) Serine-deficiency syndromes. Curr Opin Neurol 17:197–204

Dinopoulos A, Matsubara Y, Kure S (2005) Atypical variants of nonketotic hyperglycinemia. Mol Genet Metab 86:61–69

Finsterer J (2006) Central nervous system manifestations of mitochondrial disorders. Acta Neurol Scand 114:217–238

Friedman JR, Thiele EA, Wang D, Levine KB, Cloherty EK, Pfeifer HH et al (2006) Atypical GLUT1 deficiency with prominent movement disorder responsive to ketogenic diet. Mov Disord 21:241–245

Friedman SD, Shaw DWW, Ishak G, Gropman AL, Saneto RP (2010) The use of neuroimaging in the diagnosis of mitochondrial disease. Dev disabil Res rev 16:129–135

Gallagher RC, Van Hove JL, Scharer G, Hyland K, Plecko B, Waters PJ et al (2009) Folinic acid-responsive seizures are identical to pyridoxine-dependent epilepsy. Ann Neurol 65:550–556

Gospe SM Jr (2010) Neonatal vitamin-responsive epileptic encephalopathies. Chang Gung Med J 33:1–12

Gropman AL, Summar M, Leonard JV (2007) Neurological implications of urea cycle disorders. J Inherit Metab Dis 30:865–879

Grünewald S, Imbach T, Huijben K, Rubio-Gozalbo ME, Verrips A, de Klerk JB et al (2000) Clinical and biochemical characteristics of congenital disorder of glycosylation type Ic, the first recognized endoplasmic reticulum defect in N-glycan synthesis. Ann Neurol 47:776–781

Haberlandt E, Canestrini C, Brunner-Krainz M, Möslinger D, Mussner K, Plecko B et al (2009) Epilepsy in patients with propionic acidemia. Neuropediatrics 40:120–125

Haeuptle MA, Hennet T (2009) Congenital disorders of glycosylation: an update on defects affecting the biosynthesis of dolichol-linked oligosaccharides. Hum Mutat 30:1628–1641

Hamosh A, Scharer G, Van Hove J (2009) Glycine encephalopathy. In: Pagon RA, Bird TD, Dolan CR, Stephens K, Adam MP (eds) GeneReviewse (Internet). University of Washington, Seattle, pp 1993–2002. (Bookshelf ID: NBK1357)

Hoffmann GF, Schmitt B, Windfuhr M, Wagner N, Strehl H, Bagci S et al (2007) Pyridoxal 50-phosphate may be curative in earlyonset epileptic encephalopathy. J Inherit Metab Dis 30:96–99
Jaeken J (2002) Genetic disorders of gamma-aminobutyric acid, glycine, and serine as causes of epilepsy. J Child Neurol 17:84–87
Joshi SN, Fathalla M, Koul R, Maney MA, Bayoumi R (2010) Biotin responsive seizures and encephalopathy due to biotinidase deficiency. Neurol India 58:323–324
Kang HC, Lee YM, Kim HD (2013) Mitochondrial disease and epilepsy. Brain Dev 35(8):757–761
Klepper J (2008) Glucose transporter deficiency syndrome (GLUT1DS) and the ketogenic diet. Epilepsia 49:46–49
Klepper J, Leiendecker B (2007) GLUT1 deficiency syndrome-2007 update. Dev Med Child Neurol 49:707–716
Klepper J, Engelbrecht V, Scheffer H, van der Knaap MS, Fiedler A (2007) GLUT1 deficiency with delayed myelination responding to ketogenic diet. Pediatr Neurol 37:130–133
Korein J, Sansaricq C, Kalmijn M, Honig J, Lange B (1994) Murple syrup urine disease: clinical, EEG and plasma aminoacid correlations with a theoretical mechanism of acute neurotoxicity. Int J Neurosci 79:21–45
Kraoua I, Sedel F, Caillaud C, Froissart R, Stirnemann J, Chaurand G et al (2011) A French experience of type 3 Gaucher disease: phenotypic diversity and neurological outcome of 10 patients. Brain Dev 33(2):131–139
Kuo MF, Wang HS (2002) Pyridoxal phosphate-responsive epilepsy with resistance to pyridoxine. Pediatr Neurol 26:146–147
Leuzzi V (2002) Inborn errors of creatine metabolism and epilepsy: clinical features, diagnosis, and treatment. J Child Neurol 17(3):89–97
Leuzzi V1, Mastrangelo M, Battini R, Cioni G (2013) Inborn errors of creatine metabolism and epilepsy. Epilepsia 54(2):217–227
Martynyuk AE, Ucar DA, Yang DD, Norman WM, Carney PR, Dennis DM et al (2007) Epilepsy in phenylketonuria: a complex dependence on serum phenylalanine levels. Epilepsia 48:1143–1150
Matern D, Lehnert W, Gibson KM, Korinthenberg R (1996) Seizures in a boy with succinic semialdehyde dehydrogenase deficiency treated with vigabatrin (gamma-vinyl-GABA). J Inherit Metab Dis 19:313–318
McClelland VM, Bakalinova DB, Hendriksz C, Singh RP (2009) Glutaric aciduria type 1 presenting with epilepsy. Dev Med Child Neurol 51:235–239
Mignot C, Doummar D, Maire I (2006) French type 2 Gaucher disease study group. Type 2 Gaucher disease: 15 new cases and review of the literature. Brain Dev 28(1):39–48
Mills PB, Surtees RAH, Champion MP, Beesley CE, Dalton N, Scambler PK et al (2005) Neonatal epileptic encephalopathy caused by mutations in the PNPO gene encoding pyridox (am)ine 5′-phosphate oxidase. Hum Mol Genet 14:1077–1086
Mills PB, Footitt EJ, Mills KA, Tuschl K, Aylett S, Varadkar S et al (2010) Genotypic and phenotypic spectrum of pyridoxine-dependent epilepsy (ALDH7A1 deficiency). Brain 133:2148–2159
Milone M, Massie R (2010) Polymerase gamma 1 mutations: clinical correlations. Neurologist 16:84–91
Mole SE, Williams RE (2010) Neuronal Ceroid-Lipofuscinoses. In: Pagon RA, Bird TC, Dolan CR, Stephens K (eds) Gene reviews 2010. University of Washington, Seattle, pp 1993–2001
Monaghan TS, Delanty N (2010) Lafora disease: epidemiology, pathophysiology and management. CNS Drugs 24:549–561
Montini G, Malaventura C, Salviati L (2008) Early coenzyme Q10 supplementation in primary coenzyme Q10 deficiency. N Engl J Med 358:2849–2850
Nabbout R, Dulac O (2008) Epileptic syndromes in infancy and childhood. Curr Opin Neurol 21:161–166
Nasrallah F, Feki M, Kaabachi N (2010) Creatine and creatine deficiency syndromes: biochemical and clinical aspects. Pediatr Neurol 42:163–171

Neumaier-Probst E, Harting I, Seitz A, Ding C, Kolker S (2004) Neuroradiological findings in glutaric aciduria type I (glutaryl-CoA dehydrogenase deficiency). J Inherit Metab Dis 27:869–876

Overweg-Plandsoen WC, Groener JE, Wang D, Onkenhout W, Brouwer OF, Bakker HD et al (2003) GLUT-1 deficiency without epilepsy—an exceptional case. J Inherit Metab Dis 26:559–563

Papetti L, Parisi P, Leuzzi V, Nardecchia F, Nicita F, Ursitti F, Marra F, Paolino MC, Spalice A (2013) Metabolic epilepsy: an update. Brain Dev 35(9):827–841

Pascual JM, Campistol J, Gil-Nagel A (2008) Epilepsy in inherited metabolic disorders. Neurologist 14:2–14

Pearl PL (2009) New treatment paradigms in neonatal metabolic epilepsies. J Inherit Metab Dis 32:204–213

Pearl PL, Gibson KM (2004) Clinical aspects of the disorders of GABA metabolism in children. Curr Opin Neurol 17:107–113

Pearl PL, Taylor JL, Trzcinski S, Sokohl A (2007) The pediatric neurotransmitter disorders. J Child Neurol 22:606–616

Pearl PL, Vezina LG, Saneto RP, McCarter R, Molloy-Wells E, Heffron A et al (2009) Cerebral MRI abnormalities associated with vigabatrin therapy. Epilepsia 50:184–94

Pigeon N, Campeau PM, Cyr D, Lemieux B, Clarke JT (2009) Clinical heterogeneity in ethylmalonic encephalopathy. J Child Neurol 24:991–296

Pindolia K, Jordan M, Wolf B (2010) Analysis of mutations causing biotinidase deficiency. Hum Mutat 31:983–991

Plecko B, Stöckler S (2009) Vitamin B6 dependent seizures. Can J Neurol Sci 36:73–77

Rahman S, Clarke CF, Hirano M (2012) 176th ENMC International Workshop: diagnosis and treatment of coenzyme Q10 deficiency. Neuromuscul Disord 22:76–86

Rahman S, Footitt EJ, Varadkar S, Clayton PT (2013) Inborn errors of metabolism causing epilepsy. Dev Med Child Neurol 55(1):23–36

Rossi S, Daniele I, Bastrenta P, Mastrangelo M, Lista G (2009) Early myoclonic encephalopathy and nonketotic hyperglycinemia. Pediatr Neurol 241:371–374

Rotstein M, Engelstad K, Yang H, Wang D, Levy B et al (2010) Glut1 deficiency: inheritance pattern determined by haploinsufficiency. Ann Neurol 68:955–8

Scharer G, Brocker C, Vasiliou V, Creadon-Swindell G, Gallagher RC, Spector E et al (2010) The genotypic and phenotypic spectrum of pyridoxine-dependent epilepsy due to mutations in ALDH7A1. J Inherit Metab Dis 33:571–581

Schneider SA, Paisan-Ruiz C, Garcia-Gorostiaga I, Quinn NP, Weber YG, Lerche H et al (2009) GLUT1 gene mutations cause sporadic paroxysmal exercise-induced dyskinesias. Mov Disord 24:1684–8

Sévin M, Lesca G, Baumann N, Millat G, Lyon-Caen O, Vanier MT et al (2007) The adult form of Niemann–Pick disease type C. Brain 130(Pt 1):120–133

Seashore MR (2009) The organic acidemias: an overview. In: Pagon RA, Bird TD, Dolan CR, Stephens K, Adam MP (eds) GeneReviewse (Internet). University of Washington, Seattle, pp 1993–2001. (Bookshelf ID: NBK1134)

Seidner G, Alvarez MG, Yeh JI, O'Driscoll KR, Klepper J, Stump TS et al (1998) GLUT-1 deficiency syndrome caused by haploinsufficiency of the blood–brain barrier hexose carrier. Nat Genet 18:188–191

Singh S, Ganesh S (2009) Lafora progressive myoclonus epilepsy: a meta-analysis of reported mutations in the first decade following the discovery of the EPM2A and NHLRC1 genes. Hum Mutat 30:715–723

Sparks SE, Krasnewich DM (2011) Congenital disorders of glycosylation overview. In: Pagon RA, Bird TD, Dolan CR, Stephens K, Adam MP (eds) GeneReviewse (Internet). University of Washington, Seattle, pp 1993–2005. (Bookshelf ID: NBK1332)

Spiegel EK, Colman RF, Patterson D (2006) Adenylosuccinate lyase deficiency. Mol Genet Metab 89:19–31

Steele PE, Tang PH, DeGrauw AJ, Miles MV (2004) Clinical laboratory monitoring of coenzyme Q10 use in neurologic and muscular diseases. Am J Clin Pathol 121:113–120

Steinberg SJ, Raymond GV, Braverman NE, Moser AB (2003) Peroxisome biogenesis disorders, zellweger syndrome spectrum. In: Pagon RA, Bird TD, Dolan CR, Stephens K, Adam MP (eds) GeneReviewse (Internet). University of Washington, Seattle, pp 1993–2003. (Bookshelf ID:1448)

Stephenson DJ, Bezman L, Raymond GV (2000) Acute presentation of childhood adrenoleukodystrophy. Neuropediatrics 31:293–297

Suls A, Dedeken P, Goffin K, Van Esch H, Dupont P, Cassiman D et al (2008) Paroxysmal exercise-induced dyskinesia and epilepsy is due to mutations in SLC2A1, encoding the glucose transporter GLUT1. Brain 131:1831–1844

Summar ML (2005). Urea cycle disorders overview. In: Pagon RA, Bird TD, Dolan CR, Stephens K, Adam MP (eds) GeneReviewse (Internet). University of Washington, Seattle, pp 1993–2002. (Bookshelf ID: NBK1217)

Suzuki Y, Yang X, Aoki Y, Kure S, Matsubara Y (2005) Mutations in the holocarboxylase synthetase gene HLCS. Hum Mutat 26:285–290

Tabatabaie L, Klomp LW, Berger R, de Koning TJ (2010) L-serine synthesis in the central nervous system: a review on serine deficiency disorders. Mol Genet Metab 99:256–262

Takahashi Y, Suzuki Y, Kumazaki K, Tanabe Y, Akaboshi S, Miura K et al (1997) Epilepsy in peroxisomal diseases. Epilepsia 38(2):182–8

Thoene J, Wolf B (1983) Biotinidase deficiency in juvenile multiple carboxylase deficiency. Lancet 2:398

Tiranti V, Viscomi C, Hildebrandt T, Di Meo I, Mineri R, Tiveron C et al (2009) Loss of ETHE1, a mitochondrial dioxygenase, causes fatal sulfide toxicity in ethylmalonic encephalopathy. Nat Med 15:200–205

Tümer Z, Møller LB (2010) Menkes disease. Eur J Hum Genet 18:511–518

Tylki-Szymańska A, Vellodi A, El-Beshlawy A, Cole JA, Kolodny E (2010) Neuronopathic Gaucher disease: demographic and clinical features of 131 patients enrolled in the International Collaborative Gaucher Group Neurological Outcomes Subregistry. J Inherit Metab Dis 33(4):339–346

Van Gosen L (2008) Organic acidemias: a methylmalonic and propionic focus. J Pediatr Nurs 23:225–233

Vannucci SJ, Maher F, Simpson IA (1997) Glucose transporter proteins in brain: delivery of glucose to neurons and glia. Glia 21:2–21

Verhoeven NM, Salomons GS, Jakobs C (2005) Laboratory diagnosis of defects of creatine biosynthesis and transport. Clin Chim Acta 361:1–9

Wang IJ, Chu SY, Wang CY, Wang PJ, Hwu WL (2003) Maple syrup urine disease presenting with neonatal status epilepticus: report of one case. Acta Paediatr Taiwan 44:246–248

Wang D, Pascual JM, De Vivo D (2009) Glucose transporter type 1 deficiency syndrome. In: Pagon RA, Bird TD, Dolan CR, Stephens K, Adam MP (eds) GeneReviewse (Internet). University of Washington, Seattle, pp 1993–2002. (Bookshelf ID: NBK1430)

Wolf B (2011a) Biotinidase deficiency. In: Pagon RA, Bird TC, Dolan CR, Stephens K (eds) Gene reviews 2011. University of Washington, Seattle, pp 1993–2000

Wolf B (2011b) The neurology of biotinidase deficiency. Mol Genet Metab 104:27–34

Wolf NI, Bast T, Surtees R (2005) Epilepsy in inborn errors of metabolism. Epileptic Disord 7:67–81

Wolf NI, García-Cazorla A, Hoffmann GF (2009a) Epilepsy and inborn errors of metabolism in children. J Inherit Metab Dis 32:609–617

Wolf NI, Rahman S, Schmitt B, Taanman JW, Duncan AJ, Harting I et al (2009b) Status epilepticus in children with Alpers disease caused by POLG1 mutations: EEG and MRI features. Epilepsia 50:1596–1607

Zempleni J, Hassan YI, Wijeratne SS (2008) Biotin and biotinidase deficiency. Expert Rev Endocrinol Metab 3:715–724

Zhongshu Z, Weiming Y, Yukio F, Cheng-I Ning Z, Zhixing W (2001) Clinical analysis of West syndrome associated with phenylketonuria. Brain Dev 23:552–557

New Insights into Mechanisms Underlying Generalized Reflex Seizures

Edoardo Ferlazzo, Domenico Italiano, Sara Gasparini, Giovanbattista Gaspare Tripodi, Tiziana D'Agostino and Umberto Aguglia

Abstract "Apparently generalized" reflex seizures usually occur in the setting of idiopathic generalized epilepsies. Several animal, neurophysiological and neuroimaging evidences strongly suggest that such reflex seizures should be considered as focal with quick secondary generalization through cortico-reticular or cortico-cortical pathways. The aim of this article is to highlight mechanisms underlying "apparently generalized" reflex seizures provoked by intermittent light stimulations, reading, thinking and praxis.

Introduction

Reflex seizures are epileptic events triggered by specific motor, sensory or cognitive stimulations.

Currently, reflex epilepsies are defined as specific syndromes "in which all epileptic seizures are precipitated by sensory stimuli" (Engel 2001), excluding therefore reflex seizures occurring in the setting of focal and generalized conditions that are also associated with spontaneous seizures. Simple reflex seizures are evoked by simple, unstructured sensory stimuli. Complex reflex seizures are triggered by relatively elaborate stimuli, often with an emotional component, involving integration of higher cortical function. Latency from stimulus onset to the clinical seizure or evoked epileptic EEG activity is typically longer (minutes) in complex than simple reflex seizures (seconds) and, therefore, diagnosis can be challenging in the former. Eating, music, proprioceptive or somatosensory stimuli and hot water usually provoke focal seizures and will not be discussed in this review.

U. Aguglia (✉) · E. Ferlazzo · S. Gasparini · G. G. Tripodi · T. D'Agostino
Regional Epilepsy Centre, "Bianchi-Melacrino-Morelli" Hospital, Reggio Calabria, Italy
e-mail: u.aguglia@unicz.it

E. Ferlazzo · S. Gasparini · U. Aguglia
Department of Medical and Surgical Sciences, Magna Graecia University, Catanzaro, Italy

D. Italiano
Department of Clinical and Experimental Medicine, University of Messina, Messina, Italy

The aim of this article is to highlight mechanisms underlying "apparently generalized" reflex seizures provoked by intermittent light stimulations, reading, thinking and praxis.

Seizures Induced by Visual Stimuli

Reflex seizures provoked by visual stimulation, especially flashing lights, are the commonest and well known. Photosensitive patients present seizures when exposed to environmental flicker (such as sun shining through trees and discotheque lights), or with more complex stimuli such as television and videogames. Seizures are commonly characterized by subtle eyelid myoclonus with or without impairment of consciousness, symmetric or asymmetric jerks of the arms or absence seizures. If exposure to stimulus persists, generalized tonic–clonic seizures (GTCS) may occur (Kasteleijn-Nolst Trenité 1989, 2012). The EEG usually shows a photoparoxysmal response (PPR) with intermittent photic stimulation (IPS); usually between 10 and 30 flashes per sec. Photosensitivity is more frequent in women and has an important genetic component with likely autosomal dominant inheritance with reduced penetrance. No major gene has been identified (Waltz and Stephani 2000; Stephani et al. 2004). Several loci probably underlie the trait, possibly with subtle differences in phenotypic expression. An interaction of several susceptibility genes and environmental factors cannot be ruled out (de Kovel CG et al. 2010).

Photosensitivity can occur in different epilepsy syndromes, the most frequent association being with idiopathic generalized epilepsies (IGE), especially with juvenile myoclonic epilepsy (JME) in which it is reported in 40–90% of patients (Appleton et al. 2000). A continuum phenotypic spectrum, ranging from the generalized epilepsies to the idiopathic photosensitive occipital epilepsy has been postulated by Taylor et al. (2013). Shared genetic determinants between these two entities are likely to contribute to the complex inheritance pattern of photosensitivity. Visual stimuli can also induce focal seizures (Guerrini et al. 1994, 1995). Some subjects have idiopathic focal photosensitive occipital seizures (Guerrini et al. 1995).

Pattern sensitivity is characterized by seizures provoked by viewing environmental patterns such as escalator steps, striped wallpaper or clothing. Almost all these patients show a PPR to IPS.

Eye-closure sensitivity is a specific type of visual sensitive epilepsy in which brief, mainly generalized epileptiform changes appear in EEG within 2–4 s after closing the eyes. It is more common in females and may overlap with photosensitivity, even if eye-closure sensitivity is independent phenomenon. It may be seen in different epilepsy syndromes including Childhood Absence Epilepsy, Juvenile Absence Epilepsy (JAE), JME, Jeavons Syndrome (for review see Striano et al. 2009; Italiano et al. 2014). The paroxysmal activity underlying this phenomenon is unknown but it could be related to a mechanism of alpha rhythm augmentation (Sevgi et al. 2007).

Patients with all types of visually-induced seizures may self-induce attacks by compulsively repeated eye rolling and eyelid flicker movement, gazing at the sun ("sunflower epilepsy") or a bright light and moving one hand in front of the eyes (Tassinari et al. 1998). These patients may be either intellectually disabled or healthy photosensitive individuals. Other psychiatric conditions as obsessive-compulsive disorder may be present. Absences and myoclonic jerks are the most common seizure types in self-induction (Belcastro and Striano 2014).

After the 1997 Pokemon incident (Ishida et al. 1998), it was found that rapid changes of blue and red color frames elicit PPR more frequently than monochromatic ones, and that sensitivity to specific sequences of colors at certain frequencies may also play an important role in the generation of seizures (Tobimatsu et al. 1999). Animal and human data suggest that photosensitive patients have a predisposition to develop PPR due to hyperexcitability of the visual cortex (Ferlazzo et al. 2005). When appropriate stimuli reach the striate and parastriate areas activating a sufficient and critical amount of cortical tissue and inducing synchronization of neuronal activity, a local epileptic discharge is produced; the latter would rapidly involve the cortico-reticular or cortico-cortical pathways with propagation from the parieto-occipital areas and a generalized epileptic discharge (Ferlazzo et al. 2005).

Treatment involves a combination of preventive and pharmacological measures (Covanis et al. 2004). Avoidance of the stimulus (for example by looking away or covering one eye), watching television from a distance of at least 2 m and using a remote control to change channels are often effective precautions. Wearing a particular type of blue lens, named Z1, was found to be highly effective in controlling PPR in most of photosensitive patients (Capovilla et al. 2006). Viewing LCD/plasma screen is strongly suggested since they use a transistor for each pixel allowing the pixel to keep its state and they do not manifest flickering.

The most useful drugs are represented by valproate (VPA) (85% of patients becoming seizure-free); benzodiazepines (clobazam or clonazepam) and ethosuximide can be given as second choice. Newer antiepileptic drugs such as lamotrigine (LTG) and levetiracetam may also be used (Italiano et al. 2014)

Seizures Induced by Non-Verbal Cognitive Stimuli

Seizures induced by thinking have been reported in response to mental activity such as calculation, or playing chess or similar games (Goossens et al. 1990). Seizures usually start during adolescence and consist of generalized tonic–clonic seizures, myoclonic jerks with or without absences and absences. Rare spontaneous seizures may occur in 76% of patients. The EEG shows generalized epileptic discharges in 68% of patients. Photosensitivity may occur. Activation by mathematics or spatial tasks is found in 72%. Although there seems to be no mendelian inheritance, the family history is similar to that of patients with IGE (Goossens et al. 1990). The clinical pattern is also suggestive of IGE, especially JME or JAE. Unlike photosensitive patients, avoiding the triggering stimuli in this condition is not possible, and the majority of these patients have seizures usually controlled by VPA or CNZ (Goossens et al. 1990).

Wilkins et al. (1982) stressed the importance of the spatial components of the task in inducing epileptiform abnormalities: complex multiplication and division with remainder were epileptogenic, while addition, subtraction and simple multiplication and division, thought to involve fewer spatial components, were not.

Inoue et al. (1994) reviewed 79 patients with seizures provoked by higher brain functions and emphasized the role of a motor component in seizure activation. They introduced the term 'praxis-induced epilepsy' for patients whose seizures are provoked by 'contemplating complicated spatial task in a sequential fashion, making a decision and practically responding by using part of the body'. This was also stressed by Matsuoka et al. (2000) who studied the effects of higher mental activity on the EEG of 480 patients with epilepsy, monitored during cognitive tasks. Thirty-eight patients (36 having IGE) had triggered generalized or bilateral epileptic discharges often associated with bilateral myoclonic jerks or absences. Cognitive tasks requiring the use of the hands, such as writing, were found to be more epileptogenic than higher mental activities not requiring hand movement, such as mental calculation (Matsuoka et al. 2000). Thus, thinking in a non-verbal way seems to be the essential triggering element in this form of reflex seizures.

Reading Epilepsy

Reading epilepsy (RE) is a rare syndrome characterized by involuntary jaw jerks that occur only while reading, and that may progress to a generalized tonic–clonic seizure (Ramani 1998). Seizures characterized by alexia, associated with focal EEG abnormalities, have been reported (Koutroumanidis et al. 1998). Family history of seizures occur in about 41% of subjects. Age of onset is usually in adolescence and young adulthood with a slight preponderance in men. Interictal EEG is normal in 80% of patients, spontaneous spike and wave discharges are present in 11% and temporal paroxysmal discharges in 5%. Reading provokes epileptiform discharges in 77% of patients. Those abnormalities may be bilateral and symmetrical (32%), bilateral but asymmetrical (38%) or focal (30%). Lateralization is more frequent to the language-dominant hemisphere preferentially over the temporo-parietal region.

Autosomal inheritance with incomplete penetrance overlapping with a genetic background for IGE was proposed for some families (Daly and Forster 1975). A recent combined EEG/fMRI study showed that precipitation of reading-induced seizures involves the activation of dominant motor and premotor areas, striatum and mesiotemporal/limbic areas (Salek-Haddadi et al. 2009).

The mechanism by which reading precipitates seizures is still obscure. The most relevant factor seems to be the transformation of the linguistic material from graphemes into language (Wolf et al. 1992). In many patients, reading aloud was more activating than silent reading. Some authors stressed the importance of emo-

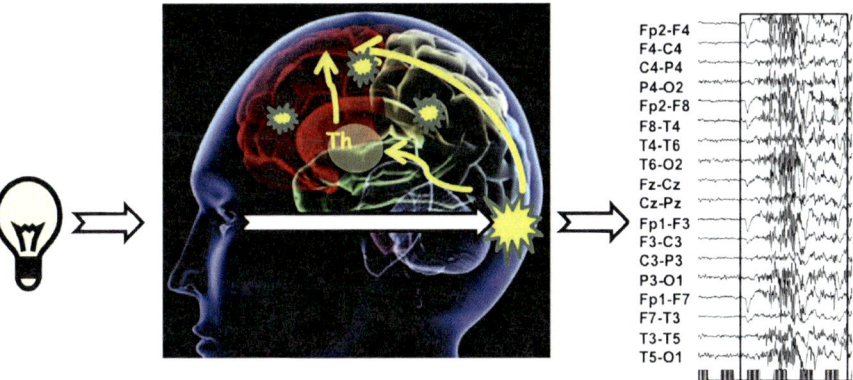

Fig. 1 *Example of "generalized" reflex seizures.* Photic stimulation induces an initial occipital lobe epileptic discharge in a patient with JME. Diffuse or multifocal hyperexcitable areas accounts for quick secondary generalization through cortico-thalamic or cortico-cortical pathways, with the final appearance of a generalized epileptic discharge. *Th* = thalami. (Reprinted from Italiano et al. 2014, with permission from Elsevier)

tional involvement with the text (Critchley et al. 1959/1960) or its comprehension (Kartsounis 1988). However nonsense text or foreign languages may also be effective (Wolf 1992).

Most patients have seizures well controlled by VPA or CNZ, and only a few decide to prevent convulsive seizures by stopping reading as soon as seizures begin (Wolf et al. 1994).

Conclusions

"Generalized" reflex seizures usually occur in the setting of IGEs. Several animal, neurophysiological and neuroimaging evidences strongly suggest that "generalized" reflex seizures should be considered as focal with quick secondary generalization through cortico-reticular or cortico-cortical pathways. IGE-patients with "generalized" reflex seizures may have focal (i.e. occipital lobe in photosensitivity) or multifocal (i.e. dominant fronto-temporal lobes on reading epilepsy) areas of hyperexcitability that, when appropriately stimulated, give rise to an epileptic activity that quickly generalizes (Fig. 1; Italiano et al. 2014).

References

Appleton R, Beirne M, Acomb B (2009) Photosensitivity in juvenile myoclonic epilepsy. Seizure 9:108–111

Belcastro V, Striano P (2014) Self-induction seizures in sunflower epilepsy: a video-EEG report. Epileptic Disord 16:93–95

Capovilla G, Gambardella A, Rubboli G, Beccaria F, Montagnini A, Aguglia U et al (2006) Suppressive efficacy by a commercially available blue lens on PPR in 610 photosensitive epilepsy patients. Epilepsia 47:529–533

Covanis A, Stodieck SR, Wilkins AJ (2004) Treatment of photosensitivity. Epilepsia 45:40–45

Critchley M, Cobb W, Sears TA (1959/1960) On reading epilepsy. Epilepsia 1:416–417

Daly RF, Forster FM (1975) Inheritance of reading epilepsy. Neurology 25:1051–1054

de Kovel CG, Pinto D, Tauer U, Lorenz S, Muhle H, Leu C et al (2010) Whole-genome linkage scan for epilepsy-related photosensitivity: a mega-analysis. Epilepsy Res 89:286–294

Engel J (2001) A proposed diagnostic scheme for people with epileptic seizures and epilepsy: report of the ILAE Task Force on Classification and Terminology. Epilepsia 42:796–803

Ferlazzo E, Zifkin BG, Andermann E, Andermann F (2005) Cortical triggers in generalized reflex seizures and epilepsies. Brain 128:700–710

Goossens L, Andermann F, Andermann E, Remillard GM (1990) Reflex seizures induced by calculation, card or board games, and spatial tasks: a review of 25 patients and delineation of the epileptic syndrome. Neurology 40:1171–1176

Guerrini R, Ferrari AR, Battaglia A, Salvadori P, Bonanni P (1994) Occipitotemporal seizures with ictus emeticus induced by intermittent photic stimulation. Neurology 44:253–259

Guerrini R, Dravet C, Genton P, Bureau M, Bonanni P, Ferrari AR et al (1995) Idiopathic photosensitive occipital lobe epilepsy. Epilepsia 36:883–891

Inoue Y, Seino M, Kubota H, Yamakaku K, Tanaka M, Yagi K (1994) Epilepsy with praxis-induced seizures. In: Wolf P (ed) Epileptic seizures and syndromes. John Libbey, London, pp 81–91

Ishida S, Yamashita Y, Matsuishi T, Ohshima M, Ohshima H, Kato H et al (1998) Photosensitive seizures provoked while viewing "Pocket Monsters," a made-for-television animation program in Japan. Epilepsia 39:1340–1344

Italiano D, Ferlazzo E, Gasparini S, Spina E, Mondello S, Labate A, Gambardella A, Aguglia U (2014) Generalized versus partial reflex seizures: a review. Seizure 23:512–520

Kartsounis LD (1988) Comprehension as the effective trigger in a case of primary reading epilepsy. J Neurol Neurosurg Psychiatry 51:128–130

Kasteleijn-Nolst Trenité DG (1989) Photosensitivity in epilepsy. Electrophysiological and clinical correlates. Acta Neurol Scand Suppl 125:1–149

Kasteleijn-Nolst Trenité DG (2012) Provoked and reflex seizures: surprising or common? Epilepsia 53:105–113

Koutroumanidis M, Koepp MJ, Richardson MP, Camfield C, Agathonikou A, Ried S, Papadimitriou A, Plant GT, Duncan JS, Panayiotopoulos CP (1998) The variants of reading epilepsy. A clinical and video-EEG study of 17 patients with reading-induced seizures. Brain 121:1409–1427

Matsuoka H, Takahashi T, Sasaki M, Matsumoto K, Yoshida S, Numachi Y et al (2000) Neuropsychological EEG activation in patients with epilepsy. Brain 123:318–330

Ramani V (1998) Reading epilepsy. In: Zifkin BG, Andermann F, Beaumanoir A, Rowan AJ (eds) Reflex epilepsies and reflex seizures. Advances in neurology, vol 75. Lippincott-Raven, Philadelphia, pp 241–262

Salek-Haddadi A, Mayer T, Hamandi K, Symms M, Josephs O, Fluegel D et al (2009) Imaging seizure activity: a combined EEG/EMG-fMRI study in reading epilepsy. Epilepsia 50:256–264

Sevgi EB, Saygi S, Ciger A (2007) Eye closure sensitivity and epileptic syndromes: a retrospective study of 26 adult cases. Seizure 16:17–21

Stephani U, Tauer U, Koeleman B, Pinto D, Neubauer BA, Lindhout D (2004) Genetics of photosensitivity (photoparoxysmal response): a review. Epilepsia 45:19–23

Striano S, Capovilla G, Sofia V, Romeo A, Rubboli G, Striano P, Trenité DK (2009) Eyelid myoclonia with absences (Jeavons syndrome): a well-defined idiopathic generalized epilepsy syndrome or a spectrum of photosensitive conditions? Epilepsia 50(Suppl 5):15–19

Tassinari CA, Rubboli G, Rizzi R, Gardella E, Michelucci R (1998) Self-induction of visually induced seizures. Adv Neurol 75:179–192

Taylor I, Berkovic SF, Scheffer IE (2013) Genetics of epilepsy syndromes in families with photosensitivity. Neurology 80:1322–1329

Tobimatsu S, Zhang YM, Tomoda Y, Mitsudome A, Kato M (1999) Chromatic sensitive epilepsy: a variant of photosensitive epilepsy. Ann Neurol 45:790–793

Waltz S, Stephani U (2000) Inheritance of photosensitivity. Neuropediatrics 31:82–85

Wilkins AJ, Zifkin B, Andermann F, McGovern E (1982) Seizures induced by thinking. Ann Neurol 11:608–612

Wolf P (1992) Reading epilepsy. In: Roger J, Bureau M, Dravet C, Dreifuss FE, Perret A, Wolf P (eds) Epileptic syndromes in infancy, childhood and adolescence, 2nd edn. John Libbey, London, pp 281–298

Wolf P, Berkovic S, Genton P, Binnie C, Anderson VE, Draguhn A (1994) Regional manifestation of idiopathic epilepsy. In: Wolf P (ed) Epileptic seizures and syndromes. John Libbey, London, pp 265–281

Current Status and Future Prospective of Neuroimaging for Epilepsy

F. Caranci, F. D'Arco, A. D'Amico, C. Russo, F. Briganti, M. Quarantelli and E. Tedeschi

Abstract Although the diagnosis of epilepsy remains mainly clinical, Magnetic Resonance Imaging (MRI) plays a crucial role in the detection of lesions that can cause epilepsy, with high impact on the diagnostic work-up as well as on therapeutic planning. Morphologic MR imaging is still the main technique for identifying lesions responsible for the epilepsy, providing images with high spatial resolution, excellent soft-tissue contrast, and multiplanar view. Quantitative MR image analysis (segmentation, voxel-based morhometry), based on 3D T1-weighted images, offers an objective means of analyzing MR images thereby improving the capability of detecting subtle lesions, often interpreted as negative by qualitative assessment of the morphologic MR imaging. Diffusion tensor imaging allows the quantification of water molecules diffusion and characterizes the degree and direction of anisotropy. Areas of abnormal diffusion, responsible for epilepsy, may be related to occult dysgenesis, or to acquired damage, resulting in neuronal loss, gliosis, and extracellular space expansion; these changes often result in reduced anisotropy and in an increase in mean diffusivity. Magnetic resonance spectroscopy provides information about the biochemical environment of the brain, thereby helping in lateralizing the epilepsy focus. Functional MR imaging is used for lateralizing language functions, and also for surgical planning predicting functional deficits following epilepsy surgery. The interpretation of MR data should always be done by a neuroradiologist expert in the field of epilepsy imaging, trying to correlate the images with clinical and electrophysiological data.

F. Caranci (✉) · F. D'Arco · A. D'Amico · C. Russo · F. Briganti · E. Tedeschi
Department of Advanced Biomedical Sciences, Neuroradiology Unit,
Federico II University, Naples, Italy
e-mail: ferdinandocaranci@libero.it

M. Quarantelli
Institute of Biostructures and Bioimaging, National Research Council, Naples, Italy

© Springer International Publishing Switzerland 2015
P. Striano (ed.), *Epilepsy Towards the Next Decade*,
Contemporary Clinical Neuroscience, DOI 10.1007/978-3-319-12283-0_7

Introduction

Modern neuroimaging techniques have had a major impact on our understanding of epilepsy. They provide high degree anatomical resolution and functional/metabolic information about the epileptic lesion, contributing to the proper classification of several epileptic disorders.

Magnetic Resonance Imaging (MRI) is a powerful tool in identifying epileptogenic tumors (gangliogliomas, DNETs, hypothalamic hamartoma, pleomorphic xanthoastrocytoma, etc.), vascular malformation (cavernous hemangiomas), severe developmental causes of intractable epilepsy (hemimegalencephaly, schizencephaly) and other syndromes that can lead to intractable seizures (neurocutaneous syndromes, such as Sturge-Weber syndrome and tuberous sclerosis). By using modern MR systems, that provide high image resolution and multiparametric reconstructions, it is often possible to recognize also small anatomical substrates responsible for epilepsy such as focal cortical malformations (focal cortical dysplasia, heterotopia, polymicrogyria) and hippocampal sclerosis. These are the most common causes of epilepsy, but they can be depicted only with a dedicated MR protocol; the sensitivity of MR imaging for such small structural abnormalities also depends on the experience of the interpreting physician (Widjaja and Raybaud 2008).

However, MRI is not always able to detect structural abnormalities in patients with seizures. Considering large case series of patients with intractable epilepsy, MRI showed a sensitivity from 82 to 86 % in identifying the epileptogenic substrate (Scott et al. 1999, Berg et al. 2000), while in children with a new diagnosis of epilepsy MRI detected epileptogenic substrates in only 13 % of cases (Bronen et al. 1996).

In recent years, new advanced MR imaging such as diffusion tensor imaging (DTI), magnetic resonance spectroscopy (MRS), quantitative MR imaging, functional MR imaging (fMRI), together with nuclear medicine procedures (positron emission tomography—PET and single-photon emission computed tomography-SPECT) have increased the sensitivity in the diagnosis of epileptic anatomical substrates and the knowledge of the mechanisms of seizures; moreover, in pre-surgical evaluation of epilepsy, the combined use of multiple imaging modalities for precise localization of the epileptogenic focus provides a better planning and ensures better surgical outcome.

The aim of this chapter is to provide an overview of structural and advanced imaging of epilepsy focusing on the best protocol tailored to the clinical diagnosis; in the second part, the MR appearance of the most important causes of epilepsy will be schematically described.

Imaging Modalities

Computed Tomography

Computed tomography (CT) uses ionizing radiation and can generate excellent hard tissue imaging contrast with moderately good soft tissue resolution. The advantages

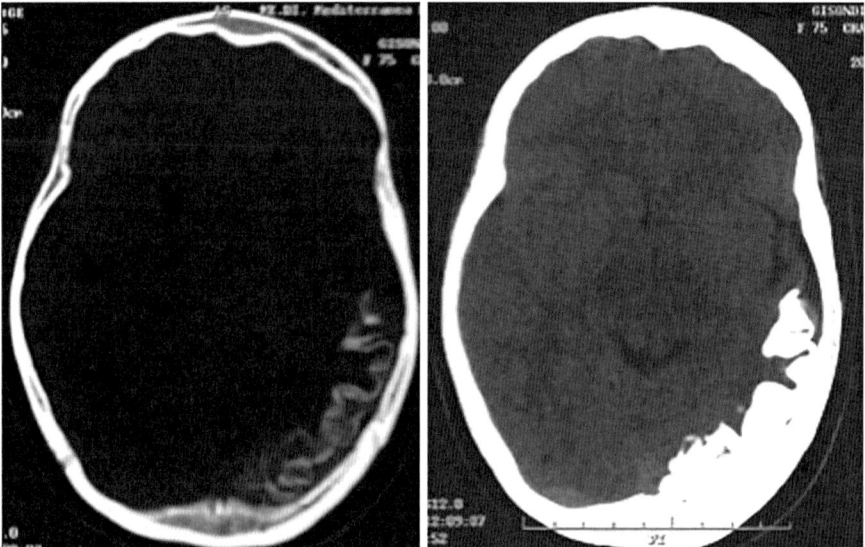

Fig. 1 CT scan of a patient with Sturge-Weber syndrome. Cortical and subcortical "*tram-track*" calcifications in the *left* posterior hemisphere are evident

of CT are wide accessibility, low cost, high speed (last CT generation can obtain a complete brain scan in few seconds) and thus it is considered a reliable brain imaging modality for most patients especially in emergency. CT can easily identify hemorrhage, infarctions, gross malformations, ventricular system pathologies, and lesions with underlying calcification (Trishit Roy and Alak 2011).

The sensitivity of CT in patients with epilepsy has not been found to be higher than 30% in unselected patient populations (Hankey et al. 1989; Wyllie et al. 1989; Gastaut and Gastaut 1979). This is due to the low resolution of CT for detecting mesial temporal lobe sclerosis (the most common epileptogenic substrates in adults) and temporal epileptogenic tumors (such as gangliogliomas or DNETs).

The use of CT for patients with epilepsy has been greatly diminished by the availability of MRI. Even if sometimes CT scan is used in neonates and infants following a pathological ultrasound (Hankey et al. 1989; Wyllie et al. 1989), it is always better to use, if available, MRI (under sedation) to better characterize the brain abnormalities.

Moreover, the role of CT in the diagnosis of tuberous sclerosis, Sturge-Weber syndrome (Fig. 1), and other pathologies with intracranial calcifications remains complementary, because MRI provides more information (noncalcified tubers).

CT can be still considered the technique of choice in the postoperative evaluation of patients undergoing surgery for uncontrolled seizures, because it can rapidly detect early complications of surgery such as hemorrhage, hydrocephalus, and major structural changes (Trishit Roy and Alak 2011).

Table 1 Epilepsy MRI protocol.

GE 3D T1	Isotropic voxel < 1 mm
Axial FLAIR & T2	2–4 mm slice thickness
Coronal FLAIR/HR T2	In temporal lobe epilepsy: coronal sections perpendicular to the long axis of hippocampus, axial sections parallel to the long axis of hippocampus
3D FLAIR	Suspect of cortical dysplasia
Axial or coronal GE T2*/axial SWI	Suspect of cavernous hemangioma, occult calcified lesions
DWI/ADC	Water restriction: cytotoxic edema
Contrast-enhanced T1	Brain tumors, Sturge-Weber syndrome

Structural Magnetic Resonance Imaging

MRI is the imaging procedure of choice in the investigation of patients with epilepsy. The advantages of MRI include the use of nonionizing radiation, high sensitivity and higher specificity than CT, multiplanar imaging capability, improved contrast of soft tissue, and high anatomical resolution. The sensitivity of MRI in detecting abnormalities in patients with epilepsy is strictly dependent on the type of pathologic substrate of the epilepsy, on the MRI protocol used, and on the experience of the interpreting physician (Widjaja and Raybaud 2008). Clinical data and electroencephalographic (EEG) findings should always guide the interpretation of MR images.

An optimal MRI protocol in an epileptic patient should use sequences with the minimum slice thickness, covering the whole brain and providing a high contrast resolution (that is the ability to distinguish between differences in intensity in an image) (Opplet 2006).

The use of 3 Tesla MR scanners, providing a better image resolution, improves the evaluation of patients with focal epilepsy when compared to 1.5 Tesla scanners (Mueller et al. 2011), with an important impact in the correct presurgical identification of the epilectic focus and thus in the treatment.

A dedicated MRI protocol should include (De Cocker et al. 2012, Table 1):

1. **3D T1 images with ≤1.5 mm slice thickness,** useful in assessing cortical thickness and gray/white matter interface (Fig. 2) and reformattable in any orthogonal or nonorthogonal planes (Widjaja and Raybaud 2008). In order to better evaluate the cortical thickening in case of focal cortical dysplasia (FCD), some authors suggest to add curvilinear reformatting using 3D data (Widjaja and Raybaud 2008; Montenegro et al. 2002). The complexity of the brain gyration is not perfectly studied using the orthogonal planes only, since the plan of analysis can be oblique to some gyri, thereby leading to apparent cortical thickening and thus to a false-positive diagnosis of FCD. In these cases, a curvilinear reconstruction can be useful to clarify the nature of a suspect area of cortical malformation detected on orthogonal planes.

Fig. 2 3D T1 weighted image reformatted on coronal plane, with excellent *gray/white* matter contrast

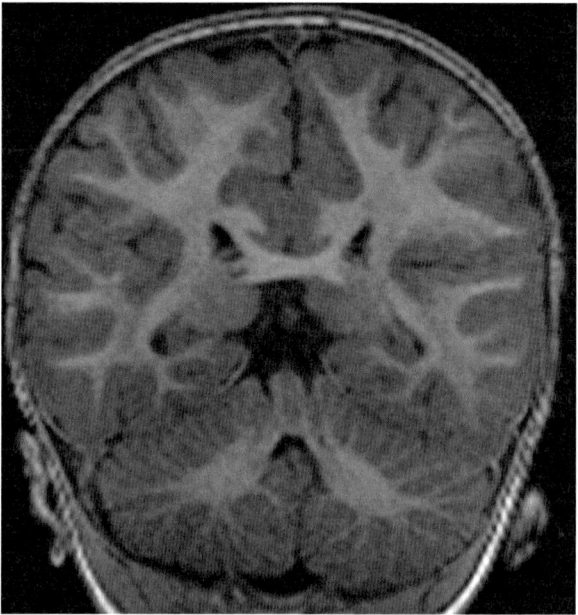

In <2 year-old children, 3D T2 w images are preferred to better evaluate the white matter because of the still incomplete myelination;
2. **T2 and FLAIR images** in axial and coronal planes, necessary to detect subtle cortical and subcotical hyperintensities (focal cortical dysplasia, small tumors, gliosis, hippocampal sclerosis); it is important to note that the best way to detect the presence of hippocampal sclerosis is to orientate the axial T2 and FLAIR sections parallel to the hippocampal axis, and the coronal slices perpendicular to it;
3. **T2* gradient echo or susceptibility weighted images (SWI) images,** helpful for identifying hemoglobin breakdown products as in post-traumatic changes and cavernomas, or when searching calcifications in tuberous sclerosis, Sturge-Weber syndrome and gangliogliomas;
4. **Contrast-enhanced T1 w images** (at least in two orthogonal planes) in case of brain tumors or Sturge-Weber syndrome;
5. **Diffusion weighted imaging** for the detection of foci of diffusion restriction consistent with cytotoxic edema or increased cellularity.

Some authors have proposed the use of **Double Inversion Recovery pulse sequence (DIR)** in case of suspected hippocampal sclerosis (Li et al. 2011). This sequence provides two different inversion pulses, which attenuates the cerebrospinal fluid (CSF) as well as the whole white matter, thus achieving a superior delineation between gray and white matter (Wattjes et al. 2007); it is considered superior to conventional MR sequences in the evaluation of subtle intensity changes in hippocampal sclerosis (Li et al. 2011).

Fetal MRI is a relatively new technique that provides increased diagnostic accuracy in the evaluation of fetal brain. This technique allows to detect small brain malformations very early during pregnancy, even not detectable by ultrasonography (US): a case of abnormal cortex and heterotopic gray matter has been described in a 24 weeks old fetus (Iaccarino et al. 2009).

Quantitative MR Imaging

Pathological examination of surgical specimens obtained in patients with intractable focal epilepsy and in whom the qualitative assessment of structural MR images was unconclusive, has shown the presence of subtle lesions such as loss of neurons, gliosis and microdysgenesis (cortical dyslamination and cytoarchitectural abnormalities) (Widjaja and Raybaud 2008; Al Sufiani and Ang 2012).

Quantitative volumetric analysis of MR images, usually performed by 3D T1-weighted images, can increase the sensitivity of MR in detecting epileptogenic foci especially in case of mesial temporal lobe epilepsy (TLE).

Volume reduction in hippocampus and in extrahippocampal regions has been demonstrated in mesial TLE (Guimaraes et al. 2007): a recent meta-analysis of voxel-based morphometry (VBM) studies on unilateral refractory temporal lobe epilepsy (Widjaja et al. 2012) showed significant reductions in ipsilateral mesiotemporal structures and in bilateral thalamus in both refractory left TLE and refractory right TLE. Bilateral abnormalities of frontal lobe and right cingulate gyrus were also found in the refractory left TLE patients, whereas right insular atrophy was found in the refractory right TLE group. Thus, quantitative MR imaging can depict the presence and laterality of hippocampal atrophy in TLE with accuracy rates, that may exceed those achieved with visual inspection of clinical MR imaging studies. Quantitative MR imaging can therefore enhance standard visual analysis, providing a useful and viable means for translating volumetric analysis into clinical practice (Farid et al. 2012).

Volumetric abnormalities affecting the thalamus and the frontal lobe were also found in generalized epilepsy (Widjaja et al. 2012). Moreover, in children with new-onset seizures a significant reduction in cortical thickness has been reported (Widjaja et al. 2012), while no significant differences in hippocampal and thalamic volumes were observed, suggesting that structural changes in cortical gray matter may predispose the patients to seizures while other changes can be due to the cumulative effects of recurrent seizures.

Diffusion Tensor Imaging (DTI)

Diffusion MRI is a MR modality that allows the mapping of the diffusion process of molecules of water in human tissues.

In the brain, water diffusion is restricted by myelin, membranes and macromolecules; in the white matter, the diffusion is mainly parallel to the white matter tracts and minimally perpendicular to them. This can explain the concept of **anisotropy** that expresses the asymmetric diffusion of water molecules in three dimensions.

Diffusion Tensor Imaging (DTI) allows the calculation of the degree and of the direction of anisotropy (Widjaja and Raybaud 2008).

From a mathematical point of view, DTI is a model of an ellipsoid that has a principal long axis and two more small axes. These three axes are perpendicular to each other and cross at the center point of the ellipsoid: the axes in this setting are called **eigenvectors** and the measures of their lengths **eigenvalues**; eigenvectors are the directions of the diffusion while each eigenvalue represents the magnitude of diffusion along each axis.

From a practical point of view, some diffusion parameters can be calculated in order to obtain information about the microstructural features of brain tissue: (1) the mean diffusivity (MD), that provides an evaluation of the magnitude of the diffusion motion in a voxel region and is measured in square millimeters per seconds; (2) the fractional anysotropy (FA), that is a scalar value between 0 and 1 that describes the values of the anisotropy. A FA value of 0 represents absence of anisotropy (the diffusion is isotropic) as in a perfect sphere, while a values of 1 represents maximal anisotropic diffusion.

Most acquired lesions and malformations of cortical development cause microstructural changes in the brain (reduction in cell denisity, impairment of normal myelin architecture, expansion of extracellular space), leading to reduced FA and increased MD (Widjaja and Raybaud 2008).

Several studies based on DTI in epileptic patients showed abnormal FA and MD.

Reduced FA was found in the white matter both ipsilateral and contralateral to the seizure focus (Widjaja et al. 2014); reduced FA and increased MD was found adjacent to cortical malformations visible on MR (Eriksson et al. 2001) and also beyond the margins of malformations detectable on MR imaging (Widjaja and Raybaud 2008) 'consistent with the EEG changes extending beyond the boundaries of focal cortical dysplasia visible on imaging, allowing a better surgical outcome' or distant from visible malformation of cortical development (Dumas de la Roque et al. 2005). Finally, significantly decreased FA values were observed in the cerebellum of patients with generalized tonic-clonic seizures (Li et al. 2010). The areas of abnormal diffusion detected using DTI in epileptic patients are thought to be caused by damage of the brain microstructure (neuronal loss, extracellular space expansion and gliosis) related to dysgenesis, acquired injuries or secondary to repeated seizures. For these reasons, DTI is considered a powerful tool in studying the anatomical substrates of epilepsy as well as the microstructural changes related to seizures. However the complexity of the analysis of the DTI data and the time-consuming post-processing limit the use of DTI parameters in routinary MR studies.

Finally, it is possible to obtain virtual 3D maps of eloquent white matter tracts (such as cortico-spinal tracts, optic radiations and arcuate fasciculus) by using DTI raw data (diffusion tensor tractography, Fig. 3); this technique is very useful in preoperative evaluation of epileptogenic lesions since it allows to localize white matter tracts and to assess their spatial relationship to the lesions (Lee et al. 2013).

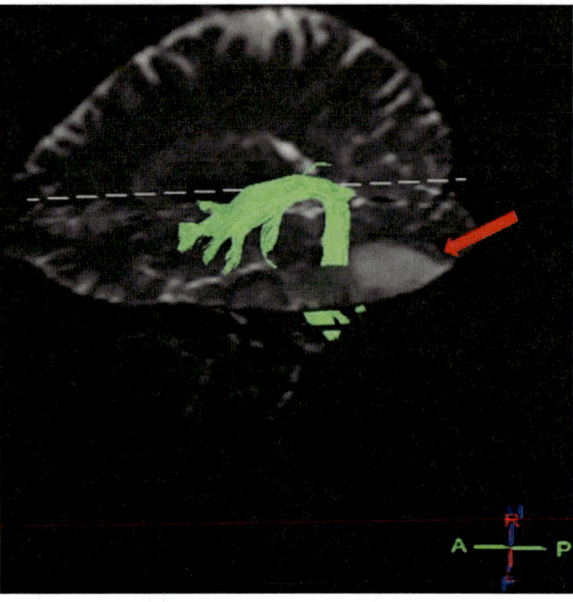

Fig. 3 Diffusion tensor tractography. 3D reconstruction of the left arcuate fasciculus (*green*) and its relationship with an epileptogenic tumor of the posterior temporal lobe (*red arrow*)

Magnetic Resonance Spectroscopy (MRS)

Magnetic resonance spectroscopy (MRS) is an analytical technique capable of evaluating the biochemical environment of the brain, and used to complement structural MRI in the characterization of tissues.

In clinical practice MRI uses signals from hydrogen protons (H^1) to determine the relative concentrations of target brain metabolites (Fig. 4). The most important brain metabolites assessed by MRS are: N-acetylaspartate (NAA), choline (Cho), creatine (Crea), lipids (Lip), lactate (Lac), glutamate-glutamine (Glx), Alanine (Ala) and myoinositol (Myo).

NAA is normally present in neurons and in axons and thus reflects the number of functioning neurons. Decrease in NAA levels indicates neuronal tissue loss or damage and axonal integrity loss.

Cho is a major component of cell membranes and the concentration of this metabolite provides information about cell density, membrane turnover and degree of myelination. An increase in Cho concentration indicates increase in cell production and/or membrane breakdown (brain tumors, demyelination).

Crea is a marker of cellular metabolism; it can be reduced in some pathological conditions such as lack of blood supply, but it can also be increased in response to cranio-cerebral trauma. However, because the concentration of Crea is relatively constant and is considered the most stable cerebral metabolite, it can be used as internal reference for calculating metabolite ratios.

Lac is not normally visible in MRS; when present, it indicates anaerobic metabolism (ischemia, hypoxia, necrotic brain tumors).

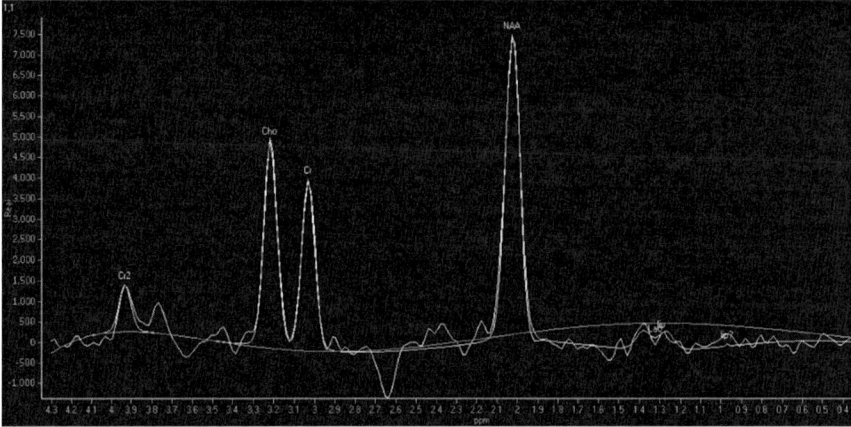

Fig. 4 Normal MRS of the brain (*single voxel, long TE*), placed in the white matter of the parietal lobe in a normal subject. Note the normal peak of NAA, higher than the Cho peak, and the absence of the pathological peaks of Lac and Lip

Lip are components of cell membranes normally not visualized on MRS. Lip peaks can be seen when there is cellular membrane breakdown or necrosis such as in metastases or primary malignant tumors.

Glx concentration reflects the glutamate concentration in glutamatergic neurons.

Ala is thought to play a role in citric acid cycle; increase of Ala concentration occurs in oxidative metabolism defects and in meningiomas.

Myo is a glial marker because is present almost exclusively in astrocytes. Myo may represent a marker of myelin degradation: elevated Myo is found in inflammation, gliosis, astrocytosis and in Alzheimer's disease.

Abnormalities on MR spectroscopy, such as reduced NAA and increased Cho levels, ipsilateral to EEG focus in patients with normal MRI findings (Widjaja and Raybaud 2008), have been used for lateralizing TLE: these results are consistent with loss of neurons, increased glial cells and neuronal dysfunction that are the pathological substrates of epilepsy. Another finding in TLE is high level of Glx, ipsilateral to the seizure onset, also in patient with "normal" MR (Savic et al. 2000).

For these reasons, MRS can play an adjunctive role in the presurgical evaluation of medically refractory TLE: by using reduced NAA/Cr ratios as a marker of neuronal loss, MRS agreed with the lateralization determined by EEG better than MR imaging volumetry alone (Caruso et al. 2013).

On the other hand, in extratemporal lobe epilepsy MRS was less specific in lateralizing the epilepsy focus (Widjaja and Raybaud 2008; Striano et al. 2008).

Abnormal metabolite concentrations have been found in patients with focal cortical dysplasia (FCD), correlating with the frequency of seizures, while normal NAA ratios have been found in patients with polymicrogyria and gray matter heterotopia (Widjaja and Raybaud 2008).

It is important to highlight that in case of FCD an abnormal spectrum is expected, because this is a malformation secondary to abnormal neuronal proliferation and differentiation.

Conversely, in case of polymicrogyria the neurons are more mature compared to FCD because this is a malformation related to abnormal cortical organization, while in gray matter heterotopias, a malformation due to abnormal migration, there is an increased number of neurons in an abnormal location. These features can explain the normal MRS findings in case of polymicrogyria and heterotopia.

Functional MR Imaging (fMRI)

Functional MRI (fMRI) is based on the quantification of the increased blood flow in areas of increased neuronal activity: this leads to a decreased oxyhemoglobin/deoxyhemoglobin ratio that is detectable using fMRI sequences (so called "BOLD" effect).

This imaging technique can be used for mapping neuronal activity during visual, language and sensorimotor activities in presurgical planning for epileptogenic lesions.

The understanding of the relationship between the areas of major neuronal activity and the epileptogenic lesion is useful to predict the post-surgical outcome and it is used for this purpose together with the location of main white matter tracts obtained using tractography.

In particular, a major role of fMRI is in lateralizing language functions in case of lesions that are close to language areas of the brain and in showing sensorimotor areas in case of lesions close to perirolandic cortex as this can predict possible language or sensorimotor deficits after surgery (Fig. 5).

In order to elicitate the BOLD effect numerous motor, sensory, verbal fluency and language comprehension tasks have been proposed and are widely used in clinical practice.

fMRI has also been used to probe the integrity of the functional connections among different brain regions, which can be detected when a subject is not performing an explicit task (resting-state fMRI, RS-fMRI) (Biswal 2012). Resting-state networks (RSNs) are the brain regions that exhibit spontaneous low-frequency synchronous fluctuations and represent brain functional connectivity. In particular, region-wise functional connectivity among the frontal, parietal, and temporal lobe, is normally present across the so-called Default-Mode Network (DMN, Fig. 6).

On the basis of PET, SPECT and EEG studies, epilepsy has been postulated to be a disorder of neuronal networks (Widjaja et al. 2013). Several studies have evaluated RSNs in patients with TLE epilepsy, generalized seizures, absence epilepsy and frontal lobe epilepsy revealing abnormal connectivity in different RSNs and negative correlation with epilepsy duration (Widjaja et al. 2013; Luo et al. 2011).

Therefore resting-state fMRI seems to be a promising technique to understand the pathological changes in neuronal connectivity in patients with different forms of epilepsy.

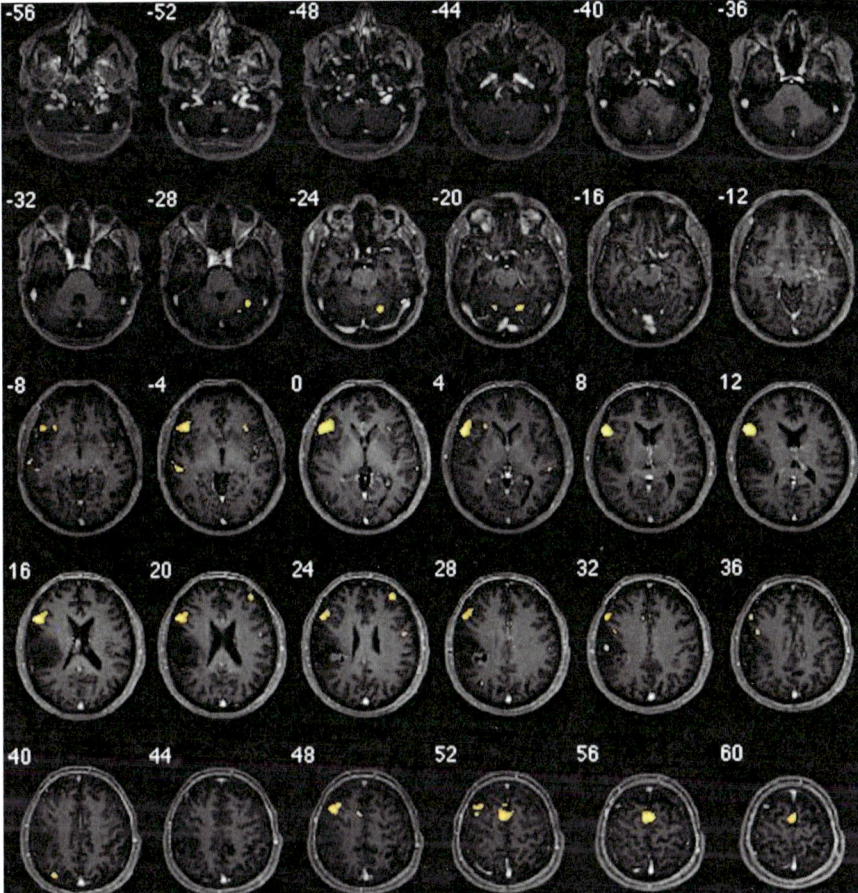

Fig. 5 fMRI (language task) after surgical excision of an epileptogenic tumor. Close relationship between the post-surgical cavity and the activation of the Broca's area, in the *left* inferior frontal gyrus

Positron Emission Tomography (PET)

The role of interictal PET with fluorodeoxyglucose (FDG-PET) is to determine the lateralization of the epileptic focus in the presurgical assessment of intractable epilepsy (Widjaja and Raybaud 2008). The spatial resolution of this technique is around 4 mm, therefore the images should be viewed side-by-side with MRI, or, even better, co-registered with MRI (Chugani et al. 1990).

The area of most severe glucose hypometabolism is tipically located in the site of ictal onset; however, the volume of hypometabolism is often widespread. This is the reason why FDG-PET hypometabolism can be used to lateralize the side of seizure onset and to determine the prognosis for complete seizure control only when it correlates with EEG recordings, MRI and/or clinical data (Widjaja and Raybaud 2008).

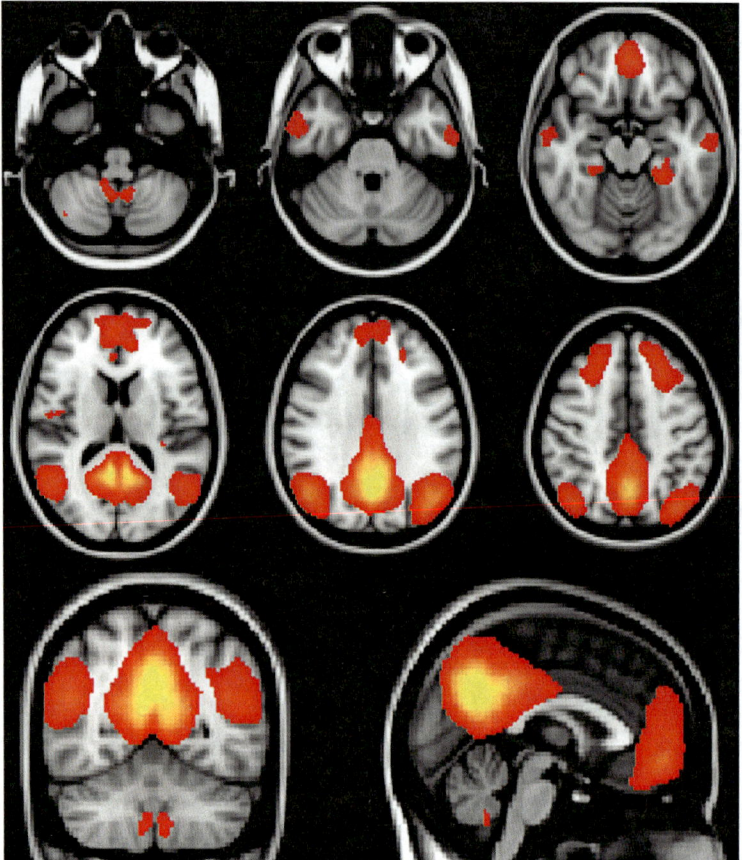

Fig. 6 Color map of the DMN as detected by Independent Component Analysis of the Resting-State fMRI data acquired at 3T in a group of 20 healthy subjects. The main components of the DMN in the inferior parietal regions, posterior cingulate and adjacent precuneus, as well as medial temporal and prefrontal cortices can be appreciated as *yellow-red* areas

In mesial TLE, interictal studies show hypometabolic areas in the epileptogenic regions in approximately 80% of patients; the changes, however, are more extensive than the structural and EEG abnormalities and may involve the lateral temporal lobe, ipsilateral frontal lobe, ipsilateral parietal lobe and basal ganglia (epileptic network) (Widjaja and Raybaud 2008; Pittau et al. 2014).

In extratemporal lobe epilepsy, interictal PET-FDG studies are less useful, especially if the MRI is normal and the scalp EEG is nonfocal. However, PET has been reported to be more sensitive in neonates and infants with focal seizures because of a possible developmental malformation. This is particularly the case in patients with infantile spasms and focal features on EEG. PET has also improved the understanding of the pathophysiology of infantile spasms by demonstrating activation of cortical regions, brainstem, and lenticular nuclei (Chugani et al. 1990).

The mechanisms underlying this hypometabolism in the epileptogenic cortex are still unresolved: it is thought that FDG-PET distribution reflects mainly synaptic activity, rather than cellular loss (Pittau et al. 2014).

In conclusion, the high sensitivity of MRI in detecting the anatomical substrates of epilepsy has diminished the role of FDG-PET in the presurgical investigation of such patients. Nevertheless, when MRI is normal, PET can be indicated to aid in localization or can be used as a guide for reviewing MRI in search for subtle overlooked cortical dysplasia or other epileptic substrates.

Single Photon Emission Computed Tomography (SPECT)

SPECT has been utilized in patients with epilepsy in the past decades, mainly using diffusible tracers for assessing brain perfusion (such as hexamethylpropylene amine oxime or ethyl cysteinate dimer). The main role of SPECT is to localize the epileptogenic zone when imaging and other non-invasive exams are unable to identify the site of seizure onset (Widjaja and Raybaud 2008).

Numerous studies using dynamic and static SPECT techniques in the ictal and interictal state have been published.

The ictal SPECT examination can identify focal hyperperfusion, while the interictal examination demonstrates hypoperfusion in the corresponding epileptogenic region. However, the diagnostic value of the interictal SPECT alone is poor, and a combined ictal/interictal SPECT study should be performed, possibly with subtraction of interictal images from the ictal ones and then coregistration with MR images, in order to achieve a better anatomical definition of the site of the seizure (Widjaja and Raybaud 2008).

Imaging of Epileptogenic Diseases

Mesial Temporal Sclerosis

The most useful MR sequences for detecting mesial temporal sclerosis (MST) are **coronal IR and/or 3D T1 weighted images,** that show a shrunken hippocampus associated with a widening of adjacent temporal horn and choroid fissure, and **coronal FLAIR (or T2)** that shows an increased hyperintensity and loss of the internal architecture of the involved hippocampus (Fig. 7a). Using 3D T1 weighted images it is also possible to detect the atrophy of the fornix and of the mamillary body ipsilateral to the affected hippocampus (Osborn 2013; Caranci et al. 2007; Iaccarino et al. 2009).

MRS typically shows a reduction of NAA (Fig. 7b) related to the neuronal loss both in hippocampal and extrahippocampal regions, while Cho and Cr are unchanged (Mueller et al. 2011).

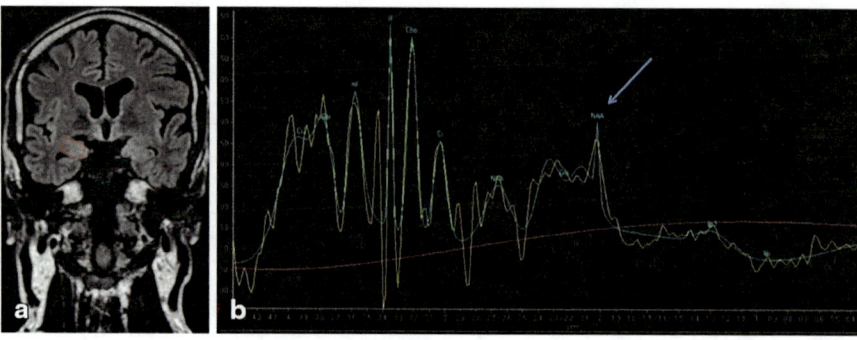

Fig. 7 Mesial temporal sclerosis. Coronal FLAIR image **a** shows mild atrophy and hyperintensity of the right hippocampus (*red box*); MRS **b** demonstrates a reduction of the NAA peak (*blue arrow*)

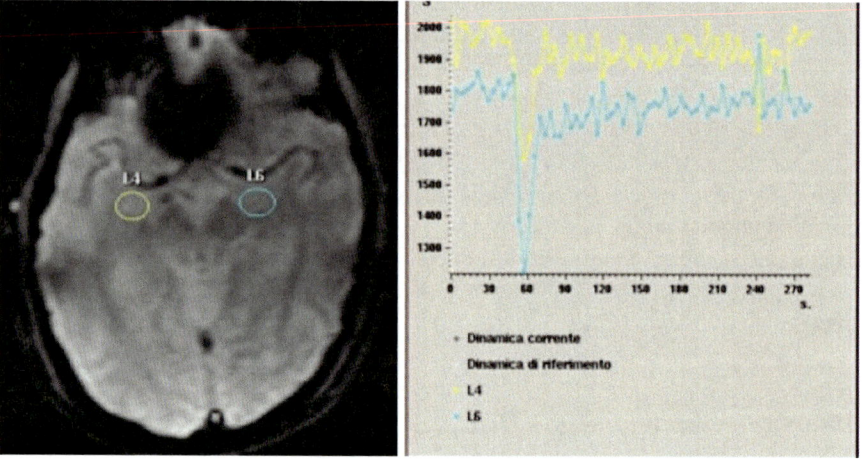

Fig. 8 Mesial temporal sclerosis. Quantitative PWI analysis in the same patient of Fig. 7 shows reduced perfusion in the right hippocampus (*yellow*) compared with the controlateral side (*blue*)

DWI shows increased diffusivity (high ADC values), while perfusion-weighted images show an interictal hypoperfusion of the affected side compared with the contralateral (Wolf et al. 2001; Fig. 8).

Volumetric analysis of hippocampi, parahippocampal giri and entorhinal cortex is used to quantify temporal lobe abnormalities in comparison with healthy controls; this is very useful in case of bilateral MST (Keller and Roberts 2007).

Finally, fMRI can be used in pre-operative planning to assess the language lateralizaton and the risk of memory disorders.

FDG-PET tipically shows hypometabolism in the affected temporal lobe in the interictal phase and is a very sensitive imaging modality for diagnosis of MTS (Osborn 2013).

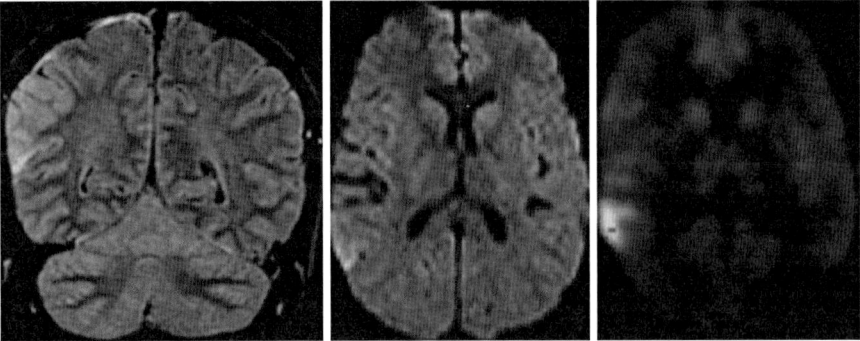

Fig. 9 Status Epilepticus. Coronal FLAIR image (**a**) showing hyperintensity and swelling of the temporo-occipital cortex; axial DWI (**b**) and FDG-PET (**c**) showing regional diffusion restriction and hypermetabolism, respectively

The differential diagnosis is with diseases that can cause seizures and FLAIR/T2 hyperintensity in the temporal lobe, including status epilepticus, characterized by gyral swelling instead of shrinking, and DWI/ADC restriction, and temporal lobe low grade glioma, that also causes mass effect instead of volume loss.

Status Epilepticus

MR findings of status epilepticus (SE) include T2/FLAIR hyperintensity and gyral swelling involving the gray matter (cortex, thalami, hippocampi) in a non-vascular distribution with sparing of both subcortical and deep white matter (Fig. 9).

Contrast enhanced T1 weighted images usually do not show any enhancement, only sometimes a variable enhancement of the gray matter; the acute phase is typically also characterized by restricted diffusion on DWI (low ADC values), hyperperfusion on PWI and hypermetabolism on FDG-PET.

Most of the MR findings in SE normalize in few days, however some patients can have permanent damage such as cortical laminar necrosis and brain atrophy (Osborn 2013; Milligan et al. 2009).

The major differential diagnosis is with cerebral ischemia, that typically shows a vascular distribution and involves both gray and white matter.

Focal Cortical Dysplasia

MR features of focal cortical dysplasia (FCD) depend on its type. The most common is FCD type II b, located most often at the bottom of a sulcus and histopathologically characterized by altered cortical layering, dysmorphic neurons and balloon cells. MRI of FCD II b shows: (1) focal cortical thickening (Fig. 10); (2) blurred interface between gray and white matter; (3) subcortical T2/FLAIR hyperintensity;

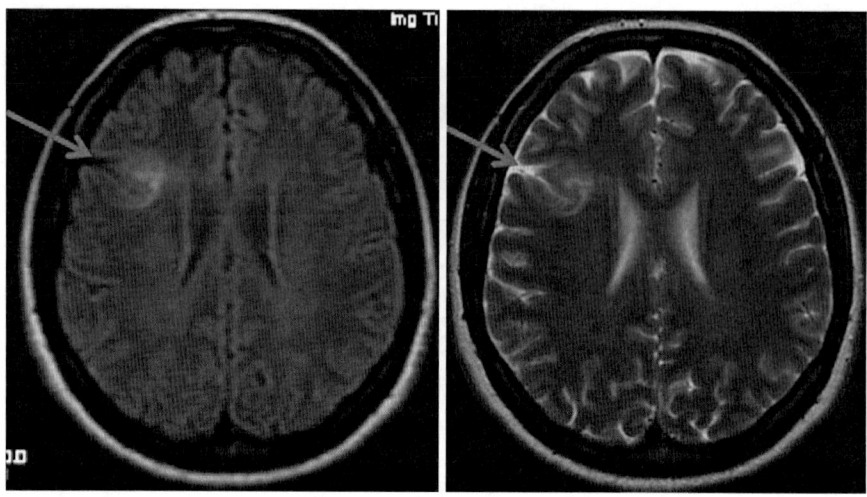

Fig. 10 Focal Cortical Dysplasia. Axial FLAIR **a** and T2 **b** images show thickening of the cortex at the bottom of a frontal sulcus with subcortical white matter hyperintensity (*arrows*)

(4) "transmantle sign" (not always present), i.e. a stripe of T2/FLAIR hyperintensity extending from the subcortical area to the margin of the ventricle due to a defect of neuronal migration (Fig. 11; Colombo et al. 2009).

CT scan is only able to detect the presence of calcifications associated with FDC (possible but rare).

The differential diagnosis of FCD includes cortically-based tumors associated with seizures such as dysembryoplastic neuroepithelial tumors (DNET), gangliogliomas and oligodendrogliomas.

MRS has been proposed as a reliable tool for differentiating FCD from a neoplasm, showing a reduced NAA/Cr ratio and no elevation in Cho/NAA (Widjaja and Raybaud 2008; Caruso et al. 2013) in the former.

Other advanced techniques can be used in the assessment of FDC: PWI shows increased perfusion in the area of FCD (Wintermark et al. 2013), while DTI can demonstrate microstructural abnormalities of the white matter extending beyond the main lesion seen on MRI (Widjaja and Raybaud 2008; Fonseca Vda et al. 2012).

Polymicrogyria and schizencephaly

Polymicrogyria (PMG) is a malformation due to abnormal late neuronal migration and cortical organization. PMG can be unilateral o bilateral and is located more often around the sylvian fissure (particularly in its posterior part), although it has been reported in any part of the cerebral cortex (Barkovich 2010).

MR findings of PMG include at least three possible aspects of the cortex (Barkovich 2010):

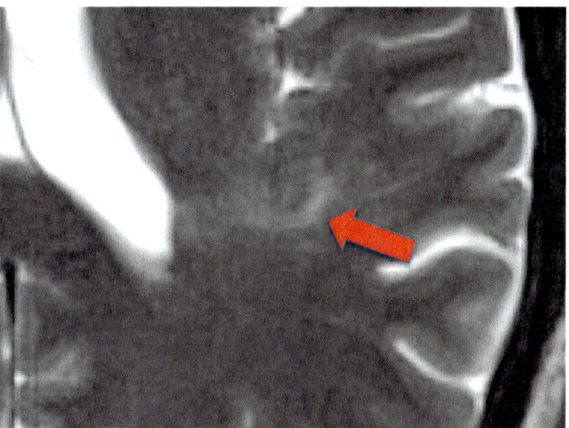

Fig. 11 "Transmantle" sign. T2w image shows a hyperintense stripe extending from the subcortical area to the margin of the ventricle due to a defect of neuronal migration (*red arrow*)

1. cortical surface with multiple small, delicate gyri;
2. thick and bumpy cortex;
3. "paradoxically" smooth cortex.

Moreover, an irregular gray/white matter interface is always present (Fig. 12a).

Sulci at the level of PMG are shallow or flat, and T2w and post-contrast T1w images can show enlarged pial veins overlying PMG (Barkovich 2007).

MRS shows relatively normal NAA ratios (Widjaja and Raybaud 2008). Using DTI it is possible to demonstrate abnormalities in the white matter underlying the polymicrogyric areas (Fig. 12b; Bonilha et al. 2007).

CT can show periventricular calcifications in case of PMG associated with congenital cytomegalovirus infection.

FDG-PET in the interictal phase can show hypometabolism in the PMG and surrounding areas (Fig. 12c; Barkovich 2007).

Schizencephaly is a disorder of neuronal migration and cortical organization, consisting in an abnormal hemispheric cleft that connects the ventricle with the subarachnoid space. Typical feature of schizencephaly is dysplastic gray matter lining its lips (PMG, pachigyria or normal-sized gyri).

Heterotopic Gray Matter

Heterotopic foci of gray matter are collections of nerve cells in abnormal locations secondary to the arrest of normal migration of neurons along the radial path between the ventricular walls (ependyma) and the subcortical regions (Bentivoglio et al. 2003).

Heterotopic gray matter may have different morphologies, but the most common are (Guerrini et al. 2006):

1. subependymal (periventricular nodular heterotopia, PVNH);
2. focal subcortical (FSH).

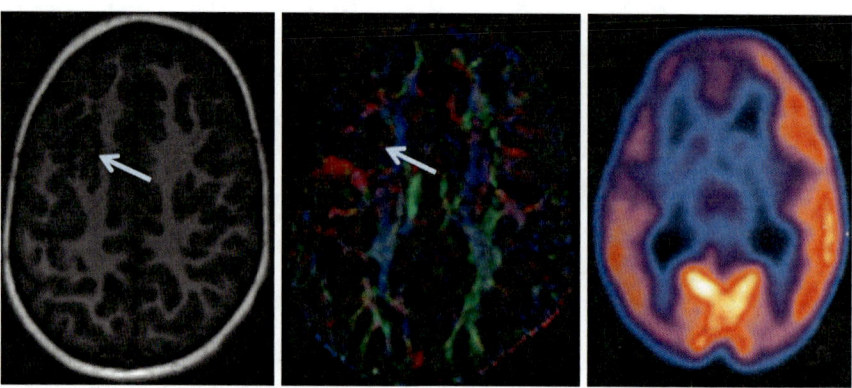

Fig. 12 Polymicrogyria. Axial 3D T1w (**a**) and DTI color map (**b**) show PMG in the left frontal lobe with associated abnormalities in the underlying white matter (*arrows*). FDG-PET (**c**) shows area of hypometabolism in the left frontal lobe

In both types, CT may rarely show dysplastic calcifications.

On MRI sequences heterotopic gray matter foci are always isointense to gray matter with distinct or blurred margins. They do not enhance after administration of contrast agents (Barkovich and Raybaud 2012).

PVNH are usually smooth, round or ovoid masses sometimes exophytic, protruding into the adjacent lateral ventricle (Wieck et al. 2005; Fig. 13).

Focal subcortical heterotopia are usually large and heterogeneous masses, which may appear as multinodular or composed of swirling, curvilinear bands of gray matter, extending from the cortex to the ventricles, and often containing blood vessels and CSF (Lim et al. 2005).

MRS may show elevated creatine and choline with normal NAA (Marsh et al. 1996), but the NAA/Cr ratio may be variable, ranging from normal to low (Li et al. 1998).

PWI, in some cases, can identify areas of hyperperfusion, and fMRI usually shows activated areas responding to stimuli just like the normal cortex (Lange et al. 2004), suggesting the participation of heterotopic cortex to integrated functional networks (Guerrini et al. 2004).

FDG-PET may show hypermetabolism or glucose uptake similar to the normal cortex (Barkovich 2007).

PVNH can be differentiated from subependymal hamartomas occurring in tuberous sclerosis, which are irregular in shape with the long axis perpendicular to the ventricular walls, are iso-hypointense on MRI sequences to white matter and enhance after intravenous infusion of contrast agents.

FSH must be differentiated from tumors, which enlarge the affected hemisphere, and are associated with normal overlying cortex and surrounding edema (Barkovich and Raybaud 2012).

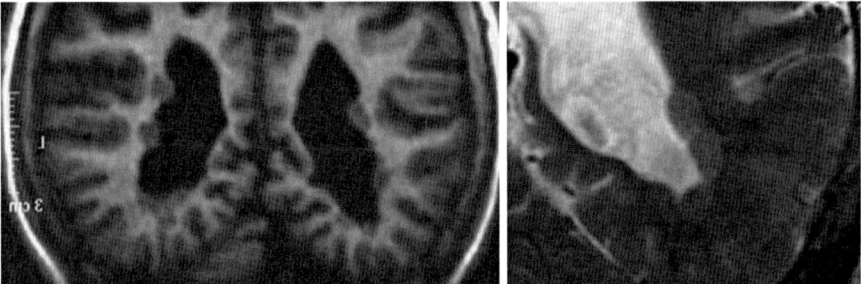

Fig. 13 T1 3D and enlarged T2 images show heterotopic gray matter nodules of subependymal heterotopia (PVNH)

Lissencephalies

The term lissencephaly means "smooth brain" and refers to the paucity of gyral and sulcal development on the surface of the brain. In lissencephaly, the cerebral cortex is abnormally thick, usually measuring 10–15 mm (Francis et al. 2006).

Lissencephalies include:

1. classical lissencephaly (type I lissencephaly);
2. subcortical band heterotopia (SBH).

They are part of a malformative complex, the agyria-pachygyria-band spectrum.

Classic lissencephaly may be complete or incomplete with absent (agyria) or decreased (pachygyria) surface convolutions, respectively (Sicca et al. 2003).

CT scan rarely shows small midline septal calcifications (Barkovich 2007).

MRI findings include (Sicca et al. 2003; Barkovich 2007):

1. smooth brain surface;
2. diminished white matter;
3. shallow and vertically oriented Sylvian fissure, configuring a figure-of-eight appearance of the cerebrum on axial images;
4. the cell-sparse zone, which is a zone of white matter separating a thin outer cortical layer from a thick deeper cortical layer.

Additional abnormalities include: hypoplasia of the corticospinal tracts, heterotopia of the inferior olives, and mild dysplasia of the cerebellar cortex.

On T1w and T2w sequences it is possible to distinguish gray and white matter layers and to identify the cell-sparse zone as areas appearing hypointense in T1w hyperintense in T2w (Barkovich 2007).

SBH is the mildest form of classic lissencephaly. In SBH, the cerebral convolutions appear either normal or mildly broad, but just beneath the cortical ribbon a thin band of white matter separates the cortex from the bands of gray matter (Fig. 14). Band heterotopia may be complete or partial. In T2w sequences, it is sometimes possible to identify foci of hyperintensity in white matter, associated with a poor motor outcome (Gleeson et al. 2000; Mai et al. 2003).

Fig. 14 Coronal T1 weigthed shows subcortical band heterotopia.

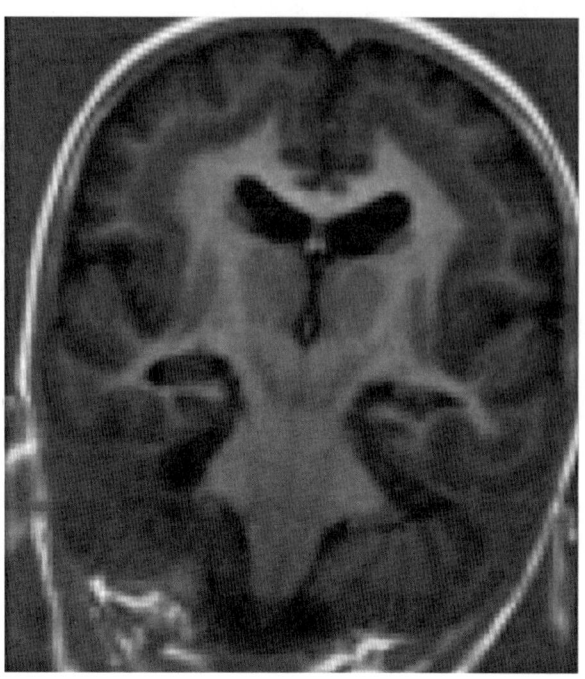

MRS may identify decreased levels of NAA in the affected cortex (Barkovich 2007). FDG-PET may show glucose uptake similar to or greater than normal cortex (De Volder et al. 1994).

Differential diagnosis include lissencephaly variants (without cell-sparse zone) and bilateral and diffuse subependymal heterotopia.

Tuberous Sclerosis

Tuberous sclerosis, or tuberous sclerosis complex (TSC), belongs to the group of the phakomatoses and involves primarily the central nervous system, the skin and the kidney (Luat et al. 2007).

The characteristic brain lesions are (Jahodova et al. 2014):

1. cortical tubers;
2. subependymal nodules (SEN);
3. subependymal giant cell astrocytomas (SEGAs);
4. white matter abnormalities.

Cortical tubers are the most characteristic lesions of tuberous sclerosis and they are most commonly supratentorial, appearing as pyramidal-shaped gyral expansions.

On CT scans, cortical tubers in neonates and infants appear hyperdense, while in children and adults they may be difficult to identify if they are not calcified (Kalantari and Salamon 2008).

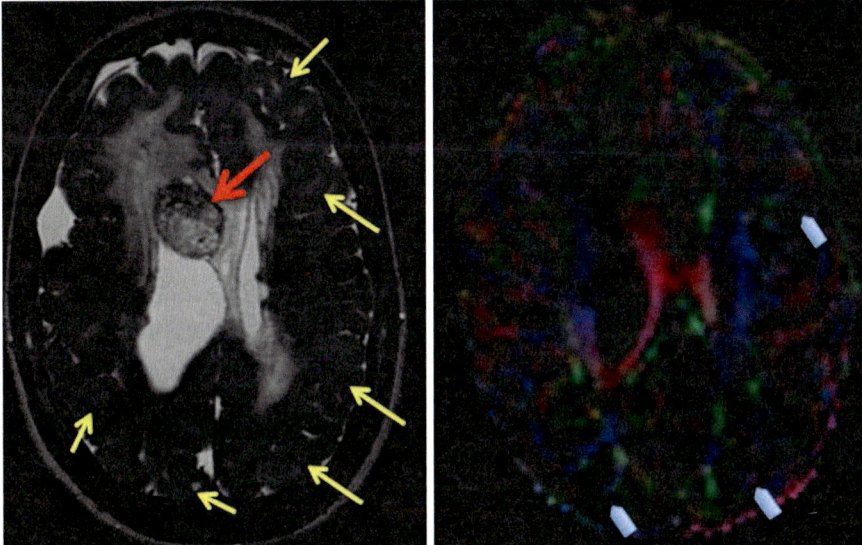

Fig. 15 Axial T2 weigthed and DTI FA color map show subependymal giant cell astrocytoma (SEGA) in the right foramen of Monro (*red arrow*) surrounded by oedema, and multiple cortical tubers (*yellow arrows*). In the DTI FA color map, decreased anisotropy of the tubers (*arrowheads*), compared with normal white matter, can be appreciated

The MRI appearance also changes with ages, as white matter myelinates, and cortical tubers may have different imaging patterns called "gyral core" and "sulcal island". In neonates they are hyperintense in T1w and hypointense in T2w sequences compared to unmyelinated white matter. In older infants and children, they become hypointense in T1w and hyperintense in T2w and FLAIR sequences, the latter being the most sensitive sequence for the detection of tubers in children and in adults (Kalantari and Salamon 2008).

When cortical tubers calcify, they may appear hyperintense on T1w images. Generally, they do not enhance after administration of contrast medium (Jahodova et al. 2014; Kalantari and Salamon 2008).

On DWI and DTI, they have increased diffusivity and decreased anisotropy compared to normal white matter (Peters et al. 2013 Fig. 15).

MRS has been proposed as a reliable tool for the differential diagnosis of cortical tubers versus neoplasms, especially when cortical tubers are solitary. In cortical tubers, MRS shows normal to slightly elevated levels of choline and slightly diminished levels of NAA, while neoplasms have marked elevation of choline and marked diminution of NAA, with an increased peak of myo-inositol (Kalantari and Salamon 2008; Jahodova et al. 2014). However, low-grade tumors may have MRS findings similar to tubers. In these cases, PWI may be helpful: most cortical tubers are generally hypoperfused compared to gray matter (Jahodova et al. 2014).

SEN are the most common brain hamartomas in tuberous sclerosis. They are more frequently located along the ventricular surface of the caudate nucleus.

On CT scans they may be difficult to identify in infants, but become more easily and progressively detected as they calcify (Barkovich 2007).

On MRI, their appearance changes as the surrounding white matter myelinates. In neonates they are hyperintense on T1w and hypointense on T2w sequences, while they become more isointense to gray matter with age and are better identified on T1w images. If they calcify, they may appear hypointense on T2*w GE or susceptibility-weighted images. After intravenous administration of contrast material, SEN show variable enhancement; they also have increased diffusivity and reduced fractional anisotropy compared to surrounding white matter (Luat et al. 2007).

SEGA is an enlarged subependymal nodule, usually located near the foramen of Monro. Their incidence in TSC is 5–10% and they may result in a clinical presentation of hydrocephalus, as they tend to enlarge (Barkovich 2007).

On imaging studies, SEGAs are identified by the demonstration of tumor growth on serial studies or by the development of hydrocephalus associated with tumor near the foramen of Monro. They tend to grow into the ventricle and rarely infiltrate the adjacent nervous tissue: in this case, they frequently have degenerated into high-grade neoplasms (Barkovich 2007).

The major differential diagnosis of TSC include subependymal heterotopia, TORCH and Taylor type cortical dysplasia.

Sturge-Weber Syndrome

Sturge-Weber Syndrome (SWS) is a sporadic phakomatosis characterized by angiomatosis involving the face (port-wine stain), the choroid of the eye and the leptomeninges. Venous occlusion and ischemia lead to angiomatosis with cortical calcium deposition and atrophy (Barkovich 2007).

CT findings include gyral and subcortical calcifications, occurring exclusively in brain areas near the angioma (Barkovich 2007).

MRI in patients with Sturge-Weber can show (Juhász et al. 2007b):

1. atrophy;
2. high signal on T2WI due to gliosis;
3. low signal in areas with calcifications;
4. leptomeningeal enhancement (Fig. 16).

On MRI, in neonatal cases, the white matter appears hypointense on T2w images in the affected hemisphere because of "accelerated" myelin maturation. The affected hemisphere becomes progressively atrophic with hyperpneumatization of the paranasal sinuses, thick diploe, and enlarged ipsilateral choroid plexus (Juhász et al. 2007b).

Calcifications may be identified on SWI or T2*w GE images as a thin ribbon of low signal intensity adjacent to the cerebral cortex in the affected areas. SWI sequences appears superior also in assessing the enlarged transmedullary veins, the pathological periventricular veins and the cortical gyriform abnormalities (Wu et al. 2011).

Current Status and Future Prospective of Neuroimaging for Epilepsy 131

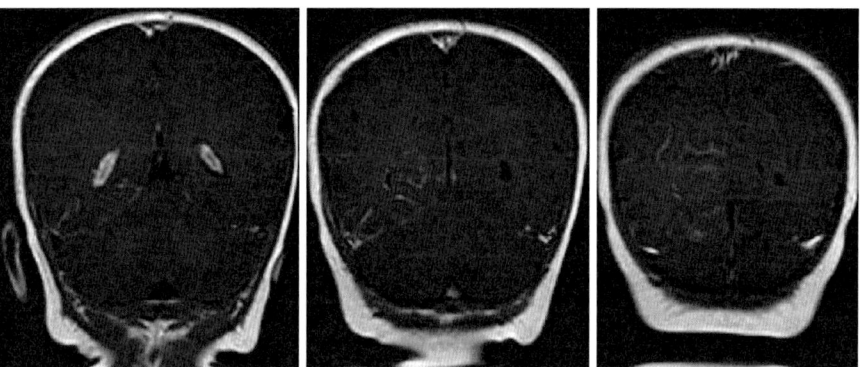

Fig. 16 Coronal contrast-enhanced T1 of leptomeningeal angiomatosis involving the right occipital lobe, a feature consistent with Sturge-Weber Syndrome

After gadolinium administration, subcortical, leptomeningeal, and choroidal serpentine T1 hyperintensity is the rule, due to the enhancement of the pial angiomatous malformation mainly localized in the occipital lobes (Barkovich 2007; Juhász et al. 2007a).

PWI may be useful in detecting cerebral hypoperfusion of the brain underlying the enhancing pial angioma, due to impaired venous drainage (Miao et al. 2011).

MRS of the affected brain regions reveals reduced NAA, elevated choline and lactate (Batista et al. 2008).

FDG-PET shows hypermetabolism in the affected area in the early stages with subsequent hypometabolism, and it may be useful in surgical planning of cortical resection for intractable seizures (Juhász et al. 2007a).

For the differential diagnosis, it is important to consider leptomeningeal enhancement and other neurocutaneous syndromes.

Hemimegalencephaly

Hemimegalencephaly is a rare disease characterized by hamartomatous growth of one cerebral hemisphere or part of it.

Associated findings include macrocephaly and somatic hemihypertrophy of the body.

On CT and MRI, part or all of a hemisphere may be affected (Flores-Sarnat 2002).

The most typical findings are (Kamiya et al. 2013):

1. enlarged hemisphere with ipsilateral ventriculomegaly (Fig. 17);
2. abnormal gyral pattern with broad gyri, shallow sulci, blurring of the cortical-white matter junction and cortical thickening;
3. thickened cortex with a wide spectrum of abnormalities, such as lissencephaly, pachygyria or polymicrogyria;

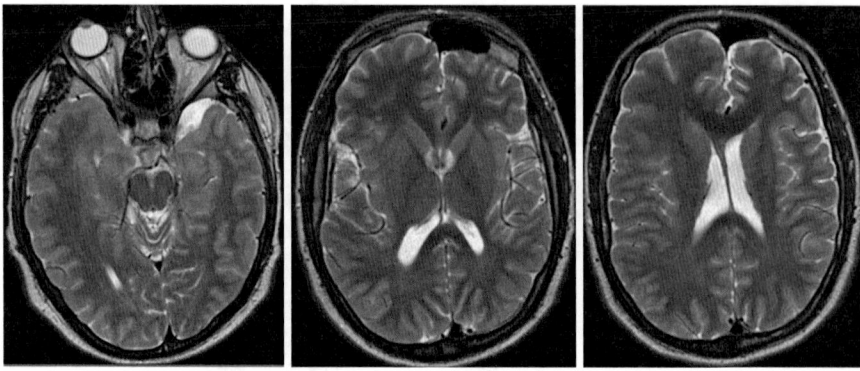

Fig. 17 Axial T2 weighted images showing right hemimegalencephaly

4. abnormal white matter, frequently hypodense on CT scans and heterogeneous on T2w MRI images, due to heterotopia and dysplastic neurons;
5. resultant midline shift.

In the late stage, the involved hemisphere may be atrophic due to constant seizure activity.

MRS shows reduction of NAA and glutamate peaks in the affected white matter, correlated to a diminished metabolic activity on FDG-PET (Flores-Sarnat 2002).

Ganglioglioma

Gangliogliomas are rare tumors, accounting for 3% of all pediatric brain neoplasms, and 6–7% of supratentorial tumors in children. Affected patients usually present in their second decade of life (Barkovich and Raybaud 2012).

Ganglioglioma typically involves the cortex of the cerebral lobes, especially the temporal lobe (85% of cases), and is the most common tumor associated with temporal lobe epilepsy (Barkovich and Raybaud 2012).

The most typical imaging findings are (Barkovich and Raybaud 2012):

1. presentation as a cyst with enhancing mural nodule (although it may be entirely solid);
2. calcifications in up to 50%.

On CT scan, gangliogliomas appear as hypodense, well circumscribed lesions with calcifications and little edema, tipically located within the cerebral cortex. Solid portions of the tumor may appear isodense, hypodense or mixed, with variable contrast enhancement. Erosion of the adjacent inner table of the calvarium may occur when the ganglioglioma is located peripherally (Gelabert-González et al. 2011).

On MRI, the ganglioglioma may have sharply or poorly defined margins. It may be solid, cystic, cystic with mural nodule or multi-cystic. It is usually hyperintense on T2w sequences and heterogeneous on T1w. The solid portion of the lesion may enhance (Barkovich 2012).

Differentiation from DNET and pleomorphic xanthoastrocytoma is difficult and calcifications are important distinguishing factors (Raz et al. 2012).

Dysembryoplastic Neuroepithelial Tumor (DNETs)

DNET are benign lesions of the cerebral cortex and are the cause of 20 % of the cases of medically refractory epilepsy in children and young adults (Ozlen et al. 2010).

The key findings on imaging are (Barkovich and Raybaud 2012):

1. swollen gyrus;
2. bubbly cystic appearance;
3. wedge-shaped and pointing toward the ventricle;
4. usually no or only little enhancement;
5. association with focal cortical dysplasia.

On CT studies, DNETs are quite well-demarcated, lobulated cortical lesions hypodense to white matter. Calcifications are present in one third of the cases and erosion of the inner table of the calvarium may be observed (Ozlen et al. 2010).

On MRI, DNET in typical cases present as a bubbly mass which expands the affected gyri. The bubbly cystic appearance is seen as small cyst-like intratumoral structures that are very hyperintense on T2w sequences and hypointense on T1w sequences, with a variable signal on FLAIR images (from hypointense to hyperintense). Diffusivity is elevated compared to the normal gray and white matter. Contrast enhancement is seen in 20–40 % of the cases and is usually patchy (Barkovich and Raybaud 2012).

MRS shows no significant difference of the metabolite ratios from normal cortex, while PWI reveals lower CBV than normal cortex (Ozlen et al. 2010; Barkovich and Raybaud 2012).

Furthermore, DNETs are metabolically inactive tumors with no significant glucose uptake on FDG-PET (Barkovich and Raybaud 2012).

Hypothalamic Hamartoma (HH)

Hypothalamic hamartoma is also known as diencephalic or tuber cinereum hamartoma.

It represents nonneoplastic congenital grey matter heterotopia in the region of tuber cinereum of the hypothalamus. It is seen in infants presenting with epilepsy (gelastic type) and precocious puberty (Mittal et al. 2013; Chung et al. 2012).

A clinical-topographical classification by Valdueza et al. (Valdueza et al. 1994) distinguished 4 types of HH:

1. Ia hamartomas, small lesions with a peduncolated attachment to the tuber cinereum, generally asymptomatic;
2. Ib hamartomas, small peduncolated masses attached to a mammillary body, usually associated to precocious puberty;

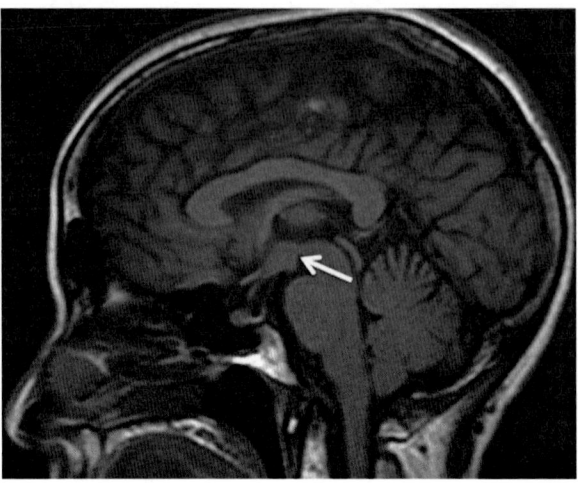

Fig. 18 Sagittal T1 weighted image shows a II A hypothalamic hamartoma (*arrow*)

3. IIa hamartomas, sessile masses (larger than 1.5 cm in diameter) attached to the floor of the third ventricle and mammillary bodies, typically presenting with gelastic seizures;
4. IIb hamartomas, large sessile masses (larger than 1.5 cm in diameter) distorting the walls and the floor of the third ventricle, associated with mental and behavioral disorders in addition to gelastic and mixed epilepsy.

CT studies are sensitive only in detecting large hamartomas, which appear as homogeneous, hypodense rounded masses, without contrast enhancement (Chung et al. 2012).

MRI represents the gold standard imaging modality. The MRI appearance is that of a well defined, round or ovoid mass lying within the hypothalamus (Fig. 18). HH are isointense on T1w sequences compared to gray matter and isointense to slightly hyperintense on T2w images. No enhancement is seen after contrast administration (Amstutz et al. 2006).

Reports in the literature suggest that on MRS the myo-inositol is higher and NAA lower than in the adjacent thalamus, probably due to astrogliosis (Martina et al. 2003).

References

Al Sufiani F, Ang LC (2012) Neuropathology of temporal lobe epilepsy. Epilepsy Res Treat. vol 2012, Article ID 624519

Amstutz DR, Coons SW, Kerrigan JF et al (2006) Hypothalamic hamartomas: correlation of MR imaging and spectroscopic findings with tumor glial content. AJNR Am J Neuroradiol 27(4)794–798

Barkovich AJ (2010) Current concepts of polymicrogyria. Neuroradiology 52:479–487

Barkovich AJ, Raybaud C (2012) Pediatric neuroimaging-Fifth edition. Lippincott Williams & Wilkins, Philadelphia

Barkovich AJ et al (2007) Diagnostic imaging—pediatric neuroradiology. Amyrsis

Batista CE, Chugani HT, Hu J et al (2008) Magnetic resonance spectroscopic imaging detects abnormalities in normal-appearing frontal lobe of patients with Sturge-Weber syndrome. J Neuroimaging 18(3):306–313

Bentivoglio M, Tassi L, Pech E et al (2003) Cortical development and focal cortical dysplasia. Epileptic Disord 5(Suppl 2):S27–34

Berg AT, Testa FM, Levy SR, Shinnar S (2000) Neuroimaging in children with newly diagnosed epilepsy: a community-based study. Pediatrics 106:527–532

Biswal BB (2012) Resting state fMRI: a personal history. Neuroimage 62(2):938–944

Bonilha L, Halford J, Rorden C et al (2007) Microstructural white matter abnormalities in nodular heterotopia with overlying polymicrogyria. Seizure 16(1):74–80

Bronen RA, Fulbright RK, Spencer DD et al (1996) Refractory epilepsy: comparison of MR imaging, CT, and histopathologic findings in 117 patients. Radiology 201:97–105

Caranci F, Bartiromo F, Cirillo L et al (2007) Thalamic changes in mesial temporal sclerosis: a limbic system pathology. Neuroradiol J 218–223

Caruso PA, Johnson J, Thibert R et al (2013) The use of magnetic resonance spectroscopy in the evaluation of epilepsy. Neuroimag Clin N Am 23:407–424

Chugani HT, Shields WD, Shewmon DA et al (1990) Infantile spasms: I. PET identifies focal cortical dysgenesis in cryptogenic cases for surgical treatment. Ann Neurol 7(4):406–413

Chung EM, Biko DM, Schroeder JW et al (2012) From the radiologic pathology archives: precocious puberty: radiologic-pathologic correlation. Radiographics 32(7):2071–2099

Colombo N, Salamon N, Raybaud C et al (2009) Imaging of malformations of cortical development. Epileptic Disord 11(3):194–205

De Cocker L, D'Arco F, Demaerel P, Smithuis R (2012) Role of MRI in epilepsy. The Radiology Assistant. www.radiologyassistant.nl/en. Accessed 1 Sept 2012

De Volder AG, Gadisseux JF, Michel CJ et al (1994) Brain glucose utilization in band heterotopia: synaptic activity of "double cortex". Pediatr Neurol 11:290–294

Dumas de la Roque A, Oppenheim C, Chassoux F et al (2005) Diffusion tensor imaging of partial intractable epilepsy. Eur Radiol 15:279–285

Eriksson SH, Rugg-Gunn FJ, Symms MR et al (2001) Diffusion tensor imaging in patients with epilepsy and malformations of cortical development. Brain 124:617–626

Farid N, Girard HM, Kemmotsu N et al (2012) Temporal lobe epilepsy: quantitative MR volumetry in detection of hippocampal atrophy. Radiology 26(2):542–550

Flores-Sarnat L (2002) Hemimegalencephaly: part 1. Genetic, clinical, and imaging aspects. J Child Neurol 17(5):373–384

Fonseca Vde C, Yasuda CL, Tedeschi GG et al (2012) White matter abnormalities in patients with focal cortical dysplasia revealed by diffusion tensor imaging analysis in a voxelwise approach. Front Neurol 121

Francis F, Meyer G, Fallet-Bianco C et al (2006) Human disorders of cortical development: from past to present. Eur J Neurosci 23:877–893

Gastaut H, Gastaut JL (1979) Computerized transverse axial tomography in epilepsy. Epilepsia 17:325–336

Gelabert-González M, Amo JM, Arcos Algaba A et al (2011) Intracranial gangliogliomas. A review of a series of 20 patients. Neurologia 26(7):405–415

Gleeson JG, Luo RF, Grant PE et al (2000) Genetic and neuroradiological heterogeneity of double cortex syndrome. Ann Neurol 47:265–269

Guerrini R, Marini C (2006) Genetic malformations of cortical development. Exp Brain Res 173(2):322–333

Guerrini R, Mei D, Sisodiya S et al (2004) Germline and mosaic mutations of FLN1 in men with periventricular heterotopia. Neurology 63:51–56

Guimaraes CA, Bonilha L, Franzon RC et al (2007) Distribution of regional gray matter abnormalities in a pediatric population with temporal lobe epilepsy and correlation with neuropsychological performance. Epilepsy Behav 11:558–566

Hankey G, Davies L, Gubbay SS (1989) Long term survival with early childhood intracerebral tumours. J Neurol Neurosurg Psychiatry 52:778–781

Iaccarino C, Tedeschi E, RapanÃ A et al (2009) Is the distance between mammillary bodies predictive of a thickened third ventricle floor? J Neurosurg 110(5):852–857

Jahodova A, Krsek P, Kyncl M et al (2014) Distinctive MRI features of the epileptogenic zone in children with tuberous sclerosis. Eur J Radiol 83(4):703–709

Juhász C, Batista CE, Chugani DC et al (2007a) Evolution of cortical metabolic abnormalities and their clinical correlates in Sturge-Weber syndrome. Eur J Paediatr Neurol Sep; 11(5):277–284

Juhász C, Haacke EM, Hu J et al (2007b) Multimodality imaging of cortical and white matter abnormalities in Sturge-Weber syndrome. AJNR Am J Neuroradiol 28(5):900–906

Kalantari BN, Salamon N (2008) Neuroimaging of tuberous sclerosis: spectrum of pathologic findings and frontiers in imaging. AJR Am J Roentgenol 190(5):W304–309

Kamiya K, Sato N, Saito Y et al (2013) Accelerated myelination along fiber tracts in patients with hemimegalencephaly. J Neuroradiol 30. pii: S0150-9861(13)00084-9. doi:10.1016/j.neurad.2013.08.005. (Epub ahead of print)

Keller SS, Roberts N (2007) Voxel-based morphometry of temporal lobe epilepsy: an introduction and review of literature. Epilepsia 49(5):741–757

Lange M, Winner B, Muller JL et al (2004) Functional imaging in periventricular nodule heterotopy caused by a new Filamin A mutation. Neurology 62:151–152

Lee MJ, Kim HD, Lee JS, Kim DS, Lee SK (2013) Usefulness of diffusion tensor tractography in pediatric epilepsy surgery. Yonsei Med J 54(1):21–27

Li LM, Cendes F, Bastos AC et al (1998) Neuronal metabolic dysfunction in patients with cortical developmental malforamtions. A proton magnetic resonance spectroscopic imaging study. Neurology 50:755–759

Li Y, Du H, Xie B et al (2010) Cerebellum abnormalities in idiopathic generalized epilepsy with generalized tonic-clonic seizures revealed by diffusion tensor imaging. PLoS One 5(12):e15219

Li Q, Zhang Q, Sun H, Zhang Y, Bai R (2011) Double inversion recovery magnetic resonance imaging at 3 T: diagnostic value in hippocampal sclerosis. J Comput Assist Tomogr 35(2):290–293

Lim CC, Yin H, Loh NK, Chua VG et al (2005) Malformations of cortical development: high-resolution MR and diffusion tensor imaging of fiber tracts at 3T. AJNR Am J Neuroradiol 26(1):61–64

Luat AF, Makki M, Chugani HT (2007) Neuroimaging in tuberous sclerosis complex. Curr Opin Neurol 20(2):142–150

Luo C, Li Q, Lai X et al (2011) Altered functional connectivity in default mode network in absence epilepsy: a resting-state FMRI study. Hum Brain Mapp 32:438–449

Mai R, Tassi L, Cossu M et al (2003) A neuropathological, stereo-EEG, and MRI study of subcortical band heterotopia. Neurology 60:1834–1838

Marsh l, Lim Ko, Sullivan EV et al (1996) Proton magnetic resonance spectroscopy of a gray matter heterotopia. Neurology 47:1571–1574

Martina DD, Seegerb U, Rankea MB, Groddb W (2003) MR imaging and spectroscopy of a tuber cinereum hamartoma in a patient with growth hormone deficiency and hypogonadotropic hypogonadism. AJNR Am J Neuroradiol 24:1177–1180

Miao Y, Juhász C, Wu J et al (2011) Clinical correlates of white matter blood flow perfusion changes in Sturge-Weber syndrome: a dynamic MR perfusion-weighted imaging study. AJNR Am J Neuroradiol 32(7):1280–1285

Milligan TA, Zamani A, Bromfield E (2009) Frequency and patterns of MRI abnormalities due to status epilepticus. Seizure 18(2):104–108

Mittal S, Mittal M, Montes JL et al (2013) Hypothalamic hamartomas. Part 1. Clinical, neuroimaging, and neurophysiological characteristics. Neurosurg Focus 34(6):E6

Montenegro MA, Li LM, Guerreiro MM et al (2002) Focal cortical dysplasia: improving diagnosis and localization with magnetic resonance imaging multiplanar and curvilinear reconstruction. J Neuroimaging 12:224–230

Mueller SG, Ebel A, Barakos J et al (2011) Widespread extrahippocampal NAA/(Cr + Cho) abnormalities in TLE with and without mesial temporal sclerosis. J Neurol 258(4):603–612

Opplet A (2006) Imaging systems for medical diagnostics. fundamentals, technical solutions and applications for systems applying ionizing radiation, nuclear magnetic resonance and ultrasound. Siemens, 2006

Osborn AG (2013). Osborn's brain: imaging, pathology, and anatomy. Amyrsis Salt Lake City, Utah

Ozlen F, Gunduz A, Asan Z et al (2010) Dysembryoplastic neuroepithelial tumors and gangliogliomas: clinical results of 52 patients. Acta Neurochir (Wien) 152(10):1661–1671

Peters JM, Taquet M, Prohl AK et al (2013) Diffusion tensor imaging and related techniques in tuberous sclerosis complex: review and future directions. Future Neurol 8(5):583–597

Pittau F, Grouiller F, Spinelli L et al (2014) The role of functional Neuroimaging in pre-surgical epilepsy evaluation. Front Neurol 5:31

Raz E, Zagzag D, Saba L et al (2012) Cyst with a mural nodule tumor of the brain. Cancer Imaging 12:237–244

Savic I, Thomas AM, Ke Y et al (2000) In vivo measurements of glutamine + glutamate (Glx) and Nacetyl aspartate (NAA) levels in human partial epilepsy. Acta Neurol Scand 102:179–188

Scott CA, Fish DR, Smith SJ et al (1999) Presurgical evaluation of patients with epilepsy and normal MRI: role of scalp video-EEG telemetry. J Neurol Neurosurg Psychiatry 66:69–71

Sicca F, Kelemen A, Genton P et al (2003) Mosaic mutations of the LIS1 gene cause subcortical band heterotopia. Neurology 61:1042–1046

Striano P, Caranci F, Di Benedetto R et al (2009) ^{1}H MR spectroscopy indicates prominent cerebellar dysfunction in benign adult familial myoclonic epilepsy. Epilepsia 50(6):1491–1497

Trishit Roy T, Alak Pandit (2011) Neuroimaging in epilepsy. Ann Indian Acad Neurol 14(2):78–80

Valdueza JM, Cristante L, Dammann O et al (1994) Hypothalamic hamartomas: with special reference to gelastic epilepsy and surgery. Neurosurgery 34(6):949–958

Wattjes MP, Lutterbey GG, Gieseke J et al (2007) Double Inversion Recovery brain imaging at 3T: diagnostic value in the detection of multiple sclerosis lesions. AJNR Am J Neuroradiol 28:54–59

Widjaja E, Raybaud C (2008) Advances in neuroimaging in patient with epilepsy. Neurosurg Focus 25(3):E3

Widjaja E, Zarei S, Mahmoodabadi CG et al (2012) Reduced cortical thickness in children with new-onset seizures. AJNR Am J Neuroradiol 33:673–677

Widjaja E, Zamyadi M, Raybaud C et al (2013) Abnormal functional network connectivity among resting-state networks in children with frontal lobe epilepsy. AJNR Am J Neuroradiol 4:2386–2392

Widjaja E, Kis A, Go C et al (2014) Bilateral white matter abnormality in children with frontal lobe epilepsy. Epilepsy Res 108(2):289–294

Wieck G, Leventer RJ, Squier WM et al (2005) Periventricular nodular heterotopia with overlying polymicrogyria. Brain 128(Pt 12):2811–2821

Wintermark P, Lechpammer M, Warfield SK et al (2013) Perfusion imaging of focal cortical dysplasia using Arterial Spin Labeling: correlation with histopathological vascular density. J Child Neurol 28(11):1474–1482

Wolf RL, Alsop DC, Levy-Teis I et al (2001) Detection of mesial temporal lobe hypoperfusion in patients with temporal lobe epilepsy by use of arterial spin labeled perfusion MR imaging. AJNR Am J Neuroradiol 22(7):1334–1341

Wu J, Tarabishy B, Hu J et al (2011) Cortical calcification in Sturge-Weber syndrome on MRI-SWI: relation to brain perfusion status and seizure severity. J Magn Reson Imaging 34(4):791–798

Wyllie E, Rothner AD, Luders H (1989) Partial seizures in children. Clinical features, medical treatment, and surgical considerations. Pediatr Clin North Am 36:343–364

The Complex Relationship Between Epilepsy and Headache and the Concept of Ictal Epileptic Headache

Pasquale Parisi

Key Points

- Headache/migraine is often associated with epilepsy in children, as a pre-ictal, ictal, post-ictal or inter-ictal phenomenon.
- Epidemiological aspects of the co-morbidity between epilepsy and headache are clearly different in children and adults.
- Headache as a symptom, with migraine characteristics and/or tension-type headache characteristics, may be the only clinical ictal manifestation of an epileptic seizure: this condition is now classified as "ictal epileptic headache".
- In particular, according to published criteria, the term "ictal epileptic headache" must be used in cases of a headache/migraine attack as the sole clinical ictal symptom of epileptic origin, confirmed by ictal-EEG recording and clinical-EEG responsiveness to intravenous antiepileptic drugs.
- EEG is not recommended as a routine examination for children diagnosed with headache, but is mandatory and must be carried out promptly in cases of prolonged migraine/headache that does not respond to antimigraine drugs and in which epilepsy is suspected.
- This is not a marginal question, because these possible, isolated, non-motor, ictal manifestations (i.e. ictal epileptic headache) should be taken into account before declaring an epileptic patient as "seizure free" so as to be able to suspend anticonvulsant therapy safely.

P. Parisi (✉)
Child Neurology, Pediatric Headache & Sleep Disorders Centre, Chair of Pediatrics, NESMOS Department, Faculty of Medicine and Psychology, Sapienza University, Via Di Grottarossa, 1035–1039, 00189 Rome, Italy
e-mail: pasquale.parisi@uniroma1.it

Summary

The possible link/comorbidity (causal or not causal) between epilepsy and headaches has been a topic of debate for over a hundred years, ever Since William Richard Gowers's time. In recent decades, new data have emerged in favor of a non-random relationship between these two entities. They are both characterized by transient attacks of altered brain function with a clinical, pathophysiological, genetic and therapeutic overlap, and may thus mimic each other. Indeed, the clinical distinction between headache and epilepsy may make the differential diagnosis a highly challenging task. Both are common and often co-morbid, with headache attacks in epilepsy being temporally related to the occurrence of epileptic seizures as a pre-ictal, ictal, post-ictal or inter-ictal event. Yet, they are both paroxysmal and chronic neurological disorders that share many clinical and epidemiological aspects, and they may both present with visual, cognitive, sensitive-sensorial and motor signs/symptoms; these neurophysiologicl phenomena arise from the cerebral cortex and are modulated by sub-cortical connections. Even from an epidemiological point of view, data in the literature regarding the co-morbidity between headache and epilepsy appear to be quite distinct in children. What makes this scenario even more variegated and complex are new data suggesting that a headache may, in some cases, even be the only ictal manifestation of an epileptic seizure. The latter condition, the so-called "ictal epileptic headache", is a new entity that has recently been cited in the new classification of headache disorders (ICHD-III), whose diagnostic criteria have very recently been defined and published. The data that have led, over the last decade, to the proposed "diagnostic criteria" for "ictal epileptic headache" are reported below. In this regard, it is crucial to stress that the authors who proposed the diagnostic criteria for ictal epileptic headache, have deliberately and consciously chosen "criteria" that underestimate the phenomenon so as to avoid spreading panic among both patients and physicians, who tend to be reluctant to accept this concept because of the stigma attached to the diagnosis of epilepsy. In the future, once this concept (i.e. "headache" as the sole ictal epileptic manifestation) has been more widely accepted, it will hopefully be possible to propose "different and less restrictive" criteria than those recently published.

Abstract Since William Richard Gowers' time, for over a hundred years, it has been discussing the possible link/comorbidity (causal or not causal) between epilepsy and headaches. During the latest decades, new data have emerged in favor of a non-random relationship between these two entities. They are both characterized by transient attacks of altered brain function with a clinical, pathophysiological, common genetic factors and therapeutic overlap, thus, they may also mimic each other. In fact, the clinical distinction between headache and epilepsy can be so difficult to make the differential diagnosis sometimes highly problematic. Both are common, often co-morbid and, in this latter case, headache attacks can be temporally related with the occurrence of epileptic seizures as pre-ictal, ictal, post-ictal or inter-ictal phenomenon. Yet, they are both paroxysmal and chronic neurological disorders that share many clinical and epidemiological aspects, and they may, both, present with visual, cognitive, sensitive-sensorial and motor signs/symptoms; these neurophysiologic phenomenon arise from the cerebral cortex and are modulated by sub-cortical connections.

Even from an epidemiological point of view data from the Literature about the co-morbidity between headache and epilepsy seem to be clearly different in children.

In addition, to make this scene even more variegated and complex, new data, supporting the possibility that a headache may even be, in some cases, "*the only ictal manifestation of an epileptic seizure*", became available. This latter condition, the so-called "Ictal Epileptic Headache", is a new entity that has recently been quoted in the new classification of headache disorders (ICHD-III, third edition, published in July 2013 in Cephalalgia), whose diagnostic criteria have also been suggested and published very recently.

Here it have been reported, in their essential aspects, the available data that during the latest decade have led to propose "diagnostic criteria" for "Ictal Epileptic Headache". In this regards, in particular, it is crucial to stress that the Authors who proposed the diagnostic criteria for the "Ictal Epileptic Headache", have "deliberately" and "consciously" chose to formulate "criteria" that underestimate the phenomenon rather than to spread the "panic" among patients and physicians who are reluctant to accept this concept because the stigma attached to the diagnosis of epilepsy. In the future, when this concept (an "headache" as sole ictal epileptic manifestation) will be "metabolized", we will be able to propose "different and less restrictive" criteria than those recently published.

Keywords Ictal epileptic headache (IEH) · Epilepsy · Migraine headache · Autonomic seizures · Autonomic status epilepticus · Criteria for IEH diagnosis

Epilepsy and Headache: Diagnostic Challenges and Their Reasons

Clinical Issues

Both occipital seizures and migraine/headache may be characterized by the presence of a transitory visual disturbance that follows headache and other autonomic symptoms. A mis-diagnosis of visual seizures as migraine with visual aura is frequent and costly. The main factor that contributes to such an error is that the description of a visual hallucination is often limited to terms such as scintillations, fortification spectrum, teichopsia, phosphenes, and variations of these signs (Panayiotopoulos 1987, 1999a, b, c). Elementary visual hallucinations of occipital seizures are usually different from the visual aura of migraine. "Ictal" elementary visual hallucinations are defined according to color, shape, size, location, movement, speed of appearance, duration, frequency and associated symptoms of progression. Elementary visual hallucinations are mainly colored and circular, develop within seconds, and are brief in duration (2–3 min); they often appear on the periphery of a temporal visual hemi-field, becoming larger and multiplying in the course of the seizure, frequently moving horizontally toward the other side (Panayiotopoulos 1987, 1999a, b, c). Significantly, post-ictal headache, which is often indistinguishable from migraine, occurs in more than half of patients, even after brief visual seizures. Post-ictal headache frequently occurs 3–15 min after the seizure ends, in a situation known in migraine as the "asymptomatic interval". Thus, occipital seizures often generate migraine-like attacks, i.e. an "epilepsy-migraine sequence" (Panayiotopoulos 1987, 1999a, b, c). In migraine, visual aura usually starts as a flickering, uncolored, zigzag line in the center of the visual field and affects the central vision. It gradually spreads over 4 min toward the periphery of one hemi-field, and usually lasts < 30 min, often leaving a scotoma. The total duration of visual auras is less than 60 min. Furthermore, migraine visual aura rarely occurs daily, and is not associated with non-visual ictal occipital symptoms, such as eye and head deviation and repetitive eyelid closures. Less typical features of migraine visual aura, such as spots, circles and beads, with or without colors, maybe experienced during migraine visual aura, though they are rarely dominant. Clustering of other symptoms, such as those described above, betray their migraine nature (Panayiotopoulos 1987, 1999a, b, c).

Electroencephalographic Issues

Children with a definitive diagnosis of epilepsy may not have epileptic EEG abnormalities recordable from the scalp (for example, as may occur in some cases of "nocturnal frontal lobe epilepsy", Nobili 2007), whereas EEG abnormalities that resemble paroxysmal abnormalities (usually present during an epileptic seizure) may be recorded during a migraine attack.

In general, EEG is of less diagnostic importance in the study of patients suffering from migraine. However, the temporal and spatial pattern of EEG anomalies in headache is usually extremely different from that observed during a real epileptic seizure (Andermann 1987; Andermann and Zifkin 1998). In the few cases of ictal epileptic headaches reported in the literature, the patients had headache/migraine as the only manifestation of a non-convulsive epileptic seizure, diagnosed on the basis of an EEG recording alone (Parisi et al. 2007, 2012a, b, 2013a, b; Parisi 2009a, b, 2011; Parisi and Kasteleijn-Nolst Trenité 2010). Intravenous administration of AEDs in these patients was able to control the seizure, as demonstrated by means of a scalp EEG as well as resolution of the headache symptoms (Parisi et al. 2012a, b). However, some isolated cases of "ictal epileptic headache" reported in the literature were incidentally detected when the critical focus was being recorded by means of deep electrodes (Laplante et al. 1983; Fusco et al. 2011). Therefore, although the EEG recording may not prove very useful as a screening instrument for migraine, it does play a fundamental role in pediatric patients who have headache/migraine symptoms that do not respond to commonly-used anti-migraine drugs (Parisi et al. 2012a, b).

Furthermore, the ictal EEG abnormalities recorded in IEH patients display aspecific features; indeed, various ictal EEG patterns have been recorded during migraine-like complaints in both symptomatic and idiopathic cases (Belcastro et al. 2011a, b; Parisi et al. 2012a, b).

Epidemiological Issues: Beyond the Controversial Evidence

An "editorial" dedicated exclusively to these epidemiological aspects, to their possible biases and to the underestimation potentially related particularly to pediatric age has recently been published (Belcastro et al. 2012).

The prevalence and incidence of co-morbid epilepsy and headache in the general population, including all stages of life, vary (Olafsson et al. 2005). Indeed, headache predominates in males before puberty, whereas it is more common in females afterwards. By contrast, epilepsy predominates in males at all ages (Olafsson et al. 2005). The age at the peak incidence of these two conditions also differs. The peak age for migraine occurs during the working years (Lyngberg et al. 2005), whereas for epilepsy it is under the age of one year and over the age of 60 years (Hauser et al. 1993). The fact that data in the literature on this topic are somewhat conflicting (Lipton et al. 1994a, b; Andermann 1987; Tonini et al. 2012) may be attributed to the co-occurrence (synergistic and/or divergent) of confounding variables adopted in the different sampling methods and study designs. These conflicting results may partly be explained by differences in the target populations, study design, age range, methods, by inclusion criteria that are limited to referral patients with epilepsy or tertiary headache centers, by the lack of appropriate control groups, and/or by different or ill-defined diagnostic criteria (Tonini et al. 2012). Thus, these studies cannot easily be compared with one another.

Although there is not yet any conclusive evidence of a real causal relationship between the two disorders, we must bear in mind that the comorbidity of headache and epilepsy is, beyond any doubt, different in children and in adults. Children are more likely to have an autonomic symptomatology in both epilepsy and headache attacks (Fogarasi et al. 2006; Kasteleijn-Nolst Trenité and Parisi 2012). Moreover, they may have isolated, long-lasting ictal autonomic manifestations, while ictal autonomic manifestations (in both epilepsy and headache) in adults are usually associated, whether simultaneously or sequentially, with other motor or sensory ictal signs and symptoms (Fogarasi et al. 2006; Kasteleijn-Nolst Trenité and Parisi 2012). Furthermore, it should be borne in mind, despite the limited number of studies in the literature (Yamamane et al. 2004; Piccinelli et al. 2006; Toldo et al. 2010), that the framework assumes markedly different shapes in the pediatric population.

Among 50 children with epilepsy, Yamamane et al. (2004) found that 46% had headache, and that 10 (43.5%) out of the 23 headache sufferers had migraine. Most patients with headache were older than 10 years (54.5%) and had idiopathic epilepsy (65.2%). In some specific childhood epilepsy syndromes, migraine/headaches appear to be more prevalent (Kinast et al. 1982; Bladin 1987; Andermann and Zifkin 1998; Panayiotopoulos 1999b; Yankovsky et al. 2005; Wirrell and Hamiwka 2006; Clarke et al. 2009). The best known of these syndromes are benign occipital epilepsy of childhood with occipital paroxysms and benign rolandic epilepsy; remarkably, in the majority of the cases reported (95%), the headache started in the same year as, or after, the diagnosis of epilepsy (Andermann and Zifkin 1998; Panayiotopoulos 1999b; (Ito et al. 1996, 2003, 2004; Leniger et al. 2001; Toldo et al. 2010; Verrotti et al. 2011a).

Piccinelli et al. (2006) found EEG interictal abnormalities in 16 (12.8%) out of 137 children and adolescents with headache, particularly in those with migraine with aura. Another intriguing issue is the comorbidity of headache in patients with idiopathic epilepsy of infancy. Indeed, it is widely known that patients with epilepsy with rolandic or occipital paroxysms, or even those without seizures, have concomitant migraine in up to 60% of cases (Panayiotopoulos 1999a, b, c; Tonini et al. 2012). When they investigated a large, consecutive, pediatric population of 1795 patients with headache under 18 years of age diagnosed at a headache center, Toldo et al. (2010) found a strong association between migraine and epilepsy. In that study, migraineurs displayed a risk for epilepsy that was 3.2 times higher than that for tension-type headache, with no significant difference between migraine with and without aura. Migraineurs affected by focal epilepsies had a risk for cryptogenic epilepsy that was three times higher than that for idiopathic epilepsy. Recently, adolescents with any headache type were reported (Lateef et al. 2012) to have significantly higher rates of epilepsy, as previously confirmed by Baca et al. (2011), who found the comorbidity of migraine and epilepsy in 15% of the children they studied. Interestingly, Colombo et al. (2011), whose data confirmed previous findings (Sasmaz et al. 2004) showing that almost 36% of parents of children with headache are unaware of the headache, stressed that pediatric headache is still under-diagnosed and is not adequately considered as a health problem either by the medical community or in social settings: on the one hand, this indicates the extent to which headache is underestimated, on the other, it confirms that the clinical picture in co-morbid cases is dominated by the diagnosis of epilepsy.

In recent years, following several reported cases of headache as the sole manifestation of an epileptic seizure (Parisi et al. 2007, 2008a, b, 2012a, b; Parisi 2009a, b; Parisi and Kasteleijn-Nolst Trenité 2010; Belcastro et al. 2011a, b; Verrotti et al. 2011b), the term ictal epileptic headache has been proposed to identify these events (Parisi 2011, 2012a, b). In particular, ictal epileptic headache is recognized as a headache ("as the sole ictal manifestation" and without presenting a "specific" clinical picture of migraine, migraine with aura or tension-type headache), lasting from seconds to days, with evidence of ictal epileptiform EEG discharges, which immediately resolve upon administration of intravenous antiepileptic medications (Parisi et al. 2012a).

In conclusion, although controversial epidemiological data in adults are often used as evidence that no association exists between these two conditions, some studies, particularly those conducted in pediatric populations, point to the comorbidity of headache and epilepsy. Thus, further studies on larger pediatric populations are warranted to definitively confirm this comorbidity (Belcastro et al. 2012).

Shared Pathophysiology, Classification and Genetic Aspects

Common Pathways, Substrates and Genetics

Many studies support the hypothesis of excessive neocortical cellular excitability as the main pathological mechanism underlying the onset of attacks in both diseases (Somjen 2001; Berger et al. 2008). Indeed, since hypo- and hyper-excitation in migraine occur sequentially as rebound phenomena (during a spreading depression), the term "dys-excitability" may better describe migraine pathophysiology than hyper-excitability (Somjen 2001; Berger et al. 2008; Tottene et al. 2009, 2011; Fabricius et al. 2008; Parisi et al. 2012a, b). Cortical Spreading Depression (CSD), which may be considered the link between headache and epilepsy (Moskowitz et al. 1993; Bolay et al. 2002; Ayata et al. 2006; Ayata 2010; Parisi et al. 2008a, b; Parisi 2009a, b; Eikermann-Haerter & Ayata 2010; Belcastro et al. 2011a, b; Zhang et al. 2011; Parisi et al. 2012a; Ghadiri et al. 2012), is characterized by a slowly propagating wave of sustained strong neuronal depolarization that generates transient intense spike activity, followed by neural suppression, which may last for minutes. The depolarization phase is associated with an increase in regional cerebral blood flow, whereas the phase of reduced neural activity is associated with a reduction in blood flow. On the other hand, CSD activates the trigeminovascular system, inducing the cascade release of numerous inflammatory molecules and neurotransmitters, which results in pain during the migraine attack (Zhang et al. 2011, Parisi et al. 2012a, b).

There is emerging evidence from both basic and clinical neurosciences that cortical spreading depression and an epileptic focus may facilitate each other, though to different extents (Parisi et al. 2008; Parisi 2009a, b; Belcastro et al. 2011a, b, c; Parisi et al. 2012a). When a certain threshold is reached, the onset and propagation of neuronal depolarisation are triggered in both CSD and a seizure. The required threshold is presumed to be lower for CSD than for a seizure, which would explain why it is far more likely to observe an epileptic patient who presents a peri-ictal headache than a migraine patient who presents an epileptic seizure (Verrotti et al. 2011b, c; Parisi et al. 2012a, b).

Once the cortical event has started, spreading subsequently depends on the size of the onset zone, velocity, semiology and type of propagation (Parisi et al. 2008; Parisi 2009b). Moreover, the onset of both CSD and that of the epileptic seizure may facilitate each other (Berger et al. 2008; Fabricius et al. 2008; Zhang et al. 2011; Parisi et al. 2012b), with these two phenomena possibly being triggered by more than one pathway converging upon the same destination: depolarization and hyper-synchronization (Parisi et al. 2008, 2012a; Parisi 2009a,b; Belcastro et al. 2011a, b; Ghadiri et al. 2012). The triggering causes, which may be environmental or individual (whether genetically determined or not), result in a flow of ions that mediate CSD through neuronal and glial cytoplasmic bridges rather than through interstitial spaces, as instead usually occurs in the spreading of epileptic seizures (Gigout et al. 2006; Parisi et al. 2008; Parisi 2009b; Tamura et al. 2011; Tottene et al. 2009, 2011). As mentioned above, the threshold required for the onset of CSD is likely to be lower than that required for the epileptic seizure. In this regard, a "migraleptic" event would be unlikely to occur (Belcastro et al. 2011a; Verrotti et al. 2011b; Parisi et al. 2012b).

As regards the role of "photosensitivity" in this topic, it should be stressed that flashes, phosphenes and other positive or negative visual manifestations are often part of the clinical picture in both headache and occipital epilepsy (Wendorff et al. 2005; Kasteleijn-Nolst Trenité and Parisi 2012). Moreover, intermittent photic stimulation (IPS) induces "flashes and phosphenes" as well as migraine/headache and seizures. Moreover, IPS may induce photoparoxysmal EEG responses (PPR), migraine and epileptic seizures (Kasteleijn-Nolst Trenité and Parisi 2012). Although occipital lobe epilepsy already has much in common with migraine (visual aura, positive and negative ictal signs, and autonomic disturbances such as pallor and vomiting), the photosensitive variant of occipital epilepsy, and photosensitive epilepsy in general, share even more similarities, such as a higher prevalence in women (female/male ratio 3:2) and a sensitivity to flickering light stimuli and striped patterns that induce attacks (Kasteleijn-Nolst Trenitè et al. 2010a, b).

Both migraine and epilepsy have an important genetic component, with strong evidence pointing to a shared genetic basis between headache and epilepsy emerging from clinical/EEG and genetic studies on Familial Hemiplegic Migraine (FHM) (Haglund and Schwartzkroin 1990; De Fusco et al. 2003; Vanmolkot et al. 2003; Kors et al. 2004; Dichgans et al. 2005; De Vries et al. 2009; Tottene et al. 2009, 2011; Gambardella and Marini 2009; Van Den Maagdenberg et al. 2010; Riant et al. 2010; Pietrobon 2010; Uchitel et al. 2012). Errors in the same gene may be associated with migraine in some cases and with epilepsy in others. Recent data suggest shared genetic substrates and phenotypic–genotypic correlations with mutations in some ion transporter genes, including CACNA1A, ATP1A2 and SCN1A (Vanmolkot et al. 2003; Kors et al. 2004; Dichgans et al. 2005; De Vries et al. 2009; Tottene et al. 2009; Gambardella and Marini 2009; Van Den Maagdenberg et al. 2010; Riant et al. 2010; Pietrobon 2010). Other genetic findings pointing to a link between migraine and epilepsy have been published. They include mutations on SLC1A3, a member of the solute carrier family that encodes excitatory amino acid transporter 1, and 57 POLG58, C10 and F259, which encode mitochondrial DNA polymerase and helicase twinkle (Tzoulis et al. 2006; Lonnqvist et al. 2009).

In addition, glutamate metabolism (Jen et al. 2005), serotonin metabolism (Johnson and Griffiths 2005), dopamine metabolism (Chen 2006) and ion channel (sodium, potassium and chloride) function might be impaired in both epilepsy and migraine (Steinlein 2004; Pietrobon 2010). In particular, it is likely that voltage-gated ion channels play a critical role in the pathways associated with migraine and epilepsy (Vanmolkot et al. 2003; Kors et al. 2004; Dichgans et al. 2005; De Vries et al. 2009; Tottene et al. 2009; Gambardella and Marini 2009; Van Den Maagdenberg et al. 2010; Riant et al. 2010; Pietrobon 2010).

Temporal Relationship and Classification Issues

Seizures and epilepsy syndromes are classified according to guidelines of the ILAE (Panayiotopoulos 2012), and headaches according to the International Classification of Headache Disorders (ICHD). The current version, the ICHD-III, was published in Cephalalgia in July 2013 (Headache Classification Committee of the International Headache Society 2013).

Seizure-related headaches may be peri-ictal (40–75 %) or inter-ictal headaches (25–60 %) (Verrotti et al. 2011a; Cai et al 2008; Ito et al 1996, 2003, 2004). Peri-ictal headache can be divided into pre-ictal, ictal and post-ictal headache; among the peri-ictal forms, the "post-ictal" is, without doubt, the most frequently reported (15–50 %), probably because both children and adults tend to remember the events more easily once the seizure has resolved (post-ictal). (Verrotti et al. 2011a; Cai et al. 2008; Ito et al. 1996, 2003, 2004). "Ictal Headache" (i.e. a "headache" with either migraine or tension-type characteristics associated with other sensory–motor–autonomic–psychiatric ictal epileptic manifestations) occurs in as few as 3–5 % of cases (Forderreuter et al. 2002).

The latest edition of the International Headache Classification (ICHD-III) (Headache Classification Committee of the International Headache Society 2013) makes a distinction between three entities (Table 1): (1) epilepsy induced by migraine with aura or "migraine-triggered seizure" (previously referred to as migralepsy) (code 1.4.4); (2) epileptic migraine (code 7.6.1); (3) post-convulsive headache (code 7.6.2). Diagnostic criteria for the new entity ictal epileptic headache (IEH) (Table 1) have recently been proposed (Parisi et al. 2012a) and more recently cited in the Appendix of the new ICHD-3 edition (Headache Classification Committee of the International Headache Society 2013).

Hemicrania Epileptica or "Epileptic Migraine" (code ICHD-3 7.6.1)

This condition, despite being very rare, has been included in the ICHD Classification and confirmed in the new ICHD-III classification. The diagnostic criteria are: (a) headache lasting seconds-to-minutes, with migraine features that satisfy criteria C and D; (b) patient presenting a partial seizure; (c) the headache develops together with the seizure and is homo-lateral to the ictal event; (d) headache resolves immediately after the convulsion.

Table 1 Current ICHD-3 classification of headache-related seizures and proposed criteria for ictal epileptic headache (IEH)

Current ICHD-3 classification of headache-related seizures
Migraine-triggered seizure (Migralepsy) (1.4.4)
Diagnostic Criteria:
A. Migraine fulfilling criteria for 1.2 ("*Migraine with aura*")
B. A seizure fulfilling diagnostic criteria for one type of epileptic attack occurs during or within 1 h of a migraine aura
Hemicrania Epileptica or "Epileptic Migraine" (7.6.1)
Diagnostic Criteria:
A. Headache lasting seconds to minutes, with features of migraine fulfilling criteria C and D
B. The patient is having a partial epileptic seizure
C. Headache develops synchronously with the seizure and is ipsilateral to the ictal discharge
D. Headache resolves immediately after the seizure
Post-ictal Headache (7.6.2)
Diagnostic Criteria:
A. Headache with features of "*tension-type headache*" or, in a patient with migraine of "*migraine headache*" and fulfilling criteria C and D
B. The patient has had a partial or generalized epileptic seizure
C. Headache develops within three hours of the seizure
D. Headache resolves within 72 h of the seizure
Proposed criteria for ictal epileptic headache (IEH)
Diagnostic criteria A-D must all be fulfilled to make a diagnosis of "IEH"
A. Headache[a] lasting minutes, hours or days;
B. Headache that is ipsilateral or contralateral to lateralised ictal epileptiform EEG discharges (if EEG discharges are lateralised);
C. Evidence of epileptiform (localised[b], lateralised or generalised) discharges on scalp EEG concomitantly with headache; different types of EEG anomalies may be observed (generalized spike-and-wave or polyspike-and-wave, focal or generalised rhythmic activity or focal sub-continuous spikes or theta activity that may be intermingled with sharp waves) with or without photoparoxysmal response (PPRs)
D. Headache resolves immediately (within minutes) of i.v. antiepileptic drug administration

[a] A specific headache pattern is not required (Migraine with or without aura, or tension-type headache are all admitted)
[b] Any localisation (frontal, temporal, parietal, occipital) is admitted

Post-ictal Headache (code ICHD-3 7.6.2)

Headache with migraine characteristics that manifests itself in about 50 % of patients after a convulsive epileptic seizure. The criteria of ictal headache in the ICHD classification are: (1) tension headache, or migraine that satisfies criteria C and D; (2) the patient presents a partial or general epileptic seizure; (3) headache develops within 3 h of seizure onset; (4) headache resolves within 72 h of the convulsion. Post-ictal headache, though often associated with symptomatic epilepsy, is a frequent event in idiopathic occipital epilepsy in children.

Epilepsy Induced by Migraine with Aura ("migralepsy") (code ICHD-3 1.4.4)

According to the ICHD-III, migralepsy is a recognized complication of migraine (Table 1).

It does not exist in the international classifications of epilepsy (ILAE).

The term migralepsy, which was first used in 1960 (Lennox and Lennox 1960) to define a condition of "opthalmic migraine with associated nausea and vomiting followed by symptoms characteristic of epilepsy", was reintroduced in 1993 by Marks and Ehrenberg (1993). However, the term migralepsy used to refer to a temporal sequence of a migraine with aura attack that evolves into tonic-clonic seizures was widely criticized by many authors, and patients classified as having migralepsy were subsequently defined as having epileptic seizures of the occipital lobe. Unfortunately, there is no clear EEG documentation of cases of migralepsy showing a critical surge of the scalp EEG in patients in whom migraine with aura is followed by a tonic-clonic seizure. Despite the skepticism expressed by various authors regarding the concept of "migralepsy", this clinical condition was inserted in the ICHD-2 classification (Headache Classification Committee of the International Headache Society 2004) as a complication of migraine (code 1.5.5), whereas in the latest revision of the ICHD-3 classification (Headache Classification Committee of the International Headache Society 2013), this condition is codified as epilepsy induced by migraine with aura (code 1.4.4), the term migralepsy being omitted. According to the ICHD-III criteria (Headache Classification Committee of the International Headache Society 2013), epilepsy induced by migraine with aura is defined as an epileptic fit that manifests itself within an hour of a migraine with aura attack in the absence of other causes. Although epilepsy and migraine are among the most common neurological illnesses, this event is very rare in children. Indeed, Sances et al. (2009) recently demonstrated that of the 50 cases of migralepsy mentioned in the literature, only two patients satisfied the migralepsy criteria according to ICHD 2 (Headache Classification Committee of the International Headache Society 2004)

Sunset of the "Migralepsy" Concept

Most reported cases of "migralepsy" do not allow a meaningful and unequivocal migraine-epilepsy sequence to be detected (Sances et al. 2009; Verrotti et al. 2011b, c). There are approximately 50 cases of potential migralepsy reported in the literature (Sances et al. 2009). The majority of these cases have been the subject of criticism by various authors because the diagnosis in the majority of the patients is uncertain for the following reasons: the information available is not clear (38%), the cases do not fulfil the current ICHD-III criteria (30 %), the diagnosis is questionable (28 %). Indeed, most previous reports of "migralepsy" may have been occipital seizures imitating migraine with aura (Verrotti et al. 2011a, b, c; Barrè et al. 2008).

It has recently even been suggested (Belcastro et al. 2011a; Parisi et al. 2012b) that many of the "published" migralepsy cases may be an "ictal epileptic headache" followed by other sensori-motor or autonomic ictal signs/symptoms. Indeed, although unequivocal epileptiform abnormalities in patients with paroxysmal sensations or behavioural changes usually point to a diagnosis of epilepsy, the lack of clear ictal epileptic spike-wave activity is frequent in autonomic epilepsies, such as Panayiotopoulos syndrome (Koutroumanidis 2007), or even in frontal lobe epilepsy (Nobili 2007). In such cases, ictal epileptic EEG activity might be recorded as unspecific slow wave abnormalities without any spike activity (Belcastro et al. 2011a; Parisi et al. 2012b). Interestingly, there may, on rare occasions, be an isolated epileptic headache that has no associated ictal epileptic manifestations or scalp EEG abnormalities but whose ictal epileptic origin can be demonstrated by depth electrode studies, even purely by chance (Laplante et al. 1983; Fusco et al. 2011).

This misunderstanding has perpetuated the concept of "migralepsy" since the 1960s (Lennox and Lennox 1960) to the detriment of the entity we now define as ictal epileptic headache. We believe that this has led to ictal epileptic headache being severely underestimated, on the one hand, and to migralepsy being clearly overestimated, on the other (Parisi et al. 2012b, 2013a, b).

The Rise of the "Ictal Epileptic Headache" Concept: "A Long and Winding Road...."

Sir Gowers's famous book, published in 1907 (Gowers 1907), first stated that "migraine is in the borderland of epilepsy", and in an epoch before electroencephalography (EEG), Gowers stated: "...the most frequent relation of migraine to epilepsy is as source of error;....in extremely rare instances one affection may develop while the other goes on".

More than 100 years later, we can firmly state that on occasion "migraine itself may be epilepsy" (Parisi et al. 2007, 2008a, b, 2012a, b, 2013a, b). The overlap between these two conditions is partial or complete, not always synchronous (i.e. mainly a peri-ictal phenomenon), but in some cases (whose number is probably largely underestimated) "the headache represents the only ictal phenomenon": we recently named this condition "ictal epileptic headache" (Parisi et al. 2012a). Since 1971, fewer than 20 IEH cases diagnosed according to proposed criteria (Table 1, Parisi et al. 2012b) have been reported (Grossman et al. 1971; Walser and Isler 1982; Laplante et al. 1983; Isler et al. 1987; Niedermeyer 1993; Marks and Ehrenberg 1993; Walker et al. 1995; Ghofrani et al. 2006; Parisi et al. 2007; Piccioli et al. 2009; Perucca et al. 2010; Belcastro et al. 2011c; Fusco et al. 2011; Italiano et al. 2011; Fanella et al. 2012; Cianchetti et al. 2013). Verrotti et al. recently published 16 other potential ictal epileptic headache cases from a large multicentre neuropediatric sample (Verrotti et al. 2011b), stressing the concept of "probable underestimate phenomenon" (Parisi et al. 2012a; Parisi et al. 2013a, b).

Nonetheless, this belief goes back a long way. Indeed, it is ever since the 1950s that cases have been reported in German (Heyck and Hess 1955), English (Nymgard 1956) and Italian (Lugaresi 1955; Morocutti and Vizioli 1957) literature in which it has been suggested that "headache" may actually be "an epileptic headache" and "… may even be the only clinical manifestation of idiopathic epilepsy" (Morocutti and Vizioli 1957). Thus, the concept of "ictal headache" dates from a long time ago (Lugaresi 1955; Heyck and Hess 1955; Nymgard 1956; Morocutti and Vizioli 1957). However, the term migralepsy was subsequently coined in the 1960s (Lennox and Lennox 1960) and has permeated the epilepsy and headache culture ever since.

With regard to migralepsy cases reported in the literature, recent articles (Sances et al. 2009; Verrotti et al. 2011a, b, c) have provided a clear demonstration of the inadequacy of both the ICHD-2 and ICHD-3 definitions of this condition. Following the introduction of the migralepsy concept by Lennox and Lennox (1960), an increasing number of ictal epileptic headaches have been reported since the 1970s (Grossman et al. 1971; Walser and Isler 1982; Laplante et al. 1983; Isler et al. 1987; Niedermeyer 1993; Marks and Ehrenberg 1993; Walker et al. 1995; Ghofrani et al. 2006; Parisi et al. 2007; Piccioli et al. 2009; Perucca et al. 2010; Belcastro et al. 2011c; Fusco et al. 2011; Italiano et al. 2011; Fanella et al. 2012; Cianchetti et al. 2013). It has also been suggested (Parisi and Kasteleijn-Nolst Trenité 2010; Belcastro et al. 2011a; Verrotti et al. 2011a, b; Striano et al. 2011, 2012; Parisi et al. 2012a, 2013a, b) that the migralepsy sequence may not exist at all and that the initial part of the "migralepsy sequence" may merely be an "ictal epileptic headache" (Parisi et al. 2012a) followed by other ictal autonomic, sensory, motor or psychic features.

It should be borne in mind that cortical and subcortical areas appear to be hierarchically divided according to how likely they are to develop CSD, with the occipital lobe appearing to be the most likely area (Verrotti et al. 2011a, b; Parisi et al. 2012b).

How can CSD and epileptic discharges facilitate each other, though to a varying extent? In other words, why may the onset of an epileptic seizure facilitate the onset of CSD to a greater extent that CSD facilitates the onset of an epileptic seizure (Belcastro et al. 2011a; Parisi et al. 2012b)? Some experimental and clinical data in the literature discuss this topic. The most interesting data on genetic defects leading to both epilepsy and migraine are related to FHM, as stated above (Haglund and Schwartzkroin 1990; De Fusco et al. 2003; Vanmolkot et al. 2003; Kors et al. 2004; Dichgans et al. 2005; De Vries et al. 2009; Tottene et al. 2009; Gambardella and Marini, 2009; Van Den Maagdenberg et al. 2010; Riant et al. 2010; Pietrobon 2010; Escayg and Goldin2010; Uchitel et al. 2012).

With regard to the concept of "cortex dys-excitability" in migraine subjects, new advances now support this point of view (Berger et al. 2008; Fabricius et al. 2008; Tottene et al. 2009, 2011; Faragauna et al. 2010; De Souza et al. 2011). Indeed, if we consider the specific polysynaptic inhibitory sub-circuit involving fast-spiking (FS) inter-neurons and pyramidal cells (PC) that have been investigated in FHM1 mice (Berger et al. 2008; Fabricius et al. 2008; Tottene et al. 2009, 2011; Faragauna et al. 2010; De Souza et al. 2011), the gain in function following glutamate release at the recurrent synapses between pyramidal cells would certainly increase network excitation; by contrast, the gain in function following glutamate release at the PC-FS synapses would lead to enhanced recruitment of inter-neurons and enhanced inhibition. Although this analysis is restricted to a specific sub-circuit, it does raise the important point that the differential effect of FHM1 mutations on excitatory and inhibitory neurotransmission may produce over-excitation in certain brain conditions, while leaving the excitation-inhibition balance within physiological limits in others. This would explain the episodic nature of the disease with alternate hyper-excitation and hypo-excitation in the same subject at different times (thus supporting the dis-excitability concept in migraine subjects) (Pinto et al. 2005; Berger et al. 2008; Fabricius et al. 2008; Tottene et al. 2009, 2011; Faragauna et al. 2010; De Souza et al 2011; Tamura et al. 2011; Escayg and Goldin 2010; Uchitel et al. 2012).

A possible explanation for the clearly different extent to which CSD and epileptic seizures facilitate each other is that although these two conditions are triggered by similar mechanisms, their evolution is different depending on whether the neuronal hyperactivity and consequent increase in (K+) exceed a critical level that causes self-regeneration of the depolarization; according to this hypothesis, CSD may be defined as "a poorly-controlled seizure" in which (K+) regulation is completely disrupted. Indeed, local neuronal hyperactivity that progressively recruits a synchronous discharge via recurrent excitatory collaterals and (K+) accumulation has been hypothesized to initiate epileptic discharge in slice models (Pinto et al. 2005). CSD, experimentally induced in rats, increases cortico-cortical evoked responses and strongly induces "brain-derived neurotrophic factor" with synaptic potentiation in vivo (Faraguna et al. 2010), while the induction of a "long-term potentiation-like" (LTP-like) phenomenon by CSD receives support from experimental evidence. In vivo data that lend support to the idea of a CSD-induced LTP-like phenomenon also exist (De Souza et al. 2011). Another recent and intriguing finding regarding CSD propagation is the model according to which interstitial (K) diffusion initiates the positive feedback cycle that ignites CSD in adjacent dendrites, and which is in contrast to the hypothesis that CSD spreads through gap junctions. In particular, according to this hypothesis, the opening of the gap junctions would not be required for CSD propagation, but is instead required for extracellular homeostasis following CSD (Tamura et al. 2011). A causative link between enhanced glutamate release and CSD facilitation has been demonstrated by means of an in vitro model of CSD (Tottene et al. 2009). The synapse-specific effect of FHM1 mutations points to the disruption of the excitation-inhibition balance and neuronal hyperactivity as the bases for episodic vulnerability to CSD ignition in migraine. This finding provides direct evidence that the gain in function following glutamate release at synapses onto pyramidal cells is likely to facilitate experimental CSD in FHM1 mutant mice, and thus provides novel insights into the controversial mechanisms of CSD initiation and propagation.

These data are consistent with and support a model of CSD initiation, in which activation of pre-synaptic voltage-gated Ca channels, and the consequent release of glutamate from recurrent cortical pyramidal cell synapses and activation of NMDA receptors, are key components of the positive feedback cycle that ignites CSD. In this regard, the role of different voltage-gated Ca2+ channels in CSD has recently been investigated (Tottene et al. 2011). After blockade of either the P-/Q-type Ca2+ channels or the NMDA receptors, CSD cannot be induced in wild-type mouse cortical slices. By contrast, the blockade of N- or R-type Ca2 channels only has a slight inhibitory effect on the CSD threshold and velocity of propagation. These findings support a model according to which the initiation and propagation of the CSD involved in migraine require the influx of Ca2+ through pre-synaptic P-/Q-type Ca2+ channels, which in turn releases glutamate from the recurrent cortical pyramidal cell synapses and activates NMDA receptors (Tottene et al. 2009, 2011).

The temporal and spatial associations between CSD and seizures have been studied by means of electro-corticographic (ECoG) recordings in patients with acutely injured cerebral cortex (Fabricius et al. 2008). The authors reported clinically overt seizures in only one patient, while each of the patients with both CSD and seizures displayed one of the following four different patterns of interaction between CSD and seizures: (a) in four patients, CSD was immediately preceded by prolonged seizure activity; (b) in three patients, the two phenomena were separated in time, with multiple CSDs being replaced by ictal activity; (c) in one patient, seizures appeared to trigger repeated CSDs at the adjacent electrode; (d) in two patients, ongoing repeated seizures were interrupted whenever CSD occurred. These four patterns were consistent within recordings from the same patient, but were different in each of the patients.

Patients 3 and 4 described by Fabricius et al. (2008) are particularly interesting as seizure activity in these two subjects spread from electrode to electrode at the same slow speed as CSD, but preceded it by several minutes. This is noteworthy because seizure activity under other conditions spreads much faster than CSD. To better understand the relevance of this finding, it should be borne in mind that a Ferrari can be driven at the top speed of a Fiat 500, though not vice versa. This example may help to understand why the onset of an epileptic seizure facilitates the onset of CSD to a greater degree than CSD facilitates the onset of an epileptic seizure. Indeed, a Ferrari is usually driven on fast roads, such as highways (myelinic), whereas a Fiat 500 tends to be driven on minor roads (amyelinic), though it must be stressed that a Ferrari can easily be driven on roads (amyelinic) usually taken by a Fiat 500, while the reverse is not true. According to these reflections, it is noteworthy that the patterns recorded by Fabricius et al. (2008) were consistent within the same patient, but differed between patients: highways (myelinic) and minor roads (amyelinic) within the same patient do not usually vary to any great extent, at least not over a relatively short period of time.

Yet another important finding reported by Fabricius et al. (2008), which confirmed our hypothesis (Parisi et al. 2008, 2012a, b, 2013a, b; Belcastro et al. 2011a, 2013; Kasteleijn-Nolst Trenité et al. 2010), is that CSD in their sample was encountered more often than seizures, as demonstrated by the fact that there were twice as many patients with CSD/peri-infarct depolarization alone than with CSD/peri-infarct depolarization plus seizures. Moreoever, 10 of the 11 patients with seizure activity also had CSD, while clinical overt seizures were only observed in 1 of the 11 patients, and seizures were not suspected on clinical grounds in any of the remaining 10 patients.

Interestingly, in the so-called IEH case reports (Grossman et al. 1971; Walser and Isler 1982; Laplante et al. 1983; Isler et al. 1987; Niedermeyer 1993; Marks and Ehrenberg 1993; Walker et al. 1995; Ghofrani et al. 2006; Parisi et al. 2007; Piccioli et al. 2009; Perucca et al. 2010; Belcastro et al. 2011c; Fusco et al. 2011; Italiano et al. 2011; Fanella et al. 2012; Cianchetti et al. 2013), patients were both idiopathic (whether photosensitive or not) and symptomatic; moreover, they also often presented a clinical history (personal and/or familial) of epilepsy and migraine. Intermittent photic stimulation evokes headache in patients with a positive photo-paroxysmal response, who may also have visually induced seizures (Table 1) (Parisi et al. 2012b). With regard to the EEG abnormalities recorded in "ictal epileptic headache" cases (Grossman et al. 1971; Walser and Isler 1982; Laplante et al. 1983; Isler et al. 1987; Niedermeyer 1993; Marks and Ehrenberg 1993; Walker et al. 1995; Ghofrani et al. 2006; Parisi et al. 2007; Piccioli et al. 2009; Perucca et al. 2010; Belcastro et al. 2011c; Fusco et al. 2011; Italiano et al. 2011; Fanella et al. 2012; Cianchetti et al. 2013), the same wide spectrum of different EEG patterns (spike-wave activity, "theta" or even "delta" shape, without any spike activity) associated with both CSD and/or seizures were also confirmed "in vivo" by electrocorticography (Fabricius et al. 2008).

Drawbacks: The Current "Ictal Epileptic Headache" Definition Will Inevitably Underestimate the Phenomenon

The proposed IEH criteria are reported in Table 1. Nonetheless, we wish to stress that the IEH criteria inevitably underestimate the number of cases with "ictal epileptic headache" events. Besides highlighting the strengths of "our published criteria" (Parisi et al. 2012a), we would also like to point out "their inevitable drawbacks".

Headache and epilepsy classifications have so far ignored each other. In the ILAE classification, headache is considered exclusively as a possible semiological ictal phenomenon that is included among the "non-motor" features. In particular, headache is described as a "cephalic" sensation, but is not considered as the sole ictal expression of an epileptic seizure. Moreover, headache is not classified as a "pain" (among the "somatosensory" features) or "autonomic" sensation, whereas signs of involvement of the autonomic nervous system, including cardiovascular, gastrointestinal, vasomotor and thermoregulatory functions, are classified as autonomic features. Despite still being considered a controversial issue, we must consider that headache pain may actually originate in the terminal nervous fibers ("vasomotor") in cerebral blood vessels; consequently, headache should be classified as an "autonomic" sensation in the ILAE Glossary and Terminology (Berg et al. 2010). It may thus be possible to interpret headache as the sole expression of an epileptic seizure and classify it as an autonomic seizure.

To explain why headache may be the sole ictal epileptic symptom, Parisi et al. (2012a, b, 2013a, b) previously suggested that an autonomic seizure (i.e. in IEH) remains purely autonomic if ictal neuronal activation of non-autonomic cortical areas fails to reach the symptomatogenic threshold, as previously described for other ictal autonomic manifestations in Panayiotopoulos syndrome (Koutromanidis 2007). In this regard, Parisi et al. have suggested that ictal epileptic headache may be considered an autonomic form of epilepsy, like Panayiotopoulos syndrome, and cases with long-lasting ictal epileptic headache episodes may accordingly even fulfil the criteria that allow them to be considered as "autonomic status epilepticus" (Ferrie et al. 2007).

In addition, it has been suggested (see all Papers by Parisi et al. from 2008 to 2013) that the social stigma attached to epilepsy may explain a general reluctance (Parisi 2009a) (not only in the general public, but even among physicians) to recognize the growing number of documented cases of IEH. Another noteworthy point is that while unequivocal epileptiform abnormalities usually point to a diagnosis of epilepsy, the lack of clear epileptic spike-and-wave activity is frequent in other ictal autonomic manifestations, as well as in patients with a deep epileptic focus arising, for example, from the orbitomesial frontal zone (Nobili 2007). In such cases, ictal epileptic EEG activity may be recorded from the scalp or exclusively by means of deep stereo-EEG recording, sometimes purely by chance (Laplante et al. 1983; Fusco et al. 2011).

Yet another point that deserves attention is the lack of a clear, repetitive EEG headache-associated pattern owing to the fact that ictal EEG recording in such patients does not yield a specific EEG picture. Indeed, different EEG patterns have been recorded during migraine-like complaints in both symptomatic and idiopathic cases (Grossman et al. 1971; Walser and Isler 1982; Laplante et al. 1983; Isler et al. 1987; Niedermeyer 1993; Marks and Ehrenberg 1993; Walker et al. 1995; Ghofrani et al. 2006; Parisi et al. 2007; Piccioli et al. 2009; Perucca et al. 2010; Belcastro et al. 2011c; Fusco et al. 2011; Italiano et al. 2011; Fanella et al. 2012; Cianchetti et al. 2013). Moreover, when EEG anomalies are recorded, no specific cortical correlations emerge (e.g. focal frontal, parietal, temporal, occipital and primary or secondary generalized), as has also been reported (thus confirming our hypothesis) for autonomic manifestations in children affected by Panayiotopoulos syndrome. Lastly, the criteria we propose do not offer the possibility of confirming all suspected cases of ictal epileptic headache by means of intravenous anticonvulsant administration, just as it is not always possible for other types of epileptic seizures; indeed, although there appears to be a clinical response in the vast majority of published cases affected by autonomic seizures as well as in ictal epileptic headache, we cannot be sure whether intravenous anticonvulsant drug administration has the ability to stop a seizure.

For all the afore-mentioned reasons, we firmly believe that the diagnosis of IEH (even according to our proposed new criteria) will remain an underestimated phenomenon owing, in particular, to:

a. the psychosocial stigma attached to this disease;
b. the fact that IEH cannot always be detected from the scalp by means of EEG recording;
c. IEH is rarely refractory to i.v. antiepileptic drug administration, as can instead occur in other type of seizures.

Conclusions

A clear clinical picture of IEH appears to be difficult to obtain. Since its epileptic nature can only be documented by means of ictal EEG recording and simultaneous intravenous antiepileptic drug administration, it is difficult to obtain firm conclusions regarding the frequency of IEH in epidemiological studies. Headache/migraine of epileptic origin must always be suspected in pediatric patients who do not respond to treatment with anti-migraine drugs in order to promptly perform an EEG and thus make a correct diagnosis.

Moreover, ictal epileptic headache may not have the characteristics of migraine with or without aura, or those of a tension headache; indeed, any "type" of headache may be defined as an ictal epileptic headache in the presence of symptoms associated with ictal EEG anomalies (whether focal or generalized) that resolve immediately after i.v. administration of an anticonvulsant drug.

On the basis of current knowledge and clinical experiences, "migralepsy" or a migraine-triggered seizure is highly unlikely to exist. We thus believe that these terms should be removed from the Appendix of International Headache Disorders Classifications until clear evidence is provided of the existence of such conditions.

Ictal Epileptic Headache criteria (Table 1) should be used to classify the rare events in which headache may represent the sole ictal epileptic manifestation. "These findings further highlight the important role of EEG recording in patients with headache, which has traditionally been opposed by fierce ancestral adversity (Parisi 2009a) against the possible link between headache and epilepsy".

We as a group thus suggest that the term ictal epileptic headache be maintained for cases in which headache is "the sole ictal manifestation", and that the term "ictal headache" be maintained for cases in which the headache, whether brief or long-lasting, is merely part of a more complex seizure including ictal manifestations that are either sequential or overlapping (sensory–motor, psychiatric or other non-autonomic manifestations). In our opinion, this distinction is crucial, as has been explained in detail in this chapter, owing to the markedly different prognosis it entails. In fact, this is not a marginal question, because these possible, isolated, non-motor, ictal manifestations (i.e. ictal epileptic headache) should be taken into account before declaring an epileptic patient as "seizure free" so as to be able to suspend anticonvulsant therapy safely.

In conclusion, by applying the ictal epileptic headache criteria proposed here (Table 1) to a large pediatric population in the future, we should be able to understand whether ictal epileptic headache is a marginal phenomenon or is, instead, an underestimated event that deserves greater attention.

Acknowledgments My interest and my commitment to research in the field of pediatric neurology have been made possible only by the patience, trust and love of my parents, Maria and Alfonso, my son, Eduardo Alfonso, and my wife, Loredana, to whom I will forever be indebted, as well as by the support of all the people I have met over the years.

References

Anderman F (1987) Clinical features of migraine-epilepsy syndrome. In: Andermann F, Lugaresi E (eds) Migraine and epilepsy. Butterworth Publishers, Boston, pp 20–89

Andermann F, Zifkin B (1998) The benign occipital epilepsies of childhood: an overview of the idiopathic syndromes and of the relationship to migraine. Epilepsia 39:9–23

Ayata C (2010) Cortical spreading depression triggers migraine attack: pro. Headache 50(4):725–730

Ayata C, Jin H, Kudo C, Dalkara T, Moskowitz MA (2006) Suppression of cortical spreading depression in migraine prophylaxis. Ann Neurol 59(4):652–661

Baca CB, Vickrey BG, Caplan R, Vassar SD, Berg AT (2011) Psychiatric and medical comorbidity and quality of life outcomes in childhood-onset epilepsy. Pediatrics 128(6):e1532–e1543

Barré M, Hamelin S, Minotti L, Kahane P, Vercueil L (2008) Epileptic seizure and migraine visual aura: revisiting migralepsy. Rev Neurol (Paris) 164:246–252

Belcastro V, Striano P, Kasteleijn-Nolst Trenite DGA, Villa MP, Parisi P (2011a) Migralepsy, hemicrania epileptica post-ictal headache and "ictal epileptic headache": a proposal for terminology and classification revision. J Headache Pain 12:289–294

Belcastro V, Striano P, Parisi P (2011b) Seizure or migraine? The eternal dilemma. Comment on: "recurrent occipital seizures misdiagnosed as status migrainosus". Epileptic Disord 13(4):456

Belcastro V, Striano P, Pierguidi L, Calabresi P, Tambasco N (2011c) Ictal epileptic headache mimicking status migrainosus: EEG and DWI-MRI findings. Headache 51:160–162

Belcastro V, Striano P, Parisi P. (2012) "Ictal epileptic headache": beyond the epidemiological evidence. Epilepsy Behav 25(1):9–10

Belcastro V, Striano P, Parisi P (2013) From migralepsy to ictal epileptic headache: the story so far. Neurol Sci 34(10):1805–1807

Berg AT, Berkovic SF, Brodie MJ et al (2010) Revised terminology and concepts for organization of seizures and epilepsies: report of the ILAE Commission on Classification and Terminology, 2005–2009. Epilepsia 51:676–685

Berger M, Speckmann EJ, Pape HC, Gorji A (2008) Spreading depression enhances human neocortical excitability in vitro. Cephalalgia 28:558–562

Bladin PF (1987) The association of benign rolandic epilepsy with migraine. In: Andermann F, Lugaresi E (eds) Migraine and epilepsy. Butterworth Publishers, Boston, pp 145–52

Bolay H, Reuter U, Dunn AK, Huang Z, Boas DA, Moskowitz MA (2002) Intrinsic brain activity triggers trigeminal meningeal afferents in migraine model. Nat Med 8:136–142

Cai S, Hamiwka LD, Wirrell EC (2008) Peri-ictal headache in children: prevalence and character. Pediatr Neurol 39:91–96

Chen SC (2006) Epilepsy and migraine: the dopamine hypotheses. Med Hypotheses 66:466–472

Cianchetti C, Pruna D, Porcu L, Peltz MT, Ledda MG (2013) Pure epileptic headache and related manifestations: a video-EEG report and discussion of terminology. Epileptic Disord 15(1):84–92

Clarke T, Baskurt Z, Strug LJ et al (2009) Evidence of shared genetic risk factors for migraine and rolandic epilepsy. Epilepsia 50:2428–2433

Colombo B, Dalla Libera D, De Feo D, Pavan G, Annovazzi PO, Comi G (2011) Delayed diagnosis in pediatric headache: an outpatient Italian survey. Headache 51(8):1267–1273

De Fusco MM, Silvestri L et al (2003) Haploinsufficiency of ATP1A2 encoding the Na+/K+ pump a2 subunit associated with familial hemiplegic migraine type 2. Nat Genet 33:192–196 de Souza TK, e Silva MB, Gomes AR, de Oliveira HM, Moraes RB, de Freitas Barbosa CT et al (2011) Potentiation of spontaneous and evoked cortical electrical activity after spreading depression: in vivo analysis in wellnourished and malnourished rats. Exp Brain Res 214:463–469

De Vries B, Frants RR, Ferrari MD, van den MAM (2009) Molecular genetics of migraine. Hum Genet 126:115–132

Dichgans M, Freilinger T, Eckstein G, Babini E, Lorenz-Depiereux B, Biskup S (2005) Mutations in the neuronal voltage-gated sodium channel SCN1A in familial hemiplegic migraine. Lancet 366:371–377

Eikermann-Haerter K, Ayata C (2010) Cortical spreading depression and migraine. Curr Neurol Neurosci Rep 10(3):167–173

Escayg A, Goldin AL (2010) Critical review and invited commentary: sodium channel SCN1A and epilepsy: mutations and mechanisms. Epilepsia 51:1650–1658

Fabricius M, Fuhr S, Willumsen L, Dreier JP, Bhatia R, Boutelle MG et al (2008) Association of seizures with cortical spreading depression and peri-infarct depolarisations in the acutely injured human brain. Clin Neurophysiol 119(9):1973–1984

Fanella M, Fattouch J, Casciato S, Lapenta L, Morano A, Egeo G et al (2012) Ictal epileptic headache as "subtle" symptom in generalized idiopathic epilepsy. Epilepsia 53(4):e67–e70

Faraguna U, Nelson A, Vyazovskiy VV, Cirelli C, Tononi G (2010) Unilateral cortical spreading depression affects sleep need and induces molecular and electrophysiological signs of synaptic potentiation in vivo. Cereb Cortex 20:2939–2947

Ferrie CD, Caraballo R, Covanis A et al (2007)Autonomic status epilepticus in Panayiotopoulos syndrome and other childhood and adult epilepsies: a consensus view. Epilepsia 48(6):1165–1172

Fogarasi A, Janszky J, Tuxhorn I (2006) Autonomic symptoms during childhood partial epileptic seizures. Epilepsia 47:584–588.

Forderreuther S, Henkel A, Noachtar S et al (2002) Headache associated with epileptic seizures: epidemiology and clinical characteristics. Headache 42:649–655

Fusco L, Specchio N, Ciofetta G, Longo D, Trivisano M, Vigevano (2011) Migraine triggered by epileptic discharges in a Rasmussen's encephalitis patient after surgery. Brain Dev 33:597–600

Gambardella A, Marini C (2009) Clinical spectrum of SCN1A mutations. Epilepsia 50:20–23

Ghadiri MK, Kozian M, Ghaffarian N, Stummer W, Kazemi H, Speckmann EJ et al (2012) Sequential changes in neuronal activity in single neocortical neuron after spreading depression. Cephalalgia 32(2):116–124

Ghofrani M, Mahvelati F, Tonekaboni H (2006) Headache as a sole manifestation in nonconvulsive status epilepticus. J Child Neurol 21:981–983

Gigout S, Louvel J, Kawasaki H, D'Antuono M, Armand V, Kurcewicz I (2006) Effects of gap junction blockers on human neocortical synchronization. Neurobiol Dis 22:496–508

Gowers WR (1907) The border-land of epilepsy, Chapter V. Reprint 1995. Arts & Boeve, Nijmegen, pp 76–102

Grossman RM, Abramovich I, Lefebvre AB (1971) Epileptic headache: report of a case with EEG recorded during the crisis. Arq Neuropsiquiatr (San Paulo) 29:198–206

Haglund MM, Schwartzkroin PA (1990). Role of Na-K pump potassium regulation and IPSPs in seizures and spreading depression in immature rabbit hippocampal slices. J Neurophysiol 63(2):225–239

Hauser AW, Annegers JF, Anderson EV, Kurlan LT (1993) The incidence of epilepsy and unprovoked seizure in Rochester, Minnesota, 1935–1984. Epilepsia 34:453–468

Heyck H, Hess R (1955) Vasomotoric headaches as symptom of masked epilepsy. Schweiz Med Wochenschr J Suisse Med 85(24):573–575

International Headache Society (2004) The international classification of headache disorders: 2nd edition. Cephalalgia 24(1):9–160

International Headache Society. (2013) The international classification of headache disorders: third edition. Cephalalgia 33(9):629–808

Isler H, Wieser HG, Egli M (1987) Hemicrania epileptica: synchronous ipsilateral ictal headache with migraine features. In: Andermann F, Lugaresi E (eds) Migraine and epilepsy. Butterworth Publishers, Boston, pp 249–263

Italiano D, Grugno R, Calabro' RS, Bramanti P, Di Maria F, Ferlazzo E (2011) Recurrent occipital seizures misdiagnosed as status migrainosus. Epileptic Disord 13:197–201

Ito M, Schachter SC (1996) Frequency and characteristics of interictal headaches in patients with epilepsy. J Epilepsy 9:83–86

Ito M, Adachi N, Nakamura F et al (2003) Multi-center study on post-ictal headache in patients with localization-related epilepsy. Psychiatry Clin Neurosci 57:385–389

Ito M, Adachi N, Nakamura F et al (2004) Characteristics of postictal headache in patients with partial epilepsy. Cephalalgia 24(1):23–28

Jen JC, Wan J, Palos TP (2005) Mutations in the glutamate transporter EAAT1 causes episodic ataxia, hemiplegia, and seizures. Neurology 65:529–534

Johnson MP, Griffiths LR (2005) A genetic analysis of serotonergic biosynthetic and metabolic enzymes in migraine using a DNA pooling approach. J Hum Genet 50:607–610

Kasteleijn-Nolst Trenité DGA, Verrotti A, Di Fonzo A et al (2010a) Headache, epilepsy and photosensitivity: how are they connected? J Headache Pain 11:469–476

Kasteleijn-Nolst TDGA, Cantonetti L, Parisi P (2010b) Visual stimuli, photosensitivity and photosensitive epilepsy. Chapter 94. In: Shorvon S, Guerrini R, Andermann F (eds) Common and uncommon causes of epilepsy. Cambridge University, Cambridge

Kasteleijn-Nolst Trenité DGA, Parisi P (2012) Migraine in the borderland of epilepsy: "migralepsy" an overlapping syndrome of children and adults? Epilepsia 53(Suppl 7):20–25

Kinast M, Lueders H, Rothner AD et al (1982) Benign focal epileptiform discharges in childhood migraine. Neurology 32:1309–1311

Kors EE, Melberg A, Vanmolkot KR, Kumlien E, Haan J, Raininko R, Flink R, Ginjaar HB, Frants RR, Ferrari MD, van den Maagdenberg AM (2004) Childhood epilepsy, familial hemiplegic migraine, cerebellar ataxia, and a new CACNA1A mutation. Neurology 63(6):1136–1137

Koutroumanidis M (2007) Panayiotopoulos syndrome: an important electroclinical example of benign childhood system epilepsy. Epilepsia 48:1044–1053

Laplante P, Saint-Hilaire JM, Bouvier G (1983) Headache as an epileptic manifestation. Neurology 33:1493–1495

Lateef TM, Cui L, Nelson KB, Nakamura EF, Merikangas KR (2012) Physical comorbidity of migraine and other headaches in US adolescents. J Pediatr 161(2).308–313

Leniger T, Isbruch K, Driesch S, Diener HC, Hufnagel A (2001) Seizure-associated headache in epilepsy. Epilepsia 42:1176–1179

Lennox WG, Lennox MA (1960) Epilepsy and related disorders. Little Brown, Boston, p 451

Lipton RB, Silberstein SD (1994) Why study the comorbidity of migraine? Neurology 44(10, Suppl 7):S4–S5

Lipton RB, Ottman R, Ehrenberg BL, Hauser WA (1994a) Comorbidity of migraine: the connection between migraine and epilepsy. Neurology 44(Suppl 7):S28–S32

Lonnqvist T, Paeteau A, Valanne L, Pihko H (2009) Recessive twinkle mutations cause severe epileptic encephalopathy. Brain 132:1553–1562

Lugaresi E (1955) EEG investigations in monosymptomatic headache in infants. Riv Neurol 25(4):582–588

Lyngberg AC, Rasmussen BK, Jørgensen T, Jensen R (2005) Incidence of primary headache: a Danish epidemiologic follow-up study. Am J Epidemiol 161(11):1066–1073

Marks DA, Ehrenberg BL (1993) Migraine-related seizures in adults with epilepsy, with EEG correlation. Neurology 43:2476–2483

Morocutti C, Vizioli R (1957) Episodes of paroxysmal headache as the only clinical manifestation of idiopathic epilepsy. Riv Neurol 27(4):427–430

Moskowitz MA, Nozaki K, Kraig RP (1993) Neocortical spreading depression provokes the expression of C-fos proteinlike immunoreactivity within trigeminal nucleus caudalis via trigeminovascular mechanisms. J Neurosci 13:1167–1177

Niedermeyer E (1993) Migraine-triggered epilepsy. Clin EEG 24:37–43

Nobili L (2007) Nocturnal frontal lobe epilepsy and non-rapid eye movement sleep parasomnias: differences and similarities. Sleep Med Rev 11:251–254

Nymgard K. (1956) Epileptic headache. Acta Psych Neurol Scand: Suppl 108:291–300

Olafsson E, Ludvigsson P, Gudmundsson G, Hesdorffer D, Kjartansson O, Hauser WA (2005) Incidence of unprovoked seizures and epilepsy in Iceland and assessment of the epilepsy syndrome classification: a prospective study. Lancet Neurol 4(10):627–634

Panayiotopoulos CP (1987) Difficulties in differentiating migraine and epilepsy based on clinical EEG findings. In: Andermann F, Lugaresi E (eds) Migraine and epilepsy. Butterworth Publishers, Boston, pp 142–151

Panayiotopoulos CP (1999a) Visual phenomena and headache in occipital epilepsy: a review, a systematic study and differentiation from migraine. Epileptic Disord 1:205–216

Panayiotopoulos CP (1999b) Differentiating occipital epilepsies from migraine with aura, acephalic migraine and basilar migraine. In: Panayiotopoulos CP (ed) Benign childhood partial seizures and related epileptic syndromes. John Libbey & Company Ltd, London, pp 281–302

Panayiotopoulos CP (1999c) Elementary visual hallucination, blindness, and headache in idiopathic occipital epilepsy: differentiation from migraine. J Neurol Neurosurg. Psychiatry 66:536–540

Panayiotopoulos CP (2012) The new ILAE report on terminology and concepts for the organization of epilepsies: critical review and contribution. Epilepsia 53(3):399–404

Parisi P (2009a) Who's still afraid of the link between headache and epilepsy? Some reactions to and reflections on the article by Marte Helene Bjørk and co-workers. J Headache Pain 10(5):327–329

Parisi P (2009b) Why is migraine rarely, and not usually, the sole ictal epileptic manifestation? Seizure 18(5):309–312

Parisi P (2011) Comments on the article by Fusco L. et al. entitled "Migraine triggered by epileptic discharges in a Rasmussen's encephalitis patient after surgery". Brain Dev 33(8):704–705

Parisi P, Kasteleijn-Nolst Trenite DGA. (2010) "Migralepsy": a call for revision of the definition. Epilepsia 51(5):932–933

Parisi P, Kasteleijn-Nolst TDG, Piccioli M et al (2007) A case with atypical childhood occipital epilepsy "Gastaut type": an ictal migraine manifestation with a good response to intravenous diazepam. Epilepsia 48:2181–2186

Parisi P, Piccioli M, De Sneeuw S, de Sneeuw S, de Kovel C, van Nieuwenhuizen O, Buttinelli C, Villa MP, Kasteleijn-Nolst Trenité DG (2008a) Redefining headache diagnostic criteria as epileptic manifestation? Cephalalgia 28:408–409

Parisi P, Piccioli M, Villa MP, Buttinelli C, Kasteleijn-Nolst Trenéte DGA (2008b) Hypothesis on neurophysiopathological mechanisms linking epilepsy and headache. Med Hypoth 70:1150–1154

Parisi P, Striano P, Kasteleijn-Nolst Trenite DGA et al (2012a) "Ictal Epileptic Headache": recent concepts for new classifications criteria. Cephalalgia 32(9):723–724

Parisi P, Striano P, Negro A, Martelletti P, Belcastro V (2012b) Ictal epileptic headache: an old story with courses and appeals. J Headache Pain 13(8):607–613

Parisi P, Striano P, Belcastro V (2013a) The crossover between headache and epilepsy. Expert Rev Neurother 13(3):231–233

Parisi P, Striano P, Verrotti A, Villa MP, Belcastro V (2013b) What have we learned about ictal epileptic headache? A review of well-documented cases. Seizure 22(4):253–258

Perucca P, Terzaghi M, Manni R (2010) Status epilepticus migrainosus: clinical, electrophysiologic, and imaging characteristics. Neurology 75:373–374

Piccioli M, Parisi P, Tisei P, Villa MP, Buttinelli C, Kasteleijn-Nolst Trenité DGA (2009) Ictal headache and visual sensitivity. Cephalalgia 29:194–203

Piccinelli P, Borgatti R, Nicoli F et al (2006) Relationship between migraine and epilepsy in paediatric age. Headache 46:413–421

Pietrobon D (2010) Biological science of headache channels. Handb Clin Neurol 97:73–83

Pinto DJ, Patrick SL, Huang WC, Connors BW (2005) Initiation, propagation and termination of epileptiform activity in rodent neocortex in vitro involve distinct mechanisms. J Neurosci 25(36):8131–8140

Riant F, Ducros A, Ploton C, Banbance C, Depienne C, Tournie-Lasserve E (2010) De novo mutations in ATP1A2 and CACNA1A are frequent in early-onset sporadic hemiplegic migraine. Neurology 75:967–972

Sances G, Guaschino E, Perucca P, Allena M, Ghiotto N, Manni R (2009) Migralepsy: a call for revision of the definition. Epilepsia 50:2487–2496

Sasmaz T, Bugdayci R, Ozge A, Karakelle A, Kurt O, Kaleagasi H (2004) Are parents aware of their schoolchildren's headaches? Eur J Public Health 14(4):366–368

Somjen GG (2001) Mechanisms of spreading depression and hypoxic spreading depression-like depolarization. Physiol Rev 81:1065–1096

Steinlein OK (2004) Genetic mechanisms that underlie epilepsy. Nat Rev Neurosci 5:400–408

Striano P, Belcastro V, Parisi P (2011) Status epilepticus migrainosus: clinical, electrophysiologic, and imaging characteristics. Neurology 76:761

Striano P, Belcastro V, Parisi P (2012) From "migralepsy" to "ictal epileptic headache" concept. Epilepsy Behav 23(3):392

Tamura K, Alessandri B, Heimann A, Kempski O (2011) The effects of a gap-junction blocker, carbenoxolone, on ischemic brain injury and cortical spreading depression. Neuroscience 194:262–271

Toldo I, Perissinotto E, Menegazzo F et al (2010) Comorbidity between headache and epilepsy in a pediatric headache center. J Headache Pain 11:235–240

Tonini MC, Giordano L, Atzeni L (2012) Primary headache and epilepsy: a multicenter cross-sectional study. Epilepsy Behav 23:342–347

Tottene A, Conti R, Fabbro A, Vecchia D, Shapovalova M, Santello M, van den Maagdenberg AM, Ferrari MD, Pietrobon D (2009) Enhanced excitatory transmission at cortical synapses as the basis for facilitated spreading depression in Ca(v)2.1 knockin migraine mice. Neuron 61(5):762–773

Tottene A, Urbani A, Pietrobon D (2011) Role of different voltage-gated Ca2+ channels in cortical spreading depression. Channels (Austin) 5(2):110–114

Tzoulis C, Engelsen BA, Telstad W (2006) The spectrum of clinical disease caused by the A467T and W748S POLG mutations: a study of 26 causes. Brain 129:1685–1692

Uchitel OD, Inchauspe CG, Urbano FJ, Di Guilmi MN (2012) Ca(V)2.1 voltage activated calcium channels and synaptic transmission in familial hemiplegic migraine pathogenesis. J Physiol Paris 106(1–2):12–22

Van Den Maagdenberg AM, Terwindt GM, Haan J, Frants RR, Ferrari MD (2010) Genetics of headaches. Handb Clin Neurol 97:85–97

Vanmolkot KR, Kors EE, Hottenga JJ, Terwindt GM, Haan JJ, Hoefnagels WA (2003) Novel mutations in the Na+/K+-ATPase pump gene ATP1A2 associated with familial hemiplegic migraine and benign familial infantile convulsions. Ann Neurol 54:360–366

Verrotti A, Coppola G, Spalice A et al (2011a) Peri-ictal and inter-ictal headache in children and adolescents with idiopathic epilepsy: a multicenter cross-sectional study. Childs Nerv Syst 27:1419–1423

Verrotti A, Coppola G, Di Fonzo A et al (2011b) Should "migralepsy" be considered an obsolete concept? A multicenter retrospective clinical/EEG study and review of the literature. Epilepsy Behav 21(1):52–59

Verrotti A, Striano P, Belcastro C et al (2011c) Migralepsy and related conditions: advances in pathophysiology and classification. Seizure 20:271–275

Walker MC, Smith SJM, Sisodya SM, Shorvon SD (1995) Case of simple partial status epilepticus in occipital lobe epilepsy misdiagnosed as migraine: clinical, electrophysiological, and magnetic resonance imaging characteristics. Epilepsia 36:1233–1236

Walser H, Isler H (1982) Frontal intermittent rhythmic delta activity. Impairment of consciousness and migraine. Headache 22(2):74–80

Wendorff J, Juchniewicz B (2005) Photosensitivity in children with idiopathic headaches. Neurol Neurochir Pol 39(4 Suppl 1):S9–S16

Wirrell EC, Hamiwka LD (2006) Do children with benign rolandic epilepsy have a higher prevalence of migraine than those with other partial epilepsies or nonepilepsy controls? Epilepsia 47:1674–1681

Yamamane LE, Montenegro MA, Guerreiro MM (2004) Comorbidity headache and epilepsy in childhood. Neuropediatrics 35(2):99–102

Yankovsky AE, Andermann F, Bernasconi A (2005) Characteristics of headache associated with intractable partial epilepsy. Epilepsia 46:1241–1245

Zhang X, Levy D, Kainz V, Noseda R, Jakubowski M, Burstein R (2011) Activation of central trigeminovascular neurons by cortical spreading depression. Ann Neurol 69(5):855–865

Epilepsy and Immune System: A Tour Around the Current Literature

Laura Mumoli, Angelo Labate, Antonietta Coppola, Giovambattista De Sarro, Emilio Russo and Antonio Gambardella

Abstract It is widely acknowledged that immune system influences several aspects of the central nervous system. Literature data have shown that immune system and autoimmune response play an important role in the pathogenesis of several neurodegenerative/neurological diseases (i.e Parkinson's and Alzheimer's Diseases, Multiple Sclerosis). However, very recent evidences of specific antibodies found in epileptic encephalitis, the good response to immune therapy in refractory epileptic syndromes and the strong relationship between systemic autoimmune disease and epilepsy suggest a plausible role for the immune system also in paroxysmal neurological disorders. In fact, an immune hypothesis represents a new way to approach epilepsy and could contribute to clarify several unanswered questions in the next future. In this review, we analysed these points mimicking a tour around current evidences from experimental animal models to clinical suggestions.

First Stop: Background

Although it is widely accepted that epilepsy can be defined as the persistent spontaneous tendency to generate seizure that underlies a persistent brain hyperexcitability, the exact physiopathology of epileptogenesis still remains unclear (Goldberg and Coulter 2013).

E. Russo (✉) · G. De Sarro
Department of Health Sciences, University "Magna Graecia" of Catanzaro, Catanzaro, Italy
e-mail: erusso@unicz.it

L. Mumoli · A. Labate · A. Gambardella
Institute of Neurology, University "Magna Graecia" of Catanzaro, Catanzaro, Italy

A. Labate · A. Gambardella
Institute of Molecular Bioimaging and Physiology, National Research Council (IBFM-CNR), Viale Europa, 88100 Germaneto, CZ, Italy

A. Coppola
Epilepsy Centre, Department of Neuroscience, Reproductive and Odontostomatological Sciences, Federico II University, Naples, Italy

Throughout the last few decades different hypotheses spanning from genetics to environmental or infective factors have been proposed to explain this latter phenomenon but unfortunately each attempt failed to give a unique and satisfactory explanation; even though the general knowledge has been widely improved (Goldberg and Coulter 2013).

Since 1977, several reports documented the efficacy of immunomodulating therapy (e.g. corticosteroids, immunoglobulin) in refractory epilepsy and based on this clinical observation a plausible immune origin has been postulated (Peachardre et al. 1977).

Currently, the recent findings of experimental studies have shown that the brain innate immunological cells such as microglia and astrocytes are able to produce cytokines and other mediators of inflammation contributing to seizure activity (Friedman and Dingledine 2011; Granata et al. 2011). These latest discoveries together with the detection of specific autoantibodies against channels or neuronal surface proteins in epileptic syndromes (Vincent et al. 2011b; Zuliani et al. 2012) and the intriguing findings that some mutated genes (LG1, KCNQ2-KCNQ3) are the cause of some types of epileptic syndromes have contributed to the immune theory as a plausible trigger or contributor of epilepsy.

In this article, we reviewed evidences for the immune system involvement in epilepsy, mostly focusing our attention on epilepsy syndromes strongly or suggestively linked to the immune system. We first describe what is our current knowledge obtained from experimental models, secondly from clinical studies showing a close relationship between the immune system and epilepsy.

Second Stop: Immunity Involvement in Experimental Models of Epilepsy

The role of immune activation in the generation of acute seizures and epilepsy has been investigated in animal models. While the role of inflammation during or after seizures in several models has been more systematically studied (Vezzani et al. 2011b, 2012), in comparison, only few studies have been pointed to the study of the immune system response and seizures following infections or the injection of LPS (Harvey and Boksa 2012). Some interesting results have been obtained studying seizures' generation in models of bacterial meningitis involving injection of group B streptococcus in infant rats (Kim et al. 1995; Kolarova et al. 2003) and models of viral CNS infection involving administration of Theiler's murine encephalomyelitis virus in young mice (Libbey and Fujinami 2011; Libbey et al. 2011). On the other hand, an enhancement in seizure susceptibility to convulsant drugs (e.g. lithium-pilocarpine, kainic acid) or an increase in seizure number and severity following LPS has also been characterized (Arican et al. 2006; Dmowska et al. 2010; Lee et al. 2000; Mirrione et al. 2010; Russo et al. 2013, 2014). Furthermore, several research groups provided evidence that, early in postnatal life, infection/immune activation can lead to long-lasting increased seizure susceptibility in adulthood.

Stewart et al. (Stewart et al. 2010) showed that young mice (P28–35) developed spontaneous epileptic seizures 2–7 months following injection of Theiler's murine encephalomyelitis virus. Similarly, Galic et al. (2009) have shown that intracerebroventricular injection of poly (I:C) in P14 rat pups increases seizure's susceptibility at adulthood; these data supports the concept that early exposure to either bacterial or viral immunogens can contribute or can be the cause of seizures and epilepsy also in adulthood.

Inflammation consists of the production of a cascade of inflammatory mediators, as well as anti-inflammatory molecules and other molecules induced to resolve inflammation, as a response to noxious stimuli (such as infection or injury), or immune stimulation, and is designed to defend the host against pathogenic threats (Vezzani et al. 2011b). The innate/adaptive immunity have been implicated in epilepsy; microglia, astrocytes and neurons are believed to contribute to the innate immunity-type processes causing brain's inflammation (Vezzani et al. 2011b). The activation of innate immunity and the transition to adaptive immunity are mediated by a large variety of inflammatory mediators, including cytokines-polypeptides, which play a pivotal role (Nguyen et al. 2002; Vezzani et al. 2011a, c). Cytokines are released by immunocompetent and endothelial cells, as well as by glia and neurons in the CNS, thereby enabling communication between effector and target cells during an immune challenge or tissue injury (Vezzani et al. 2011b). The most extensively studied are represented by IL-1beta, TNF-alpha and IL-6 (Friedman and Dingledine 2011). The major results in the field both support the concept that inflammation causes seizures or seizures cause neuroinflammation. Regarding the latter, several evidences have been reported regarding neuroinflammation-induction in the brain following single or prolonged seizures in various models in areas directly involved in seizure activity (Devinsky et al. 2013; Dhote et al. 2007; Foresti et al. 2011; Librizzi et al. 2012; Maroso et al. 2011; Ravizza et al. 2008; Reid et al. 2013). On the other hand, several studies support the role of inflammation in seizure generation (Auvin et al. 2007, 2009, 2010a, b, 2012; Galic et al. 2012; Marchi et al. 2012; Vezzani et al. 2008), even if it has recently been reported that LPS intrahippocampal infusion inhibits the development of kindling in rats (Ahmadi et al. 2013).

Neuronal hyperexcitability may be affected by cytokines in different ways (Silveira et al. 2012). IL-1beta contribute to excitotoxicity increasing calcium influx into neurons by N-methyl-D-aspartate (NMDA) receptors activation (Viviani et al. 2003), inhibiting glutamate reuptake by astrocytes (Hu et al. 2000). In addition, application of IL-1beta to hippocampal neurons results in the block of gamma-aminobutyric acid type A (GABAA) receptor function (Wang et al. 2000). Similarly, TNF-alpha has also been implicated in GABAA transmission decrease by endocytosis of its receptors in hippocampal pyramidal cells (Stellwagen et al. 2005). Neuroanatomic, imaging, neurochemical, and mechanistic approaches in vivo have been employed, and effector molecules of the cytokine have been described (Balosso et al. 2008; Friedman and Dingledine 2011; Maroso et al. 2010; Viviani et al. 2003). The involvement of TNF-alpha in programming neuronal excitability acutely and long term has also been examined (Riazi et al. 2010). Involvement of glutamate receptors in these TNF-alpha–mediated effects is suggested by the evidenced changes in glutamate receptor subunit expression after LPS administration (Harre et al. 2008) as well as in TNF-alpha receptor subtypes knockout mice (Balosso et al. 2009), and by the ability of TNF-alpha to induce rapid changes in AMPA receptor subunit expression and function (Stellwagen et al. 2005). Finally, the recent involvement of mTOR pathway during epileptogenesis might also be considered as suggestive of an involvement of immune system (Russo et al. 2014). The role of the mTOR signaling pathway on the innate immune system and neuroinflammatory processes in the diseased brain has now begun to receive considerable attention (Maiese et al. 2013; Russo et al. 2013; Wang et al. 2013); this is despite the fact that the general immunomodulatory functions of mTOR have been widely studied, and rapamycin (RAP), a specific mTOR inhibitor, is a commonly used and successful immunosuppressant drug (Chi 2012). Recent findings support a specific interplay between neuroinflammation and mTOR pathway indicating the latter as a suitable target for the treatment of epilepsy and epileptogenesis (Curatolo and Moavero 2013; Liu et al. 2014; Russo et al. 2014).

Third Stop: Clinical Evidence of the Relationship Between Immune System and Epilepsy

Types of Epilepsy Strongly Linked with an Immune–Involvement

Rasmussen Encephalitis

The first evidence of fascinating relationship between seizures and immune-system dates back to 1958 when Theodore Rasmussen described a case of encephalitis later called Rasmussen encephalitis (RE). RE is a dramatic rare acquired disease with unfavourable prognosis that is characterized by progressive focal cortical unilateral hemispheric atrophy. Epilepsy is one of main feature of RE, especially unilateral untreatable motor partial seizure is the first sign of RE and remains for all the course of disease. It's possible to identify three clinical phases in RE: initially a prodromal stage (mean duration 7.1 months) characterized by low seizure frequency and mild hemiparesis (this phase could be absent in several cases); secondly an acute stage (RE could present directly in this stage) (mean duration 8 months) identified by high rate of simple partial untreatable motor seizure called "epilepsia partialis continua" (Bien et al. 2005; Granata et al. 2003b) reported it in 56–92 % of subjects) plus the appearance of neurological signs as progressive cognitive deterioration, hemiparesis, hemiatrophia and aphasia (if the dominant hemisphere is involved); finally, a residual stage characterized by stable neurological deficits and seizures with a lower frequency compared to acute phase. The duration of each stage is correlated with the severity of destructive processes even if there is a high rate of variability for each case.

Recently, new scientific discoveries have been achieved leading to an expansion of RE's syndrome spectrum, modifying and completing previous concepts. In fact, despite RE was considered a childhood's disease, it is currently accepted that it could occur at all ages especially in adolescent and even in adult subjects (Gambardella et al. 2008). An adulthood onset of RE is usually less severe, tends not only to progress more slowly but also more likely to respond to immunologic treatment (Hart et al. 1997; Leach et al. 1999; McLachlan et al. 1993). Furthermore, the clinical manifestations of RE related to "epilepsia partialis continua" and movement disorders as hemidystonia and hemiathetosis are emerging as adjunctive signs of RE and it seems to correlate with atrophy of the head of caudate nucleus (Gambardella et al. 2008). This latter radiological finding in addition to a progressive mono-hemispheric focal cortical atrophy and grey/white matter T2/FLAIR hyperintensities are considered one of the diagnostic neuroimaging criteria to diagnose RE (Bien et al. 2005). There are also very few reports of bilateral RE usually showing the presence of an underlying dual pathology (for example RE plus low grade tumour, cortical dysplasia, tuberosis sclerosis) (Chinchilla et al. 1994; Firlik et al. 1999; McLachlan et al. 1993; Palmer et al. 1999; Tobias et al. 2003).

It is very well-known that an early surgical exclusion of the affected hemisphere played a major role in seizure treatment (Wennberg et al. 1997) however, clinical experiences suggested that a combinations of corticosteroids, apheresis and high-dose IV immunoglobulin immunomodulatory treatment could be considered in some selective cases (Granata et al. 2003a).

The only unsolved concern regards the possible immune genesis of RE. Neuropathological studies of RE showing the presence of inflammatory infiltrates of T cells plus astrogliosis, suggest an involvement of adaptive part of immune system in RE disease (Bauer et al. 2007). In fact, histopathological-immunohistochemical studies on RE brain have shown that T lymphocytes not only are the main pathologic features of disease but are involved in neuronal death and damage through an apoptosis phase due to the release of granzyme B (Bien et al. 2002). The reasons for this attack against neurons and microglia activated by T lymphocytes is still unknown and needs to be clarified even if a viral hypothesis could be suggested but never definitely demonstrate.

Additionally, in 1990s the evidence in RE patients' serum of antibodies against glutamate receptor 3 (GluR3) has been pointed out as the possible cause for neuronal death through a damage mediated by antibodies or complement (Saiz et al. 2008).

Conversely, other studies have demonstrated that GluR3 antibodies are not present in all RE patients and are not specific of RE being observed in other epileptic syndromes, too (Mantegazza et al. 2002; Watson et al. 2004; Wiendl et al. 2001)

Limbic Encephalitis "neuronal surface antibodies syndrome"

Since upon the first description of Corsellis and colleagues (1968) limbic encephalitis (LE) was considered a rare paraneoplastic disorder with poor prognosis. The underlying immune-mediated pathogenesis was identified in a cytotoxic response of T cell induced by onco-neuronal antibodies against intracellular antigens (Tuzun and Dalmau 2007). The disorder is characterized by a subacute onset of episodic memory loss, disorientation and behavioural changes associated with seizures, hallucinations, sleep disturbance. Neuroimaging usually shows signal changes in T2-weighted images, FLAIR sequences or diffusion in medial temporal lobe (Asztely and Kumlien 2012). It is becoming frequent the evidence that LE is not a classical onco-neuronal disorder but is associated with antibodies binding cell surface called "neuronal surface antibodies" (NSAbs) (Zuliani et al. 2012). This new entity is clinically similar to paraneoplastic LE with an important exception, namely the prognosis is favourable. In fact, these types of LE respond so well to immunotherapy to determine a substantial complete recovery (Vincent et al. 2011a). Although the incidence of this disorder is not well-established, current data seems to suggest that is more frequent than all encephalitis associated with paraneoplastic antibodies (Lancaster et al. 2011).

VGKC Complex Encephalitis

This represents a new fascinating chapter where the representative form is the LE associated to antibodies against voltage gated potassium channel complex (VGKC-Ab). Vincent et al. described a series of cases as a potentially reversible form of limbic encephalitis responsive to immunotherapy. Clinical and neuropsychological features of patients with VGKC-Ab are characterized by a subacute amnesia (1–52 week), global impairment of memory with sparing general intellect, confusion, sleep disturbance, hypothermia and seizure that occur mainly in adulthood males. Typically, MRI studies revealed either unilateral or bilateral change of signal in medial temporal lobes, especially on T2 or FLAIR weighted sequences, moreover this feature could be absent in 45 % of patients at onset (Irani et al. 2008).

In almost 80 % of cases, paraneoplastic screening (including paraneoplastic antibodies) is always negative and serum VGKC-Ab ranges from 450 to 5128 pM. These values decrease (from 2 to 88 % in comparison to basal level) after treatment with steroids, plasma exchange and intravenous immunoglobulin. The decrement of VGKC-Ab is correlated with improvement of neuropsychological performances and all symptoms are broadly revertible after immunotherapy (Vincent et al., 2004). This last point represents the substantial difference between VGCK syndrome and other untreatable rapidly progressive dementia conditions such as Creutzfeld-Jakob disease (Geschwind et al. 2008).

Previous reports postulated that VGKC complex belongs to two proteins: the leucine rich glioma inactivated 1 protein (LGI1) and contactin associated protein 2 (CASPR2), but nowadays the results of several investigators indicate that these are two diverse entities from VGKC (Irani et al. 2010; Lai et al. 2010).

LGI1 is a secreted neuronal protein highly expressed in hippocampus and neocortex that interacts with a presynaptic protein called ADAM23 and postsynaptic ADAM 22, modulating presynaptic Kv1 potassium channels (Lancaster et al. 2011).

Previously, genetic studies have demonstrated that LGI1 plays an important role in epilepsy field in fact mutations of LGI1 are responsible of some autosomal dominant partial epilepsy with auditory seizures indeed the studies performed on models of transgenic mouse have shown that LGI1 mutations increase the excitatory synaptic transmission modifying dendritic morphology (Nobile et al. 2009; Zhou et al. 2009).

Despite a common substrate, the clinical spectrum correlated to LGI1 mutation is different by LE with LGI1-Ab. In fact LE with LGI1Ab is characterized by an encephalitis rapidly progressive, with two peculiar findings: hyponatraemia and antiepileptic drug refractory facio-brachial dystonic seizures, in almost all cases without surface EEG ictal pattern but always reversible after immunotherapy (Irani et al. 2008).

NMDAR-Ab Encephalitis

A distinctive entity among LE field is the encephalitis N-methyl-D-aspartate-antibody related (NMDAR-Ab). The target of this syndrome is represented by NMDA receptors (NMDARs) which are ligand-gated cation channels involved in neuronal plasticity and synaptic transmission, widely expressed in the amygdala, thalamus, hippocampus and prefrontal cortex. It is known that the pathogenic mechanism of anti NMDAR antibodies is based by a selective and reversible reduction of density of NMDAR surface protein with a subsequent decrease of synaptic NMDAR-mediated current. It has been suggested that a reduction of NMDAR activity might promote epileptogenesis through an increase in glutamatergic activity (Dalmau et al. 2011).

The spectrum of LE NMDAR-Ab differs by other LE for demographic, clinical and instrumental findings. In fact, people affected by LE NMDAR-Ab is peculiar since it is mainly represented by children and young woman. The symptomatology profile is characterized by a prodromal phase presenting as viral illness followed by psychiatric disorders (including anxiety, behavioural changes and psychosis) thereafter followed by the occurrence of seizures, alteration of consciousness and dysautonomia. Sometimes autonomic disturbance may require admission to intensive care unit for central hypoventilation and a temporary pacemaker for cardiac rhythmic alterations (Sansing et al. 2007). Although the first descriptions of LE NMDAR-Ab was related to a ovaric teratoma, recent data showed that less than 5 % of cases are related to tumours. This LE has a good response to immune-treatment including plasma exchange, steroids, intravenous immunoglobulin or combination. Some uncontrolled studies have suggested a second-line treatment with immunotherapy, ciclophosphamide, rituximab or both (Titulaer et al. 2013)

Forth Stop: Practical Evidence of Autoimmune Disease and Epilepsy

It is very well known how the incidence of seizures is very high in course of systemic autoimmune diseases (Vincent and Crino 2011).

Based on this, several authors have postulated that immune system might be involved in the pathogenesis of some forms of epilepsy and hyperexcitabilty. The forth stop of our tour will focus on autoimmune diseases that are strongly linked to seizures trying to give a key for understanding this close relationship.

Systemic Lupus Erythematous

Seizures occur in almost 10–20 % of patients affected by systemic lupus erythematous (SLE) and the prevalence of epilepsy in these patients is eight times higher than general population. Furthermore, in a significant quote of patients (5–10 %) seizures can precede the clinical onset of SLE of several years (Aarli 2000).

These seizures are mainly generalized while seizures occurring during SLE are either focal or generalized-tonic (Mackworth-Young and Hughes 1985).

An important prognostic factor in patients with a single epileptic seizure is the presence of antiphospholipid antibodies since they are correlated with a greater risk of developing new seizures (Peltola et al. 2000).

Several theories speculate that an immune-mediated damage of antinuclear antibodies that cross-react with neuronal antigens or an immune complex-mediated vasculitis could underlie epileptogenic mechanism in patients with SLE. On the other hand, there still is an important unanswered question: what is the meaning of increased level of antinuclear antibodies in patients with idiopathic epilepsy? Some authors reported that the use of antiepileptic drugs might responsible for the increment of antinuclear and antiphospholipid antibodies (Billiau et al. 2005). Although the real cause is almost unknown, epilepsy plays an important role in SLE representing one of the diagnostic criteria for its diagnosis.

Coeliac Disease

Coeliac disease (CD) is a chronic immune T cell-mediated disease against gluten and related proteins that occurs in individuals with a genetic predisposition. In the last years, an increasing number of reports have shown that CD is not only confined to gastrointestinal system, in fact other systems such as nervous system are involved in this disease. The association between CD and epilepsy is controversial and it ranges from 0.5 to 7.2 % (Ruggieri et al. 2008; Zelnik et al. 2004), but a recent meta-analysis showed that paediatric population with CD has a 2.1 fold increased risk of developing epilepsy (Lionetti et al. 2010). On the other hand, some authors reported in epileptic population a greater prevalence of CD (about 2–3 %) (Cronin et al. 1998; Emami et al. 2008; Labate et al. 2001). Recently, a population-based cohort analysis confirmed a moderate risk of epilepsy in individuals with CD reinforcing the potential role of immunologic pathogenesis in the development of epilepsy (Ludvigsson et al. 2012).

This role is also supported by the exposure to gliadin determining an immune response through the activation of T cell against transglutaminase (especially form 6 that is expressed mainly on cerebral cortex, amygdala, hippocampus, cerebellum) and the production of aggressive pro-inflammatory cytokines. Although temporal lobe epilepsy represents the most common type of epilepsy associated with CD, especially in patients with hippocampal sclerosis (Peltola et al. 2009), Labate et al. among other has reported a high rate of silent CD in partial epilepsy with occipital paroxysms, with and without cerebral calcifications and proposing a routine sierological screening of CD antibodies in this population. In fact, it is already accepted as indicator of silent or latent CD the positivity to antibodies to gliadin or transglutaminase in patients without signs of gastrointestinal involvement (Vincent and Crino 2011).

This condition is extremely common in adulthood, when CD is often asymptomatic and neurological illness, including epilepsy, is reported as the first sign preceding the diagnosis of CD (Maki and Collin 1997).

The correlation between seizure and gluten is demonstrated by the successful control of refractory seizures in patients with CD after a treatment with gluten-free diet (Harper et al. 2007; Mavroudi et al. 2005).

A recent study of Ranua et al. (Ranua et al. 2005)reported that the presence of CD antibodies did not differ between patients with epilepsy compared to control group, but the prevalence of antigliadin antibodies is higher in patients with primary generalized epilepsy, suggesting a genetic predisposition. However, several aspects remain obscure and further studies are needed to better clarify the link between immune system and epilepsy in CD patients.

Multiple Sclerosis

Multiple Sclerosis (MS) is an immune-mediated disease of the central nervous system of unknown origin (Vincent and Crino 2011).

Seizures are considered part of MS spectrum and can occur in every phase of the disease either before or after the onset of MS but mainly during relapses. In the majority of cases, seizures appear after several years of MS onset and occasionally can be the unique presenting manifestation of MS (Gambardella et al. 2003; Trouillas and Courjon 1972).

Although the prevalence of epilepsy in MS is more common than in the general population (mean value is assessed about 2.3 %) the rate is extremely heterogeneous ranging from 0.5 to 10.8 % (Drake and Macrae 1961; Ghezzi et al. 1990; Kinnunen and Wikstrom 1986; Matthews 1962).

Each type of seizure have been reported in association with MS, however, secondary generalized tonic-clonic are the most frequent types of seizures (Sander et al. 1990; Striano et al. 2003).

Moreover, several studies reported the occurrence of uncommon forms of epilepsy associated to MS as sensory partial seizures, musicogenic epilepsy and aphasic status epilepticus in particular aphasia is been noted as a predominant symptom of seizure in MS.

Although seizures might occur at any time during the course and in every form of MS (primary or secondary progressive as well as relapsing-remitting), most typically, seizures appearing during relapses are self-limiting and do not generally require antiepileptic treatment.

Several hypotheses have been proposed regarding the pathogenic mechanism underlying epilepsy in MS. Some suggested that epileptogenesis is either sustained by oedema originated by active demyelinating lesions by pro-inflammatory cytokines such as tumour necrosis factor-α, specific interleukins that are produced by activation of microglia in MS lesions.

Hashimoto Encephalopathy

Since 1966, Hashimoto encephalopathy (HE) is considered a rare steroid-responsive treatable encephalopathy associated with autoimmune thyroiditis (Hashimoto thyroiditis)(Chong et al. 2003). HE has only a gender predilection (as other autoimmune diseases, it affects 4–5 times more females than males) but it is not age dependent. There are two clinical HE-correlated subtypes: a vasculitic-like pattern (acute or subacute multiple stroke-like episodes associated with focal neurological deficit and variable degrees of cognitive and consciousness dysfunction) and a diffuse gradual cognitive impairment with dementia, neuropsychiatric symptoms and impairment of consciousness (Afshari et al. 2012).

Seizures affect 66 % of HE patients, especially in childhood when HE might be very insidious and it might be suspected when new onset unexplained deterioration is associated with refractory epilepsy (Berger et al. 2010; Vasconcellos et al. 1999).

Type of seizures reported in HE subject are focal or secondary generalized convulsions, whereas very rarely status epilepticus or absence status have been described (Ferlazzo et al. 2006; Vasconcellos et al. 1999). EEG findings are not specific and show a diffuse background slowing as reported in other encephalopathy (Chong et al. 2003; Kothbauer-Margreiter et al. 1996), a resolution of EEG pattern is correlated with clinical improvement (Henchey et al. 1995).

Brain imaging is not useful, in fact, half cases show normal MRI while the others show non specific alterations (Chong et al. 2003; Kothbauer-Margreiter et al. 1996).

Conversely, in 60–85 % of subject with HE, elevated protein concentration and less commonly lymphocytic pleocytosis is present at CSF analysis. However, two are the hallmark features of HE: firstly the detection in serum of antithyroid antibodies (but is not specific in fact are positive in 10 % of general population) independently of thyroid hormonal profile and the gravity of disease; secondly the rapid and dramatic response to high–dose to steroid-treatment with a complete recovery within 2 months. The combination of these characteristics is distinctive and exclusive of HE and allows an easy differential diagnosis by other autoimmune forms of encephalitis. Disease relapse is possible after steroids' tapering, and in some cases, a combined immunomodulatory treatment with rituximab, cyclophosphamide, immunoglobulins and plasma exchange is necessary (Vincent and Crino 2011).

How antithyroid antibodies can trigger seizures and neurological manifestations still remains unclear. Post–mortem studies showed a lymphocyte infiltration around small vessels in brain parenchyma of HE patients, thus suggesting a microvascular inflammation damage for an immune complex deposition mediated by antithyroid antibodies (Duffey et al. 2003; Nolte et al. 2000).

Nevertheless, it is unknown the mechanism by which anti-thyroids antibodies can reverse neuronal and thyroid tissue damage, actually, it is not possible to exclude other antibodies or/and other mechanism underlying HE.

Fifth Stop: Practical Evidence of Probable Dysimmune Epilepsy

In these last years, there is a growing interest about catastrophic childhood epileptic encephalophaties, to prevent neurological complications and negative cognitive influences. To achieve these purposes, it is necessary to know the real underlying causes to establish a correct early clinical and therapeutic approach to obtain better outcomes. Based on the effectiveness of immunomodulatory treatments in these epilepsies (Eriksson et al. 2001; Lousa et al. 2000; Wiendl et al. 2001) a potential role of immunity in their pathogenesis is reasonable. In this stop we focus the attention on these severe epileptic syndromes.

West Syndrome

West syndrome (WS) is an infancy epileptic syndrome with onset in the first years of life characterized by brief tonic spasms, arrest or regression of psychomotor development and a chaotic pattern on electroencephalogram called as " hypsarrhythmia". Essentially, it is possible to distinguish two subgroups in WS: symptomatic (underlying a brain damage such as neonatal asphyxia, meningoencephalitis, cerebral dysgenesis, and congenital metabolic disorders) or cryptogenic, if occurred in previously healthy children *(International League Against Epilepsy Task Force, 1989)*.

Independently by subgroups of origin, the hallmark of WS is the responsiveness to immune suppressants therapy as adrenocorticotropic hormone (ACTH) or glucocorticoids, which is responsible of cessation of spasms in more than 90% of patients (Kondo et al. 2005; Tsuji et al. 2007; Yamamoto et al. 2007).

Hence, an immune mechanism in the pathophysiology of WS is highly supported. Therefore, in order to clarify the immune pathophysiology of WS, several studies have faced the question, however, the profile of cytokine patterns produced by lymphocytes and other pro and anti-inflammatory molecules in plasma e CSF fluid of patients with WS were inconclusive (Haginoya et al. 2009; Liu et al. 2001).

Takashi Shiiharaa et al. (2010) studying peripheral blood lymphocyte subset and serum cytokine profiles have suggested that, in WS, T-cell and B-cell activation played a role and ACTH therapy may associate with T-cell inactivation, unfortunately these data were not confirmed by other reports. The efficacy of ACTH and the superiority to corticosteroids can be explained through a suppression of endogenous convulsant hormone CRH (highly expressed in WS brain) through a direct activation melanocortin receptor and a block of transcription of nuclear factor-kB (this factor is involved in inflammation and epileptogenesis)(Baram et al. 1992; Baram and Schultz 1991).

To date, even if conceivable, an immune mechanism in the pathophysiology of WS has been suggested, however, the exact mechanisms remain unknown and yet might be seriously investigated.

Landau-Kleffner Syndrome

Landau-Kleffner syndrome (LKS) is another rare developmental syndrome, characterized by acquired aphasia during early childhood epileptic seizures, behavioral problems, abnormal epileptic activity on EEG during sleep. Since the first description by Landau and Kleffner in 1957, several hypotheses were investigated to discover the pathogenesis, but of all only an immune pathologic cause seems to be more reliable. This theory was suggested by Mikati et al. and Lagae et al., following the observation of a significant improvement of language function and EEG abnormalities after treatment with corticosteroids and/or repeated IVIG infusions (Lagae et al. 1998; Mikati and Saab 2000)

Moreover Fayad et al. (1997) and then other investigators (Mikati and Saab 2000) have strengthened an immune origin through the evidence of intrathecal IgG index in liquor in patients affected by LKS observing that the rate of IgG was normalized after immunoglobulin treatment. Another evidence of immunological mechanism for epilepsy in LKS was documented by Connolly et al. (Connolly et al. 1999, 2006) through the demonstration of the presence of serum autoantibodies anti-brain endothelial cells and cell nuclei IgG in 45 % of patients affected by LKS in comparison to control. All these observations support the hypothesis that an autoimmune mechanism could be the cause of this epileptic syndrome.

Febrile Infection-Related Epilepsy Syndrome (FIRES)

An expanding interest has been focused on a new important subtype of childhood syndromes called febrile infection related epilepsy syndrome (FIRES).

Just the name of syndrome includes and defines the clinical spectrum of FIRES that is represented by a febrile infection followed, after weeks, by acute onset of an extraordinary high seizure activity most difficult to treat. As first described by van Baalen et al (2010) and Kramer et al. (2011) the population involved in this syndrome is younger than 15 and a slight male predominance. All patients had suffered an infection, mainly respiratory in the weeks prior the onset of symptoms. This syndrome has a poor prognosis and it is burdened by a mortality rate of about 9% and many patients have cognitive sequelae.

The preceding infection and the lacking evidence of infectious encephalitis support an immune-mediated patho-mechanism for FIRES. Apart from antibody-related encephalitis, alternative hypotheses grant a special role to the innate immune system even to a genetic predisposition in FIRES pathogenesis (Howell et al. 2012; Nabbout et al. 2011)

To date the failure of antibody-detection against the known neuronal antigens as well as the ineffectiveness of immunotherapy questions a role for autoantibodies in the epileptogenesis of classical FIRES (van Baalen et al. 2012).

Despite these limits, the literature data on this syndrome is growing up and an immune involvement is not possible to be excluded. In fact, (Specchio et al. 2010), the presence of severe epileptic syndrome in previously normal children preceded by fever is a highly suggesting element of an immune-mediated or inflammatory processes and it deserves further consideration.

Last Stop: Future Directions

The hypothesis that immune system might have a crucial role in epilepsy was first postulated 20 years ago. Since then, the discoveries in biochemical mediators in epilepsy models plus the possibility to identify an association with specific autoantibodies in several epileptic syndromes are strengthening the relationship between epilepsy and immunity. On the other hand, very interesting is the proven efficacy of immunotherapies such as adrenocorticotropic hormone (ACTH), corticosteroids (dexamethasone, hydrocortisone, prednisone/prednisolone, and methylprednisolone), cyclophosphamide, methotrexate, and rituximab in several forms of refractory epilepsies supporting this link.

Overall, in preclinical settings, the link between inflammation, immune system and epilepsy has to be considered determined, however, more research is warranted in order to better define possible targets for pharmacological interventions and the role played by the various mediators, which is still debated with controversial results being published.

Furthermore some antiepileptic drugs (AEDs) such as valproate, carbamazepine, phenytoin, vigabatrin, levetiracetam, and diazepam have been found to modulate the immune system activity by affecting humoral and cellular immunity (Beghi and Shorvon 2011). Based on these considerations immunity might be considered a promising and enchanting challenge to understand the epilepsy.

Ethical Publication We confirm that we have read the Journal's position on issues involved in ethical publication and affirm that this report is consistent with those guidelines.

Disclosure of Conflicts of Interest None of the authors has any conflict of interest to disclose.

References

Aarli JA (2000) Epilepsy and the immune system. Arch Neurol 57:1689–1692

Afshari M, Afshari ZS, Schuele SU (2012) Pearls & oy-sters: hashimoto encephalopathy. Neurology 78:e134–137

Ahmadi A, Sayyah M, Khoshkholgh-Sima B, Choopani S, Kazemi J, Sadegh M, Moradpour F, Nahrevanian H (2013) Intra-hippocampal injection of lipopolysaccharide inhibits kindled seizures and retards kindling rate in adult rats. Exp Brain Res 226:107–120

Arican N, Kaya M, Kalayci R, Uzun H, Ahishali B, Bilgic B, Elmas I, Kucuk M, Gurses C, Uzun M (2006) Effects of lipopolysaccharide on blood-brain barrier permeability during pentylenetetrazole-induced epileptic seizures in rats. Life Sci 79:1–7

Asztely F, Kumlien E (2012) The diagnosis and treatment of limbic encephalitis. Acta Neurol Scand 126:365–375

Auvin S, Shin D, Mazarati A, Nakagawa J, Miyamoto J, Sankar R (2007) Inflammation exacerbates seizure-induced injury in the immature brain. Epilepsia 48(Suppl 5):27–34

Auvin S, Porta N, Nehlig A, Lecointe C, Vallee L, Bordet R (2009) Inflammation in rat pups subjected to short hyperthermic seizures enhances brain long-term excitability. Epilepsy Res 86:124–130

Auvin S, Mazarati A, Shin D, Sankar R (2010a) Inflammation enhances epileptogenesis in the developing rat brain. Neurobiol Dis 40:303–310

Auvin S, Shin D, Mazarati A, Sankar R (2010b) Inflammation induced by LPS enhances epileptogenesis in immature rat and may be partially reversed by IL1RA. Epilepsia 51(Suppl 3):34–38

Auvin S, Bellavoine V, Merdariu D, Delanoe C, Elmaleh-Berges M, Gressens P, Boespflug-Tanguy O (2012) Hemiconvulsion-hemiplegia-epilepsy syndrome: current understandings. Eur J Paediatr Neurol 16:413–421

Balosso S, Maroso M, Sanchez-Alavez M, Ravizza T, Frasca A, Bartfai T, Vezzani A (2008) A novel non-transcriptional pathway mediates the proconvulsive effects of interleukin-1beta. Brain 131:3256–3265

Balosso S, Ravizza T, Pierucci M, Calcagno E, Invernizzi R, Di Giovanni G, Esposito E, Vezzani A (2009) Molecular and functional interactions between tumor necrosis factor-alpha receptors and the glutamatergic system in the mouse hippocampus: implications for seizure susceptibility. Neuroscience 161:293–300.

Baram TZ, Schultz L (1991) Corticotropin-releasing hormone is a rapid and potent convulsant in the infant rat. Brain Res Dev Brain Res 61:97–101

Baram TZ, Mitchell WG, Snead OC 3rd, Horton EJ, Saito M (1992) Brain-adrenal axis hormones are altered in the CSF of infants with massive infantile spasms. Neurology 42:1171–1175

Bauer J, Elger CE, Hans VH, Schramm J, Urbach H, Lassmann H, Bien CG (2007) Astrocytes are a specific immunological target in Rasmussen's encephalitis. Ann Neurol 62:67–80

Beghi E, Shorvon S (2011) Antiepileptic drugs and the immune system. Epilepsia 52(Suppl 3):40–44

Berger I, Castiel Y, Dor T (2010) Paediatric Hashimoto encephalopathy, refractory epilepsy and immunoglobulin treatment—unusual case report and review of the literature. Acta Paediatr 99:1903–1905

Bien CG, Bauer J, Deckwerth TL, Wiendl H, Deckert M, Wiestler OD, Schramm J, Elger CE, Lassmann H (2002) Destruction of neurons by cytotoxic T cells: a new pathogenic mechanism in Rasmussen's encephalitis. Ann Neurol 51:311–318

Bien CG, Granata T, Antozzi C, Cross JH, Dulac O, Kurthen M, Lassmann H, Mantegazza R, Villemure JG, Spreafico R, Elger CE (2005) Pathogenesis, diagnosis and treatment of Rasmussen encephalitis: a European consensus statement. Brain 128:454–471

Billiau AD, Wouters CH, Lagae LG (2005) Epilepsy and the immune system: is there a link? Eur J Paediatr Neurol 9:29–42

Chi H (2012) Regulation and function of mTOR signalling in T cell fate decisions. Nat Rev Immunol 12:325–338

Chinchilla D, Dulac O, Robain O, Plouin P, Ponsot G, Pinel JF, Graber D (1994) Reappraisal of Rasmussen's syndrome with special emphasis on treatment with high doses of steroids. J Neurol Neurosurg Psychiatry 57:1325–1333

Chong JY, Rowland LP, Utiger RD (2003) Hashimoto encephalopathy: syndrome or myth? Arch Neurol 60:164–171

Connolly AM, Chez MG, Pestronk A, Arnold ST, Mehta S, Deuel RK (1999) Serum autoantibodies to brain in Landau-Kleffner variant, autism, and other neurologic disorders. J Pediatr 134:607–613

Connolly AM, Chez M, Streif EM, Keeling RM, Golumbek PT, Kwon JM, Riviello JJ, Robinson RG, Neuman RJ, Deuel RM (2006) Brain-derived neurotrophic factor and autoantibodies to neural antigens in sera of children with autistic spectrum disorders, Landau-Kleffner syndrome, and epilepsy. Biol Psychiatry 59:354–363

Corsellis JA, Goldberg GJ, Norton AR (1968) "Limbic encephalitis" and its association with carcinoma. Brain 91:481–496

Cronin CC, Jackson LM, Feighery C, Shanahan F, Abuzakouk M, Ryder DQ, Whelton M, Callaghan N (1998) Coeliac disease and epilepsy. QJM 91:303–308

Curatolo P, Moavero R (2013) mTOR inhibitors as a new therapeutic option for epilepsy. Expert Rev Neurother 13:627–638

Dalmau J, Lancaster E, Martinez-Hernandez E, Rosenfeld MR, Balice-Gordon R (2011) Clinical experience and laboratory investigations in patients with anti-NMDAR encephalitis. Lancet Neurol 10:63–74

Devinsky O, Vezzani A, Najjar S, De Lanerolle NC, Rogawski MA (2013) Glia and epilepsy: excitability and inflammation. Trends Neurosci 36:174–184

Dhote F, Peinnequin A, Carpentier P, Baille V, Delacour C, Foquin A, Lallement G, Dorandeu F (2007) Prolonged inflammatory gene response following soman-induced seizures in mice. Toxicology 238:166–176

Dmowska M, Cybulska R, Schoenborn R, Piersiak T, Jaworska-Adamu J, Gawron A (2010) Behavioural and histological effects of preconditioning with lipopolysaccharide in epileptic rats. Neurochem Res 35:262–272

Drake WE Jr, Macrae D (1961) Epilepsy in multiple sclerosis. Neurology 11:810–816

Duffey P, Yee S, Reid IN, Bridges LR (2003) Hashimoto's encephalopathy: postmortem findings after fatal status epilepticus. Neurology 61:1124–1126

Emami MH, Taheri H, Kohestani S, Chitsaz A, Etemadifar M, Karimi S, Eshagi MA, Hashemi M (2008) How frequent is celiac disease among epileptic patients? J Gastrointestin Liver Dis 17:379–382

Eriksson K, Peltola J, Keranen T, Haapala AM, Koivikko M (2001) High prevalence of antiphospholipid antibodies in children with epilepsy: a controlled study of 50 cases. Epilepsy Res 46:129–137

Fayad MN, Choueiri R, Mikati M (1997) Landau-Kleffner syndrome: consistent response to repeated intravenous gamma-globulin doses: a case report. Epilepsia 38:489–494

Ferlazzo E, Raffaele M, Mazzu I, Pisani F (2006) Recurrent status epilepticus as the main feature of Hashimoto's encephalopathy. Epilepsy Behav 8:328–330

Firlik KS, Adelson PD, Hamilton RL (1999) Coexistence of a ganglioglioma and Rasmussen's encephalitis. Pediatr Neurosurg 30:278–282

Foresti ML, Arisi GM, Shapiro LA (2011) Role of glia in epilepsy-associated neuropathology, neuroinflammation and neurogenesis. Brain Res Rev 66:115–122

Friedman A, Dingledine R (2011) Molecular cascades that mediate the influence of inflammation on epilepsy. Epilepsia 52(Suppl 3):33–39

Galic MA, Riazi K, Henderson AK, Tsutsui S, Pittman QJ (2009) Viral-like brain inflammation during development causes increased seizure susceptibility in adult rats. Neurobiol Dis 36:343–351

Galic MA, Riazi K, Pittman QJ (2012) Cytokines and brain excitability. Front Neuroendocrinol 33:116–125

Gambardella A, Valentino P, Labate A, Sibilia G, Ruscica F, Colosimo E, Nistico R, Messina D, Zappia M, Quattrone A (2003) Temporal lobe epilepsy as a unique manifestation of multiple sclerosis. Can J Neurol Sci 30:228–232

Gambardella A, Andermann F, Shorvon S, Le Piane E, Aguglia U (2008) Limited chronic focal encephalitis: another variant of Rasmussen syndrome? Neurology 70:374–377

Geschwind MD, Tan KM, Lennon VA, Barajas RF Jr, Haman A, Klein CJ, Josephson SA, Pittock SJ (2008) Voltage-gated potassium channel autoimmunity mimicking creutzfeldt-jakob disease. Arch Neurol 65:1341–1346

Ghezzi A, Montanini R, Basso PF, Zaffaroni M, Massimo E, Cazzullo CL (1990) Epilepsy in multiple sclerosis. Eur Neurol 30:218–223

Goldberg EM, Coulter DA (2013) Mechanisms of epileptogenesis: a convergence on neural circuit dysfunction. Nat Rev Neurosci 14:337–349

Granata T, Fusco L, Gobbi G, Freri E, Ragona F, Broggi G, Mantegazza R, Giordano L, Villani F, Capovilla G, Vigevano F, Bernardina BD, Spreafico R, Antozzi C (2003a) Experience with immunomodulatory treatments in Rasmussen's encephalitis. Neurology 61:1807–1810

Granata T, Gobbi G, Spreafico R, Vigevano F, Capovilla G, Ragona F, Freri E, Chiapparini L, Bernasconi P, Giordano L, Bertani G, Casazza M, Dalla Bernardina B, Fusco L (2003b) Rasmussen's encephalitis: early characteristics allow diagnosis. Neurology 60:422–425

Granata T, Cross H, Theodore W, Avanzini G (2011) Immune-mediated epilepsies. Epilepsia 52(Suppl 3):5–11

Haginoya K, Noguchi R, Zhao Y, Munakata M, Yokoyama H, Tanaka S, Hino-Fukuyo N, Uematsu M, Yamamoto K, Takayanagi M, Iinuma K, Tsuchiya S (2009) Reduced levels of interleukin-1 receptor antagonist in the cerebrospinal fluid in patients with West syndrome. Epilepsy Res 85:314–317

Harper E, Moses H, Lagrange A (2007) Occult celiac disease presenting as epilepsy and MRI changes that responded to gluten-free diet. Neurology 68:533–534

Harre EM, Galic MA, Mouihate A, Noorbakhsh F, Pittman QJ (2008) Neonatal inflammation produces selective behavioural deficits and alters N-methyl-D-aspartate receptor subunit mRNA in the adult rat brain. Eur J Neurosci 27:644–653

Hart YM, Andermann F, Fish DR, Dubeau F, Robitaille Y, Rasmussen T, Berkovic S, Marino R, Yakoubian EM, Spillane K, Scaravilli F (1997) Chronic encephalitis and epilepsy in adults and adolescents: a variant of Rasmussen's syndrome? Neurology 48:418–424

Harvey L, Boksa P (2012) Prenatal and postnatal animal models of immune activation: relevance to a range of neurodevelopmental disorders. Dev Neurobiol 72:1335–1348

Henchey R, Cibula J, Helveston W, Malone J, Gilmore RL (1995) Electroencephalographic findings in Hashimoto's encephalopathy. Neurology 45:977–981

Howell KB, Katanyuwong K, Mackay MT, Bailey CA, Scheffer IE, Freeman JL, Berkovic SF, Harvey AS (2012) Long-term follow-up of febrile infection-related epilepsy syndrome. Epilepsia 53:101–110

Hu S, Sheng WS, Ehrlich LC, Peterson PK, Chao CC (2000) Cytokine effects on glutamate uptake by human astrocytes. Neuroimmunomodulation 7:153–159

Irani SR, Buckley C, Vincent A, Cockerell OC, Rudge P, Johnson MR, Smith S (2008) Immunotherapy-responsive seizure-like episodes with potassium channel antibodies. Neurology 71:1647–1648

Irani SR, Alexander S, Waters P, Kleopa KA, Pettingill P, Zuliani L, Peles E, Buckley C, Lang B, Vincent A (2010) Antibodies to Kv1 potassium channel-complex proteins leucine-rich, glioma inactivated 1 protein and contactin-associated protein-2 in limbic encephalitis, Morvan's syndrome and acquired neuromyotonia. Brain 133:2734–2748

Kim YS, Sheldon RA, Elliott BR, Liu Q, Ferriero DM, Tauber MG (1995) Brain injury in experimental neonatal meningitis due to group B streptococci. J Neuropathol Exp Neurol 54:531–539

Kinnunen E, Wikstrom J (1986) Prevalence and prognosis of epilepsy in patients with multiple sclerosis. Epilepsia 27:729–733

Kolarova A, Ringer R, Tauber MG, Leib SL (2003) Blockade of NMDA receptor subtype NR2B prevents seizures but not apoptosis of dentate gyrus neurons in bacterial meningitis in infant rats. BMC Neurosci 4:21

Kondo Y, Okumura A, Watanabe K, Negoro T, Kato T, Kubota T, Hiroko K (2005) Comparison of two low dose ACTH therapies for West syndrome: their efficacy and side effect. Brain Dev 27:326–330

Kothbauer-Margreiter I, Sturzenegger M, Komor J, Baumgartner R, Hess CW (1996) Encephalopathy associated with Hashimoto thyroiditis: diagnosis and treatment. J Neurol 243:585–593

Kramer U, Chi CS, Lin KL, Specchio N, Sahin M, Olson H, Nabbout R, Kluger G, Lin JJ, van Baalen A (2011) Febrile infection-related epilepsy syndrome (FIRES): pathogenesis, treatment, and outcome: a multicenter study on 77 children. Epilepsia 52:1956–1965

Labate A, Gambardella A, Messina D, Tammaro S, Le Piane E, Pirritano D, Cosco C, Doldo P, Mazzei R, Oliveri RL, Bosco D, Zappia M, Valentino P, Aguglia U, Quattrone A (2001) Silent celiac disease in patients with childhood localization-related epilepsies. Epilepsia 42:1153–1155

Lagae LG, Silberstein J, Gillis PL, Casaer PJ (1998) Successful use of intravenous immunoglobulins in Landau-Kleffner syndrome. Pediatr Neurol 18:165–168

Lai M, Huijbers MG, Lancaster E, Graus F, Bataller L, Balice-Gordon R, Cowell JK, Dalmau J (2010) Investigation of LGI1 as the antigen in limbic encephalitis previously attributed to potassium channels: a case series. Lancet Neurol 9:776–785

Lancaster E, Martinez-Hernandez E, Dalmau J (2011) Encephalitis and antibodies to synaptic and neuronal cell surface proteins. Neurology 77:179–189

Landau WM, Kleffner FR (1957) Syndrome of acquired aphasia with convulsive disorder in children. Neurology 7:523–530

Leach JP, Chadwick DW, Miles JB, Hart IK (1999) Improvement in adult-onset Rasmussen's encephalitis with long-term immunomodulatory therapy. Neurology 52:738–742

Lee SH, Han SH, Lee KW (2000) Kainic acid-induced seizures cause neuronal death in infant rats pretreated with lipopolysaccharide. Neuroreport 11:507–510

Libbey JE, Fujinami RS (2011) Neurotropic viral infections leading to epilepsy: focus on Theiler's murine encephalomyelitis virus. Future Virol 6:1339–1350

Libbey JE, Kennett NJ, Wilcox KS, White HS, Fujinami RS (2011) Interleukin-6, produced by resident cells of the central nervous system and infiltrating cells, contributes to the development of seizures following viral infection. J Virol 85:6913–6922

Librizzi L, Noe F, Vezzani A, de Curtis M, Ravizza T (2012) Seizure-induced brain-borne inflammation sustains seizure recurrence and blood-brain barrier damage. Ann Neurol 72:82–90

Lionetti E, Francavilla R, Pavone P, Pavone L, Francavilla T, Pulvirenti A, Giugno R, Ruggieri M (2010) The neurology of coeliac disease in childhood: what is the evidence? A systematic review and meta-analysis. Dev Med Child Neurol 52:700–707

Liu ZS, Wang QW, Wang FL, Yang LZ (2001) Serum cytokine levels are altered in patients with West syndrome. Brain Dev 23:548–551

Liu J, Reeves C, Michalak Z, Coppola A, Diehl B, Sisodiya SM, Thom M (2014) Evidence for mTOR pathway activation in a spectrum of epilepsy-associated pathologies. Acta Neuropathol Commun 2:71

Lousa M, Sanchez-Alonso S, Rodriguez-Diaz R, Dalmau J (2000) Status epilepticus with neuron-reactive serum antibodies: response to plasma exchange. Neurology 54:2163–2165

Ludvigsson JF, Zingone F, Tomson T, Ekbom A, Ciacci C (2012) Increased risk of epilepsy in biopsy-verified celiac disease: a population-based cohort study. Neurology 78:1401–1407

Mackworth-Young CG, Hughes GR (1985) Epilepsy: an early symptom of systemic lupus erythematosus. J Neurol Neurosurg Psychiatry 48:185

Maiese K, Chong ZZ, Shang YC, Wang S (2013) mTOR: on target for novel therapeutic strategies in the nervous system. Trends Mol Med 19:51–60

Maki M, Collin P (1997) Coeliac disease. Lancet 349:1755–1759

Mantegazza R, Bernasconi P, Baggi F, Spreafico R, Ragona F, Antozzi C, Bernardi G, Granata T (2002) Antibodies against GluR3 peptides are not specific for Rasmussen's encephalitis but are also present in epilepsy patients with severe, early onset disease and intractable seizures. J Neuroimmunol 131:179–185

Marchi N, Granata T, Ghosh C, Janigro D (2012) Blood-brain barrier dysfunction and epilepsy: pathophysiologic role and therapeutic approaches. Epilepsia 53:1877–1886

Maroso M, Balosso S, Ravizza T, Liu J, Aronica E, Iyer AM, Rossetti C, Molteni M, Casalgrandi M, Manfredi AA, Bianchi ME, Vezzani A (2010) Toll-like receptor 4 and high-mobility group box-1 are involved in ictogenesis and can be targeted to reduce seizures. Nat Med 16:413–419

Maroso M, Balosso S, Ravizza T, Iori V, Wright CI, French J, Vezzani A (2011) Interleukin-1beta biosynthesis inhibition reduces acute seizures and drug resistant chronic epileptic activity in mice. Neurother 8:304–315

Matthews WB (1962) Epilepsy and disseminated sclerosis. Q J Med 31:141–155

Mavroudi A, Karatza E, Papastavrou T, Panteliadis C, Spiroglou K (2005) Successful treatment of epilepsy and celiac disease with a gluten-free diet. Pediatr Neurol 33:292–295

McLachlan RS, Girvin JP, Blume WT, Reichman H (1993) Rasmussen's chronic encephalitis in adults. Arch Neurol 50:269–274

Mikati MA, Saab R (2000) Successful use of intravenous immunoglobulin as initial monotherapy in Landau-Kleffner syndrome. Epilepsia 41:880–886

Mirrione MM, Konomos DK, Gravanis I, Dewey SL, Aguzzi A, Heppner FL, Tsirka SE (2010) Microglial ablation and lipopolysaccharide preconditioning affects pilocarpine-induced seizures in mice. Neurobiol Dis 39:85–97

Nabbout R, Vezzani A, Dulac O, Chiron C (2011) Acute encephalopathy with inflammation-mediated status epilepticus. Lancet Neurol 10:99–108

Nguyen MD, Julien JP, Rivest S (2002) Innate immunity: the missing link in neuroprotection and neurodegeneration? Nat Rev Neurosci 3:216–227

Nobile C, Michelucci R, Andreazza S, Pasini E, Tosatto SC, Striano P (2009) LGI1 mutations in autosomal dominant and sporadic lateral temporal epilepsy. Hum Mutat 30:530–536

Nolte KW, Unbehaun A, Sieker H, Kloss TM, Paulus W (2000) Hashimoto encephalopathy: a brainstem vasculitis? Neurology 54:769–770

Palmer CA, Geyer JD, Keating JM, Gilliam F, Kuzniecky RI, Morawetz RB, Bebin EM (1999) Rasmussen's encephalitis with concomitant cortical dysplasia: the role of GluR3. Epilepsia 40:242–247

Peachardre JC, Sauvezie B, Osier C, Gilbert J (1977) Traitement does encéphalopathies épileptiques de l'enfant par les gamma globulines. Rev Electroencephalogr Neurophysiol Clin 7:443–447

Peltola JT, Haapala A, Isojarvi JI, Auvinen A, Palmio J, Latvala K, Kulmala P, Laine S, Vaarala O, Keranen T (2000) Antiphospholipid and antinuclear antibodies in patients with epilepsy or new-onset seizure disorders. Am J Med 109:712–717

Peltola M, Kaukinen K, Dastidar P, Haimila K, Partanen J, Haapala AM, Maki M, Keranen T, Peltola J (2009) Hippocampal sclerosis in refractory temporal lobe epilepsy is associated with gluten sensitivity. J Neurol Neurosurg Psychiatry 80:626–630

Ranua J, Luoma K, Auvinen A, Maki M, Haapala AM, Peltola J, Raitanen J, Isojarvi J (2005) Celiac disease-related antibodies in an epilepsy cohort and matched reference population. Epilepsy Behav 6:388–392

Ravizza T, Gagliardi B, Noe F, Boer K, Aronica E, Vezzani A (2008) Innate and adaptive immunity during epileptogenesis and spontaneous seizures: evidence from experimental models and human temporal lobe epilepsy. Neurobiol Dis 29:142–160

Reid AY, Riazi K, Campbell Teskey G, Pittman QJ (2013) Increased excitability and molecular changes in adult rats after a febrile seizure. Epilepsia 54:e45–e48

Riazi K, Galic MA, Pittman QJ (2010) Contributions of peripheral inflammation to seizure susceptibility: cytokines and brain excitability. Epilepsy Res 89:34–42

Ruggieri M, Incorpora G, Polizzi A, Parano E, Spina M, Pavone P (2008) Low prevalence of neurologic and psychiatric manifestations in children with gluten sensitivity. J Pediatr 152:244–249

Russo E, Citraro R, Donato G, Camastra C, Iuliano R, Cuzzocrea S, Constanti A, De Sarro G (2013) mTOR inhibition modulates epileptogenesis, seizures and depressive behavior in a genetic rat model of absence epilepsy. Neuropharmacology 69:25–36

Russo E, Andreozzi F, Iuliano R, Dattilo V, Procopio T, Fiume G, Mimmi S, Perrotti N, Citraro R, Sesti G, Constanti A, De Sarro G (2014) Early molecular and behavioral response to lipopolysaccharide in the WAG/Rij rat model of absence epilepsy and depressive-like behavior, involves interplay between AMPK, AKT/mTOR pathways and neuroinflammatory cytokine release. Brain Behav Immun.

Saiz A, Blanco Y, Sabater L, Gonzalez F, Bataller L, Casamitjana R, Ramio-Torrenta L, Graus F (2008) Spectrum of neurological syndromes associated with glutamic acid decarboxylase antibodies: diagnostic clues for this association. Brain 131:2553–2563

Sander JW, Hart YM, Johnson AL, Shorvon SD (1990) National General Practice Study of Epilepsy: newly diagnosed epileptic seizures in a general population. Lancet 336:1267–1271

Sansing LH, Tuzun E, Ko MW, Baccon J, Lynch DR, Dalmau J (2007) A patient with encephalitis associated with NMDA receptor antibodies. Nat Clin Pract Neurol 3:291–296

Shiihara T, Miyashita M, Yoshizumi M, Watanabe M, Yamada Y, Kato M (2010) Peripheral lymphocyte subset and serum cytokine profiles of patients with West syndrome. Brain Development 32:695–702

Silveira G, de Oliveira AC, Teixeira AL (2012) Insights into inflammation and epilepsy from the basic and clinical sciences. J Clin Neurosci 19:1071–1075

Specchio N, Fusco L, Claps D, Vigevano F (2010) Epileptic encephalopathy in children possibly related to immune-mediated pathogenesis. Brain Development 32:51–56

Stellwagen D, Beattie EC, Seo JY, Malenka RC (2005) Differential regulation of AMPA receptor and GABA receptor trafficking by tumor necrosis factor-alpha. J Neurosci 25:3219–3228

Stewart KA, Wilcox KS, Fujinami RS, White HS (2010) Development of postinfection epilepsy after Theiler's virus infection of C57BL/6 mice. J Neuropathol Exp Neurol 69:1210–1219

Striano P, Orefice G, Brescia Morra V, Boccella P, Sarappa C, Lanzillo R, Vacca G, Striano S (2003) Epileptic seizures in multiple sclerosis: clinical and EEG correlations. Neurol Sci 24:322–328

Titulaer MJ, McCracken L, Gabilondo I, Armangue T, Glaser C, Iizuka T, Honig LS, Benseler SM, Kawachi I, Martinez-Hernandez E, Aguilar E, Gresa-Arribas N, Ryan-Florance N, Torrents A, Saiz A, Rosenfeld MR, Balice-Gordon R, Graus F, Dalmau J (2013) Treatment and prognostic factors for long-term outcome in patients with anti-NMDA receptor encephalitis: an observational cohort study. Lancet Neurol 12:157–165

Tobias SM, Robitaille Y, Hickey WF, Rhodes CH, Nordgren R, Andermann F (2003) Bilateral Rasmussen encephalitis: postmortem documentation in a five-year-old. Epilepsia 44:127–130

Trouillas P, Courjon J (1972) Epilepsy with multiple sclerosis. Epilepsia 13:325–333

Tsuji T, Okumura A, Ozawa H, Ito M, Watanabe K (2007) Current treatment of West syndrome in Japan. J Child Neurol 22:560–564

Tuzun E, Dalmau J (2007) Limbic encephalitis and variants: classification, diagnosis and treatment. Neurologist 13:261–271

van Baalen A, Hausler M, Boor R, Rohr A, Sperner J, Kurlemann G, Panzer A, Stephani U, Kluger G (2010) Febrile infection-related epilepsy syndrome (FIRES): a nonencephalitic encephalopathy in childhood. Epilepsia 51:1323–1328

van Baalen A, Hausler M, Plecko-Startinig B, Strautmanis J, Vlaho S, Gebhardt B, Rohr A, Abicht A, Kluger G, Stephani U, Probst C, Vincent A, Bien CG (2012) Febrile infection-related epilepsy syndrome without detectable autoantibodies and response to immunotherapy: a case series and discussion of epileptogenesis in FIRES. Neuropediatrics 43:209–216

Vasconcellos E, Pina-Garza JE, Fakhoury T, Fenichel GM (1999) Pediatric manifestations of Hashimoto's encephalopathy. Pediatr Neurol 20:394–398

Vezzani A, Ravizza T, Balosso S, Aronica E (2008) Glia as a source of cytokines: implications for neuronal excitability and survival. Epilepsia 49(Suppl 2):24–32

Vezzani A, Aronica E, Mazarati A, Pittman QJ (2011a) Epilepsy and brain inflammation. Exp Neurol.

Vezzani A, French J, Bartfai T, Baram TZ (2011b) The role of inflammation in epilepsy. Nat Rev Neurol 7:31–40

Vezzani A, Maroso M, Balosso S, Sanchez MA, Bartfai T (2011c) IL-1 receptor/Toll-like receptor signaling in infection, inflammation, stress and neurodegeneration couples hyperexcitability and seizures. Brain Behav Immun 25:1281–1289

Vezzani A, Balosso S, Ravizza T (2012) Inflammation and epilepsy. Handb Clin Neurol 107:163–175

Vincent A, Crino PB (2011) Systemic and neurologic autoimmune disorders associated with seizures or epilepsy. Epilepsia 52(Suppl 3):12–17

Vincent A, Buckley C, Schott JM, Baker I, Dewar BK, Detert N, Clover L, Parkinson A, Bien CG, Omer S, Lang B, Rossor MN, Palace J (2004) Potassium channel antibody-associated encephalopathy: a potentially immunotherapy-responsive form of limbic encephalitis. Brain 127:701–712

Vincent A, Bien CG, Irani SR, Waters P (2011a) Autoantibodies associated with diseases of the CNS: new developments and future challenges. Lancet Neurol 10:759–772

Vincent A, Irani SR, Lang B (2011b) Potentially pathogenic autoantibodies associated with epilepsy and encephalitis in children and adults. Epilepsia 52(Suppl 8):8–11

Viviani B, Bartesaghi S, Gardoni F, Vezzani A, Behrens MM, Bartfai T, Binaglia M, Corsini E, Di Luca M, Galli CL, Marinovich M (2003) Interleukin-1beta enhances NMDA receptor-mediated intracellular calcium increase through activation of the Src family of kinases. J Neurosci 23:8692–8700

Wang S, Cheng Q, Malik S, Yang J (2000) Interleukin-1beta inhibits gamma-aminobutyric acid type A (GABA(A)) receptor current in cultured hippocampal neurons. J Pharmacol Exp Ther 292:497–504

Wang SJ, Bo QY, Zhao XH, Yang X, Chi ZF, Liu XW (2013) Resveratrol pre-treatment reduces early inflammatory responses induced by status epilepticus via mTOR signaling. Brain Res 1492:122–129

Watson R, Jiang Y, Bermudez I, Houlihan L, Clover L, McKnight K, Cross JH, Hart IK, Roubertie A, Valmier J, Hart Y, Palace J, Beeson D, Vincent A, Lang B (2004) Absence of antibodies to glutamate receptor type 3 (GluR3) in Rasmussen encephalitis. Neurology 63:43–50

Wennberg RA, Quesney LF, Villemure JG (1997) Epileptiform and non-epileptiform paroxysmal activity from isolated cortex after functional hemispherectomy. Electroencephalogr Clin Neurophysiol 102:437–442

Wiendl H, Bien CG, Bernasconi P, Fleckenstein B, Elger CE, Dichgans J, Mantegazza R, Melms A (2001) GluR3 antibodies: prevalence in focal epilepsy but no specificity for Rasmussen's encephalitis. Neurology 57:1511–1514

Yamamoto H, Fukuda M, Miyamoto Y, Murakami H, Kamiyama N (2007) A new trial liposteroid (dexamethasone palmitate) therapy for intractable epileptic seizures in infancy. Brain Dev 29:421–424

Zelnik N, Pacht A, Obeid R, Lerner A (2004) Range of neurologic disorders in patients with celiac disease. Pediatrics 113:1672–1676

Zhou YD, Lee S, Jin Z, Wright M, Smith SE, Anderson MP (2009) Arrested maturation of excitatory synapses in autosomal dominant lateral temporal lobe epilepsy. Nat Med 15:1208–1214

Zuliani L, Graus F, Giometto B, Bien C, Vincent A (2012) Central nervous system neuronal surface antibody associated syndromes: review and guidelines for recognition. J Neurol Neurosurg Psychiatry 83:638–645

Novel Molecular Targets for Drug-Treatment of Epilepsy

Vincenzo Belcastro and Alberto Verrotti

Abstract Nowadays several antiepileptic drugs (AEDs) are available for the treatment of patients with epilepsy. Nevertheless, up to 30% of patients continue to present recurrent seizures. So, the challenge for new more efficacious and better tolerated drugs is continuing. Advances in understanding of pathophysiology of epilepsy and in the physiology of ion channels and other molecular targets provide opportunities to create new and improved AEDs. Potentially interesting molecular targets include KCNQ-type K+ channels, SV2A synaptic vesicle protein, ionotropic and metabotropic glutamate receptors. The pipeline for the development of new AEDs with novel mechanisms of action is narrowing with only a few interesting compounds on the immediate horizon. In fact, only perampanel (modulates AMPA mediated neurotransmission) and brivaracetam (binds to SV2A protein and sodium channel) are likely to reach the market-place in the next 3 years. Eslicarbazepine has approved in the last year as add-on treatment for partial onset seizures.

This chapter reviews the available information on various classes of molecules that are in the pipeline for the treatment of epilepsy.

Introduction

The armamentarium to treat epilepsy includes today more than 20 drugs. These are classically distinguished as standard, traditional or first generation antiepileptic drugs (AEDs), which include phenobarbital, phenytoin, carbamazepine, valproic acid, ethosuximide and benzodiazepines, and new or second generation AEDs, which are vigabatrin, lamotrigine, felbamate, gabapentin, oxcarbazepine, tiagabine, topiramate, stiripentol, pregabalin, levetiracetam, rufinamide, zonisamide. More recently, other three compounds have been introduced, i.e. lacosamide, retigabine and eslicarbazepine acetate, also defined as third generation AEDs. However, despite

V. Belcastro (✉)
Neurology Unit, S. Anna Hospital, Como, Italy
e-mail: vincenzobelcastro@libero.it

A. Verrotti
Department of Paediatric, University of Perugia, Perugia, Italy

the therapeutic arsenal of old and new AEDs, approximately 30% of patients with epilepsy still suffer from seizures (Brodie 2010). Noteworthy, none of the AEDs that have been introduced since 1990, including those that act on newly identified targets, can be considered a 'magic bullet' that reliably cures a patient's seizures. Nevertheless, when compared with the old AEDs (developed before 1990), the new AEDs that have since been developed have provided considerable improvements in terms of safety, tolerability and pharmacokinetics (Brodie 2010). Of equal importance, the existence of more than 20 AEDs offers a broad range of therapeutic options, which may lead to better personalized medicine (Sander 2004; Perucca and Tomson 2011).

The ways for developing new AEDs and innovative therapeutic strategies are multiple and different. One possibility is the evolution of pre-existing drugs with the objective of enhancing their efficacy or eliminating troublesome side effects. The alternative strategy is the research of new compounds identified through the study of mechanisms of drug-resistance and of additional molecular targets or, of course, through the characterization of the biological activity of molecules identified through sheer serendipity (Meldrum and Rogawski 2007). Currently marketed AEDs predominantly target voltage-gated channels (e.g., alpha-subunits of voltage-gated Na+ channels, T-type voltage-gated Ca2+ channels) or influence GABA-mediated inhibition (Rogawski and Loscher 2004). Recently identified new and potentially interesting molecular targets include KCNQ-type K+ channels, SV2A synaptic vesicle protein, ionotropic and metabotropic glutamate receptors (Löscher et al. 2013).

In this chapter we review the available information on new classes of molecules that are in the pipeline for the treatment of epilepsy in the next future.

Novel Structural Analogues of Pre-existing AEDs

Many of the traditional AEDs are highly efficacious but their tolerance and safety profile can be disappointing. Therefore, the attempt to enhance the efficacy of pre-existing drugs by manipulation of their structure is a logical approach to obtain drugs with greater efficacy or better tolerability.

UCB Pharma is currently developing two levetiracetam (LEV) analogues: brivaracetam and seletracetam (ucb 44212). Compared with LEV, both compounds have higher affinity for the SV2A-binding site, which mediates the antiepileptic activity of LEV, and show much greater potency in animal models of seizures and epilepsy.

Eslicarbazepine acetate (ESL), a prodrug of eslicarbazepine (S-licarbazepine), shares with carbamazepine and oxcarbazepine the dibenzazepine nucleus bearing the 5-carboxamide substitute but is structurally different at the 10,11-position. This structural variation is expected to result in different metabolism (with once daily dosing) and improved tolerability (Verrotti et al. 2014).

Carisbamate, a compound endowed with broad-spectrum anticonvulsant activity in animal models, is a monocarbamate with some structural relation to the dicarbamate felbamate.

Brivaracetam: Mechanism of Action, Pharmacokinetic Profile, Efficacy and Safety

Brivaracetam (BRV) is a highly selective and reversible SV2A ligand with a 15- to 30-fold higher affinity than LEV in rat and human brain. At therapeutically relevant doses, BRV occupies 80–90% of SV2A within 5–15 min, representing maximal protection against seizures in animal models (Matagne et al. 2008; Gillard et al. 2011). The high anticonvulsant activity of BRV compared with LEV, observed in several animal models of epilepsy, seems not simply explained by the high affinity to SV2A but probably resides in the way BRV modulates SV2A function. In this regard, it is important to point that the exact role of SV2A in synaptic transmission is not completely clear (Matagne et al. 2008; Gillard et al. 2011). However, a strong functional correlation has been established between SV2A binding and the anticonvulsant potency for both focal and generalized seizures (Kaminski et al. 2008). In addition, BRV displays inhibitory activity at neuronal voltage-dependent sodium channels (Zona et al. 2010). How much such effect contributes to the anticonvulsant potency of BRV is not fully elucidated yet. At any rate, the sodium channel modulation represents a distinct activity of BRV compared with LEV (Margineanu et al. 2009) and might explain why BRV, and not LEV, is also active on the maximal electroshock-induced model (Matagne et al. 2008). It appears evident that further studies are needed to fully understand the neurobiology of BRV.

BRV has nearly complete bioavailability after oral administration. Single-dose studies under fasting conditions show a t_{max} of about 1 h and a dose proportional C_{max} for a dose range between 10 and 1400 mg. However, area under the curve (AUC) deviates from dose linearity above 600 mg, and high fat meals seem to slightly delay t_{max} to 3 h and to slightly decrease C_{max} of about 28% (Sargentini-Maier et al. 2007; Von Rosenstiel 2007). BRV is less than 20% protein bound with a volume of distribution of 0.6 l/kg (Rolan et al. 2008). BRV has a half-life of about 8 h and is primarily metabolized via hydrolysis of the acetamide group and CYP2C19-mediated hydroxylation, but all metabolites are not pharmacologically active (Von Rosenstiel 2007). CYP2C19 mutations affect slightly the BRV metabolism with a 30% decreased clearance which does not seem to be clinically relevant. Patients with severe renal impairment without dialysis require no major adjustments in dosage. However, subjects with severe hepatic failure present a 50–60% increase in plasma concentrations compared with healthy controls (Sargentini-Maier et al. 2012; Stockis et al. 2013). Data from add-on trials show no significant effect of BRV on plasma concentrations of concomitant AEDs such as lamotrigine, phenobarbital, phenytoin, topiramate, valproic acid or zonisamide. Preliminary data suggest no significant interactions with oral contraceptives (Bialer et al. 2013).

The efficacy of BRV as an adjunctive therapy for patients with focal epilepsy has been evaluated in three phase III studies. Two of which use a fixed-dose design and include patients with focal epilepsy only (Biton et al. 2014; Ryvlin et al. 2014), while the third study adopts a flexible dose design and also enrols patients with generalized seizures (Kwan et al. 2014). In study NO1252, the efficacy of BRV (20 mg, 50 mg and 100 mg/day) was investigated against placebo in patients aged 16–70 years with uncontrolled partial-onset seizures (POS) with/without secondary generalization, despite treatment with one to two concomitant AEDs at a stable and optimal dosage. The primary efficacy endpoint (percentage reduction over placebo in baseline-adjusted focal seizure frequency per week over the 12-week treatment period) was statistically significant only for BRV 100 mg/day (Ryvlin et al. 2014). In study NO1253, patients aged 16–70 years were randomized (1:1:1:1) to placebo or BRV 5 mg, 20 mg or 50 mg/day without up titration. The primary efficacy endpoint (percentage reduction over placebo in baseline adjusted POS frequency per week during the 12-week treatment period) was statistically significant only for BRV 50 mg (Biton et al. 2014). Finally, study NO1254 adopted a flexible dose design in adults (16–70 years) with uncontrolled epilepsy; however, up to 20% were patients with generalized epilepsies. After a prospective 4-week baseline, patients were randomized to BRV or placebo, initiated at 20 mg/day and increased, as needed, to 150 mg/day during an 8-week dose-finding period followed by another 8-week stable dose maintenance period. During the 16-week treatment period, median percentage reduction from baseline in POS frequency per week was not statistically significant, while 50% responder rate is 30.3% for BRV and 16.7% for placebo ($p=0.006$) (Kwan et al. 2014).

In general terms, all studies report no difference in the treatment-emergent adverse events (TEAEs) for patients treated with BRV compared with placebo. The most commonly reported TEAEs leading to discontinuation were psychiatric adverse events represented mainly by aggression, anxiety, irritability, depression and insomnia.

Seletracetam: Profile of a New Pyrrolidone Derivative in Animal Models of Epilepsy

Seletracetam (SEL) is a pyrrolidone derivative with a one-log-unit higher affinity for the synaptic vesicle protein 2A (SV2A) than LEV (Bennett et al. 2007). SEL shows a potent anti-seizure activity in animal models mimicking partial-onset (kindled animals) and generalized epilepsy (audiogenic seizure susceptible mice and genetic absence epilepsy rats from Strasbourg (GAERS)). In amygdala-kindled rats, SEL increased the generalized seizure threshold current and decreased the duration of the after-discharge and the seizure severity observed at the after-discharge threshold current, and generally had a much more potent effect than previously observed for LEV (Bennett et al. 2007). SEL showed no psychomimetic effects and a very high central nervous system (CNS) tolerability in both kindled and GAERS rats, markedly superior to that of LEV and other AEDs (Matagne et al. 2009). A number of in vitro studies have been conducted with whole-cell patch clamp techniques to ascertain if SEL modulates voltage- and/or ligand-operated ion channels. SEL, in contrast to another new pyrrolidone derivative, BRV, did not modify tetrodotoxin-sensitive fast Na+ currents in rat cortical neurons and did not alter persistent Na+ currents, like phenytoin, in neurons from the CAI region of rat hippocampal slices (Zona et al. 2005). Likewise, another study on rat hippocampal neurons reported that SEL is devoid of any significant effect on GABA-, glycine-, N-methyl-D-aspartate (NMDA)-, kainic acid- and AMPA-gated currents, with the exception of a minor inhibition of the plateau phase of the NMDA current (Bennett et al. 2007). Therefore, SEL does not appear to directly modulate Na+ channels as well as other key ligand-gated ion channels involved in inhibitory and excitatory neurotransmission. In a preliminary study in patients with photosensitive epilepsy, SEL was effective in suppressing light-induced electroencephalographic discharges (Perucca et al. 2007).

Eslicarbazepine Acetate: Mechanism of Action, Pharmacokinetic Profile, Efficacy and Safety

Eslicarbazepine acetate (ESL) is a new active compound that uses as the mechanism of action of blocking the voltage-gated sodium channel (VGSC) to obtain anticonvulsant activity on the CNS (Benes et al. 1999). ESL has a huge blocking power of VGSC like many AEDs that interact with ion channels and neurotransmitter receptors to block VGSC reducing excitability of the membrane. Several examples can be found such as carbamazepine, lamotrigine, oxcarbazepine and phenytoin (Rogawski 2002).

ESL has no interaction with receptors of benzodiazepines, GABA and glutamate (Ambrosio et al 2000; Ambrosio et al. 2001; Parada and Soares da Silva 2002; Cunha et al. 2002) and, like carbamazepine, it works on the inactive state of the channel but it has three times less efficacy on the stand-by state, this characteristic makes ESL something more selective on neurons that tend to be rapidly firing with respect to normal ones, this means that the probability of adverse neurological consequences is lower (Brown and El-Mallakh 2010).

ESL administration is oral, its absorption by the intestine is high. An extensive and rapid biotransformation takes place and the drug is metabolized into eslicarbazepine which delivers the pharmacological effect (Almeida et al. 2009). The process is immediate and as a consequence the plasma concentration stays low and under the quantification limit.

The bio transformation is carried out by hepatic first-pass metabolism and the metabolites in the circulation are eliminated via renal excretion (Almeida et al. 2009). One oral dose of ESL can be found more than 90% in urine as metabolites, of the remaining part 92% is eslicarbazepine that is excreted in urine too, by on third as glucuronide conjugate and two thirds as unconjugated form.

Regarding the interaction between this drug and other AEDs several studies are available in vitro and in vivo (Almeida and Soares da Silva 2007; Almeida et al. 2009; Vaz-da-Silva et al. 2009; Rocha et al. 2009). Treatment in concomitance with carbamazepine shows a significant increase of clearance of ESL and no change in exposure to carbamazepine, this indicates that ESL dose may need to be increased. Treatment of patients in concomitance with phenytoin and phenobarbital did show increased clearance of eslicarbazepine too but with a significant increase of phenytoin exposure, this may suggest to increase ESL dose and decrease phenytoin dose (Almeida et al. 2009). The pharmacokinetic interaction between ESL and other AEDs is not relevant. There are other non AEDs medications that interact with ESL, the most notable of which are warfarin (Almeida and Soares da Silva 2007), simvastatin (Falcao et al. 2013a) and oral contraceptives (Falcao et al. 2013b).

The efficacy of ESL for the treatment of POS has been the object of a number of randomized clinical trials (Elger et al. 2007; Elger et al. 2009; Ben-Menachem et al. 2010). These studies addressed patients that had at least four POS monthly and that were refractory to treatment with 1–3 AEDs. Overall these studies showed that ESL single-daily doses of 800 and 1200 mg were effective and safe in patients who were refractory to treatment with one or two concomitant AEDs and displayed that the percentage of subjects who responded increased together with increase of ESL dose. A recent study on efficacy and safety of ESL analyzed the data of a population of 1049 patients enrolled in all of the three phase III studies to focus on a broader population and subpopulations (Gil-Nagel et al. 2013). The data showed that single-daily doses of 800 and 1200 mg work well as adjunctive therapy of POS, and are also well tolerated. The good outcome in efficacy and safety of these doses showed to be independent of population characteristics (gender, geographic area, epilepsy duration, age at diagnosis time, type of seizure) and the type and number of concomitant AEDs therapy. Noteworthy, the long term efficacy of single-daily doses of ESL was addressed in open label extension studies during a 1 year period. The drug was used as adjunctive treatment in adult subjects with POS in trials conducted in double-blind and placebo controlled. The dose for the trial was of 800 mg once-daily for 4 weeks, after that the dose could be changed raising or lowering it in the range 400 and 1200 mg without altering the doses of concomitant AEDs (Halasz et al. 2010; Hufnagel et al. 2013). In all these open label extension studies ESL, in single-daily doses, showed to be effective in reducing the frequency of seizures and well tolerated as an adjunctive long-term therapy in adults with refractory POS (Halasz et al. 2010; Hufnagel et al. 2013).

ESL is currently approved by EMEA in the European Union for adjunctive use in partial epilepsy in adults at the daily dosage of 400–1200 mg/day. ESL is not recommended below 18 years and only few data about efficacy and safety of this drug in pediatric patients are available. A low-dose tablet formulation (200 mg) and ad oral suspension formulation (50 mg/ml) were developed. Currently there is only one published trial which analyzed pharmacokinetics, efficacy and safety of this drug in pediatric population (Almeida et al. 2008). In each age group (2–6 years, 7–11 years and 12–17 years) three different dosages were studied. The study demonstrated a clear dose-related reduction in seizure frequency with good tolerability.

Regarding the safety, the analysis performed in the multicenter studies previously illustrated showed that the incidence of TEAEs increased with the increase in the dose of ESL, both for all TEAEs as well as for those considered at least possibly related to treatment (Elger et al. 2009; Gil-Nagel et al. 2009; Ben-Menachem et al. 2010; Halasz et al. 2010; Gil-Nagel et al. 2013). The incidence of TEAEs was more marked during the early treatment phase. In fact, from day 42 to the end of the studies there was only a little difference between ESL and placebo groups of each trial (Gil-Nagel et al. 2009; Ben-Menachem et al. 2010). TEAEs were usually mild to moderate in intensity, while the incidence of serious TEAEs was very low and similar in each of the ESL treatment groups (Elger et al. 2007; Elger et al. 2009; Gil-Nagel et al. 2009; Gil-Nagel et al. 2013; Ben-Menachem et al. 2010; Halasz et al. 2010). The most common adverse events with an incidence >2% were: dizziness, somnolence, nausea, diplopia, headache, vomiting, abnormal coordination, blurred vision, vertigo and fatigue. The incidence of rash, which is the most common idiosyncratic reaction with all AEDs (Zaccara et al. 2007), occurred in approximately 1% in all phase III studies (Elger et al. 2009; Gil-Nagel et al. 2009; Ben-Menachem et al. 2010), while this incidence has been reported to be up to 10% in subjects treated with oxcarbazepine (Shorvon 2000) and 11% in subjects treated with carbamazepine (Mattson et al. 1992). The incidence of behavior or psychiatric adverse events is low and there was no case of suicide or suicide attempt.

Carisbamate

Carisbamate (Ortho-McNeil Janssen, Titusville, NJ, USA), a compound endowed with broad-spectrum anticonvulsant activity in animal models, is a monocarbamate with some structural relation to the dicarbamate felbamate. The mechanism of action contributing to the broad-spectrum anticonvulsant activity has not been fully elucidated but it is of considerable interest that the drug displays antiepileptogenic and neuroprotective activity when administered repeatedly after lithium-pilocarpine status epilepticus in rats (Francois et al. 2005). After an initial study with positive results against POS at doses of more than 200 mg/day (Faught et al. 2008), three placebo-controlled adjunctive-therapy trials assessing doses of 200, 400, 800 and 1200 mg/day found inconsistent evidence of only modest efficacy (Sperling et al. 2010). As a result, carisbamate's development in epilepsy has been discontinued.

Novel Investigated AEDs: Focus on the Photosensitivity Epilepsy Model

The photosensitivity epilepsy model provides a means of assessing the effects of potential AEDs in patients in a controlled laboratory setting. Photosensitivity describes the ability to produce an epileptiform electroencephalography (EEG) response evoked by intermittent photic stimulation (IPS). This EEG pattern is called a photoparoxysmal response (PPR). Administration of approved and experimental AEDs, as single or repeated doses, clearly diminishes or even abolishes the response to IPS (Binnie et al. 1986). A standardized method for eliciting PPR in response to IPS has been devised to quantify the effects of promising new therapies for epilepsy (Kasteleijn-Nolst Trenité et al. 2012).

This clinical photosensitivity epilepsy proof-of-concept model has been applied to determine and evaluate a potentially effective AED. In fact, this assay allows investigators to establish that the compound under study is penetrating to the CNS compartment and engaging the target channel or receptor under study by suppressing the epileptiform EEG discharges. Currently, these studies refer as phase II studies.

JNJ-26489112

JNJ-26489112 [(S)-N-(6-Chloro-2,3-dihydrobenzo(1,4)dioxin-2-yl)methyl)sulfamide] is a centrally active, broad spectrum investigational anticonvulsant having in vitro activity at multiple CNS targets including N-methyl-D-aspartate(NMDA), kainate, gamma-aminobutyric acid (GABA), TypeII Na, KCNQ and N-type calcium channels. In addition, JNJ-26489112 demonstrates modest activity at the 5-HT-2creceptor, dopamine transporter, and dopamine and serotonin uptake sites (Di Prospero et al. 2014). Although the precise mechanism of action of JNJ-26489112 is unknown, it has demonstrated significant anticonvulsant activity in a wide range of pre-clinical seizure models. In fact, JNJ-26489112 has anticonvulsant activity against audiogenic, electrically-and chemically induced seizures. By limiting seizure spread and elevating seizure threshold, JNJ-26489112 is more effective pre clinically than marketed anti-epileptic drugs in several severe seizure models including metrazol-induced seizures and hippocampal kindling. Based on its activity in these models, JNJ-26489112 may be useful in treating generalized tonic-clonic, complex partial and absence seizures; in addition, it might also be effective in pharmaco-resistant epilepsy (Di Prospero et al. 2014).

In a recent small study, JNJ-26489112 showed a positive response at all investigated doses with complete suppression of the IPS induced PPR (Di Prospero et al. 2014).

ICA-105665

ICA-105665, identified by Icagen, Inc. (Durham, NC, U.S.A.), is a novel small molecule that opens Kv7.2/7.3 and Kv7.3/7.5 potassium channels (Roeloffs et al. 2008), also known as KCNQ2/3 and KCNQ3/5 channels. The compound has demonstrated broad spectrum antiseizure activity in multiple animal models including maximal electroshock, 6 Hz seizures, pentylenetetrazole, and electrical kindling at doses from <1–5 mg/kg (Roeloffs et al. 2008). ICA-105665 was well-tolerated in both a single ascending dose study in healthy volunteers at doses up to 400 mg and in healthy volunteers and patients with epilepsy when administered twice daily in a 7-day repeat dose study at total daily doses up to 600 mg (Kasteleijn-Nolst Trenité et al. 2013). Moreover, ICA-105665 reduced the PPR in patients with photosensitivity epilepsy at single doses of 100, 400, and 500 mg (Kasteleijn-Nolst Trenité et al. 2013). Interestingly ICA-105665 is the first activator of neuronal Kv7 potassium channels tested for activity in the photosensitive epilepsy model in humans. In fact, retigabine another activator of Kv7 potassium channels (Rogawski and Bazil 2008), is effective for the treatment of partial seizures in humans and has been approved for use in Europe and the United States, but it has never been assessed in patients with photosensitive epilepsy or in patients with IGE. The reduction of PPR in patients with photosensitivity epilepsy provides evidence of CNS penetration by ICA-105665, and preliminary evidence that engagement with neuronal Kv7 potassium channels has antiseizure effects.

The most common TEAEs following ICA-105665 administration included dizziness, somnolence, ataxia, and tremor. Both efficacy and emergence of dizziness appeared to correlate with increasing plasma concentrations of ICA-105665.

Pitolisant (BF2.649)

The potential benefit of pitolisant (BF2.649), a histamine 3 receptor (H3R) antagonist, has been evaluated in different seizure models in rats and mice, predictive for generalized and partial types of seizures. The occurrence and duration of EEG discharges and seizures were determined in: (i) the genetic absence epilepsy rats from Strasbourg (GAERS), where BF2.649 (20 mg/kg, i.p.) significantly decreased both the number and cumulated durations of spike-and-wave discharges; (ii) the maximal electroshock test, where a complete suppression of EEG epileptiform activity and seizures was observed in mice treated with BF2.649; (iii) the kainate-induced hippocampal seizures in mice, where pitolisant is extremely effective to reduce the cumulated duration of hippocampal discharges.

Recently, the pharmacodynamic effect of pitolisant was tested in patients with epilepsy using the photosensitivity proof of concept model. A total of 14 adult patients were studied for 3 days to evaluate the effect of a single oral dose of pitolisant on EEG photosensitivity ranges (Kasteleijn-Nolst Trenité et al. 2013). In this study, a statistically significant suppressive effect (standardized photosensitive response [SPR] reduction) for 20-, 40-, or 60-mg doses of pitolisant was seen in 9/14 (64%) patients of whom 6/14 (43%) showed abolition of the response to IPS. Patients on the highest dosage (60 mg) showed the strongest effect with an effect lasting up to 28 h (Kasteleijn-Nolst Trenité et al. 2013).

AMPA Receptors as Novel Targets for Antiepileptic Drugs

Alpha-Amino-3-hydroxy-5-methyl-4-isoxazolepropionic acid (AMPA) receptors are key mediators of seizure spread in the CNS and represent promising targets for AEDs. There is emerging evidence that AMPA receptors may play a role in epileptogenesis and in seizure-induced brain damage (Russo et al. 2012). This evidence suggests that AMPA receptor antagonists could have broad utility in epilepsy therapy. Competitive and noncompetitive AMPA receptor antagonists are broad-spectrum anticonvulsants in animal seizure models. There is evidence that these antagonists can potentiate the antiseizure activity of NMDA receptor antagonists and conventional AEDs (Rogawski 2013). This evidence suggests that the preferred use of AMPA receptor antagonists may be in combination therapies. AMPA receptors are distributed widely in the CNS and are present in all areas relevant to epilepsy, including the cerebral cortex, amygdala, and thalamus. Although there are regional differences in the expression of the four subunits, GluA1, GluA2, and GluA3 are the most abundant subunits in the forebrain, with the exception of some thalamic nuclei, where GluA4 is also abundant. Approximately 80 % of AMPA receptors at excitatory synapses on CA1 hippocampal pyramidal neurons are GluA1/GluA2 heteromers (Rogawski 2013).

BGG492 (Selurampanel), a Competitive AMPA/Kainate Antagonist in Clinical Development for the Treatment of Epilepsy

BGG492 shows anticonvulsant activity in several animal models of epilepsy, including electroshock and chemically-induced seizures in rodents, WAG/Rij rats (a genetic model of absence epilepsy), the rat amygdala kindling model (indicating a potential anti-epileptogenic effect), and in fully kindled rats (a model of therapy-resistant partial seizures in human) (Russo et al. 2012). It is well understood that properties required for high affinity at AMPA receptors are contrary to those required for oral bioavailability. As a compromise, BGG492 has moderate binding affinity for rat and human AMPA receptors, but > 100-fold selectivity with regards to the glycine-binding site of NMDA receptors and no significant affinity in a 150-target safety panel (Russo et al. 2012). BGG492 is only metabolized to a limited extent, and does not inhibit CYP450 enzymes. Its very favourable safety profile is evidence by a lack of cardiovascular, phototoxic or teratogenic potential, as well as by results of in vivo toxicology studies in rats, dogs and monkeys, where only minor and reversible effects were observed. Dose-limiting adverse effects were related to the classical signs of exaggerated pharmacology for AMPA/kainate receptor antagonism, mostly ataxia and decreased locomotor activity. BGG492 is currently in Phase II development for epilepsy but clinical trials are ongoing also for migraine and pain.

A phase II study on the effect of BGG492 on PPR in patients with photosensitive epilepsy has been recently completed (ClinicalTrials.gov Identifier: NCT00784212).

NS-1209

NS-1209 is a water-soluble AMPA antagonist that shows high efficacy against status epilepticus induced by electrical stimulation of the amygdala or by subcutaneous administration of kainic acid in rats (Pitkanen et al. 2007). It also displayed some neuroprotective activity against status-induced hippocampal neurodegeneration (Perucca 2009). Clinical testing in the treatment of recurrent seizures has been initiated but no published data are available. Moreover, at the present time, no clinical trials in epilepsy are registered. Conversely, a Phase II controlled study in neuropathic pain has been published (Gormsen et al. 2009).

Perampanel: Pharmacokinetic Profile, Efficacy and Safety

Perampanel, a non-competitive selective antagonist of the AMPA glutamate receptors, exerts a direct influence on post-synaptic glutamatergic transmission (Rogawski 2011).

Investigation of this agent took place in two different neurological diseases: drug-resistant partial-onset epilepsies and Parkinson's disease (Rascol et al. 2012; Zaccara et al. 2013). The studies on patients with drug-resistant POS were carried out on little less than 1700 patients that were recruited in phase II and phase III (French et al. 2012; French et al. 2013; Krauss et al. 2012a).

After oral administration, absorption of perampanel is rapid and almost complete. Bioavailability was found to be complete with low systemic clearance after oral administration, consistent with a low first-pass metabolism. Fasting conditions do not affect the extent of absorption, but do slow drug absorption. After multiple oral dosing both the C_{max} and the AUC increase proportionally with dose. C_{max} of perampanel is reached within approximately 1 h. In humans, perampanel is 95% bound to plasma proteins (Franco et al. 2013). The average $t_{1/2}$ of perampanel without concomitant inducer AEDs is approximately 105 h, the longest among the new generation AEDs, allowing once-daily dosing. Perampanel is extensively metabolised by primary oxidation mediated by CYP3A followed by glucuronidation. About 30% of an orally administered dose of radio-labeled perampanel is found in the urine and 70% in the faeces, primarily as a mixture of oxidative and conjugated metabolites (Franco et al. 2013). The pharmacokinetics of a single 1 mg dose of perampanel were examined in subjects with mild to moderate hepatic impairment. The mean apparent clearance of perampanel was lower in both mildly and moderately impaired subjects compared to demographically matched healthy subjects. In view of these pharmacokinetic properties of perampanel a dose reduction in hepatically impaired subjects should be considered. To date, no data have been reported on the perampanel pharmacokinetics in subjects with renal impairment (Franco et al. 2013).

The efficacy of perampanel was demonstrated in three phase III, randomized, placebo-controlled, double-blind, multicentre trials (studies 304, 305, and 306).

Studies 304 and 305 compared once-daily perampanel doses of 8 and 12 mg with placebo (French et al., 2012; 2013) and study 306 (Krauss et al. 2012a) compared once daily perampanel doses of 2, 4 and 8 mg with placebo. Design of trials scheduled a 6-week baseline, a starting dose of 2 mg/day and weekly increments of 2 mg/day to the target dose. The titration phase was followed by a 13 weeks maintenance period. In Study 304 patients were randomized to receive once daily treatment with perampanel 8 mg ($n=133$), perampanel 12 mg ($n=134$) or placebo ($n=121$), with a 19-week double-blind phase (6-week titration and 13-week maintenance period). The 50% responder rates compared to placebo for the intention-to-treat (ITT) population were 37.6% with 8 mg ($p=0.0760$), 36.1% with 12 mg ($p=0.0914$) versus 26.4% with placebo. The median percentage change in seizure frequency for the ITT population was: -26.3% with 8 mg ($p=0.0261$) and -34.5% with 12 mg ($p=0.0158$) versus -21.0% with placebo (French et al. 2012). In Study 305 patients on a stable regimen of 1–3 AEDs were randomized to receive once daily treatment with perampanel 8 mg ($n=129$), perampanel 12 mg ($n=121$) or placebo ($n=136$) in a 19-week double-blind treatment (6-week titration and 13-week maintenance period). The 50% responder rates compared to placebo for the ITT population were 33.3% with 8 mg ($p=0.0018$), 33.9% with 12 mg ($p<0.001$) versus 14.7% with placebo. The median percent change in seizure frequency for the ITT population were: -30.5% with 8 mg ($p<0.001$) and -17.6% with 12 mg ($p=0.011$) versus -9.7% with placebo (French et al. 2013). In Study 306 patients on a stable regimen of 1–3 AEDs were randomized to receive once daily treatment with perampanel 2 mg ($n=180$), perampanel 4 mg ($n=172$), perampanel 8 mg ($n=169$) or placebo ($n=185$) in a 19 week double-blind add-on treatment (6-week titration and 13-week maintenance period). The 50% responder rates of perampanel compared to placebo for the ITT population were 20.6% with 2 mg ($p=$ns), 28.5% with 4 mg ($p=0.013$) and 34.9% with 8 mg ($p<0.001$) versus 17.9% with placebo. The median percentage change in seizure frequency for the ITT population was -13.6% with perampanel 2 mg ($p=$ns), -23.3% with perampanel 4 mg ($p=0.003$) and -30.8% with perampanel 8 mg ($p<0.001$) versus -10.7% with placebo (Krauss et al. 2012a).

A good tolerability profile of perampanel emerges from a recent meta-analysis of randomized controlled trials which demonstrated that AEs were not significantly more frequent during perampanel treatment in respect to placebo (Krauss et al. 2012b). The two main AEs associated with perampanel were dizziness and ataxia. Moreover, perampanel has been associated with somnolence. Noteworthy, no AEs clearly related to cognition have been associated with the treatment with perampanel while the drug was associated to weight gain, the only non-neurological AE of this drug. Idiosyncratic AEs haven't been associated with perampanel since rash appeared in a number of patients so small that it was considered non-significant. Perampanel is metabolised partly by reactive metabolites and these are responsible of mediation of most immune-mediated idiosyncratic drug reactions (European Public Assessment Reports). These facts are reassuring but cannot exclude other rare idiosyncratic AEs because the number of subjects included in the RCTs was insufficient to detect them (Zaccara et al. 2007; 2013).

Perampanel is approved in Europe and the US as adjunctive therapy for adults with focal seizures with or without secondary generalization.

General Remarks and Conclusions

There is a remarkable array of new chemical entities in the current AED development pipeline. In some cases, the compounds were synthesized in an attempt improve upon the activity of marketed AEDs. In other cases, the discovery of antiepileptic potential was largely serendipitous. Entry into the pipeline begins with the demonstration of activity in one or more animal screening models. Results from testing in a panel of such models provide a basis to differentiate agents and may offer clues as to the mechanism. Target activity may then be defined through cell-based studies, often years after the initial identification of activity. Some pipeline compounds are believed to act through conventional targets, whereas others are structurally novel and may act by novel mechanisms. A variety of AEDs that may act through novel targets are also in clinical development. However, so far, the mechanism of action of an antiepileptic compound has been rarely useful to predict its efficacy. Moreover, some of the new agents have been already evaluated in various non-epileptic conditions, such as neuropathic pain, migraine, and Parkinson's disease and it is even more difficult to predict their impact of epilepsy treatment.

The ideal AED would be effective well-tolerated, easy to take, devoid of significant drug interactions. However, mechanisms underlying epileptogenesis are multiple and complex, and is probably a utopian ideal to realize "the gold standard" AED. Nevertheless, the challenge for new more efficacious, more specific, and better tolerated drugs is continuing and a better knowledge of mechanisms underlying epilepsy should represent the guide for future research. The ultimate goal of should be not only to render seizure-free the patients but also allow to improve quality of life of individuals and reduce costs of medical care. It remains to be demonstrated whether any of the AEDs in the pipeline will reach these objectives.

References

Almeida L, Soares da Silva P (2007) Eslicarbazepine acetate (BIA 2-093). Neurotherapeutics 4:88–96

Almeida L, Minciu I, Nunes T, Falcão A, Magureanu SA, Soares-da-Silva P (2008) Pharmacokinetics, efficacy, and tolerability of eslicarbazepine acetate in children and adolescent with epilepsy. J Clin Pharmacol 48:966–977

Almeida L, Bialer M, Soares da Silva P (2009) Eslicarbazepine acetate. In: Shorvon S, Perucca E, Engel J (eds) The treatment of Epilepsy, 3rd edn. Blackwell, Oxford, pp 485–498

Ambrosio AF, Silva AP, Araujo I, Malva JO, Soares-da-Silva P, Carvalho AP et al (2000) Neurotoxic/neuroprotective profile of carbamazepine, oxcarbazepine and two new putative antiepileptic drugs, BIA 2-093 and BIA 2-024. Eur J Pharmacol 406:191–201

Ambrosio AF, Silva AP, Malva JO, Soares-da-Silva P, Carvalho AP, Carvalho CM (2001) Inhibition of glutamate release by BIA 2-093 and BIA 2-024, two novel derivates of carbazepine, due to blockade of sodium but not calcium channel. Biochem Pharmacol 61:1271–1275

Benes J, Parada A, Figueiredo AA, Alves PC, Freitas AP, Learmonth DA et al (1999) Anticonvulsivant and sodium channel-blocking properties of novel 10,11-dihydro-5H-dibenz(b,f)azepine-5-carboxamide derivative. J Med Chem 42:2582–2587

Ben-Menachem E, Gabbai AA, Hufnagel A, Maia J, Almeida L, Soares-da-Silva P (2010) Eslicarbazepine acetate as adjunctive therapy in adult patients with partial epilepsy. Epilesy Res 89:278–285

Bennett B, Matagne A, Michel P, Leonard M, Cornet M, Meeus MA et al (2007) Seletracetam (UCB 44212). Neurother 4(1):117–122

Bialer M, Johannessen SI, Levy RH, Perucca E, Tomson T, White HS (2013) Progress report on new antiepileptic drugs: a summary of the Eleventh Eilat Conference (EILAT XI). Epilepsy Res 103(1):2–30

Binnie CD, Kasteleijn-Nolst Trenité DGA, De Korte RA (1986) Photosensitivity as a model for acute antiepileptic drug studies. Electroencephalogr Clin Neurophysiol 63:35–41

Biton V, Berkovic SF, Abou-Khalil B, Sperling MR, Johnson ME, Lu S (2014) Brivaracetam as adjunctive treatment for uncontrolled partial epilepsy in adults: a phase III randomized, double-blind, placebo-controlled trial. Epilepsia 55(1):57–66

Brodie MJ (2010) Antiepileptic drug therapy the story so far. Seizure 19(10):650–655

Brown ME, El-Mallakh RS (2010) Role of eslicarbazepine in the treatment of epilepsy in adult patients with partial-onset seizures. Ther Clin Risk Manag 6:103–109

Cunha RA, Coelho JE, Costenla AR, Lopes LV, Parada A, de Mendonça A et al (2002) Effects of carbazepine and novel 10,11-dihydro-5H-dibenz(b,f)azepine-5-carboxamide derivatives on synaptic transmission in rat hippocampal slices. Pharmacol Toxicol 90:208–213

Di Prospero NA, Gambale JJ, Pandina G, Ford L, Girgis S, Moyer JA et al (2014) Evaluation of JNJ-26489112 in patients with photosensitive epilepsy: A placebo-controlled, exploratory study. Epilepsy Res 108(4):709–716

Elger C, Bialer M, Cramer JA, Maia J, Almeida L, Soares-da-Silva P (2007) Eslicarbazepine acetate: a double blind, add-on, placebo-controlled exploratory trial in adult patients with partial-onset seizures. Epilepsia 48:497–504

Elger C, Halasz P, Maia J, Almeida L Soares-da-Silva P (2009) BIA-2093–301 Investigators Study Group. Efficacy and safety of eslicarbazepine acetate as adjunctive treatment in adults with refractory partial-onset seizures: a randomized, double-blind, placebo-controlled, parallel-group phase 3 study. Epilepsia 50:454–463

Falcao A, Pinto R, Nunes T, Soares-da-Silva P (2013a) Effect of repeated administration of eslicarbazepine acetate on the pharmacokinetics of simvastatin in healthy subjects. Epilepsy Res 106(1–2):244–249

Falcao A, Vaz-da-Silva M, Gama H, Nunes T, Almeida L, Soares-da-Silva P (2013b) Effect of eslicarbazepine acetate on the pharmacokinetics of a combined ethinylestradiol/levonorgestrel oral contraceptive in healty women. Epilepsy Res 105:368–376

Faught E, Holmes GL, Rosenfeld WE, Novak G, Neto W, Greenspan A et al (2008) Randomized, controlled, dose ranging trial of carisbamate for partial-onset seizures. Neurology 71:1586–1593

Franco V, Crema F, Iudice A, Zaccara G, Grillo E (2013) Novel treatment options for epilepsy: focus on perampanel. Pharmacol Res 70(1):35–40

Francois J, Ferrandon A, Koning E, Nehlig A (2005) A new drug RWJ 333369 protects limbic areas in the lithium-pilocarpine model of epilepsy and delays or prevents the occurrence of spontaneous seizures. Epilepsia 46(Suppl 8):269–270

French JA, Krauss GL, Biton V, Squillacote D, Yang H, Laurenza A et al (2012) Adjunctive Perampanel for refractory partial-onset seizures: randomized phase III study 304. Neurology 79:589–596

French JA, Krauss GL, Steinhoff BJ, Squillacote D, Yang H, Kumar D et al (2013) Evalutation of adjunctive Perampanel in patients with refractory partial-onset seizures: results of randomized global phase III study 305. Epilepsia 54:117–125

Gillard M, Fuks B, Leclercq K, Matagne A (2011) Binding characteristics of brivaracetam, a selective, high affinity SV2A ligand in rat, mouse and human brain: relationship to anti-convulsant properties. Eur J Pharmacol 664(1–3):36–44

Gil-Nagel A, Lopes-Lima J, Almeida L, Maia J Soares-da-Silva P (2009) BIA-2093–303 Investigators Study Group. Efficacy and safety of 800 and 1200 mg eslicarbazepine acetate as adjunctive treatment in adults with refractory partial-onset seizures. Acta Neurol Scand 120:281–287

Gil-Nagel A, Elger C, Ben-Menachem E, Halász P, Lopes-Lima J, Gabbai AA et al (2013) Efficacy and safety of eslicarbazepine acetate as add-on treatment in patients with focal onset seizures: integrated analysis of pooled data from double-blind phase III clinical studies. Epilepsia 54:98–107

Gormsen L, Finnerup NB, Almqvist PM, Jensen TS (2009) The efficacy of the AMPA receptor antagonist NS1209 and lidocaine in nerve injury pain: a randomized, double-blind, placebo-controlled, three-way crossover study. Anesth Analg 108(4):1311–1319

Halasz P, Cramer JA, Hodoba D, Członkowska A, Guekht A, Maia J et al (2010) Long term efficacy of eslicarbazepine acetate: results of a 1-year open-label extension study in partial-onset seizures in adults with epilepsy. Epilepsia 51:1963–1969

Hufnagel A, Ben-Menachem E, Gabbai AA, Falcão A, Almeida L, Soares-da-Silva P (2013) Long-term safety and efficacy of eslicarbazepine acetate as adjunctive therapy in the treatment of partial-onset seizures in adults with epilepsy: results of a 1-year open-label extension study. Epilepsy Res 103:262–269

Kaminski RM, Matagne A, Leclercq K, Gillard M, Michel P, Kenda B et al (2008) SV2A protein is a broad-spectrum anticonvulsant target: functional correlation between protein binding and seizure protection in models of both partial and generalized epilepsy. Neuropharmacology 54(4):715–720

Kasteleijn-Nolst Trenité D Rubboli G, Hirsch E, Martins da Silva A, Seri S, Wilkins A et al (2012) Methodology of photic stimulation revisited: updated European algorithm for visual stimulation in the EEG laboratory. Epilepsia 53:16–24

Kasteleijn-Nolst Trenité DG, Biton V, French JA, Abou-Khalil B, Rosenfeld WE, Diventura B et al (2013) Kv7 potassium channel activation with ICA-105665 reduces photoparoxysmal EEG responses in patients with epilepsy. Epilepsia 54(8):1437–1443

Krauss GL, Serratosa JM, Villanueva V, Endziniene M, Hong Z, French J et al (2012a) Randomized phase 3 study 306: adjunctive Perampanel for refractory partial-onset seizures. Neurology 78:1408–1415

Krauss GL, Bar M, Biton V, Klapper JA, Rektor I, Vaiciene-Magistris N et al (2012b) Tolerability and safety of Perampanel: two randomized dose-escalation studies. Acta Neurol Scand 125:8–15

Kwan P, Trinka E, Van Paesschen W, Rektor I, Johnson ME, Lu S (2014) Adjunctive brivaracetam for uncontrolled focal and generalized epilepsies: results of a phase III, double-blind, randomized, placebo-controlled, flexible-dose trial. Epilepsia 55(1):38–46

Löscher W, Klitgaard H, Twyman RE, Schmidt D (2013) New avenues for anti-epileptic drug discovery and development. Nat Rev Drug Discov 12(10):757–756

Margineanu DG, Klitgaard H (2009) Brivaracetam inhibits spreading depression in rat neocortical slices in vitro. Seizure 18(6):453–456

Matagne A, Margineanu DG, Kenda B, Michel P, Klitgaard H (2008) Anti-convulsive and anti-epileptic properties of brivaracetam (ucb 34714), a high-affinity ligand for the synaptic vesicle protein, SV2A. Br J Pharmacol 154(8):1662–1671

Matagne A, Margineanu DG, Potschka H, Löscher W, Michel P, Kenda B et al (2009) Profile of the new pyrrolidone derivative seletracetam (ucb 44212) in animal models of epilepsy. Eur J Pharmacol 614(1–3):30–37

Mattson RH, Cramer JA, Collins JF (1992) A comparison of valproate with carbamazepine for the treatment of partial seizures and secondary generalized tonic-clonic seizures in adults. N Eng J 327:765–771

Meldrum BS, Rogawski MA (2007) Molecular targets for antiepileptic drug development. Neurother 4(1):18–61

Parada A, Soares da Silva P (2002) The novel anticonvulsivant BIA2-093 inhibits transmitter release during opening of voltage-gated sodium channels: a comparison with carbamazepine and oxcarbazepine. Neurochem Int 40:435–440

Perucca E (2009) What is the promise of new antiepileptic drugs in status epilepticus? Focus on brivaracetam, carisbamate, lacosamide, NS-1209, and topiramate. Epilepsia 50(Suppl 12):49–50

Perucca E, Tomson T (2011) The pharmacological treatment of epilepsy in adults. Lancet Neurol 10(5):446–456

Perucca E, French J, Bialer M (2007) Development of new antiepileptic drugs: challenges, incentives, and recent advances. Lancet Neurol 6(9):793–804

Pitkanen A, Mathiesen C, Ronn LC, Møller A, Nissinen J (2007) Effect of novel AMPA antagonist, NS1209, on status epilepticus. An experimental study in rat. Epilepsy Res 74(1):45–54

Rascol O, Barone P, Behari M, Emre M, Giladi N, Olanow CW et al (2012) Perampanel in Parkinson disease fluctuations: a double-blind randomized trial with placebo and entacapone. Clin Neuropharmacol 35:15–20

Rocha JF, Vaz-da-Silva M, Almeida L, Falcão A, Nunes T, Santos AT et al (2009) Effect of eslicarbazepine acetate on the pharmacokinetics of metformin in healthy subjects. Int J Clin Pharmacol Ther 7:255–261

Roeloffs R, Wickenden AD, Crean C, Werness S, McNaughton-Smith G, Stables J et al (2008) In vivo profile of ICA-27243 [N-(6-chloro-pyridin-3-yl)-3,4-difluorobenzamide], a potent and selective KCNQ2/Q3 (Kv7.2/Kv7.3) activator in rodent anticonvulsant models. J Pharmacol Exp Ther 326:818–828

Rogawski MA (2002) Principles of antiepileptic drug action. In Levy RH, Mattson RH, Meldrum BS, Perucca E (eds) Antiepileptic drugs, 5th ed. Lippincot Williams & Wilkins, Philadelphia, pp 3–22

Rogawski MA (2011) Revisiting AMPA receptors as an antiepileptic drug target. Epilepsy Curr 11:56–63

Rogawski MA (2013) AMPA receptors as a molecular target in epilepsy therapy. Acta Neurol Scand 197:9–18

Rogawski MA, Bazil CW (2008) New molecular targets for antiepileptic drugs: alpha (2) delta, SV2A, and K(v)7/KCNQ/Mpotassium channels. Curr Neurol Neurosci Rep 8:345–352

Rogawski MA, Loscher W (2004) The neurobiology of antiepileptic drugs. Nat Rev Neurosci 5:553–564

Rolan P, Sargentini-Maier ML, Pigeolet E, Stockis A (2008) The pharmacokinetics, CNS pharmacodynamics and adverse event profile of brivaracetam after multiple increasing oral doses in healthy men. Br J Clin Pharmacol 66(1):71–75

Russo E, Gitto R, Citraro R, Chimirri A, De Sarro GB (2012) New AMPA antagonists in epilepsy. Expert Opin Investig Drugs 21(9):1371–1389

Ryvlin P, Werhahn KJ, Blaszczyk B, Johnson ME, Lu S (2014) Adjunctive brivaracetam in adults with uncontrolled focal epilepsy: results from a double-blind, randomized, placebo-controlled trial. Epilepsia 55(1):47–55

Sander JW (2004) The use of antiepileptic drugs–principles and practice. Epilepsia 45(S6):28–34

Sargentini-Maier ML, Rolan P, Connell J, Tytgat D, Jacobs T, Pigeolet E et al (2007) The pharmacokinetics, CNS pharmacodynamics and adverse event profile of brivaracetam after single increasing oral doses in healthy males. Br J Clin Pharmacol 63(6):680–688

Sargentini-Maier ML, Sokalski A, Boulanger P, Jacobs T, Stockis A (2012) Brivaracetam disposition in renal impairment. J Clin Pharmacol 52(12):1927–1933

Shorvon S (2000) Oxcarbazepine: a review. Seizure 9:75–79

Sperling MR, Greenspan A, Cramer JA, Kwan P, Kälviäinen R, Halford JJ et al (2010) Carisbamate as adjunctive treatment of partial onset seizures in adults in two randomized, placebo controlled trials. Epilepsia 51:333–343

Stockis A, Sargentini-Maier ML, Horsmans Y (2013) Brivaracetam disposition in mild to severe hepatic impairment. J Clin Pharmacol 53(6):633–641

Vaz-da-Silva M, Costa R, Soares E, Maia J, Falcão A, Almeida L et al (2009) Effects of eslicarbazepine acetate on the pharmacokinetics of digoxin in healthy subjects. Fundam Clin Pharmacol 23:504–514

Verrotti A, Loiacono G, Rossi A, Zaccara G (2014) Eslicarbazepine acetate: an update on efficacy and safety in epilepsy. Epilepsy Res 108(1):1–10

Von Rosenstiel P (2007) Brivaracetam (UCB 34714). Neurother 4(1):84–87

Zaccara G, Franciotta D, Perucca P (2007) Idiosyncratic adverse reaction to antiepileptic drugs. Epilepsia 48:1223–1244

Zaccara G, Giovannelli F, Cincotta M, Verrotti A, Grillo E (2013) The adverse profile of Perampanel: meta-analysis of randomized controlled trials. Eur J Neurol 20:1204–1211

Zona C, Niespodziany I, Pieri M, Klitgaard H, Margineanu DG (2005) Seletracetam (ucb 44212), a new pyrrolidone derivative, lacks effect on Na+ currents in rat brain neurons in vitro. Epilepsia 46(Suppl 8):116

Zona C, Pieri M, Carunchio I, Curcio L, Klitgaard H, Margineanu DG (2010) Brivaracetam (ucb 34714) inhibits Na(+) current in rat cortical neurons in culture. Epilepsy Res 88(1):46–54

Reproductive Hormones in Epilepsy Therapy: From Old Promises to New Hopes

Alberto Verrotti, Giovanni Prezioso, Claudia D'Egidio and Vincenzo Belcastro

Abstract A significant mutual interaction between sex hormones and the central nervous system has been reported by several studies in the past years.

This paper aims to discuss the genomic and electrophysiological effects of androgens, estrogens and progestogens on neurons, their influence on seizure frequency and antiepileptic drugs metabolism, and, conversely, the hormonal changes and reproductive dysfunction in patients with epilepsy.

In conclusion, the correlations between Polycystic Ovary Syndrome (PCOS), reduced effectiveness of contraceptives and antiepileptic drugs have also been mentioned.

Epilepsy represents one of the most frequent neurological diseases worldwide, affecting nearly 1% of the population. It is characterized by the chronic recurrence of seizures in an unpredictable fashion. The prevention of seizures is the primary goal of epilepsy treatment. This requires the use of anti-epileptic drugs (AEDs) for the majority of patients.

Biological, experimental, and clinical studies for a number of years reported a noticeable interaction between epilepsy, AEDs and sex hormones.

Not only seizures and AEDs can induce significant changes in sex hormones regulation, sexual development, and reproductive functions, but also seizures and AEDs metabolism are strongly influenced by steroid hormones and their derivatives.

A. Verrotti (✉)
Department of Pediatrics, University of Perugia, Perugia, Italy
e-mail: averrotti@unich.it

G. Prezioso · C. D'Egidio
Department of Pediatrics, University of Chieti, Chieti, Italy

V. Belcastro
Neurology Unit, Department of Medicine, S. Anna Hospital, Como, Italy

Sex Hormones Role on CNS

Androgens, estrogens, and progestogens play a key role in shaping neural activity since prenatal life, moreover they exert a regulatory action on neurotransmitters and their receptors in neurons. Sex hormones interact with tissues, like the brain, by means of both receptor mediated and receptor-independent pathways. The first mechanism starts with nuclear receptor dimerization after hormone binding, thereafter the steroid-receptor complex in turn binds to and regulates the transcription of sex steroid responsive genes or post-transcriptional proteins. (Keefe 2002; Reddy 2009; Finocchi and Ferrari 2011) Each steroid receptor isoform is linked with different adapter protein and different response elements. Non-genomic effects of sex steroids have also been reported. (Keefe 2002) These effects are usually faster than nuclear receptor-mediated ones. Indeed, some steroids seem to conditioning electrophysiologic neuronal activity by interacting with high affinity on specific binding sites of the main neurotransmitter receptors, like GABA-A and NMDA receptors, effectively acting as neural stimulants or depressant. For instance, some of these hormones exert an agonist action on the GABA-A receptor even more potent than benzodiazepines and barbiturates, clearly affecting the epileptogenic activity. (Keefe 2002; Reddy 2009)

Estrogens show evident proconvulsant and epileptogenic properties, even if in sometimes they may exhert protective and anticonvulsant effects. Several studies have found estradiol (E2) both to enhance NMDA receptor activity and to temporarily decrease GABA inhibition. Moreover it appears to decrease GABA synthesis. There is also emerging evidence of hippocampus hyperexcitability due to estrogen and brain derived neurotrophic factor interactions (Foldvary-Schaefer et al. 2004; Erel and Guralp 2011; Finocchi and Ferrari 2011; Guille et al. 2008)

In contrast, clinical studies reported marked anticonvulsant qualities of progesterone, in particular of its 5α-reduced metabolites, like allopregnanolone. It has been demonstrated that glial and neuronal cells of the cerebral cortex and subcortical white matter are even able to synthesize by themselves from cholesterol these metabolites (neurosteroids). (Finocchi and Ferrari 2011; Reddy 2009; Guille et al. 2008). Neurosteroids activate the GABA receptor both directly, by binding two distinct sites different from the benzodiazepine or barbiturate sites, and indirectly, through a longer-term action on progesterone nuclear receptors. (Reddy 2009)

Nevertheless, proconvulsant neurosteroid acting as NMDA receptor agonists and GABA-A receptor antagonists have been identified (i.e. pregnenolone sulfate). (Reddy 2009)

In regards to androgens, testosterone affects neural activity depending on its conversion to androstanediol, which has anticonvulsant and GABA-A agonist property, and E2 with the aforementioned proconvulsant activity. (Reddy 2009; Keefe 2002; Hamed 2008)

Table 1 Three types of catamenial epilepsy. (modified from Reddy 2009)

Subcategories of catamenial epilepsy	Main characteristics
C1—perimenstrual	Due to the rapid decrease in anticonvulsant progesterone and neurosteroids blood level during days −3 to +3 which lead to descreased GABA inhibition
	Incidence: 71% of women with regular ovulatory cycles
C2—periovulatory	Due to the proconvulsant estrogen peak not balanced by an adequate level of neurosteroids in days 10–13. This requires a decrease in GABA inhibition and a marked excitation
	Incidence: 71% of women with regular ovulatory cycles
C3—inadequate luteal	Due to the insufficient brain levels of neurosteroids resulting from progesterone metabolism in anovulatory cycles. The final result is both a marked estradiol excitation and a persistent low GABA inhibition during days
	Incidence: 78% of women with anovulatory cycles (luteal phase)

Effects on Seizures-Catamenial Epilepsy

There is clinical evidence that interaction between steroid hormones and the CNS influences seizure susceptibility in epileptic patients. Being this especially evident in women, whose remarkable hormonal changes during puberty as well as pregnancy and menopause affect seizure frequency and AED metabolism. Moreover menopausal transition seems to have an effect on seizure susceptibility with a seizure increase of about 30% in women with epilepsy in perimenopause and a tendency to decrease after menopause, but there is no consensus on these findings (Erel and Guralp 2011; Røste et al. 2008).

Another clear proof of the role of sex hormones in seizure exacerbation is the catamenial epilepsy phenomenon in women. Catamenial epilepsy (CE) derives from the Greek word *katamenios*, which means "monthly", and it refers to a cyclical variation in seizures frequency in relation to menstrual cycle phases in women affected by epilepsy (genetic, structural, metabolic or of unknown cause), especially temporal-lobe epilepsy (Foldvary-Schaefer et al. 2004; Guille et al. 2008; Reddy and Rogawski, 2009; Pennell 2009; Pack 2010; Finocchi and Ferrari 2011). The incidence varies from 10 to 70% due to the lack of an unambiguous definition and to methodological differences among studies (Reddy 2009). A significant contribution to the study of CE was made by Herzog et al. (Herzog et al. 1997), who proposed the most accepted definition: a twofold increase in average daily frequency during a phase of menstrual cycle. They also described three subcategories of CE: C1 during the perimenstrual period, probably due to the withdrawal of allopregnanolone and other inibitory neurosteroids at the time of menstruation, C2 in the periovulatory phase, owing to an increase in estrogen concentration, defined above as proconvulsant, and C3, typical of the inadequate luteal phase of women with anovulatory cycles (see Table 1 and Fig. 1). A concurrent involvement in CE of AEDs blood level fluctuations and of the possible changes in water, pH, and electrolyte cannot

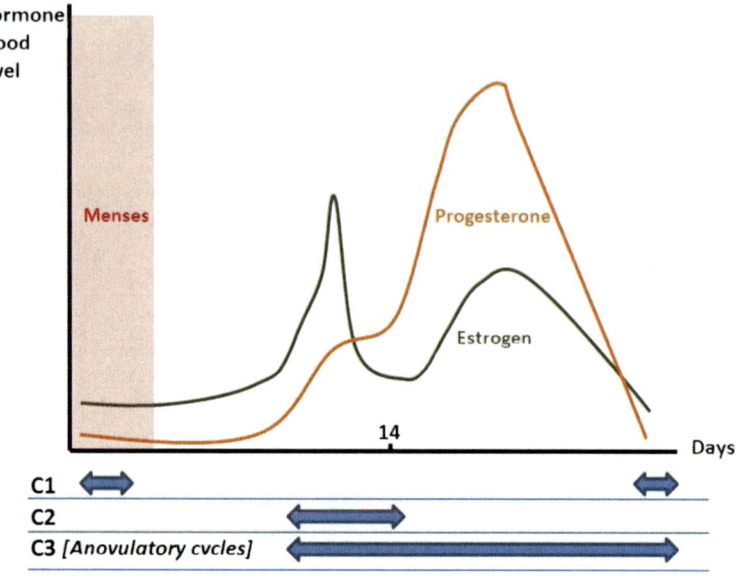

Fig. 1 Seizures distribution within the three patterns of catamenial epilepsy

be definitely ruled out, although hormonal oscillation maintains a key etiologic role (Pack 2010). A correct diagnosis of CE requires both careful compilation and evaluation of menstrual and seizure diaries as well as characterization of cycle type and duration (Reddy 2009; Foldvary-Schaefer et al. 2004). It is not rare to find women with CE presenting with state-dependant pharmacoresistance or intractable seizures (Reddy and Rogawski 2009). The first-line therapy includes the AEDs and usually requires cyclic dosage adjustments or supplement with other AEDs. Adjunctive hormonal treatment with progesterone or estrogen antagonists has been reported to augment seizure control in appropriate CE patients (Pack 2010; Herzog 1995, 1999). Natural progesterone, mainly converted to the anticonvulsant allopregnanolone, has been demonstrated to be effective in women with focal epilepsy and CE, even though it showed endocrine and CNS side effects (Herzog 1986, 1995, 1999; Pack 2010; Reddy 2009). Synthetic progestogen like Medroxyprogesterone is only partially converted to active neurosteroids, resulting in moderate improvement in seizure frequency (Zimmerman et al. 1973; Mattson et al. 1984). Nonetheless it provokes menses interruption and consequently reproductive disturbances in long-term therapy. On the contrary the use of combined oral contraceptives has a questionable effectiveness (Guille et al. 2008). Antiestrogens (clomiphene citrate), androgen, or synthetic gonadotropin-releasing hormone therapies have a limited utility as they show several adverse events (Foldvary-Schaefer et al. 2004). Great interest has been recently engendered about the use of Ganaxolone in CE. Ganaxolone is a synthetic analogue of allopregnanolone with potent positive action on GABA-A receptors resulting in anticonvulsant effect (Reddy 2009; Pack 2010; Guille et al.

2008; McAuley et al. 2001). Non-hormonal therapy of CE includes Acetazolamide, a carbonic anhydrase inhibitor with a broad spectrum of efficacy on seizures. The mechanism of action on seizure reduction is unclear, maybe a diuretic effect is implied. Since Acetazolamide is subject to tolerance as much as significant adverse events, it is usually administered intermittently (Reddy 2009; Ross 1958; Lim et al. 2001; Pack 2010).

Anyway, further investigation is needed in the matter of CE therapy given that data currently available largely belong to small non randomized studies and empirical evidence (Reddy 2009).

Effects of Seizures on Sex Functions

Epilepsy itself may directly influence the hypothalamic-pituitary axis (Herzog et al. 1986; Isojärvi et al. 2005; Scharfman et al. 2008) Preclinical investigations suggested that the involvement of medial temporal lobe regions in epilepsy may alter sex hormone secretion and reproductive functions. A dysregulation of GnRH pulsatility and therefore of LH/FSH ratio in epileptic women and of testosterone/LH ratio in epileptic men has been described (Verrotti et al. 2011; Drislane et al. 1994; Ciampani et al. 2005; Morell 2003; Herzog 2008; Hamed 2008; Fawley et al. 2006). Experimental studies reported an interesting correlation with laterality, even though datas are not univocal (Pack 2010; Quigg et al. 2009). Unilateral left-sided temporolimbic discharges seem to increase the pulse frequencies of GnRH secretion, resulting in a higher occurrence with PCOS, while right-sided temporolimbic discharges may decrease GnRH pulse frequency and are more commonly associated with sexual dysfunction (hypotalamic amenorrhea, functional hyperprolactinemia, infertility and premature menopause in women; decreased libido, abnormal semen analysis and reduced fertility in men with epilepsy) (Verrotti et al. 2011; Herzog et al. 1986; Morrell et al. 2005; Quigg et al. 2009; Harden 2008; Duncan et al. 2009).

Furthermore, some studies draw attention to the increased risk of premature ovarian failure and perimenopausal symptoms in women with epilepsy (Klein et al. 2001; Harden 2003).

Mutual Interactions Between Sex Hormones and AEDs

There is increasing evidence of AEDs contribution to sex dysfunction in epileptic patients. Indeed, studies suggest that AEDs may variably influence both the steroid hormones metabolism and their binding proteins (Hamed 2008; Bauer et al. 2002). Enzyme-inducing AEDs (EIAEDs)—such as phenobarbital (PB), phenytoin (PHT), and carbamazepine (CBZ)—rather than non-EIAEDs (NEIAEDs) have been known to play a critical role in steroid hormones abnormalities (see Table 2). EIAEDs can

Table 2 EIAEDs and NEIAEDs. (modified from Reddy 2009)

EIAEDs	NEIAEDs
Carbamazepine	Clobazam
Oxcarbazepine	Clonazepam
Phenobaribatal	Ethosuximide
Methylphenobarbital	Mesuximide
Phenobarbital sodium	Valproic Acid[a]
Phenytoin	Lamotrigine
Fosphenytoin sodium	Gabapentin
Felbamate	Pregabalin
Topiramate	Vigabatrin
Primidone	Tiagabine
	Zonisamide
	Sultiame
	Beclamide
	Levetriacetam
	Rufinamide
	Stripentol

[a] weak CYP inducer

induce hepatic cytochrome P450 (CYP450), a system of mixed oxidative enzymes which metabolizes AEDs to a more water-soluble form. As the CYP450 is involved in steroids metabolism, its induction implies a faster hormone clearance (Isojärvi et al. 2005). In addition, EIAEDs were found to increase serum sex hormone-binding globulin (SHBG) levels in patients with epilepsy, reducing biologically active steroid hormones like DHEAS, T, free androgen index, and E2 (Pennell 2009; Erel and Guralp 2011; Verrotti et al. 2011). This may result in impotence and decreased fertility in men, hyperandrogenism and menstrual disorders in women. EIAEDs may also correlate with hypogonadotropic hypogonadism, by inhibiting LH secretion, and with decreasing libido and potency, by increasing the conversion of testosterone to E2 (Hamed 2008).

Anyway also NEIAEDs have been found to be involved in hormonal dysregulation. Valproate (VPA), in particular, seems to be associated with high serum concentrations of T, androstenedione and DHEAS, and with increase in levels of LH and LH/FSH ratio, leading to polycystic ovary syndrome, hyperandrogenism and amenorrhea in women, reduced feritlity and sperm abnormalities in morphology, count and motility in men (Verrotti et al. 2011; Isojärvi et al. 1993; Prabhakar et al. 2007; Rauchenzauner et al. 2010; Xiaotian et al. 2013; Morrell et al. 2005; Chen et al. 1992). Even if not yet clear, VPA may directly alter androgen production in ovaries or it may act indirectly, by inhibiting steroid hormones metabolism and thereby increasing serum androgen levels. It is noticeable that hyperandrogenism has been evidenced more frequently in women who have gained weight during VPA therapy and in those who started treatment before the age of twenty.

Low prevalence of reproductive disorders has been reported during therapy with new AEDs like oxcarbazepine and lamotrigine (LTG), but no data are available for a number of other new AEDs. Besides, women on LTG, previously treated with VPA, seems to show a marked improvement in reproductive dysfunction (Isojärvi et al. 1998).

PCOS and AEDs

Among reproductive dysfunctions, PCO and PCOS appears to be particularly common in epileptic women (10–25 versus 5–6% of health women) (Herzog et al. 2003). Even though PCOS has been detected in non-treated epileptic patients, several studies highlighted an increased incidence during AED treatment, in particular with VPA, but the evidence still remains controversial (Isojärvi et al. 1993, 1995, 1996, 2001; Morrell et al. 2002, 2008; Betts et al. 2003; Prabhakar et al. 2007). Special attention should be directed to weight gain and insulin resistance, which can be common adverse effects of vigabatrin, carbamazepine and gabapentin other than VPA, because they have been reported as important risk factors for PCOS (Verrotti et al. 2011). Elevated serum leptin, impaired adipokine regulation, and low serum IGFBP-1 levels have been linked to VPA-related obesity in women with epilepsy (Belcastro et al. 2013; Greco et al. 2005; Gungor et al. 2007; Verrotti et al. 2011, 1999; Rauchenzauner et al. 2008). Moreover, VPA seems also able to stimulate insulin secretion both indirectly, by increasing free fatty acids levels in blood through albumin binding competition and directly by stimulating pancreatic b-cells and inhibiting glucose transporter protein type 1 (GLUT-1) activity (Luef et al. 2002, 2009, 2003; Wong et al. 2005; Belcastro et al. 2013; Evans et al. 2003). Finally, the inhibition of sympathetic nervous system, the impairment of insulin signal transduction pathway and the direct influence on the ovary via aromatase inhibition are other proposed mechanism of action of VPA (Verrotti et al. 2011). It has been demonstrated that all of these metabolic abnormalities predispose also to metabolic syndrome, which is frequently detected among VPA overweight patients as a consequence of the excess fat mass (Belcastro et al. 2013; Verrotti et al. 2010; see Fig. 2).

On the other hand, EIAEDs like phenitoine and CBZ may prevent from the development of PCOS, by reducing free T levels.

Contraceptive and AEDs

EIAEDs may also interfere with oral contraceptive pills (OCPs) efficacy similarly by reducing the biological active hormone concentrations and allowing ovulations. In order to reduce the risk of unplanned pregnancies it is recommended to

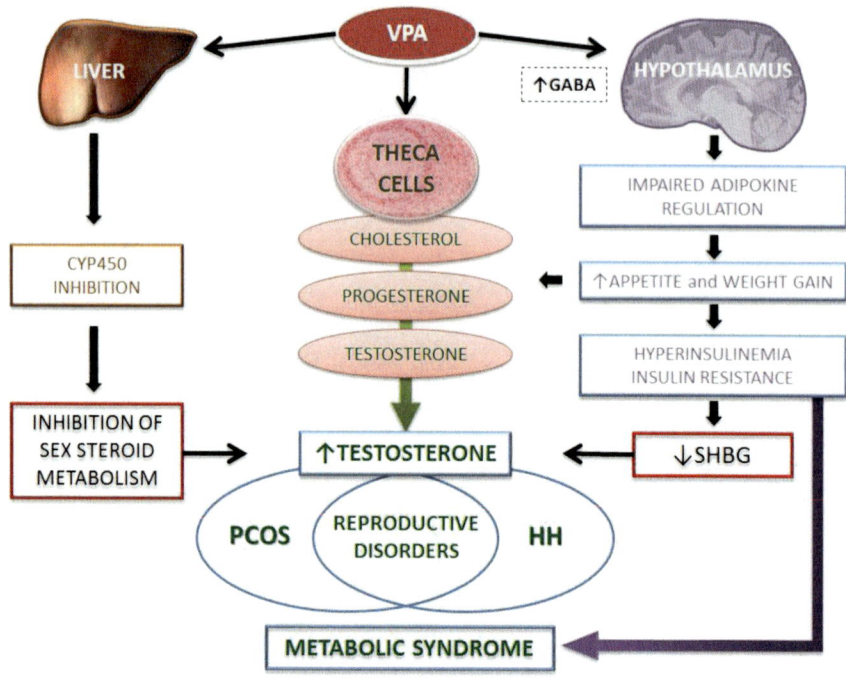

Fig. 2 Hormonal and metabolic alterations in VPA therapy. (modified from Verrotti et al.2011)

prescribe an OCP with higher doses of ethinyl estradiol (≥ 50 micrograms) and to use additional non-hormonal forms of contraception (Erel and Guralp 2011; Bartoli et al. 1997; Pennell 2009; Guberman 1999).

Refrences

Bartoli A, Gatti G, Cipolla G, Barzaghi N, Veliz G, Fattore C, Mumford J, Perucca E (1997) A double-blind, placebo-controlled study on the effect of vigabatrin on in vivo parameters of hepatic microsomal enzyme induction and on the kinetics of steroid oral contraceptives in healthy female volunteers. Epilepsia 38:702–707

Bauer J, Isojärvi JI, Herzo AG, Reuber M, Polson D, Taubøll E, Genton P, van der Ven H, Roesing B, Luef GJ, Galimberti CA, van Parys J, Flügel D, Bergmann A, Elger CE (2002) Reproductive dysfunction in women with epilepsy: recommendations for evaluation and management. J Neurol Neurosurg Psychiatry 73:121–125

Belcastro V, D'Egidio C, Striano P, Verrotti A (2013) Metabolic and endocrine effects of valproic acid chronic treatment. Epilepsy Res 107:1–8

Betts T, Yarrow H, Dutton N, Greenhill L, Rolfe T (2003) A study of anticonvulsant medication on ovarian function in a group of women with epilepsy who have only ever taken one anticonvulsant compared with a group of women without epilepsy. Seizure 12:323–329

Chen SS, Shen MR, Chen TJ, Lai SL (1992) Effects of antiepileptic drugs on sperm motility of normal controls and epileptic patients with long-term therapy. Epilepsia 33:149–153

Ciampani M, Verrotti A, Chiarelli F (2005) Sex hormones in patients with epilepsy-hormonal changes in epileptic men and women taking antiepileptics. Horm Metab Res 37:184–188

Drislane FW, Coleman AE, Schomer DL, Ives J, Levesque LA, Seibel MM, Herzog AG (1994) Altered pulsatile secretion of luteinizing hormone in women with epilepsy. Neurology 44:306–310

Duncan S, Talbot A, Sheldrick R, Caswell H (2009) Erectile function, sexual desire, and psychological well-being in men with epilepsy. Epilepsy Behav 15:351–357

Erel T, Guralp O (2011) Epilepsy and menopause. Arch Gynecol Obstet 284:749–755

Evans JL, Goldfine ID, Maddux BA, Grodsky GM. (2003) Are oxidative stress-activated signaling pathways mediators of insulin resistance and beta-cell dysfunction? Diabetes 52:1–8

Fawley JA, Pouliot WA, Dudek FE (2006) Epilepsy and reproductive disorders: the role of the gonadotropin-releasing hormone network. Epilepsy Behav 8:477–482

Finocchi C, Ferrari M (2011) Female reproductive steroids and neuronal excitability. Neurol Sci 32(Suppl 1):S31–35

Foldvary-Schaefer N, Harden C, Herzog A, Falcone T (2004) Hormones and seizures. Cleve Clin J Med 71(Suppl 2):S11–18

Greco R, Latini G, Chiarelli F, Iannetti P, Verrotti A (2005) Leptin, ghrelin, and adiponectin in epileptic patients treated with valproic acid. Neurology 65:1808–1809

Guberman A (1999) Hormonal contraception and epilepsy. Neurology 53:S38–40

Guille C, Spencer S, Cavus I, Epperson CN (2008) The role of sex steroids in catamenial epilepsy and premenstrual dysphoric disorder: implications for diagnosis and treatment. Epilepsy Behav 13:12–24

Gungor S, Yücel G, Akinci A, Tabel Y, Ozerol IH, Yologlu S (2007) The role of ghrelin in weight gain and growth in epileptic children using valproate. J Child Neurol 22:1384–1388

Hamed SA (2008) Neuroendocrine hormonal conditions in epilepsy: relationship to reproductive and sexual functions. Neurologist 14:157–169

Harden CL (2003) Menopause and bone density issues for women with epilepsy. Neurology 61:S16–S22

Harden CL (2008) Sexual dysfunction in women with epilepsy. Seizure 17:131–135

Herzog AG (1986) Intermittent progesterone therapy of partial complex seizures in women with menstrual disorders. Neurology 36:1607–1610

Herzog AG (1995) Progesterone therapy in women with complex partial and secondary generalized seizures. Neurology 45:1660–1662

Herzog AG (1999) Progesterone therapy in women with epilepsy: a 3-year follow-up. Neurology 52:1917–1918

Herzog AG (2008) Disorders of reproduction in patients with epilepsy: primary neurological mechanisms. Seizure 17:101–110

Herzog AG, Seibel MM, Schomer DL Vaitukaitis JL, Geschwind N (1986) Reproductive endocrine disorders in women with partial seizures of temporal lobe origin. Arch Neurol 43:341–346

Herzog AG, Klein P, Ransil BJ (1997) Three patterns of catamenial. Epilepsia 38:1082–1088

Herzog AG, Coleman AE, Jacobs AR, Klein P, Friedman MN, Drislane FW, Ransil BJ, Schomer DL (2003) Interictal EEG discharges, reproductive hormones, and menstrual disorders in epilepsy. Ann Neurol 54:625–637

Isojärvi JI, Laatikainen TJ, Pakarinen AJ, Juntunen KT, Myllylä VV (1993) Polycystic ovaries and hyperandrogenism in women taking valproate for epilepsy. N Engl J Med 329:1383–1388

Isojärvi JI, Laatikainen TJ, Pakarinen AJ, Juntunen KT, Myllylä VV (1995) Menstrual disorders in women with epilepsy receiving carbamazepine. Epilepsia 36:676–681

Isojärvi JI, Laatikainen TJ, Knip M, Pakarinen AJ, Juntunen KT, Myllylä VV (1996) Obesity and endocrine disorders in women taking valproate for epilepsy. Ann Neurol 39:579–584

Isojärvi JI, Rättyä J, Myllylä VV, Knip M, Koivunen R, Pakarinen AJ, Tekay A, Tapanainen JS (1998) Valproate, lamotrigine, and insulinmediated risks in women with epilepsy. Ann Neurol 43:446–451

Isojärvi JI, Taubøll E, Pakarinen AJ, van Parys J, Rättyä J, Harbo HF, Dale PO, Fauser BC, Gjerstad L, Koivunen R, Knip M, Tapanainen JS (2001) Altered ovarian function and cardiovascular risk factors in valproate-treated women. Am J Med 111:290–296

Isojärvi JI, Taubøll E, Herzog AG (2005) Effect of antiepileptic drugs on eproductive endocrine function in individuals with epilepsy. CNS Drugs 19:207–223

Keefe DL (2002) Sex hormones and neural mechanisms. Arch Sex Behav 31:401–403

Klein P, Serje A, Pezzullo JC (2001) Premature ovarian failure in women with epilepsy. Epilepsia 42:1584–1589

Lim LL, Foldvary N, Mascha E, Lee J (2001) Acetazolamide in women with catamenial epilepsy. Epilepsia 42:746–749

Luef G, Abraham I, Trinka E, Alge A, Windisch J, Daxenbichler G, Unterberger I, Seppi K, Lechleitner M, Krämer G, Bauer G. (2002) Hyperandrogenism, postprandial hyperinsulinism and the risk of PCOS in a cross sectional study of women with epilepsy treated with valproate. Epilepsy Res 48:91–102

Luef G, Lechleitner M, Bauer G, Trinka E, Hengster P (2003) Valproic acid modulates islet cell insulin secretion: a possible mechanism of weight gain in epilepsy patients. Epilepsy Res 55:53–58

Luef G, Rauchenzauner M, Waldmann M, Sturm W, Sandhofer A, Seppi K, Trinka E, Unterberger I, Ebenbichler CF, Joannidis M, Walser G, Bauer G, Hoppichler F, Lechleitner M (2009) Nonalcoholic fatty liver disease (NAFLD), insulin resistance and lipid profile in antiepileptic drugtreatment. Epilepsy Res 86:42–47

Mattson RH, Cramer JA, Caldwell BV, Siconolfi BC (1984) Treatment of seizures with medroxyprogesterone acetate: preliminary report. Neurology 34:1255–1258

McAuley JW, Moore JL, Reeves AL, Flyak J, Monaghan EP, Data J (2001) A pilot study of the neurosteroid ganaxolone in catamenial epilepsy: clinical experience in two patients. Epilepsia 42:85–85

Morell MJ (2003) Reproductive and metabolic disorders in women with epilepsy. Epilepsia 44:11–20

Morrell MJ, Giudice L, Flynn KL, Seale CG, Paulson AJ, Doñe S, Flaster E, Ferin M, Sauer MV (2002) Predictors of ovulatory failure in women with epilepsy. Ann Neurol 52:704–711

Morrell MJ, Flynn KL, Doñe S, Flaster E, Kalayjian L, Pack AM (2005) Sexual dysfunction, sex steroid hormone abnormalities, and depression in women with epilepsy treated with antiepileptic drugs. Epilepsy Behav 6:360–365

Morrell MJ, Hayes FJ, Sluss PM, Adams JM, Bhatt M, Ozkara C, Warnock CR, Isojärvi J (2008) Hyperandrogenism, ovulatory dysfunction, and polycystic ovary syndrome with valproate versus lamotrigine. Ann Neurol 64:200–211

Pack AM (2010) Implications of hormonal and neuroendocrine changes associated with seizures and antiepileptic drugs: a clinical perspective. Epilepsia 51(Suppl 3):150–153

Pennell PB (2009) Hormonal aspects of epilepsy. Neurol Clin 27:941–965

Prabhakar S, Sahota P, Kharbanda PS, Siali R, Jain V, Lal V, Khurana D (2007) Sodium valproate, hyperandrogenism and altered ovarian function in Indian women with epilepsy: a prospective study. Epilepsia 48:1371–1377

Quigg M, Smithson SD, Fowler KM, Sursal T, Herzog AG (2009) Laterality and location influence catamenial seizure expression in women with partial epilepsy. Neurology 73:223–227

Rauchenzauner M, Haberlandt E, Scholl-Bürgi S, Karall D, Schoenherr E, Tatarczyk T, Engl J, Laimer M, Luef G, Ebenbichler CF (2008) Effect of valproic acid treatment on body composition, leptin and the soluble leptin receptor in epileptic children. Epilepsy Res 80:142–149

Rauchenzauner M, Bitsche G, Svalheim S, Tauboll E, Haberlandt E, Wildt L, Rostasy K, Luef G (2010) Effects of levetiracetam and valproic acid monotherapy on sex-steroid hormones in prepubertal children—results from a pilot study. Epilepsy Res 88:264–268

Reddy DS (2009) The role of neurosteroids in the pathophysiology and treatment of catamenial epilepsy. Epilepsy Res 85:1–30

Reddy DS, Rogawski MA (2009) Neurosteroid replacement therapy for catamenial epilepsy. Neurotherapeutics 6:392–401

Ross IP (1958) Acetazolamide therapy in epilepsy. Lancet 2:1308–1309

Røste LS, Taubøll E, Svalheim S, Gjerstad L (2008) Does menopause affect the epilepsy? Seizure 17:172–175

Scharfman HE, Kim MK, Hintz TM, MacLusky NJ (2008) Seizures and reproductive function: insights from female rats with epilepsy. Ann Neurol 64:687–697

Verrotti A, Basciani F, Morresi S, de Martino M, Morgese G, Chiarelli F (1999) Serum leptin changes in epileptic patients who gain weight after therapy with valproic acid. Neurology 53:230–232

Verrotti A, Manco R, Agostinelli S, Coppola G, Chiarelli F (2010) The metabolic syndrome in overweight epileptic patients treated with valproic acid. Epilepsia 51:268–273

Verrotti A, D'Egidio C, Mohn A, Coppola G, Parisi P, Chiarelli F (2011) Antiepileptic drugs, sex hormones, and PCOS. Epilepsia 52:199–211

Wong HY, Chu TS, Lai JC, Fung KP, Fok TF, Fujii T (2005) Sodium valproate inhibits glucose transport and exacerbates Glut1-deficiency in vitro. J Cell Biochem 96:775–785

Xiaotian X, Hengzhong Z, Yao X, Zhipan Z, Daoliang X, Yumei W (2013) Effects of antiepileptic drugs on reproductive endocrine function, sexual function and sperm parameters in Chinese Han men with epilepsy. J Clin Neurosci 20:1492–1497

Zimmerman AW, Holder DT, Reiter EO, Dekaban AS (1973) Medroxyprogesterone acetate in the treatment of seizures associated with menstruation. J Pediatr 83:959–963

Neuromodulation for the Treatment of Drug-Resistant Epilepsy

Pantaleo Romanelli and Alfredo Conti

Abstract Surgical neuromodulation for epilepsy refers to procedures involving the electrical stimulation of cortical, diencephalic, cerebellar and peripheral targets (such as the vagus nerve). Stereotactic radiosurgery also provides a neuromodulatory approach, affecting the discharging behavior of epileptic neurons in absence of evident target necrosis. Cortical transections or Multiple Subpial Transections (MST) are a non-resective technique useful to treat epileptogenic foci located in eloquent cortex. Electrical stimulation, stereotactic radiosurgery, and MST are emerging procedures for the treatment of medically refractory epilepsy in patients not amenable to resective surgery due to inability to map the focus, presence of multiple epileptogenic foci and/or involvement of eloquent cortex. Radiosurgery can also be offered to patients ineligible for invasive surgery for a variety of medical contraindications.

Introduction

Epilepsy is the most common chronic neurological disorder affecting approximately 1 % of the population worldwide. About one-third of the patients with epilepsy respond poorly to medical therapy with the initial response being highly predictive of long-term seizure control (Go and Snead 2008; Kuzniecky and Devinsky 2007). Patients who are poor responders to established medical treatments are exposed to the long-term medical and social sequelae of intractable seizures leading to multiple hospital admissions and chronic disability. Patients with uncontrolled generalized

---

P. Romanelli (✉)
Centro Diagnostico Italiano, via Saint Bon 20, Milano, Italy
e-mail: radiosurgery2000@yahoo.com

A. Conti
Department of Neurosurgery, University of Messina, AOU Policlinico "G. Martino" via Consolare Valeria 1, 98125 Messina, Italy
e-mail: alfredo.conti@unime.it

© Springer International Publishing Switzerland 2015
P. Striano (ed.), *Epilepsy Towards the Next Decade*,
Contemporary Clinical Neuroscience, DOI 10.1007/978-3-319-12283-0_12

tonic-clonic seizures also have a substantially increased risk of death either due to trauma, sudden unexplained cardiorespiratory arrest, or sudden unexplained death in epilepsy (SUDEP) (Go and Snead 2008; Kuzniecky and Devinsky 2007; Cascino 2004; Tolstykh and Cavazos 2013).

Identification of the epileptic focus and subsequent surgical resection is the gold standard for the treatment of medically refractory seizures. Resective surgery can be performed when a concordant localization of the seizure focus is obtained based on seizure semiology, EEG and MRI, with further support provided by advanced techniques such as magnetic resonance spectroscopy (MRS), magnetoenecephalography (MEG), ictal and interictal SPECT, and PET.

Mesial temporal lobe resection provides excellent outcomes in terms of seizure freedom with rare serious complications while neocortical epilepsy has lower, but still appealing rates of post-surgical seizure freedom (Jette and Wiebe 2013).

However, a large number of refractory patients cannot be selected for resective surgery due to the inability to identify and localize the epileptogenic focus or to the overlap between the epileptogenic focus with eloquent neo-cortex (sensorimotor, speech areas) or archicortex (dominant hippocampus), which cannot be resected without inducing serious neurologic sequelae. Seizure foci located in close proximity to eloquent cortex can be safely resected but accurate cortical mapping is mandatory (Bauman et al. 2005; Devinsky et al. 2003). Detection of multiple independent seizure foci is generally considered a contraindication to resective surgery, but in highly selected cases, multiple resections guided by staged cortical mapping can still provide positive seizure control (Devinsky et al. 2003; Romanelli et al. 2001, 2002; Bauman et al. 2008).

Neuromodulation is an emerging surgical option employed to treat patients when resective surgery is contraindicated (Anderson et al. 2009; Romanelli et al. 2012; Saillet et al. 2009). It is commonly reserved for patients who are not candidates for resective surgery due to poor localization of the epileptic zone, spatial co-localization of ictal focus and eloquent cortex, presence of multiple independent ictal foci or seizure origin located in deep brain regions. In such circumstances surgical approaches to entirely and effectively remove the epileptogenic tissue carry a relevant risk of new neurological deficits (the typical case of Hypothalamic Hamartomas [HH]).

Neuromodulation is a non-resective approach to the treatment of epilepsy that has historically included the delivery of electrical stimulation to selected brain regions or to the vagus nerve. Other non-resective approaches such as stereotactic radiosurgery and multiple subpial transections, which affect the pathophysiology of an epileptic focus without removing it, can also be included in this definition. This paper provides a concise review of the following neuromodulatory treatments: vagus nerve stimulation (VNS), deep brain stimulation (DBS), the new responsive neurostimulation devices, stereotactic radiosurgery, and multiple subpial transections.

Vagus Nerve Stimulation

Vagus nerve stimulation is the most common neuromodulation procedure for the treatment of epilepsy. It is usually offered to patients with severe medically-refractory epilepsy who are not eligible for or not responsive to microsurgical resection. Vagus nerve stimulation is characterized by low surgical risk and lack of significant toxicity and has been approved for the treatment and prevention of refractory seizures in adults and adolescents by the European and US regulatory boards in 1994 and 1997, respectively.

According to Cyberonics, the manufacturer of the VNS, the device has been implanted in over 65,000 patients worldwide (http://eu.cyberonics.com/en/vns-therapy/healthcare-professionals/implanted-components). Despite the extensive clinical use, a satisfactory explanation of the mechanism of action of VNS in epilepsy is not yet available. Modulation of the brainstem, thalamic insular and limbic targets exerted by the afferent branches of the vagus nerve is involved in the anti-epileptic effect of VNS, but a coherent framework explaining why the patient's response is so widely unpredictable is still missing. The development of implant strategies based on improved patient response prediction is an essential step needed for the progress of VNS.

Vagus nerve stimulation placement requires a delicate but quick microsurgical procedure on the left vagus nerve (selected to minimize vegetative side-effects). The nerve is identified and dissected by spreading apart the carotid artery medially and the jugular vein laterally. Three coils are looped around the left vagus nerve in order to minimize vegetative side effects. From a patient perspective, VNS implant surgery is usually performed on an outpatient basis, or at times, patients may need to stay in the hospital for one day. Complication rates are usually minimal, but stimulation-induced side effects such as dysphonia (20–30%) and cough (6%) are relatively common. These side effects are usually transient and can be managed by the reduction of stimulation intensity (Ben-Menachem et al. 1999; Uthman et al. 2004). Other side effects such as sleep apnea or cardiopulmonary dysfunction are very rare (Papacostas et al. 2007).

The long term safety and efficacy of VNS in the adjunctive treatment of epilepsy have been widely reported over the years both in adults and children (Baaj et al. 2008; Ben-Menachem et al. 1999; Benifla et al. 2006; Uthman et al. 2004). Reduction of seizure frequency following VNS implantation occurs over time, with further improvement in seizure reduction recorded 1 year after VNS implantation (Mapstone 2008). The number of patients achieving complete seizure freedom after VNS is very small ($\approx 2\%$), while an approximate 40% of patients achieving a substantial reduction (more than 50%) of seizure frequency and intensity is commonly reported (Mapstone 2008). A noticeable improvement in social, affective and behavioral scores in association with the ability to reduce the medical therapy load has been reported in responsive patients (Baaj et al. 2008; Ben-Menachem et al. 1999; Benifla et al. 2006; Uthman et al. 2004). Overall, VNS is a safe and well-tolerated therapy providing improved seizure control to some patients. The main shortcomings of VNS therapy are the unpredictability of response, cost of the device, and the need for additional minor surgeries to replace the battery. A better understanding of the mechanism of action of VNS is needed to pave the way for improved, targeted patient selection.

Deep Brain Stimulation

Deep brain stimulation (DBS) is a minimally invasive neuromodulation technique of proven efficacy in the treatment of movement disorders such as Parkinson's disease. Treatment of medically refractory epilepsy is one of the many emerging DBS applications. As for VNS, the DBS mechanism of action in the suppression of epileptic activity is not entirely clear. The insertion of a DBS lead within a neural network acts to modulate seizures either through direct electrical stimulation or by a microlesional effect, or some combination of both (Van Roost et al. 2007; Stacey and Litt 2008). The anterior or centromedian nucleus of the thalamus, subthalamic nucleus, caudate nucleus, hippocampus, hypothalamus, and cerebellum are among the DBS explored targets (Ellis and Stevens 2008). The cerebellum was one of the earliest structures studied for stimulation in epilepsy patients, though only modest efficacy has been reported with modulation of this target alone (Fountas et al. 2010; Davis and Emmonds 1992; Cooper et al. 1976). The caudate nucleus has been explored as well as a brain stimulation target to improve epilepsy (Chkhenkeli and Chkhenkeli 1997): low frequency stimulation induced some improvement in the frequency of generalized seizures, but this has not generalized to firm conclusions about the true overall efficacy of the treatment.

The anterior (ANT) nucleus of the thalamus is the most attractive and explored DBS target due to its central position within the limbic system. The main output pathway of the limbic system proceeds from the hippocampus to the mammillary body, which innervates the ANT through the mammillo-thalamic tract. Within the ANT, massive projections to the cingulate gyrus and to the frontal lobe justify its definition as the "pacemaker for the cortex" and its crucial role in the genesis of generalized seizures (Mirski and Ferrendelli 1986). Bilateral ANT DBS is a widely explored procedure that is facilitated by relatively easy surgical targeting. Widely different clinical indications and outcomes have been reported (Andrade et al. 2006; Hodaie et al. 2002; Kerrigan et al. 2004; Lee et al. 2006; Osorio et al. 2005; Upton et al. 1985; Velasco et al. 2001). Overall, the procedure has been found to be relatively safe and well-tolerated while the seizure control rate is quite variable. A large multicenter randomized double-blinded trial was performed on over 100 subjects, which demonstrated an efficacy on seizure control similar to that offered by VNS (Fisher et al. 2010). Median decline in seizures was 40.5% in the stimulated group and 14.5% in the control group, while seizure intensity was improved as well. After 2 years of chronic high-frequency and high-intensity stimulation, 54% of patients noted seizure reductions while 14 patients (12.7%) were seizure-free for at least 6 months. Overall, these are encouraging results; however, the cost of bilateral DBS implantation and frequent changes of battery are known considerations. Relatively high stimulation intensities were needed, requiring frequent changes of battery, which is most likely related to the large volume of the ANT. Current DBS design was developed in order to stimulate relatively small neural volumes. ANT has an overwhelmingly larger volume than that of common DBS targets such as the subthalamic nucleus (STN), the ventrolateral thalamic nuclei or the Globus Pallidus pars interna. Insertion of multiple leads or new designs are likely to be necessary in order to provide an adequate ANT stimulation coverage.

The medial thalamus, namely the centromedian nucleus (CM), was explored as a DBS target by Velasco and colleagues (Velasco et al. 2001). This nucleus has a much smaller size than the ANT, though much larger than STN, and is likely to be more responsive to DBS than ANT. CM projects widely to the neocortex and research suggests that modulation of this nucleus exerts an antiepileptic effect mediated by cortical desynchronization (Velasco et al. 2001). There are mixed reports ranging from modest to significant improvements in seizure control for the treatment of focal epilepsy with secondary generalization (Velasco et al. 1987) and Lennox-Gastaut syndrome (Velasco et al. 2001; Shimizu and Maehara 2000) using DBS in the CM.

The subthalamic nucleus (STN) is the main DBS target for movement disorders. In addition, STN DBS has been explored also as treatment for refractory epilepsy. The antiepileptic effect of STN DBS is thought to be mediated by the inhibition of the excitatory effect of the substantia nigra pars reticulata (SNr) on gabaergic neurons in the dorsal midbrain anticonvulsant zone (DMAZ), placed in the deep layers of the superior culliculus (Gale 1986; Iadarola and Gale 1982).The extensive STN projections to sensorimotor cortex could also exert a crucial role on the control of tonic and clonic seizures. The first successful treatment of refractory epilepsy by DBS in a child with cortical dysplasia was reported by Benabid et al. (Benabid et al. 2001). Seizure reduction ranging between 60% and 80% in refractory focal epilepsies has been subsequently reported by various groups (Chabardes et al. 2002; Dinner et al. 2002; Vonck et al. 2003; Shon et al. 2005). A 50% seizure reduction rate has also been observed in progressive myoclonic epilepsy (PME) (Vesper et al. 2007). Overall, similarly to CM DBS, the number of cases reported so far is small and heterogeneous and no prospective trials have been performed. On the other hand, STN is the most explored DBS target: DBS placement here is facilitated by the unique electrophysiological STN signature found when performing microelectrode recording. Therefore a precise assessment of DBS placement (usually done within the sensorimotor domain) is much easier here than in ANT or CM, where no electrophysiologic signatures are available. STN is likely to represent a useful target to treat seizures characterized by a predominant motor component. Further studies are clearly needed to assess the role of STN DBS in the treatment of refractory epilepsy but also to inform the process of DBS thalamic (ANT, CM) versus basal ganglia (STN) target selection in general.

The hippocampus and amygdala are two of the most epileptogenic areas of the brain, and thus are the most common origins of refractory epilepsy. Microsurgical resection of the amygdalo-hippocampal complex is also the most common surgical procedure for epilepsy and is associated with post-operative resolution of seizures in up 75% of patients (Schramm 2008). DBS placement within the hippocampus has been used experimentally for poor candidates for resective surgery, such as those with bilateral ictal localization, or in whom preoperative neurophysiological findings (e.g., WADA test) predict a significant potential for memory loss. The elongated shape of the hippocampus fits very well with the insertion of DBS leads but the stimulation volume is rather large and the current DBS design is surely not optimized for this indication. Test stimulation is performed after DBS placement as continuous high-frequency square-wave pulses. A subclavicular internal pulse generator is then placed and connected via an extension cable to the implanted leads, providing chronic stimulation if a substantial reduction of interictal spike activity during a period of acute stimulation is witnessed (Boon et al. 2007; Van Roost et al. 2007). No major surgical complications have been reported, but high intensity stimulation may induce reversible memory deficts (Boex et al. 2011). A microlesional effect has been also described. Seizure outcomes are quite variable within and across the studies (Boon et al. 2007; Van Roost et al. 2007; Boex et al. 2011; McLachlan et al. 2010).

Hippocampal DBS is a procedure of great interest to treat mesiotemporal lobe epilepsy (MTLE), especially when resective surgery is not indicated. However, experience is limited, reported seizure outcomes are widely variable, and optimal stimulation protocols have not yet been defined. This is also true for other neuromodulation procedures for epilepsy: the field is promising but much more work is needed to assess the overall utility, role and impact of DBS, as well as the best criteria for patient selection.

Responsive Stimulation

The main limitation of current DBS is the inability to adapt the stimulation field and intensity to changing conditions such as those typical of an epileptic focus. Responsive stimulation aims to predict and respond to seizures, suppressing epileptiform activity by delivering stimulation in response to electrocorticographic activity (Litt 2003; Sun et al. 2008).

Responsive stimulation requires preliminary cortical mapping and then acts directly on the seizure focus. Stimulation is not continuous as is the case in DBS (closed-loop stimulation), but is provided only at specific times to abort oncoming seizures, thus potentially reducing the likelihood of functional disruption or habituation due to continuous treatment (Litt and Echauz 2002). The electrodes delivering open-loop stimulation can record the electrical activity of the target and stimulate "on demand", in a pacemaker-wise fashion. Electrocorticographic activity over the target region is continuously monitored: when abnormal activity is detected, electrical stimulation is delivered to abort the epileptic discharge and restore normal electrical rhythm. In essence, the device recognizes specific brain discharges indicating a high risk for evolution into clinical seizures and acts to abort their propagation.

A pilot trial of four patients with refractory seizures treated with responsive cortical stimulation showed suppression of clinical seizures with resolution of electrographic seizure activity (Kossoff et al. 2004). Responsive cortical stimulation using the RNS System (NeuroPace, Mountain View, CA), provides a continuous electrocorticographic analysis triggering the stimulation when ictal or pre-ictal activity is detected, was studied in patients with refractory partial-onset seizures (Morrell 2011). This large, prospective, multicenter, randomized double-blind trial has been performed on 191 adults undergoing placement of subdural or depth electrodes over 1 or 2 selected seizure foci. Overall, the procedure was safe and well-tolerated. A 37.9 % reduction in mean seizure frequency in the treatment group compared with a 17.3 % reduction in the sham group ($p = 0.012$) was found during a 12-week period. After 2 years follow-up, 46 % of patients had at least a 50 % reduction in their mean seizure frequency. Verbal functioning, visuospatial ability, and memory showed some improvement as well. These results again appear grossly comparable to those of VNS. Again, further work is needed in order to identify appropriate clinical and patient selection characteristics to ensure the best clinical outcomes.

Stereotactic Radiosurgery

Stereotactic radiosurgery (SRS) is a strikingly effective approach to treat a variety of brain tumors but is also an excellent tool to treat functional brain disorders such as trigeminal neuralgia. The attractiveness of a non-invasive treatment devoid of the risks and discomforts associated with open surgery and general anesthesia as well as the excellent results obtained explain the exponential growth of SRS over the last decade (Niranjan et al. 2012). Treatment of epilepsy is an emerging application of SRS: it was initially observed that secondary epilepsy improves after radiosurgical treatment of the underlying lesion. Following this observation, radiosurgery was proposed as a primary treatment for idiopathic localization-related epilepsy (Romanelli and Anschel 2006). Most reports focus on the treatment of secondary epilepsy, caused by intrinsic brain tumors, arteriovenous malformations (AVMs) or cavernomas, with a smaller number of recent papers describing the SRS in the treatment of epilepsy related to mesial temporal sclerosis or hypothalamic hamartoma (Regis et al. 2002b; Schrottner et al. 2002; Romanelli et al. 2008; Regis et al. 2004b). In most cases, no evident neuroradiologic or histologic changes can be appreciated over the epileptic region following treatment with SRS, which is in agreement with the fact that the doses received are usually too low to induce radionecrosis (Regis et al. 1996; Romanelli and Anschel 2006; Srikijvilaikul et al. 2004). Irradiation of epileptogenic cortex using non-necrotizing doses is associated with modulation of neurotransmitters and reduction of pathological discharges while normal neuronal activity is preserved (Regis et al. 2002a). As a non-invasive procedure, SRS provides an attractive treatment option for those cases where non-invasive seizure mapping can be achieved. Low-dose irradiation of seizure foci located in eloquent regions using magnetoencephalography (MEG)-guided SRS is a good example of a thoroughly non-invasive mapping and treatment approach providing a safe and effective procedure for highly challenging or inoperable lesions (Kurita et al. 2001; Stefan et al. 1998). The refinement and availability of non-invasive mapping techiques for refractory epilepsy is clearly a condition sine qua non needed for the evaluation of radiosurgical treatments for epilepsy. The rapid evolution of MR imaging and the introduction of high density surface EEG (Ramon and Holmes 2013) are likely to provide the needed support for an increasing use of SRS in the treatment of epilepsy in future years.

Treatment of mesial temporal lobe epilepsy (MTLE) by stereotactic radiosurgery was first reported by Regis et al. (Regis et al. 2004b). In a recent prospective multicenter pilot trial, two different radiosurgery doses were compared and the overall seizure remission rate was 69% during the third follow-up year after treatment, which is comparable to that reported for resective temporal lobectomy (Chang et al. 2010). The main shortcoming of SRS in the treatment of mesial temporal lobe epilepsy is the long delay to appreciate a visible effect on seizures.

Furthermore, SRS appears to be an excellent treatment option for hypothalamic hamartomas (HH), which are typically associated with severe medically-refractory epilepsy (Arita et al. 1998; Regis et al. 2004a; Rosenfeld 2011; Schulze-Bonhage et al. 2004; Selch et al. 2005). The efficacy of SRS for the treatment of severe epilepsy induced by HH is strictly bound to the delivery of adequate doses, with the best results obtained by prescribing a marginal dose of > 17 Gy. The post-operative course may be characterized by an increased number of seizures, sometimes with a large number over only several days. The duration of this period is brief, days or weeks, and rarely more than 1 month and it is followed by a gradual overall improvement (Regis et al. 2007; Romanelli et al. 2008). So far, no major neurological complications following SRS have been reported; this is of great importance considering the complexity and overall risks of open surgical approaches. Surgical and radiosurgical treatments can be easily integrated in patients with large HH. In such cases, a surgical debulking procedure could be followed by radiosurgery delivered to the unresectable epileptogenic intrahypothalamic component.

A large amount of retrospective data is available regarding seizure outcomes following the treatment of arteriovenous malformations (AMVs) with SRS. Pollock and coworkers (Pollock et al. 1994) retrospectively studied 67 patients with smaller AVMs (Spetzler-Martin Grade I or II) that were potentially resectable by surgery. All subjects had been first offered surgical resection, and had declined. Two patients were lost to follow-up, leaving 65 subjects who had data collected for a minimum of 24 months (mean 35 months). Thirty-one patients (47.7%) had experienced symptomatic seizures prior to radiosurgery. Mean AVM volume was 3.1 cm^3, with a mean marginal dose of 21 Gy (maximum dose: 36 Gy). Sixteen patients (52%) had seizure frequency reduced to < 1 seizure per year, while 15 had no change in seizure control. No patients developed new seizures following treatment. There was a 7.7% hemorrhage rate and a 3% mortality rate (due to hemorrhage) within 8 months following radiosurgery; however there was no risk of bleeding if there was total obliteration or subtotal obliteration with patency of early draining veins only.

A retrospective review of 40 pediatric cases of AVMs treated by a multimodality approach and followed for a mean of 38.7 months has been reported by Hoh et al. (Hoh et al. 2000). Ten patients with AVM-related seizures were treated with proton beam radiotherapy if lesions were located in eloquent areas or had a particular pattern of venous drainage. Mean dose was 15.9 Gy to a mean volume of 9.9 cm^3. Nine out of ten patients (90%) became seizure free following treatment. It should be noted that AVM embolization was also performed concurrently and seizure outcome was not analyzed in detail to assess the respective weight of radiosurgery versus embolization in the outcome of seizure freedom. The authors report no radiosurgery-related complications or morbidity. One patient had a hemorrhage after radiosurgery, before the AVM had been obliterated.

A further retrospective study published by the same group in 2002 reports on 141 patients with AVMs and seizures, representing 33% of 424 patients treated for AVM over an eight year period (Hoh et al. 2002). Follow-up data have been available for 110 patients out of 141.These patients were treated with a multimodality, multidisciplinary approach including various combinations of surgery, radiosurgery, and embolization. Mean follow-up period was 34.8 months. Those who received radiosurgery were treated with proton beam therapy with a mean dose of 15.5 Gy to a mean volume of 7.7 cm^3. These investigators identified the following pre-treatment risk factors for the development of symptomatic seizures with AVM: male gender, age <65 years, AVM size >3 cm, and temporal lobe AVM location. Of the 110 patients treated with the multiple modalities, 73 (66%) were seizure-free (Engel Class I), 11 (10%) were Class II, 1 (0.9%) was Class III, and 22 (20%) were Class IV. Treatment-specific analysis revealed that surgery had the highest number of patients with Class I outcome (81%), followed by embolization (50%), and then radiosurgery (43%). However, if the AVM was completely obliterated, then all treatments yielded the same percentage of patients with a Class I outcome. The following factors were associated with a Class I outcome: short seizure history, associated intracranial hemorrhage, generalized tonic-clonic seizure type, deep and posterior fossa AVM location, surgical resection, and complete AVM obliteration. In addition 5.7% of patients who did not have pre-treatment symptomatic seizures, subsequently developed seizures.

Hadjipanayis (Hadjipanayis et al. 2001) selected thirty-three patients with AVMs of the precentral gyrus for retrospective study out of a group of 770 patients treated Gamma Knife radiosurgery for AVMs. Median AVM volume was 3 cm^3 and the median dose to the margin was 20 Gy. Twenty-seven (87%) of the patients had initially presented with seizures. After a mean follow-up of 54 months, there was a 63% rate of seizure freedom. The remaining 37% continued to experience seizures. The following variables showed no relationship to seizure control: marginal dose, volume, Spetzler-Martin grade, gender, age, and perimotor versus motor cortex location. Two patients (6%) developed new bleeding at 12 and 23 months post-radiosurgery, resulting in one death (3%). Hand weakness occurred in one subject with an 8.4 cm^3 target volume and another patient developed subjective sensory, visual, and auditory deficits. This adverse effect profile compares favorably with most of the surgical resection series involving a similar population.

Schauble et al. (Schauble et al. 2004) identified 70 patients with AVM-associated seizures who had been treated by Gamma Knife radiosurgery. 10.3 cm^3 average tissue volume was treated with a prescription isodose average of 18 Gy. Patients were evaluated at 1, 2, and 3 years post treatment. However, a potentially significant bias was introduced by the fact that patients requiring retreatment before year 1, 2 or 3 were excluded from evaluation at all further time points. Therefore, the most poorly responding patients were excluded from long-term follow-up. Sixty-five patients were studied at the 1-year time point and 51 at the 3-year time point. At 1 and 3 years, 74 and 78% respectively, had an excellent outcome (non-disabling seizures only). Seizure-free rates at years 1 and 3 were 45% and 51% respectively. Factors predictive of a favorable outcome included low seizure frequency before radiosurgery, presence of generalized tonic-clonic seizure, and smaller size AVM. There was a 5% mortality rate from AVM-related hemorrhages post radiosurgery. In addition, one patient died due to "worsening edema." The total post-radiosurgery hemorrhage rate is not reported.

Seizure improvement has been also reported after SRS for cavernous malformations (CM). Over a 17-year period, 95 patients were treated for CMs by proton beam therapy at Massachusetts General Hospital. This group was retrospectively analyzed with an average follow-up of 5.4 years by Amin-Hanjani et al. (Amin-Hanjani et al. 1998). The patients had been treated with radiosurgery if there were symptomatic CMs in locations where the risk of surgical morbidity was deemed unacceptable or based on failed previous attempts at surgical resection. Median volume treated was 3.1 cm^3, with a median marginal dose of 16.5 Gy (90% isodose). During the first 2 years after treatment, there was a 22.4% per year per lesion hemorrhage rate (17.4% pretreatment), however after 2 years, the rate decreased to 4.5%. In addition to a 16.3% rate of permanent neurological disability, there was a 3.2% mortality rate due to hemorrhage within the first 2 years after radiosurgery. Eighteen of the subjects had seizures prior to treatment. There was a significant improvement in seizure control after treatment and no patients developed new onset seizures or intractable epilepsy after treatment. Although disability increased after radiotherapy overall, the greatest risk was in those patients with deep CMs. There were no deaths among those with lobar lesions. Since seizures were prominent in those with lobar CMs, the chance for neurological deterioration was lower in patients with seizures.

Regis et al. (Regis et al. 2000) published a retrospective multicenter report of 49 patients treated for CMs with Gamma Knife radiosurgery. All patients had drug-resistant epilepsy and were followed for greater than 12 months following treatment. Mean dose at the margin of the lesion and volume were 19.17 Gy and 2.4 cm^3 respectively. Fifty-three percent of the patients became seizure-free (Engel Class I) while 20% experienced a significant decrease in number of seizures and 26% had little or no improvement. The average time to seizure remission was 4 months. Five of the patients who failed to improve following radiosurgery were treated with microsurgery. Three of these later patients became seizure free, one has rare seizures, and one has "failed completely." One patient experienced a hemorrhage 3 months following radiosurgery. Seven patients developed severe radio-induced edema. Better outcome was associated with simple partial seizures, compared with complex partial seizure type. Mesiotemporal location was also associated with a poor outcome while lateral-temporal location was associated with a good outcome.

The combination of non-invasive mapping and frameless radiosurgical treatment using the Cyberknife to treat selected cases of drug-refractory epilepsy has been recently described (Romanelli 2014).

Even if dedicated studies are rare, the selected studies discussed indicate that SRS can provide seizure improvement or freedom in selected patients with MTLE, HH, tumors, AVMs and CMs. Patients with seizure foci located near or over eloquent cortex can substantially benefit from SRS. An expanded use of SRS can be anticipated in parallel with the development of non-invasive seizure mapping techniques.

Multiple Subpial Transections

Multiple subpial transections (MST) refer to a microsurgical procedure aiming to cut the horizontal connections mediating epileptic activity diffusion from the ictal focus to adjacent and distant cortex (Blount et al. 2004; Devinsky et al. 1994; Dogali et al. 1993; Hashizume and Tanaka 1998). MST can be considered a neuromodulatory technique because the cortex generating seizures is not resected, but rather modified to prevent the development of synchronized pathological ictal discharges in the cortex (Morrell and Hanbery 1969; Sawhney et al. 1995). This procedure predates the introduction of electrical neuromodulation and radiosurgery in the treatment of epilepsy and was, for over two decades, the only surgical option to treat patients with medically-refractory epilepsy involving eloquent cortex (Blount et al. 2004; Devinsky et al. 1994; Dogali et al. 1993; Hashizume and Tanaka 1998; Morrell and Hanbery 1969; Mulligan et al. 2001; Sawhney et al. 1995; Schramm et al. 2002; Spencer et al. 2002).

The cerebral cortex is functionally organized in vertically oriented columns of neurons working as a homogeneous processing unit. The output of the columnar networks is mainly transmitted by vertical axons directed to near or far cortical regions, to the thalamus and basal ganglia, or to the brainstem and spinal cord while adjacent columns are interconnected by horizontal axons, providing the recruitment and synchronization of the critical mass of cortex needed to generate seizures (Chervin et al. 1988; Hubel and Wiesel 1962; Mountcastle 1957; Mountcastle 1997). The spread of epileptic activity follows a non-uniform horizontal spatial pattern involving the cortical layer V, which acts as the seizure trigger (Lueders et al. 1981; Telfeian and Connors 1998). MST provides a way to undercut the horizontal axons, mediating the spread of epileptogenic activity while sparing the vertically-oriented fibers subserving neurological function.

MST is generally reserved for patients undergoing extensive preoperative and intraoperative electrophysiological evaluation to map the epileptogenic foci and their precise spatial relationships with sensorimotor, language, or visual cortices. Once a precise focus localization is achieved, multiple incisions through the epileptic cortex are placed to transect the horizontal axons responsible for the propagation of seizures, while preserving the vertical axons subserving neurological functions. The procedure aims to provide selective disconnection of cortical fibers without subcortical white matter and superficial pial injuries. Preservation of the pial blood supply to the cortex is a critical factor for the success of the procedure and is facilitated by the use of special hooks to penetrate the cortex through a puncture hole made in the pia mater. The cortical transections are spaced approximately 5 mm apart and oriented perpendicular to the long axis of the selected epileptic gyrus. Multiple parallel transections made at this distance over a cortical gyrus provide an effective disconnection and parcellization of the seizure focus without injury to or disruption of the basic functions of the columns (Morrell et al. 1999). The number of transections are estimated pre-operatively and then refined based upon electrocorticography recordings during the procedure.

MST is mostly used in conjunction with surgical resection of a preoperatively mapped epileptogenic area (Blount et al. 2004; Mulligan et al. 2001; Orbach et al. 2001a; Schramm et al. 2002; Shimizu and Maehara 2000; Spencer et al. 2002; Hufnagel et al. 1997). MST can be added to microsurgical resection of non-eloquent brain regions or used alone in severe epilepsy conditions, such as epilepsia partialis continua (EPC) and Rasmussen encephalitis (Irwin et al. 2001; Molyneux et al. 1998; Morrell et al. 1995; Morrell and Hanbery 1969). MST provides a unique tool for the treatment of Landau-Kleffner syndrome (LKS) in children, an epileptic disorder characterized by speech deterioration (Irwin et al. 2001). A literature review has identified 211 patients treated with MST (Spencer et al. 2002). Late seizure recurrence has been described, and is likely due to restoration of horizontal connections over time, provides a medium for the spreading of epileptogenic activity (Orbach et al. 2001b).

A novel way to generate cortical transections equivalent to MST has been recently described (Romanelli et al. 2013). Synchrotron-generated X-ray microplanar beams (microbeams) are characterized by the ability to deliver extremely high doses of radiation to spatially restricted volumes of tissue. The minimal dose spreading outside the beam path provides an exceptional degree of protection from radio-induced damage to the neurons and glia adjacent to the microscopic slices of tissue irradiated. The preservation of cortical architecture following high-dose microbeam irradiation and the ability to non-invasively induce the equivalent of a surgical cut over the cortex is of great interest for the development of novel experimental models in neurobiology and new treatment avenues for a variety of brain disorders (Romanelli and Bravin 2011).

When synchrotron-generated microbeams were used to replicate MST, cortical transections sized 100 and 600 μm were generated over the sensorimotor cortex of naïve and epileptic rats using incident doses of, 360 and 240 Gy, respectively. Histologically evident cortical transections were found immediately and after 7 months. No neurological injury was observed. Convulsive seizure duration was drastically reduced in rats receiving local infusion of kainic acid (Romanelli et al. 2013). This novel approach combining SRS and MST holds great experimental and clinical potential, providing a non-invasive but powerful way to modulate cortical function.

Conclusions

The treatment of medically-refractory epilepsy is one of the greatest challenges of modern medicine. Microsurgical resection of the seizure focus provides the greatest chance to achieve seizure freedom, but is feasible only in patients in good medical condition with a single detectable seizure focus not involving eloquent cortex. Neuromodulation using electrical devices such as VNS or DBS, MST, and SRS provide precious adjunctive options to improve seizure control in patients with severe medically refractory epilepsy not eligible for resective surgery.

References

Amin-hanjani S, Ogilvy CS, Candia GJ, lyons S, Chapman PH (1998) Stereotactic radiosurgery for cavernous malformations: kjellberg's experience with proton beam therapy in 98 cases at the Harvard Cyclotron. Neurosurgery 42:1229–1236; discussion 1236–1238

Anderson CT, Davis K, Baltuch G (2009) An update on brain stimulation for epilepsy. Curr Neurol Neurosci Rep 9:327–332

Andrade DM, Zumsteg D, Hamani C, Hodaie M, Sarkissian S, Lozano AM, Wennberg RA (2006) Long-term follow-up of patients with thalamic deep brain stimulation for epilepsy. Neurology 66:1571–1573

Arita K, Kurisu K, Iida K, Hanaya R, Akimitsu T, Hibino S, Pant B, Hamasaki M, Shinagawa S (1998) Subsidence of seizure induced by stereotactic radiation in a patient with hypothalamic hamartoma. Case report. J Neurosurg 89:645–648

Baaj AA, Benbadis SR, Tatum WO, Vale FL (2008) Trends in the use of vagus nerve stimulation for epilepsy: analysis of a nationwide database. Neurosurg Focus 25:E10

Bauman JA, Feoli E, Romanelli P, Doyle WK, Devinsky O, Weiner HL (2005) Multistage epilepsy surgery: safety, efficacy, and utility of a novel approach in pediatric extratemporal epilepsy. Neurosurgery 56:318–334

Bauman JA, Feoli E, Romanelli P, Doyle WK, Devinsky O, Weiner HL (2008) Multistage epilepsy surgery: safety, efficacy, and utility of a novel approach in pediatric extratemporal epilepsy. Neurosurgery 62(Suppl 2):489–505

Ben-menachem E, Hellstrom K, Waldton C, Augustinsson LE (1999) Evaluation of refractory epilepsy treated with vagus nerve stimulation for up to 5 years. Neurology 52:1265–1267

Benabid AL, Koudsie A, Benazzouz A, Vercueil L, Fraix V, Chabardes S, Lebas JF, Pollak P (2001) Deep brain stimulation of the corpus luysi (subthalamic nucleus) and other targets in Parkinson's disease. Extension to new indications such as dystonia and epilepsy. J Neurol 248(Suppl 3):III37–III47

Benifla M, Rutka JT, Logan W, Donner EJ (2006) Vagal nerve stimulation for refractory epilepsy in children: indications and experience at The Hospital for Sick Children. Childs Nerv Syst 22:1018–1026

Blount JP, Langburt W, Otsubo H, Chitoku S, Ochi A, Weiss S, Snead OC, Rutka JT (2004) Multiple subpial transections in the treatment of pediatric epilepsy. J Neurosurg 100:118–124

Boex C, Seeck M, Vulliemoz S, Rossetti AO, Staedler C, Spinelli L, Pegna AJ, Pralong E, Villemure JG, Foletti G, Pollo C (2011) Chronic deep brain stimulation in mesial temporal lobe epilepsy. Seizure 20:485–490

Boon P, Vonck K, de Herdt V, Van Dycke A, Goethals M, Goossens L, Van Zandijcke M, de Smedt T, Dewaele I, Achten R, Wadman W, Dewaele F, Caemaert J, Van Roost D (2007) Deep brain stimulation in patients with refractory temporal lobe epilepsy. Epilepsia 48:1551–1560

Cascino GD (2004) Surgical treatment for epilepsy. Epilepsy Res 60:179–186

Chabardes S, Kahane P, Minotti L, Koudsie A, Hirsch E, Benabid AL (2002) Deep brain stimulation in epilepsy with particular reference to the subthalamic nucleus. Epileptic Disord 4(Suppl 3):83–93

Chang EF, Quigg M, Oh MC, Dillon WP, Ward MM, Laxer KD, Broshek DK, Barbaro NM (2010) Predictors of efficacy after stereotactic radiosurgery for medial temporal lobe epilepsy. Neurology 74:165–172

Chervin RD, Pierce PA, Connors BW (1988) Periodicity and directionality in the propagation of epileptiform discharges across neocortex. J Neurophysiol 60:1695–1713

Chkhenkeli SA, Chkhenkeli IS (1997) Effects of therapeutic stimulation of nucleus caudatus on epileptic electrical activity of brain in patients with intractable epilepsy. Stereotact Funct Neurosurg 69:221–224

Cooper IS, Amin I, Riklan M, Waltz JM, Poon TP (1976) Chronic cerebellar stimulation in epilepsy. Clinical and anatomical studies. Arch Neurol 33:559–570

Davis R, Emmonds SE (1992) Cerebellar stimulation for seizure control: 17-year study. Stereotact Funct Neurosurg 58:200–208

Devinsky O, Perrine K, Vazquez B, Luciano DJ, Dogali M (1994) Multiple subpial transections in the language cortex. Brain 117(Pt 2):255–265

Devinsky O, Romanelli P, Orbach D, Pacia S, Doyle W (2003) Surgical treatment of multifocal epilepsy involving eloquent cortex. Epilepsia 44:718–23

Dinner DS, Neme S, Nair D, Montgomery EB, Jr. Baker KB, Rezai A, Luders HO (2002) EEG and evoked potential recording from the subthalamic nucleus for deep brain stimulation of intractable epilepsy. Clin Neurophysiol 113:1391–1402

Dogali M, Devinsky O, Luciano D, Perrine K, Beric A (1993) Multiple subpial cortical transections for the control of intractable epilepsy in exquisite cortex. Acta Neurochir Suppl (Wien) 58:198–200

Ellis TL, Stevens A (2008) Deep brain stimulation for medically refractory epilepsy. Neurosurg Focus 25:E11

Fisher R, Salanova V, Witt T, Worth R, Henry T, Gross R, Oommen K, Osorio I, Nazzaro J, Labar D, Kaplitt M, Sperling M, Sandok E, Neal J, Handforth A, Stern J, Desalles A, Chung S, Shetter A, Bergen D, Bakay R, Henderson J, French J, Baltuch G, Rosenfeld W, Youkilis A, Marks W, Garcia P, Barbaro N, Fountain N, Bazil C, Goodman R, Mckhann G, Babu Krishnamurthy K, Papavassiliou S, Epstein C, Pollard J, Tonder L, Grebin J, Coffey R, Graves N (2010) Electrical stimulation of the anterior nucleus of thalamus for treatment of refractory epilepsy. Epilepsia 51:899–908

Fountas KN, Kapsalaki E, Hadjigeorgiou G (2010) Cerebellar stimulation in the management of medically intractable epilepsy: a systematic and critical review. Neurosurg Focus 29:E8

Gale K (1986) Role of the substantia nigra in GABA-mediated anticonvulsant actions. Adv Neurol 44:343–364

Go C, Snead OC 3rd (2008) Pharmacologically intractable epilepsy in children: diagnosis and preoperative evaluation. Neurosurg Focus 25:E2

Hadjipanayis CG, Levy EL, Niranjan A, Firlik AD, Kondziolka D, Flickinger JC, Lunsford LD (2001) Stereotactic radiosurgery for motor cortex region arteriovenous malformations. Neurosurgery 48:70–76; discussion 76–77

Hashizume K, Tanaka T (1998) Multiple subpial transection in kainic acid-induced focal cortical seizure. Epilepsy Res 32:389–399

Hodaie M, Wennberg RA, Dostrovsky JO, Lozano AM (2002) Chronic anterior thalamus stimulation for intractable epilepsy. Epilepsia 43:603–608

Hoh BL, Ogilvy CS, Butler WE, Loeffler JS, Putman CM, Chapman PH (2000) Multimodality treatment of nongalenic arteriovenous malformations in pediatric patients. Neurosurgery 47:346–357; discussion 357–358

Hoh BL, Chapman PH, Loeffler JS, Carter BS, Ogilvy CS (2002) Results of multimodality treatment for 141 patients with brain arteriovenous malformations and seizures: factors associated with seizure incidence and seizure outcomes. Neurosurgery 51:303–309; discussion 309–311

Hubel DH, Wiesel TN (1962) Receptive fields, binocular interaction and functional architecture in the cat's visual cortex. J Physiol 160:106–154

Hufnagel A, Zentner J, Fernandez G, Wolf HK, Schramm J, Elger CE (1997) Multiple subpial transection for control of epileptic seizures: effectiveness and safety. Epilepsia 38:678–688

Iadarola MJ, Gale K (1982) Substantia nigra: site of anticonvulsant activity mediated by gamma-aminobutyric acid. Science 218:1237–1240

Irwin K, Birch V, Lees J, Polkey C, Alarcon G, Binnie C, Smedley M, Baird G, Robinson RO (2001) Multiple subpial transection in Landau-Kleffner syndrome. Dev Med Child Neurol 43:248–252

Jette N, Wiebe S (2013) Update on the surgical treatment of epilepsy. Curr Opin Neurol 26:201–207

Kerrigan JF, Litt B, Fisher RS, Cranstoun S, French JA, Blum DE, Dichter M, Shetter A, Baltuch G, Jaggi J, Krone S, Brodie M, Rise M, Graves N (2004) Electrical stimulation of the anterior nucleus of the thalamus for the treatment of intractable epilepsy. Epilepsia 45:346–354

Kossoff EH, Ritzl EK, Politsky JM, Murro AM, Smith JR, Duckrow RB, Spencer DD, Bergey GK (2004) Effect of an external responsive neurostimulator on seizures and electrographic discharges during subdural electrode monitoring. Epilepsia 45:1560–1567

Kurita H, Suzuki I, Shin M, Kawai K, Tago M, Momose T, Kirino T (2001) Successful radiosurgical treatment of lesional epilepsy of mesial temporal origin. Minim Invasive Neurosurg 44:43–46

Kuzniecky R, Devinsky O (2007) Surgery Insight: surgical management of epilepsy. Nat Clin Pract Neurol 3:673–681

Lee KJ, Jang KS, Shon YM (2006) Chronic deep brain stimulation of subthalamic and anterior thalamic nuclei for controlling refractory partial epilepsy. Acta Neurochir Suppl 99: 87 91

Litt B (2003) Evaluating devices for treating epilepsy. Epilepsia 44(Suppl 7):30–37

Litt B, Echauz J (2002) Prediction of epileptic seizures. Lancet Neurol 1:22–30

Lueders H, Bustamante LA, Zablow L, Goldensohn ES (1981) The independence of closely spaced discrete experimental spike foci. Neurology 31:846–851

Mapstone TB (2008) Vagus nerve stimulation: current concepts. Neurosurg Focus 25:E9

Mclachlan RS, Pigott S, Tellez-Zenteno JF, Wiebe S, Parrent A (2010) Bilateral hippocampal stimulation for intractable temporal lobe epilepsy: impact on seizures and memory. Epilepsia 51: 304–307

Mirski MA, Ferrendelli JA (1986) Anterior thalamic mediation of generalized pentylenetetrazol seizures. Brain Res 399:212–223

Molyneux PD, Barker RA, Thom M, Van Paesschen W, Harkness WF, Duncan JS (1998) Successful treatment of intractable epilepsia partialis continua with multiple subpial transections. J Neurol Neurosurg Psychiatry 65: 137–138

Morrell MJ (2011) Responsive cortical stimulation for the treatment of medically intractable partial epilepsy. Neurology 77:1295–1304

Morrell F, Hanbery JW (1969) A new surgical technique for the treatment of focal cortical epilepsy. Electroencephalogr Clin Neurophysiol 26:120

Morrell F, Whisler WW, Smith MC, Hoeppner TJ, De Toledo-Morrell L, Pierre-Louis SJ, Kanner AM, Buelow JM, Ristanovic R, Bergen D et al (1995) Landau-Kleffner syndrome. Treatment with subpial intracortical transection. Brain 118(Pt 6):1529–1546

Morrell F, Kanner AM, De Toledo-Morrell L, Hoeppner T, Whisler WW (1999) Multiple subpial transection. Adv Neurol 81:259–270

Mountcastle VB (1957) Modality and topographic properties of single neurons of cat's somatic sensory cortex. J Neurophysiol 20:408–434

Mountcastle VB (1997) The columnar organization of the neocortex. Brain 120(Pt 4):701–722

Mulligan LP, Spencer DD, Spencer SS (2001) Multiple subpial transections: the Yale experience. Epilepsia 42:226–229

Niranjan A, Madhavan R, Gerszten PC, Lunsford LD (2012) Intracranial radiosurgery: an effective and disruptive innovation in neurosurgery. Stereotact Funct Neurosurg 90:1–7

Orbach D, Romanelli P, Devinsky O, Doyle W (2001a) Late seizure recurrence after multiple subpial transections. Epilepsia 42:1130–1133

Orbach D, Romanelli P, Devinsky O, Doyle W (2001b) Late seizure recurrence after multiple subpial transections. Epilepsia 42:1316–1319

Osorio I, Frei MG, Sunderam S, Giftakis J, Bhavaraju NC, Schaffner SF, Wilkinson SB (2005) Automated seizure abatement in humans using electrical stimulation. Ann Neurol 57:258–268

Papacostas SS, Myrianthopoulou P, Dietis A, Papathanasiou ES (2007) Induction of central-type sleep apnea by vagus nerve stimulation. Electromyogr Clin Neurophysiol 47:61–63

Pollock BE, Lunsford LD, Kondziolka D, Maitz A, Flickinger JC (1994) Patient outcomes after stereotactic radiosurgery for "operable" arteriovenous malformations. Neurosurgery 35:1–7; discussion 7–8

Ramon C, Holmes MD (2013) Noninvasive localization of epileptic sites from stable phase synchronization patterns on different days derived from short duration interictal scalp dEEG. Brain Topogr 26:1–8

Regis J, Kerkerian-Legoff L, Rey M, Vial M, Porcheron D, Nieoullon A, Peragut JC (1996) First biochemical evidence of differential functional effects following Gamma Knife surgery. Stereotact Funct Neurosurg 66(Suppl 1):29–38

Regis J, Bartolomei F, Kida Y, Kobayashi T, Vladyka V, Liscak R, Forster D, Kemeny A, Schrottner O, Pendl G (2000) Radiosurgery for epilepsy associated with cavernous malformation: retrospective study in 49 patients. Neurosurgery 47:1091–1097

Regis J, Bartolomei F, Hayashi M, Chauvel P (2002a) Gamma Knife surgery, a neuromodulation therapy in epilepsy surgery! Acta Neurochir Suppl 84:37–47

Regis J, Bartolomei F, Hayashi M, Chauvel P (2002b) What role for radiosurgery in mesial temporal lobe epilepsy. Zentralbl Neurochir 63:101–105

Regis J, Hayashi M, Eupierre LP, Villeneuve N, Bartolomei F, Brue T, Chauvel P (2004a) Gamma knife surgery for epilepsy related to hypothalamic hamartomas. Acta Neurochir Suppl 91:33–50

Regis J, Rey M, Bartolomei F, Vladyka V, Liscak R, Schrottner O, Pendl G (2004b) Gamma knife surgery in mesial temporal lobe epilepsy: a prospective multicenter study. Epilepsia 45:504–515

Regis J, Scavarda D, Tamura M, Villeneuve N, Bartolomei F, Brue T, Morange I, Dafonseca D, Chauvel P (2007) Gamma knife surgery for epilepsy related to hypothalamic hamartomas. Semin Pediatr Neurol 14:73–79

Romanelli P (2014) Epilepsy. Cyberknife stereotactic radiosurgery. Brain, Vol. 1. Nova Science Publishers, Hauppage

Romanelli P, Anschel DJ (2006) Radiosurgery for epilepsy. Lancet Neurol 5:613–620

Romanelli P, Bravin A (2011) Synchrotron-generated microbeam radiosurgery: a novel experimental approach to modulate brain function. Neurol Res 33:825–831

Romanelli P, Weiner HL, Najjar S, Devinsky O (2001) Bilateral resective epilepsy surgery in a child with tuberous sclerosis: case report. Neurosurgery 49:732–734; discussion 735

Romanelli P, Najjar S, Weiner HL, Devinsky O (2002) Epilepsy surgery in tuberous sclerosis: multistage procedures with bilateral or multilobar foci. J Child Neurol 17:689–692

Romanelli P, Muacevic A, Striano S (2008) Radiosurgery for hypothalamic hamartomas. Neurosurg Focus 24:E9

Romanelli P, Striano P, Barbarisi M, Coppola G, Anschel DJ (2012) Non-resective surgery and radiosurgery for treatment of drug-resistant epilepsy. Epilepsy Res 99:193–201

Romanelli P, Fardone E, Battaglia G, Brauer-Krisch E, Prezado Y, Requardt H, Le Duc G, Nemoz C, Anschel DJ, Spiga J, Bravin A (2013) Synchrotron-generated microbeam sensorimotor cortex transections induce seizure control without disruption of neurological functions. PLoS One 8:e53549

Rosenfeld JV (2011) The evolution of treatment for hypothalamic hamartoma: a personal odyssey. Neurosurg Focus 30:E1

Saillet S, Langlois M, Feddersen B, Minotti L, Vercueil L, Chabardes S, David O, Depaulis A, Deransart C, Kahane P (2009) Manipulating the epileptic brain using stimulation: a review of experimental and clinical studies. Epileptic Disord 11:100–12

Sawhney IM, Robertson IJ, Polkey CE, Binnie CD, Elwes RD (1995) Multiple subpial transection: a review of 21 cases. J Neurol Neurosurg Psychiatry 58:344–349

Schauble B, Cascino GD, Pollock BE, Gorman DA, Weigand S, Cohen-Gadol AA, Mcclelland RL (2004) Seizure outcomes after stereotactic radiosurgery for cerebral arteriovenous malformations. Neurology 63:683–687

Schramm J (2008) Temporal lobe epilepsy surgery and the quest for optimal extent of resection: a review. Epilepsia 49:1296–1307

Schramm J, Aliashkevich AF, Grunwald T (2002) Multiple subpial transections: outcome and complications in 20 patients who did not undergo resection. J Neurosurg 97:39–47

Schrottner O, Unger F, Eder HG, Feichtinger M, Pendl G (2002) Gamma-Knife radiosurgery of mesiotemporal tumour epilepsy observations and long-term results. Acta Neurochir Suppl 84:49–55

Schulze-Bonhage A, Homberg V, Trippel M, Keimer R, Elger CE, Warnke PC, Ostertag C (2004) Interstitial radiosurgery in the treatment of gelastic epilepsy due to hypothalamic hamartomas. Neurology 62:644–647

Selch MT, Gorgulho A, Mattozo C, Solberg TD, Cabatan-Awang C, Desalles AA (2005) Linear accelerator stereotactic radiosurgery for the treatment of gelastic seizures due to hypothalamic hamartoma. Minim Invasive Neurosurg 48:310–314

Shimizu H, Maehara T (2000) Neuronal disconnection for the surgical treatment of pediatric epilepsy. Epilepsia 41(Suppl 9):28–30

Shon YM, Lee KJ, Kim HJ, Chung YA, Ahn KJ, Kim YI, Yang DW, Kim BS (2005) Effect of chronic deep brain stimulation of the subthalamic nucleus for frontal lobe epilepsy: subtraction SPECT analysis. Stereotact Funct Neurosurg 83:84–90

Spencer SS, Schramm J, Wyler A, O'Connor M, Orbach D, Krauss G, Sperling M, Devinsky O, Elger C, Lesser R, Mulligan L, Westerveld M (2002) Multiple subpial transection for intractable partial epilepsy: an international meta-analysis. Epilepsia 43:141–145

Srikijvilaikul T, Najm I, Foldvary-Schaefer N, Lineweaver T, Suh JH, Bingaman WE (2004) Failure of gamma knife radiosurgery for mesial temporal lobe epilepsy: report of five cases. Neurosurgery 54:1395–1402; discussion 1402–1404

Stacey WC, Litt B (2008) Technology insight: neuroengineering and epilepsy-designing devices for seizure control. Nat Clin Pract Neurol 4:190–201

Stefan H, Hummel C, Grabenbauer GG, Muller RG, Robeck S, Hofmann W, Buchfelder M (1998) Successful treatment of focal epilepsy by fractionated stereotactic radiotherapy. Eur Neurol 39:248–250

Sun FT, Morrell MJ, Wharen RE Jr (2008) Responsive cortical stimulation for the treatment of epilepsy. Neurotherapeutics 5:68–74

Telfeian AE, Connors BW (1998) Layer-specific pathways for the horizontal propagation of epileptiform discharges in neocortex. Epilepsia 39:700–708

Tolstykh GP, Cavazos JE (2013) Potential mechanisms of sudden unexpected death in epilepsy. Epilepsy Behav 26:410–414

Upton AR, Cooper IS, Springman M, Amin I (1985) Suppression of seizures and psychosis of limbic system origin by chronic stimulation of anterior nucleus of the thalamus. Int J Neurol 19–20:223–230

Uthman BM, Reichl AM, Dean JC, Eisenschenk S, Gilmore R, Reid S, Roper SN, Wilder BJ (2004) Effectiveness of vagus nerve stimulation in epilepsy patients: a 12-year observation. Neurology 63:1124–1126

Van Roost D, Boon P, Vonck K, Caemaert J (2007) Neurosurgical aspects of temporal deep brain stimulation for epilepsy. Acta Neurochir Suppl 97:333–336

Velasco F, Velasco M, Ogarrio C, Fanghanel G (1987) Electrical stimulation of the centromedian thalamic nucleus in the treatment of convulsive seizures: a preliminary report. Epilepsia 28:421–430

Velasco F, Velasco M, Jimenez F, Velasco AL, Marquez I (2001) Stimulation of the central median thalamic nucleus for epilepsy. Stereotact Funct Neurosurg 77:228–232

Vesper J, Steinhoff B, Rona S, Wille C, Bilic S, Nikkhah G, Ostertag C (2007) Chronic high-frequency deep brain stimulation of the STN/SNr for progressive myoclonic epilepsy. Epilepsia 48:1984–1989

Vonck K, Boon P, Goossens L, Dedeurwaerdere S, Claeys P, Gossiaux F, Van Hese P, De Smedt T, Raedt R, Achten E, Deblaere K, Thieleman A, Vandemaele P, Thiery E, Vingerhoets G, Miatton M, Caemaert J, Van Roost D, Baert E, Michielsen G, Dewaele F, Van Laere K, Thadani V, Robertson D, Williamson P (2003) Neurostimulation for refractory epilepsy. Acta Neurol Belg 103:213–217

New Radiosurgical Paradigms to Treat Epilepsy Using Synchrotron Radiation

Pantaleo Romanelli, Alberto Bravin, Erminia Fardone and Giuseppe Battaglia

Abstract Synchrotron-generated X-ray microplanar beams (microbeams) are characterized by peculiar biological properties such as a remarkable tissue-sparing effect in healthy tissues including the central nervous system and, as a direct consequence, the ability to deliver extremely high doses without induction of radionecrosis. Growing experimental evidence is showing remarkable tolerance of brain and spinal cord to irradiation with microbeam arrays delivering doses up to 400 Gy with a beam width up to 0.7 mm. Submillimetric beams can be delivered following a stereotactic design bringing to the target doses in the range of hundreds of Gray without harm to the surrounding tissues. Microbeam arrays can be used to generate cortical transections or subcortical lesions, thus enabling the non-invasive modulation of brain networks. This novel microradiosurgical approach is of great interest for the treatment of a variety of brain disorders, including epilepsy.

As discussed in the neuromodulation chapter, epilepsy is the most common chronic neurological disorder affecting approximately 1 % of the population worldwide. Current estimates indicate that 20–30 % of patients with epilepsy are refractory to antiepileptic drugs (AED) (Go and Snead 2008; Kuzniecky and Devinsky 2007). These medically intractable patients are candidates for surgical treatment. The primary goal of epilepsy surgery is to resect the seizure focus without causing neurological sequelae. While mesial temporal sclerosis declares itself on neuroimaging studies, neocortical seizure foci cannot be localized precisely on the basis of surface EEG and neuroimaging, thus requiring invasive monitoring procedures. Subdural grids and strips or stereoencephalography are performed on selected patients to obtain a precise mapping of the seizure focus and of eloquent

---

P. Romanelli (✉)
Centro Diagnostico Italiano, via Saint Bon 20, Milano, Italy
e-mail: radiosurgery2000@yahoo.com

E. Fardone · A. Bravin
European Synchrotron Radiation Facility, BP220, Grenoble, France

G. Battaglia
Istituto di Ricovero e Cura a Carattere Scientifico Neuromed,
Località Camerelle, Pozzilli (IS), Italy

cortical regions. If the seizure focus does not involve eloquent cortex, then focus resection can be performed. The risk of significant morbidity is especially high when the epileptogenic zone overlaps eloquent cortex, such as speech or primary motor areas. The main limits of current resective epilepsy surgery is the needed to remove extensive parts of the brain and the inability to treat seizure foci involving eloquent cortex (Shorvon et al. 2012). Multiple subpial transections (MST) and stereotactic radiosurgery (SRS) are attractive non-resective alternative to treat seizure foci involving eloquent cortex (Romanelli et al. 2012).

Multiple subpial transections (MSTs) is a non-resective procedure developed to treat patients with severe medically-refractory seizures involving highly functional regions of the brain (Orbach et al. 2001; Mulligan et al. 2001; Mountcastle 1957; Morrell et al. 1995; Morrell et al. 1989; Morrell and Hanbery 1969). This approach has the merit to avoid the resection of the focus but there is limited experience using it to treat also seizure foci located in non-eloquent cortex or in the hippocampus (Patil and Andrews 2013; Patil et al. 2004). MSTs are usually performed by an open surgical procedure aiming to the placement of parallel cortical incisions spaced by intervals of 5 mm. Neuroanatomic studies show that the basic functional cortical unit is arranged vertically, while epileptic activity spreads horizontally. Vertical cortical incisions interrupt the horizontal axons carrying the spreading of seizure activity thus preventing seizure propagation while preserving the vertical columns subserving neuronal function. In essence, MSTs induce a selective disconnection leaving behind an intact functional cortex (Morrell et al. 1999; Telfeian and Connors 1998; Chervin et al. 1988). Stereotactic radiosurgery (SRS) is another emerging non-resective approach to treat epilepsy: narrow beams of radiation coming from different angles are directed with high precision and accuracy to a selected target, achieving a very high energy deposition within the target volume while sparing the surrounding normal tissue thanks to a rapid dose fall-off. SRS using the Gamma-Knife or the Cyberknife is an excellent tool to treat brain tumors and functional brain disorders such as trigeminal neuralgia. The attractiveness of a non-invasive treatment devoid of the risks and discomforts associated with open surgery and general anaesthesia as well as the excellent clinical outcomes explain the exponential growth of SRS over the last decade (Niranjan et al. 2012). Treatment of epilepsy is an emerging application of SRS: it was initially observed that secondary epilepsy improves after radiosurgical treatment of the underlying lesion. Following this observation, radiosurgery was proposed as a primary treatment for idiopathic localization-related epilepsy (Romanelli and Anschel 2006). The amount of energy which can be deposited by current radiosurgical techniques is limited by the size of the smallest collimator available (4 mm for the GammaKnife, 5 mm for the Cyberknife). Therefore the seizure focus can be selectively irradiated delivering doses up to 25 Gy in single session: time to achieve seizure control can be as long as 2 years. Figure 1 shows a Cyberknife treatment delivered to treat medically-refractory complex partial seizures induced by mesial temporal sclerosis. The entire hippocampus was irradiated with a prescribed dose of 23 Gy delivered to the 81% isodose. The limits of current radiosurgical treatments for epilepsy include the long waiting time to achieve response and the possibility of complications such as radio-

New Radiosurgical Paradigms to Treat Epilepsy Using Synchrotron Radiation

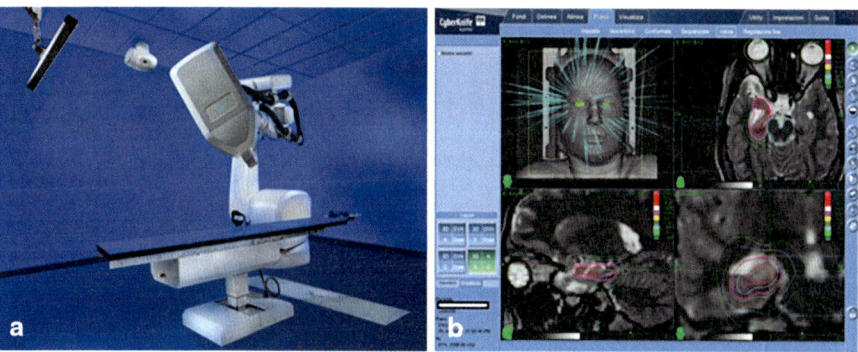

Fig. 1. **a** The Cyberknife is a robotic linac providing image-guided frameless non-isocentric beam delivery. **b** Cyberknife treatment planning for drug-refractory complex partial seizures induced by mesial temporal sclerosis. A prescribed dose of 23 Gy was delivered to the 81 % isodose

Fig. 2 European synchroton radiation facility of Grenoble. (Taken from www.esrf.eu)

induced edema and radionecrosis. Current SRS does not allow to generate in a non-invasive way the equivalent of cortical or hippocampal microsurgical transections. The delivery of higher doses to restricted tissue volumes allowing to transect the epileptogenic focus is a novel radiosurgical paradigm developed at the European Synchrotron Radiation Facility (ESRF) in Grenoble.

At the ESRF (Fig. 2), microscopic arrays of X-ray beams originating from a Synchrotron source can induce the equivalent of a microsurgical cut through the neocortex or hippocampus by delivering very intense doses of radiation (hundreds to thousands of Grays(Gy) to tissue slices of microscopic thickness (Brauer-Krisch et al. 2010; Slatkin et al. 2007; Romanelli and Bravin 2011). Synchrotron-generated microplanar beams (microbeams) are delivered as an array of parallel beam of the wanted thickness (going from 25 all the way up to 600 μm). Microbeam irradiation can deliver peak doses of several hundred Gy with doses in non-irradiated valleys limited to a few Gy. This unique irradiation modality slows, and sometimes ablates, malignant brain tumours in rodents (Laissue et al. 1998). Additionally to its high

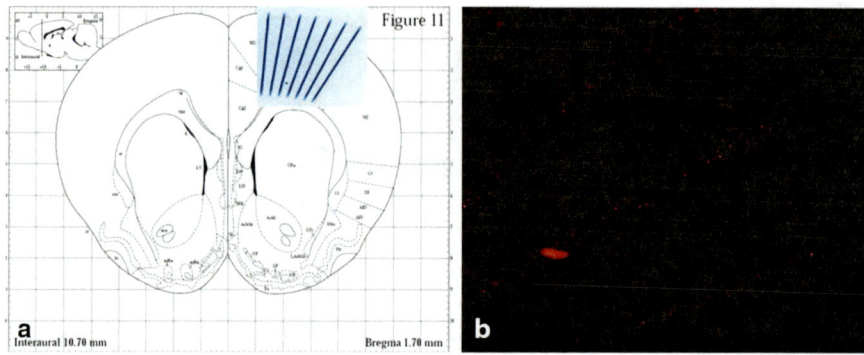

Fig. 3 a Representative picture of microbeam sensorimotor cortex transections. The brain sketch is extracted by Paxinos and Watson Atlas 1986 (Paxinos and Watson 1986). **b** Immunohistochemistry using pH2AX as marker to highlight immediate DNA double strand breaks and cell apoptosis

precision, the technique allows users to take advantage of the rapid regeneration of normal microvessels damaged in the direct paths of thin microbeams (Slatkin et al. 1995; Smilowitz et al. 2002; Blattmann et al. 2005; Van De Looij et al. 2006; Serduc et al. 2006). The preferential tumoricidal effects, or strong tumour palliation properties, are due, in part, to the lack of recovery of the tumour vasculature (Laissue et al. 1998; Zhong et al. 2003; Dilmanian et al. 2002), presumably because of structural differences between microvessels of tumour and those of the surrounding normal tissue (Denekamp et al. 1998). Microbeam radiosurgery offers great opportunities to modulate cortical function without neurological injury. An array of parallel or convergent microbeams can be used to transect epileptogenic cortex, disconnecting and parcellizing the focus through the severing of horizontal axons while maintaining the function subserved by the vertical columns (A and III B). Our group performed transections of the primary motor cortex delivering peak doses up to 360 Gy without evidence of neurological injury (Romanelli et al. 2013). The adjacent cortical columns exposed to much lower valley doses (less than 5 Gy) failed to show any histological evidence of evident tissue damage. No sign of motor deficit was found during a 6 months observation period through Rotarod® test. In a kainic-acid model of convulsive seizures originating from the sensory- motor cortex, cortical transetions over the focus induced immediate seizure control (Romanelli et al. 2013). Immunohistochemistry techniques have shown preliminary evidence of cortical neurogenesis together with the development of reactive gliosis (not visible by conventional histology) along the beam pathway (Fardone 2013). This interesting finding supports the controversial hypothesis that there is cortical neurogenesis after irradiation injury (Gould 1999; Kornack and Rakic 2001; Koketsu et al. 2003; Gould et al. 2001; Gould 2007) but also after other types of cortical injury such as ischemia (Parent et al. 2002; Kokaia et al. 2006; Thored et al. 2006), multiple sclerosis (Danilov et al. 2006), epilepsy (Scott et al. 1998; Parent et al. 2002; Zhang et al. 2014), and following focal apoptosis (Magavi et al. 2000). Further studies are ongoing to better characterise the neurogenesis process after microbeam transections both in hippocampus and neocortex before and after the exposure to microbeam irradiation (Fig. 3).

In conclusion, submillimetric transections, either placed over neocortical seizure foci or through the hippocampus, could prove to be an excellent tool to be added to the current techniques used to control seizures. The development of devices delivering submillimetric beams able to generate cortical transections might add a powerful new tool to the clinical treatment of epilepsy and, more in general, to modulate cortical functions in a wide variety of neuropsychiatric disorders.

References

Blattmann H et al (2005) Applications of synchrotron X-rays to radiotherapy. Nucl Instrum Methods Phys Res 548(1–2):17–22

Brauer-Krisch E et al (2010) Potential high resolution dosimeters for MRT. In Siu KKW (ed) 6th international conference on medical applications of synchrotron radiation, American Institute of Physics, USA, pp 89–97

Chervin RD, Pierce PA, Connors BW (1988) Periodicity and directionality in the propagation of epileptiform discharges across neocortex. J Neurophysiol 60(5):1695–1713

Danilov AI et al (2006) Neurogenesis in the adult spinal cord in an experimental model of multiple sclerosis. Eur J Neurosci 23(2):394–400

Denekamp J, Daşu A, Waites A (1998) Vasculature and microenvironmental gradients: the missing links in novel approaches to cancer therapy? Adv Enzyme Regul 38:281–299

Dilmanian FA et al (2002) Response of rat intracranial 9 L gliosarcoma to microbeam radiation therapy. Neuro Oncol 4(1):26–38

Fardone E (2013) A new application of microbeam radiation therapy (MRT) on the treatment of epilepsy and brain disorders. University Joseph Furier of Grenoble [Thesis]

Go C, Snead OC (2008) Pharmacologically intractable epilepsy in children: diagnosis and preoperative evaluation. Neurosurg Focus 25(3):E2

Gould E (1999) Neurogenesis in the neocortex of adult Primates. Science 286(5439):548–552

Gould E (2007) How widespread is adult neurogenesis in mammals? Nat Rev Neurosci 8(6):481–488

Gould E et al (2001) Adult-generated hippocampal and neocortical neurons in macaques have a transient existence. Proc Natl Acad Sci U S A 98(19):10910–10917

Kokaia Z et al (2006) Regulation of stroke-induced neurogenesis in adult brain–recent scientific progress. Cereb Cortex 16(Suppl 1):i162–i167 (New York, NY 1991)

Koketsu D et al (2003) Nonrenewal of neurons in the cerebral neocortex of adult Macaque Monkeys. J Neurosci 23(3):937–942

Kornack DR, Rakic P (2001) Cell proliferation without neurogenesis in adult primate neocortex. Science 294(5549):217–2130 (New York, NY)

Kuzniecky R, Devinsky O (2007) Surgery Insight: surgical management of epilepsy. Nat Clin Pract Neurol 3(12):673–681

Laissue JA et al (1998) Neuropathology of ablation of rat gliosarcomas and contiguous brain tissues using a microplanar beam of synchrotron-wiggler-generated X rays. Int J Cancer (Journal international du cancer) 78(5):654–660

Magavi SS, Leavitt BR, Macklis JD (2000) Induction of neurogenesis in the neocortex of adult mice. Nature 405(6789):951–955

Morrell F, Hanbery JW (1969) A new surgical technique for the treatment of focal cortical epilepsy. Electroencephalogr Clin Neurophysiol 26(1):120

Morrell F et al (1989) Multiple subpial transection: a new approach to the surgical treatment of focal epilepsy. J Neurosurg 70:231–239

Morrell F et al (1995) Landau-Kleffner syndrome. Treatment with subpial intracortical transection. Brain: A Journal of Neurology 118(Pt 6):1529–1546

Morrell F et al (1999) Multiple subpial transection. Adv Neurol 81:259–270

Mountcastle VB (1957) Modality and topographic properties of single neurons of cat's somatic sensory cortex. J Neurophysiol 20(4):408–434

Mulligan LP, Spencer DD, Spencer SS (2001) Multiple subpial transections: the Yale experience. Epilepsia 42(2):226–229

Niranjan A et al (2012) Intracranial radiosurgery: an effective and disruptive innovation in neurosurgery. Stereotact Funct Neurosurg 90(1):1–7

Orbach D et al (2001) Late seizure recurrence after multiple subpial transections. Epilepsia 42(10):1130–1133

Parent JM et al (2002) Rat forebrain neurogenesis and striatal neuron replacement after focal stroke. Ann Neurol 52(6):802–813

Patil AA, Andrews R (2013) Long term follow-up after multiple hippocampal transection (MHT). Seizure: The Journal of the British Epilepsy Association 22(9):731–734

Patil AA et al (2004) Is epilepsy surgery on both hemispheres effective? Stereotact Funct Neurosurg 82(5–6):214–221

Paxinos G, Watson C (1986) The rat brain in stereotaxic coordinates, 2nd edn. Academic Press, London

Romanelli P, Anschel DJ (2006) Radiosurgery for epilepsy. Lancet Neurol 5:613–620

Romanelli P, Bravin A (2011) Synchrotron-generated microbeam radiosurgery: a novel experimental approach to modulate brain function. Neurol Res 33(8):825–831

Romanelli P et al (2012) Non-resective surgery and radiosurgery for treatment of drug-resistant epilepsy. Epilepsy Res 99(3):193–201

Romanelli P et al (2013) Synchrotron-generated microbeam sensorimotor cortex transections induce seizure control without disruption of neurological functions. PloS One 8(1):e53549

Scott BW et al (1998) Kindling-induced neurogenesis in the dentate gyrus of the rat. Neurosci Lett 248(2):73–76

Serduc R et al (2006) In vivo two-photon microscopy study of short-term effects of microbeam irradiation on normal mouse brain microvasculature. Int J Radiat Oncol Biol Phys 64(5):1519–1527

Slatkin DN et al (1995) Subacute neuropathological effects of microplanar beams of x-rays from a synchrotron wiggler. Proc Natl Acad Sci U S A 92(19):8783–8787

Slatkin DN et al (2007) Prospects for microbeam radiation therapy of brain tumours in children. Dev Med Child Neurol 49(2):163

Smilowitz HM et al (2002) Synergy of gene-mediated immunoprophylaxis and microbeam radiation therapy for advanced intracerebral rat 9 L gliosarcomas. J Neurooncol 78(2):135–143

Telfeian AE, Connors BW (1998) Layer-specific pathways for the horizontal propagation of epileptiform discharges in neocortex. Epilepsia 39(7):700–708

Thored P et al (2006) Persistent production of neurons from adult brain stem cells during recovery after stroke. Stem Cells 24(3):739–747 (Dayton, Ohio)

Van De Looij Y et al (2006) Cerebral edema induced by Synchrotron Microbeam Radiation Therapy in the healthy mouse brain. Characterization by means of Diffusion Tensor Imaging. In Proceedings 14th Scientific Meeting International Society for Magnetic Resonance in Medicine, p 1472

Zhang L et al (2014) Hippocampal CA field neurogenesis after pilocarpine insult: the hippocampal fissure as a neurogenic niche. J Chem Neuroanat 56:45–57

Zhong N et al (2003) Response of rat skin to high-dose unidirectional x-ray microbeams: a histological study. Radiat Res 160(2):133–142

Index

Symbols
2q24.4 deletion, 2, 6
5q14.3 deletion, 2, 7
6q terminal deletion, 2, 8, 9
14q12 deletion and duplication, 2, 9
15q13.3 deletion, 2, 10

A
Androgen, 202, 204
Animal models, 19, 22, 26, 27, 164, 185, 192
Antiepileptic drugs (AED), 12, 18, 19, 51, 60, 71, 72, 85, 91, 95, 175, 192, 231
Auditory seizures, 37, 169
Autoantibodies, 36, 40–42, 164, 175
Autoimmune limbic encephalitis (LE), 35, 36, 40, 41
Autonomic seizures, 155
Autonomic status epilepticus, 155
Autosomal dominant lateral temporal epilepsy (ADLTE), 35
Autosomal dominant partial epilepsy with auditory features (ADPEAF), 36

C
Channelopathies, 16, 27
Contraceptive, 185, 188, 204
Copy number variation, 2

D
Deep brain stimulation (DBS), 214, 216
Dysembryoplastic Neuroepithelial Tumor (DNT), 55, 133

E
Electroencephalography (EEG), 84, 150, 189
Epilepsy, 1, 2, 8–10, 12, 16, 19, 25, 71, 83, 88
treatment, 195, 201

Epileptic encephalopathies, 19, 22–24
Estrogen, 202
Excitability, 15, 19, 24, 27, 145, 165

F
Functional MRI (fMRI), 62, 118

G
Ganglioglioma (GG), 49, 132
Ganglion Cell Tumors (GCTs), 52, 53, 55
Glioneuronal tumors (GNTs), 48

I
Ictal epileptic headache (IEH), 140, 142, 147, 150, 151, 154–156
Immune system, 164, 166, 170, 175
Inborn errors of metabolism, 79, 95
Intermittent light stimulations, 102
Ionotropic and metabotropic glutamate receptors, 184

K
KCNQ-type K+channels, 184

L
Leucine-rich, glioma inactivated 1 (LGI1), 27, 35–38, 42, 169
Limbic encephalitis (LE), 167

M
Magnetic Resonance Imaging (MRI), 52, 56, 59, 61, 110, 112, 119, 214
Migraine headache, 148
MR spectroscopy, 72, 117
Multiple subpial transections (MST), 223, 224, 225

N
Neuromodulation, 214, 216, 223, 231
Neurosteroids, 202, 203

P
PCOS, 205, 207
Positron Emission Tomography (PET), 119, 120, 121, 214
Praxis, 102
Progesterone, 202, 204

R
Reading, 102, 104, 105

S
SCN1A, 2, 7, 18
Seizures
  childhood onset, 92
  effects of
    in sex functions, 205
  effects on
    catamenial epilepsy, 203, 205
  induced by non-verbal cognitive stimuli, 103, 104
  induced by visual stimuli, 102, 103
  infantile onset, 82, 83
  neonatal onset, 73, 75, 76

Sex functions
  effect of seizures on, 205
Sex hormones, 201, 202, 205
Single-photon emission computed tomography (SPECT), 110, 118, 121
Stereotactic radiosurgery (SRS), 214, 219, 220, 232
Surgery, 49, 60, 62, 111, 214, 218, 232
SV2A synaptic vesicle protein, 184
Synaptic transmission, 15, 41, 42, 185
Synchrotron, 224, 233

T
Temporal lobe epilepsy (TLE), 19, 114, 132, 171, 220
Thinking, 102–104
Tumors
  epidemiology and type of, 50
  neurocytic, 59

V
Vagus nerve stimulation (VNS), 214, 215

X
Xp11.22-11.23 duplication, 11, 12